A Textbook of

ENGINEERING MATHEMATICS-III

A Textbook of
ENGINEERING MATHEMATICS-III

B. Tech. II year (III Semester) of
Uttarakhand Technical University, Dehradun

H.S. Gangwar
M.Sc., Ph.D.
Professor and Head
Department of Mathematics
Shri Siddhi Vinayak Institute of Technology
Bareilly (U.P.)

A.K. Jaiswal
M.Sc., Ph.D.
Leturer
Department of Mathematics
St. Andrew's College
Gorakhpur (U.P.)

An Imprint of

NEW AGE INTERNATIONAL (P) LIMITED, PUBLISHERS

New Delhi • Bangalore • Chennai • Cochin • Guwahati • Hyderabad
Jalandhar • Kolkata • Lucknow • Mumbai • Ranchi
Visit us at **www.newagepublishers.com**

Published by New Age International (P) Ltd., Publishers
First Edition: 2011

Branches:

- 37/10, 8th Cross (Near Hanuman Temple), Azad Nagar, Chamrajpet, **Bangalore**-560 018. Tel.: (080) 26756823, Telefax: 26756820, E-mail: bangalore@newagepublishers.com
- 26, Damodaran Street, T. Nagar, **Chennai**-600 017. Tel.: (044) 24353401, Telefax: 24351463 E-mail: chennai@newagepublishers.com
- CC-39/1016, Carrier Station Road, Ernakulam South, **Cochin**-682 016. Tel.: (0484) 2377004, Telefax: 4051303. E-mail: cochin@newagepublishers.com
- Hemsen Complex, Mohd. Shah Road, Paltan Bazar, Near Starline Hotel, **Guwahati**-781 008. Tel.: (0361) 2513881. Telefax: 2543669, E-mail: guwahati@newagepublishers.com
- No. 105, 1st Floor, Madhiray Kaveri Tower, 3-2-19, Azam Jahi Road, Nimboliadda, **Hyderabad**-500 027. Tel.: (040) 24652456, Telefax: 24652457, E-mail:hyderabad@newagepublishers.com
- RDB Chambers (Formerly Lotus Cinema)106A, 1st Floor, S.N. Banerjee Road, **Kolkata**-700 014. Tel.: (033) 22273773, Telefax: 22275247, E-mail:kolkata@newagepublishers.com
- 16-A, Jopling Road, **Lucknow**-226 001. Tel.: (0522) 2209578, 4045297, Telefax: 2204098 E-mail: lucknow@newagepublishers.com
- 142C, Victor House, Ground Floor, N.M. Joshi Marg, Lower Parel, **Mumbai**-400 013. Tel.: (022) 24927869. Telefax: 24915415, E-mail: mumbai@newagepublishers.com
- 22, Golden House, Daryaganj, **New Delhi**-110 002. Tel.: (011) 23262370, 23262368, Telefax: 43551305. E-mail: sales@newagepublishers.com

---

ISBN: 978-81-224-3026-4

Rs. 225.00

C-10-08-4884

Printed in India at Parmanand Offset, Delhi.
Typeset at Aakriti Graphics, New Delhi.

PUBLISHING FOR ONE WORLD
NEW AGE INTERNATIONAL (P) LIMITED, PUBLISHERS
4835/24, Ansari Road, Daryaganj, New Delhi-110002
Visit us at **www.newagepublishers.com**

Preface

The book Engineering Mathematics is primarily written according to the unified syllabus of Mathematics of third semester of all Engineering Colleges affiliated to Uttarakhand Technical University, Dehradun and other states of India. This book also covers the B.Tech./B.E./B.Arch. second year course of other Indian Engineering Colleges.

The subject matter is presented in very systematic and logical manner. In each chapter, all concepts, definitions and the large number of examples in the best possible way have been discussed in detail and lucid manner so that the students should feel no difficulty to understand the subject.

Thorough care has been taken to eradicate errors but the perfection cannot be claimed. The readers are requested for give constructive suggestions to improve the book will be highly appreciated.

The authors in general hope that the present book will be warmly received by the students and teachers. We shall indeed be very thankful to our colleagues for recommending this book to their students.

The authors wish to express their thanks to 'New Age International (P) Limited' Publishers and the editorial department for the inspiration given by them to undertake and complete this project.

AUTHORS

Syllabus

UTTARAKHAND TECHNICAL UNIVERSITY, DEHRADUN

TAS-301/TME-301/IPE-301: MATHEMATICS III

LTP

310

Unit I: Integral Transforms **(8)**

Fourier integral, Fourier's complex integral, Fourier sine and cosine transforms and applications to simple heat transfer equations.

Z-transform and its application to solve difference equations.

Unit II: Functions of a Complex Variable-I **(9)**

Analytic functions, C-R equations and harmonic functions, Line integral in the complex plane, Cauchy's integral theorem, Cauchy's integral formula for derivatives of analytic functions, Liouville's theorem, Fundamental theorem of algebra.

Unit III: Functions of a Complex Variable-II **(8)**

Representation of a function by power series, Taylor's and Laurent's series, singularities, zeroes and poles, Residue theorem, evaluation of real integrals of type $\int_0^{2\pi} f(\cos\theta, \sin\theta)\, d\theta$ and $\int_{-\infty}^{+\infty} f(x)\, dr$, Conformal mapping and bilinear transformations.

Unit IV: Statistics and Probability **(8)**

Moments, Moment generating functions, Skewness, Kurtossis, Correlation and Regression, Binomial distribution, Poisson distribution, Normal distribution.

Unit V: Curve Fitting and Solution of Equations **(5)**

Method of least squares and curve fitting of straight line and parabola, Solution of cubic and biquadratic equations.

Contents

UNIT I

INTEGRAL TRANSFORMS

[1–99]

1.1 Integral transforms 1
1.2 Definition 1
1.3 Fourier integral theorem 2
1.4 Fourier's complex integral 3
1.5 Fourier transform 4
1.6 Fourier sine and cosine integrals 4
1.7 Fourier sine and cosine transforms 5
1.8 Properties of Fourier transforms 27
1.9 Convolution 30
Problem set 1.1 35
1.10 Parseval's identity for Fourier transforms 38
1.11 Parseval's identity for cosine transform 38
1.12 Parseval's identity for sine transform 39
1.13 Fourier transform of derivatives 42
1.14 Relationship between Fourier and Laplace transforms 43
1.15 Fourier transforms of partial derivative of a function 43
1.16 Applications of Fourier transform of heat conduction (transfer equations) 45
Problem set 1.2 52
1.17 Z-transforms 53

1.18 Definition of *Z*-transform 54
1.19 Properties of *Z*-transforms 55
1.20 Change of scale 55
1.21 Shifting property 55
1.22 Multiplication by K 57
1.23 Division by K 57
1.24 Initial value 57
1.25 Final value 58
1.26 Partial sum 58
1.27 Convolution 59
Problem set 1.3 70
1.28 Inverse *Z*-transform 71
1.29 Method of finding inverse *Z*-transforms 71
Problem set 1.4 83
1.30 Difference equation 85
1.31 Order of a difference equation 85
Problem set 1.5 98

UNIT II

FUNCTIONS OF A COMPLEX VARIABLE I

[100–155]

2.1 Complex numbers 100
2.2 Function of a complex variable 100
2.3 Continuity 101
2.4 Derivative of $f(z)$ 101
2.5 Analytic function 101
2.6 Necessary and sufficient conditions for $f(z)$ to be analytic 101
2.7 Cauchy-Reimannian C-R equations in polar coordinate 104
2.8 Derivative of *W* in polar form 104
2.9 Harmonic function 106
2.10 Orthogonal system 106
2.11 Determination of the conjugate function 108
2.12 Milne's Thomson method 108
Problem set 2.1 128

2.13 Line integral in the complex plane 130
2.14 Simple closed curve 135
2.15 Cauchy's theorem (or Cauchy's integral theorem) 136
2.16 Multi-connected region 136
2.17 Extension of Cauchy's theorem to multi-connected region 137
2.18 Cauchy's integral formula 137
2.19 Cauchy's integral formula for the derivative of an analytic function 138
2.20 Higher order derivatives 139
2.21 Cauchy's inequality 139
2.22 Liouville's theorem 140
2.23 The fundamental theorem of algebra 141
Problem set 2.2 153

UNIT III

FUNCTIONS OF A COMPLEX VARIABLE II

[156–270]

3.1 Representation of a function by power series 156
3.1.1 Absolute convergence of power series $\sum_{n=o}^{\infty} a_n z^n$ 156
3.1.2 Radius of convergence of power series 156
3.2 Taylor's series or theorem 158
3.3 Laurent's series or theorem 159
Problem set 3.1 179
3.4 Singularity 181
3.4.1 Different types of singularities 182
3.5 Limit point of zeros 183
3.6 Limit point of poles 183
Problem set 3.2 187
3.7 Residues 187
3.7.1 Residue at infinity 188
3.8 Cauchy's residue theorem 188
3.9 Method of finding the residues 189
Problem set 3.3 200
3.10 Evaluation of real integrals by contour integration 202

3.10.1 Evaluation the integrals of the type $\int_0^{2\pi} f(\cos\theta, \sin\theta)d\theta$ by contour integration where $f(\cos\theta, \sin\theta)$ is a rational function 202
Problem set 3.4 215
3.11 Evaluation of $\int_{-\infty}^{\infty} f(x)dx$ 215
Problem set 3.5 228
3.12 Indented semi-circular contour or poles lie on the real axis 229
3.13 Branch point 230
Problem set 3.6 243
3.14 Conformal representation 243
3.14.1 Transformation or mapping 243
3.14.2 Conformal transformation 244
3.15 Some general transformations 244
3.16 The linear transformation 252
3.17 Bi-linear transformation or Mobius transformation 252
3.18 Cross ratio 252
Problem set 3.7 270

UNIT IV

STATISTICS AND PROBABILITY

[271–381]

4.1 Moments 271
4.1.1 Moments about a number 272
4.1.2 Relation between μ_r and μ'_r 272
4.1.3 Moments about the origin 273
4.1.4 Relation between μ_r, μ'_r and v_r 273
4.2 Moment generating functions 274
4.3 Skewness 282
4.3.1 Types of skewness 283
4.3.2 Measurement of skewness 283
4.4 Kurtosis 284
4.4.1 Measurement of kurtosis 284
4.4.2 Types of kurtosis 284

Problem set 4.1 ... 295
4.5 Correlation ... 297
4.5.1 Various types of correlation ... 297
4.5.2 Correlation coefficient ... 298
4.6 Scattar or dot diagrams ... 298
4.7 Correlation coefficient by covariance or Karl Pearson's method ... 299
4.8 Rank correlation (or correlation by rank method) ... 309
4.9 Edward Spearman's formula for tied ranks ... 310
4.10 Regression ... 313
4.10.1 Line of regression ... 313
4.10.2 To obtain the equation of line of regression of y on x ... 313
4.10.3 To obtain the equation of line of regression of x on y ... 314
4.11 Another form of equations of lines of regression ... 318
4.12 Some properties of regression coefficients ... 321
4.12.1 To find b_{yx} and b_{xy} using assumed mean ... 326
4.13 Error of prediction ... 327
Problem set 4.2 ... 334
4.14 Probability ... 337
4.14.1 Probability distribution ... 337
4.15 Binomial distribution (or Bernoulli distribution) ... 337
4.15.1 Conditions of Binomial distribution ... 338
4.15.2 Properties of Binomial distribution ... 338
4.15.3 Mean of the Binomial distribution ... 338
4.15.4 Variance of the Binomial distribution ... 339
4.15.5 Recurrence or recursion formula for the Binomial distribution ... 339
Problem set 4.3 ... 352
4.16 Poisson distribution ... 354
4.16.1 Properties of Poisson distribution ... 354
4.16.2 Poisson distribution as a limit of the Binomial distribution ... 354
4.16.3 Mean of the Poisson distribution ... 355
4.16.4 Variance of the Poisson distribution ... 356
4.16.5 Recurrence formula for Poisson distribution ... 356
Problem set 4.4 ... 367
4.17 Normal distribution (or Gaussian distribution) ... 369
4.17.1 Standard normal distribution ... 369
4.17.2 Properties of normal distribution ... 370
4.17.3 Importance of the normal distribution ... 370
4.17.4 Area under the normal curve ... 370
Problem set 4.5 ... 380

UNIT V

CURVE FITTING AND SOLUTION OF EQUATION

[382–441]

5.1 Curve fitting 382
5.2 Principle of least squares 382
5.3 Fitting of straight line 383
5.4 Fitting of parabola 384
5.5 Change of scale 384
5.6 Fitting of an exponential curve 393
5.7 Fitting of the curve $y = ax + bx^2$ 394
5.8 Fitting of the curve $y = ax + \frac{b}{x}$ 395
5.9 Fitting of the curve $y = \frac{c_0}{x} + c_1\sqrt{x}$ 396
Problem set 5.1 404
5.10 Polynomial 406
5.10.1 Degree of polynomial 406
5.11 Descarte's rule of signs 406
5.12 Cardon's method 408
Problem set 5.2 428
5.13 Descarte's method for solving biquadratic equation 429
5.14 Ferrari's method 437
Problem set 5.3 440

UNIT I

Integral Transforms

1.1 INTEGRAL TRANSFORMS

Integral transforms are used in the solution of partial differential equations. The choice of a particular transform to be used for the solution of a differential equation depend upon the nature of the boundary conditions of the equation and the facility with which the transform $F(s)$ can be converted to give the original function $f(x)$.

1.2 DEFINITION

The integral transform $F(s)$ of a function $f(x)$ is defined as

$$I[f(x)] = F(s) = \int_a^b f(x)k(s,x)dx$$

where $k(s, x)$ is a known function of s and x, called the kernel of the transform, s is called the parameter of the transform and $f(x)$ is cal led the inverse transform of $F(s)$.

Some of the well known transforms are given below :

(i) Laplace transform. $k(s, x) = e^{-sx}$

$$\{f(x)\} = F(s) = \int_0^\infty f(x)e^{-sx}$$

(ii) Fourier complex transform. $k(s, x) = e^{isx}$

$$F\{f(x)\} = F(s) = \int_{-\infty}^\infty f(x)e^{isx}\ dx$$

(iii) Hankel transform. $k(s, x) = x\ J_n(sx)$

$$H_n\{f(x)\} = F(s) = \int_0^\infty f(x)x\ J_n\ (sx)\ dx$$

where $J_n(sx)$ is the Bessel function of the first kind of order n.

(iv) Melline transform. $k(s, x) = x^{s-1}$

$$M\{f(x)\} = F(x) = \int_0^{\infty} f(x)x^{s-1}\, dx$$

(v) Hilbert transform. $K(s, x) = \dfrac{1}{s-x}$

$$F(s) = \int_{-\infty}^{\infty} \frac{f(x)}{s-x}\, dx$$

1.3 FOURIER INTEGRAL THEOREM

This states that

$$f(x) = \frac{1}{\pi}\int_0^{\infty}\int_{-\infty}^{\infty} f(t)\cos u(t-x)\, du\, dt$$

The integral on the right hand side is called Fourier integral of $f(x)$.

Proof: We know that Fourier series of a function $f(x)$ in $(-c, c)$ is given by

$$f(x) = \frac{a_0}{2} + \sum_{n=1}^{\infty} a_n \cos\frac{n\pi x}{c} + \sum_{n=1}^{\infty} b_n \sin\frac{n\pi x}{c} \qquad \text{...(1)}$$

where a_0, a_n and b_n are given by

$$a_0 = \frac{1}{c}\int_{-c}^{c} f(t)dt,$$

$$a_n = \frac{1}{c}\int_{-c}^{c} f(t)\cos\frac{n\pi t}{c}\, dt$$

$$b_{n\,n} = \frac{1}{c}\int_{-c}^{c} f(t)\sin\frac{n\pi t}{c}\, dt$$

Substituting the values of a_0, a_n and b_n in (1), we get

$$f(x) = \frac{1}{2c}\int_{-c}^{c} f(t)dt + \sum_{n=1}^{\infty}\frac{1}{c}\int_{-c}^{c} f(t)\cos\frac{n\pi t}{c}\cos\frac{n\pi x}{c}\, dt + \sum_{n=1}^{\infty}\frac{1}{c}\int_{-c}^{c} f(t)\sin\frac{n\pi t}{c}\sin\frac{n\pi x}{c}\, dt$$

$$= \frac{1}{2c}\int_{-c}^{c} f(t)dt + \sum_{n=1}^{\infty}\frac{1}{c}\int_{-c}^{c} f(t)\left[\cos\frac{n\pi t}{c}\cos\frac{n\pi x}{c} + \sin\frac{n\pi t}{c}\sin\frac{n\pi x}{c}\right]dt$$

$$= \frac{1}{2c}\int_{-c}^{c} f(t)dt + \sum_{n=1}^{\infty}\frac{1}{c}\int_{-c}^{c} f(t)\cos\frac{n\pi}{c}(t-x)dt$$

| As $\cos(A-B) = \cos A\cos B + \sin A\sin B$

$$= \frac{1}{2c}\int_{-c}^{c} f(t)\left\{1 + 2\sum_{n=1}^{\infty}\cos\frac{n\pi}{c}(t-x)\right\}dt \qquad \text{...(2)}$$

Since cosine function are even functions *i.e.*, $\cos(-\theta) = \cos\theta$, so the expression

$$1 + 2\sum_{n=1}^{\infty}\cos\frac{n\pi}{c}(t-x) = \sum_{n=-\infty}^{\infty}\cos\frac{n\pi}{c}(t-x)$$

Therefore, (2) becomes

$$f(x) = \frac{1}{2c}\int_{-c}^{c} f(t)\left\{\sum_{n=-\infty}^{\infty} \cos\frac{n\pi}{c}(t-x)\right\}dt$$

$$= \frac{1}{2\pi}\int_{-c}^{c} f(t)\left\{\frac{\pi}{c}\sum_{n=-\infty}^{\infty} \cos\frac{n\pi}{c}(t-x)\right\}dt \qquad ...(3)$$

Now assume that c increases indefinitely, so that we may write $\frac{n\pi}{c} = u$ and $\frac{\pi}{c} = \delta u$.

i.e., when c $\to \infty$, then δu $\to$ 0 $n = \infty$ then $u = \infty$

$\therefore$ The term $\lim_{c\to\infty} \sum_{n=-\infty}^{\infty} \cos\frac{n\pi}{c}(t-x)\cdot\frac{\pi}{c} = \lim_{\delta u\to 0} \sum_{u=\infty}^{\infty} \cos u(t-x)\,\delta u$

$$\lim_{c\to\infty}\left\{\frac{\pi}{c}\sum_{n=-\infty}^{\infty} \cos\frac{n\pi}{c}(t-x)\right\} = \int_{-\infty}^{\infty} \cos u(t-x)du \qquad \left|\; \text{As } \int_a^b f(x)dx = \lim_{\delta x\to 0}\sum_{x=a}^{b} f(x)\delta x\right.$$

$$= 2\int_0^{\infty} \cos u(t-x)du \qquad ...(4)$$

Substituting (4) in (5), we obtain

$$f(x) = \frac{1}{2\pi}\int_{-\infty}^{\infty} f(t)\left\{2\int_0^{\infty} \cos u(t-x)du\right\}dt \qquad \left|\; \text{As } c\to\infty\right.$$

Thus, $\boxed{f(x) = \frac{1}{\pi}\int_0^{\infty}\int_{-\infty}^{\infty} f(t)\cos u(t-x)du\,dt.}$ **Proved.** ...(5)

Notes: The following conditions on $f(x)$ are assumed.

(*i*) $f(x)$ is defined as single-valued except at finite points in $(-c, c)$.

(*ii*) $f(x)$ is periodic outside $(-c, c)$ with period $2c$.

(*iii*) $f(x)$ and $f'(x)$ are sectionally continuous in $(-c, c)$.

(*iv*) $\int_{-\infty}^{\infty} |f(x)|\,dx$ converges, *i.e.* $f(x)$ is absolutely integrable in $(-\infty, \infty)$.

1.4 FOURIER'S COMPLEX INTEGRAL

$$f(x) = \frac{1}{2\pi}\int_{-\infty}^{\infty} e^{-iux}du\int_{-\infty}^{\infty} f(t)e^{iut}dt$$

Proof:

$\therefore$ $\int_{-\infty}^{\infty} \sin u(t-x)du = 0$ [Since, sin u $(t-x)$ is odd]

By Fourier's integral theorem, we know that

$$f(x)\frac{1}{2\pi}\int_{-\infty}^{\infty}\int_{-\infty}^{\infty} f(t)\cos u(t-x)\,du\,dt \qquad ...(1)$$

For the term $e^{iu(t-x)}$ we add the integral $\frac{1}{2\pi}\int_{-\infty}^{\infty}\int_{-\infty}^{\infty} f(t)\cdot \sin u(t-x)\,du\,dt$ with multiplied by i we get

$$f(x)=\frac{1}{2\pi}\int_{-\infty}^{\infty}\int_{-\infty}^{\infty} f(t)\cos u(t-x)du\,dt+\frac{i}{2\pi}\int_{-\infty}^{\infty} f(t)dt\int_{-\infty}^{\infty} \sin u(t-x)du$$

Since the value of integral

$$\int_{-\infty}^{\infty} \sin x(t-x)\,du \;=0$$

As sin is an odd function, so there is no effect on edition of new term in fourier integral theorem

or $$f(x) = \frac{1}{2\pi}\int_{-\infty}^{\infty} f(t)dt\int_{-\infty}^{\infty} [\cos u(t-x)+i\sin u(t-x)]du$$

$$= \frac{1}{2\pi}\int_{-\infty}^{\infty} f(t)dt\int_{-\infty}^{\infty} e^{iu(t-x)}du$$

or $$\boxed{f(x)=\frac{1}{2\pi}\int_{-\infty}^{\infty} e^{-iux}du\int_{-\infty}^{\infty} f(t)e^{iut}dt} \qquad ...(2)$$

Relation (2) is called Fourier's Complex Integral.

1.5 FOURIER TRANSFORM

We have $$f(x)=\frac{1}{2\pi}\int_{-\infty}^{\infty} e^{-iux}du\int_{-\infty}^{\infty} f(t)e^{iut}dt$$

Let $$=\frac{1}{2\pi}\int_{-\infty}^{\infty} e^{-isx}ds\int_{-\infty}^{\infty} f(t)e^{ist}dt \quad (u=s) \qquad ...(1)$$

$$f(s) = \int_{-\infty}^{\infty} f(t)e^{ist}dt \qquad ...(2)$$

then $$f(x)=\frac{1}{2\pi}\int_{-\infty}^{\infty} e^{-isx}F(s)ds. \qquad ...(3)$$

The integral (2) is called Fourier transform of $f(x)$ and linoted as $F\{f(x)\}$
And integral (3) $f(x)$ is called Inverse Fourier transform of $F(s)$.

1.6 FOURIER SINE AND COSINE INTEGRALS

The Fourier sine and cosine transform defined as follows:

$$f(x)=\frac{2}{\pi}\int_{0}^{\infty} \sin ux\,du\int_{0}^{\infty} f(t)\sin ut\,dt \qquad \text{(Fourier sine integral)}$$

$$f(x)=\frac{2}{\pi}\int_{0}^{\infty} \cos ux\,du\int_{0}^{\infty} f(t)\cos ut\,dt \qquad \text{(Fourier cosine integral)}$$

Proof: We know that

$$\cos u\,(t-x) = \cos\,(ut-ux)$$

or $$\cos u\,(t-x) = \cos ut\cos ux+\sin ut\sin ux$$

Then equation (5) of article 1.3, can be written as

$$f(x)=\frac{1}{\pi}\int_0^{\infty}\int_{-\infty}^{\infty} f(t)(\cos ut\cos ux+\sin ut\sin ux)\,du\,dt$$

$$f(x)=\frac{1}{\pi}\int_0^{\infty}\int_{-\infty}^{\infty} f(t)\cos ut\cos ux\,du\,dt+\frac{1}{\pi}\int_0^{\infty}\int_{-\infty}^{\infty} f(t)\sin ut\sin ux\,du\,dt \qquad ...(1)$$

Case 1. *When* $f(t)$ *is odd*

$\therefore f(t)\cos ut$ is odd hence $\int_0^{\infty}\int_{-\infty}^{\infty} f(t)\cos ut\cos ux\,du\,dt=0$

From (1), we have

$$\therefore \boxed{f(x)=\frac{2}{\pi}\int_0^{\infty}\sin ux\,du\int_0^{\infty} f(t)\sin ut\,dt} \qquad ...(2)$$

The relation (2) is called **Fourier sine integral.**

Case 2. *When* $f(t)$ *is even*

$\therefore\ f(t)\sin ut$ is odd. so $\int_0^{\infty}\int_{-\infty}^{\infty} f(t)\sin ut\sin ux\,du\,dt=0$

From (1), we have

$$\boxed{f(x)=\frac{2}{\pi}\int_0^{\infty}\cos ux\,du\int_0^{\infty} f(t)\cos ut\,dt} \qquad ...(3)$$

The relation (3) is known as **Fourier cosine integral**.

1.7 FOURIER SINE AND COSINE TRANSFORMS

We know that the Fourier sine integral is

$$f(x)=\frac{2}{\pi}\int_0^{\infty}\sin sx\,ds\int_0^{\infty} f(t)\sin st\,dt \qquad (u=s)$$

Let $$F(s)=\int_0^{\infty} f(t)\sin st\,dt \qquad ...(1)$$

So $$f(x)=\frac{2}{\pi}\int_0^{\infty}\sin sx\,F(s)\,ds \qquad ...(2)$$

The integral (1) is called Fourier sine transform of $f(x)$ and denoted by $Fs\{f(x)\}$

And the integral (2) is called inverse Fourier sine transform of $F(s)$

Again we know that the Fourier cosine integral is

$$f(x)=\frac{2}{\pi}\int_0^{\infty}\cos sx\,ds\int_0^{\infty} f(t)\cos st\,dt \qquad (u=s)$$

Let $$F(s)=\int_0^{\infty} f(t)\cos st\,dt \qquad ...(3)$$

Then $$f(x)=\frac{2}{\pi}\int_0^{\infty}\cos sx\,dF(s)\,ds \qquad ...(4)$$

The integral (3) is called Fourier cosine integral of $f(x)$ and denoted by $Fc\ \{f(x)\}$

The integral (4), is called inverse Fourier cosine transform of $F(s)$.

Some Standard Results:

1. $\int_{-\infty}^{\infty} e^{-x^2} dx = \sqrt{\pi}$

2. $\int_0^{\infty} \frac{e^{ax} - e^{-ax}}{e^{\pi x} - e^{-\pi x}} dx = \frac{1}{2} \tan \frac{a}{2}$

3. $\int_0^{\infty} \frac{e^{ax} + e^{-ax}}{e^{\pi x} - e^{-\pi x}} dx = \frac{1}{2} \sec \frac{a}{2}$

4. $\int_0^{\infty} \frac{\sin ax}{x} dx = \frac{\pi}{2}$ if $a > 0$

Example 1: Express the function

$$f(x) = \begin{cases} 1, & \text{when } |x| \le 1 \\ 0, & \text{when } |x| > 1 \end{cases}$$

as a Fourier integral. Hence, evaluate

$$\int_0^{\infty} \frac{\sin \lambda \cos \lambda x}{\lambda} d\lambda.$$

[U.T.U. 2008]

Sol. The Fourier Integral for $f(x)$ is

$$f(x) = \frac{1}{\pi} \int_0^{\infty} \int_{-\infty}^{\infty} f(t) \cos \lambda(t - x) dt\, d\lambda$$

$$= \frac{1}{\pi} \int_0^{\infty} \int_{-1}^{1} \cos \lambda(t - x) dt\, d\lambda \qquad \left[\because f(t) = \begin{cases} 1, & -1 < t < 1 \\ 0, & \text{otherwise} \end{cases} \right]$$

$$= \frac{1}{\pi} \int_0^{\infty} \left[\frac{\sin \lambda(t - x)}{\lambda} \right]_{-1}^{1} d\lambda$$

$$= \frac{1}{\pi} \int_0^{\infty} \frac{\sin \lambda(1 - x) + \sin \lambda(1 + x)}{\lambda} d\lambda$$

$$\left| \text{As } \sin C + \sin D = 2 \sin \frac{C + D}{2} \cos \frac{C - D}{2} \right.$$

Thus $$f(x) = \frac{2}{\pi} \int_0^{\infty} \frac{\sin \lambda \cos \lambda x}{\lambda} d\lambda. \quad \textbf{Ans.}$$

or $$\int_0^{\infty} \frac{\sin \lambda \cos \lambda x}{\lambda} d\lambda = \frac{\pi}{2} f(x)$$

or $$\int_0^{\infty} \frac{\sin \lambda \cos \lambda x}{\lambda} d\lambda = \begin{cases} \frac{\pi}{2}, & \text{for } |x| < 1 \\ 0, & \text{for } |x| > 1 \end{cases}$$

For $|x| = 1$ *i.e.*, $x = \pm 1$ which is a point of discontinuity of $f(x)$, value of integral $= \dfrac{\frac{\pi}{2} + 0}{2} = \dfrac{\pi}{4}$ **Ans.**

Example 2: Using Fourier sine integral, show that

$$\int_0^\infty \left(\frac{1-\cos\pi\lambda}{\lambda}\right)\sin(x\lambda)\,d\lambda = \begin{cases} \dfrac{\pi}{2}, & \text{when } 0<x<\pi \\ 0, & \text{when } x>\pi \end{cases}$$

Sol. Let $f(x)=\begin{cases} \dfrac{\pi}{2}, & 0<x<\pi \\ 0, & x>\pi \end{cases}$

Using Fourier sine integral, we have

$$f(x)=\frac{2}{\pi}\int_0^\infty \sin\lambda x\,d\lambda \int_0^\infty f(t)\sin\lambda t\,dt$$

$$=\frac{2}{\pi}\int_0^\infty \sin\lambda x\left[\int_0^\pi \frac{\pi}{2}\sin\lambda t\,dt\right]d\lambda$$

$$=\int_0^\infty \sin\lambda x\left(-\frac{\cos\lambda t}{\lambda}\right)_0^\pi d\lambda$$

$$=\int_0^\infty \left(\frac{1-\cos\pi\lambda}{\lambda}\right)\sin(x\lambda)\,d\lambda \qquad \Big|\ \text{As } \cos 0=1$$

$$\therefore \int_0^\infty \left(\frac{1-\cos\pi\lambda}{\lambda}\right)\sin(x\lambda)\,d\lambda = f(x) = \begin{cases} \dfrac{\pi}{2}, & 0<x<\pi \\ 0, & x>\pi \end{cases}$$ **Proved.**

Example 3: Find the complex form of the Fourier integral representation of

$$f(x) = \begin{cases} e^{-kx}, & x>0 \text{ and } k>0 \\ 0, & \text{otherwise} \end{cases}$$

Sol. We know that

$$f(x)=\frac{1}{2\pi}\int_{-\infty}^{\infty} e^{-iux}\,du\int_{-\infty}^{\infty} f(t)e^{iut}\,dt \qquad \text{...(1)}$$

Putting the value $f(t)$ in (1), we get

$$f(x) = \frac{1}{2\pi}\int_{-\infty}^{\infty} e^{-iux}\,du\left[\int_0^\infty e^{-kt}e^{iut}\,dt\right]$$

$$=\frac{1}{2\pi}\int_{-\infty}^{\infty} e^{-iux}\,du\left[\int_0^\infty e^{-(k-iu)t}\,dt\right]$$

$$=\frac{1}{2\pi}\int_{-\infty}^{\infty} e^{-iux}\,du\left[\frac{e^{-(k-iu)t}}{-(k-iu)}\right]_0^\infty = \frac{1}{2\pi}\int_{-\infty}^{\infty}\frac{e^{-iux}}{k-iu}\,du.$$ **Ans.**

Example 4: Find the Fourier sine integral for

$$f(x)=e^{-\beta x} \qquad (\beta>0)$$

Hence, show that $\dfrac{\pi}{2}e^{-\beta x}=\displaystyle\int_0^\infty \frac{\lambda\sin\lambda x}{\beta^2+\lambda^2}\,d\lambda.$

Sol. The Fourier sine integral of $f(x)$ is

$$f(x)=\frac{2}{\pi}\int_0^\infty \sin\lambda x\,d\lambda\int_0^\infty f(t)\sin\lambda t\,dt \qquad ...(1)$$

Putting the value of $f(t)$ in (1), we get

$$e^{-\beta x}=\frac{2}{\pi}\int_0^\infty \sin\lambda x\,d\lambda\int_0^\infty e^{-\beta t}\sin\lambda t\,dt$$

$$=\frac{2}{\pi}\int_0^\infty \sin\lambda x\,d\lambda\left[\frac{e^{-\beta t}}{(\beta^2+\lambda^2)}(-\beta\sin\lambda t-\lambda\cos\lambda t)_0^\infty\right]$$

$$=\frac{2}{\pi}\int_0^\infty \sin\lambda x d\lambda\left[0+\frac{\lambda}{\beta^2+\lambda^2}\right]$$

$$\Rightarrow \quad e^{-\beta x}=\frac{2}{\pi}\int_0^\infty \frac{\lambda\sin\lambda x}{\beta^2+\lambda^2}d\lambda \quad \text{or} \quad \frac{\pi}{2}e^{-\beta x}=\int_0^\infty \frac{\lambda\sin\lambda x}{\beta^2+\lambda^2}d\lambda. \quad \textbf{Proved.}$$

Example 5: Using Fourier cosine integral representation of an appropriate function, show that

$$\int_0^\infty \frac{\cos wx}{k^2+w^2}dw=\frac{\pi e^{-kx}}{2k}. \qquad (x>0,\ k>0)$$

Sol. Let $f(x) = e^{-kx}$ we know that Fourier integral is

$$f(x)=\frac{2}{\pi}\int_0^\infty \cos uxdu\int_0^\infty f(t)\cos ut\,dt$$

Putting the value of $f(t)$ and replacing *u* by *w*, we get

$$e^{-kx}=\frac{2}{\pi}\int_0^\infty \cos wx\,dw\int_0^\infty e^{-kt}\cos wt\,dt \qquad \Big|\, f(t)=e^{-kt}$$

$$=\frac{2}{\pi}\int_0^\infty \cos wx\,dw\left[\frac{e^{-kt}}{k^2+w^2}\{-k\cos wt+w\sin wt\}_0^\infty\right]$$

$$=\frac{2}{\pi}\int_0^\infty \cos wx\,dw\left[0+\frac{k}{k^2+w^2}\right]=\frac{2k}{\pi}\int_0^\infty \frac{\cos wx\,dw}{k^2+w^2}$$

or $$\int_0^\infty \frac{\cos wx}{k^2+w^2}dw=\frac{\pi e^{-kx}}{2k}. \quad \textbf{Proved.}$$

Example 6: Find the Fourier sine transform of $e^{-|x|}$. Hence, evaluate $\int_0^\infty \frac{x\sin mx}{1+x^2}dx$.

Sol. In the interval $(0,\infty)$, x is always positive, therefore, $e^{-|x|}=e^{-x}$.

The Fourier sine integral of $f(x)$ is

$$F_s\{e^{-x}\}=\int_0^\infty e^{-x}\sin sx\,dx=\frac{s}{1+s^2}=F(s) \quad \text{(say)}$$

Now the inverse sine transform of $F(s)$ is e^{-x}.
Using inversion formula for the sine transform, we get

$$e^{-x}=\frac{2}{\pi}\int_0^\infty F(s)\sin sx\,ds$$

or

$$e^{-x}=\frac{2}{\pi}\int_0^\infty \frac{s}{1+s^2}\sin sx\,ds$$

Replacing x by m, we get

$$e^{-m}=\frac{2}{\pi}\int_0^\infty \frac{s}{1+s^2}\sin ms\,ds$$

$$=\frac{2}{\pi}\int_0^\infty \frac{x\sin mx}{1+x^2}dx \quad \text{| By definite integral property}$$

Hence, we get $\int_0^\infty \frac{x\sin mx}{1+x^2}dx=\frac{\pi}{2}e^{-m}$. **Ans.**

Example 7: Find the Fourier transform of

$$f(x)=\begin{cases}1-x^2, & \text{if } |x|\le 1\\ 0, & \text{if } |x|>1\end{cases}$$

and use it to evaluate

$$\int_0^\infty\left(\frac{x\cos x-\sin x}{x^3}\right)\cos\frac{x}{2}dx.$$

Sol. We have

$$f(x)=\begin{cases}1-x^2, & -1<x<1\\ 0, & |x|>1\end{cases}$$

The Fourier transform of a function $f(x)$ is given by

$$F(s)=\int_{-\infty}^\infty f(x).e^{isx}dx \quad \text{...(1)}$$

Substituting the value of $f(x)$ in (1), we get

$$F(s)=\int_{-1}^1 (1-x^2)e^{isx}dx$$

Integrating by parts, we get

$$F(s)=\left[(1-x^2)\frac{e^{isx}}{is}-(-2x)\frac{e^{isx}}{(is)^2}+(-2)\frac{e^{isx}}{(is)^3}\right]_{-1}^{1}$$

$$=\left[-2\frac{e^{is}}{s^2}+2\frac{e^{is}}{is^3}-2\frac{e^{-is}}{s^2}-2\frac{e^{-is}}{is^3}\right]$$

$$=\left[-\frac{2}{s^2}(e^{is}+e^{-is})+\frac{2}{is^3}(e^{is}-e^{-is})\right]$$

$$=\left[-\frac{2}{s^2}(2\cos s)+\frac{2}{is^3}(2i\sin s)\right]$$

$$=\frac{4}{s^3}[-s\cos s+\sin s]$$

By inversion formula for Fourier transform, we have

$$f(x)=\frac{1}{2\pi}\int_{-\infty}^{\infty}F(s)e^{-isx}\cdot ds$$

$$=\frac{1}{2\pi}\int_{-\infty}^{\infty}\frac{4}{s^3}(-s\cos s+\sin s)e^{-isx}\cdot ds$$

$$=\frac{2}{\pi}\int_{-\infty}^{\infty}\left(\frac{-s\cos s+\sin s}{s^3}\right)\cdot(\cos x-i\sin sx)\,ds$$

$$=\frac{4}{\pi}\int_{0}^{\infty}\left(\frac{-s\cos s+\sin s}{s^3}\right)\cdot\cos sx\,ds$$

| The second term is 0 by (definite integral property)

Putting $x=\frac{1}{2}$, we get

$$f(1/2)=\frac{4}{\pi}\int_{0}^{\infty}\left(\frac{-s\cos s+\sin s}{s^3}\right)\cdot\cos\frac{s}{2}\,ds$$

$$\frac{3}{4}=-\frac{4}{\pi}\int_{0}^{\infty}\left(\frac{s\cos s-\sin s}{s^3}\right)\cdot\frac{\cos s}{2}\,ds$$

| $f(1/2)=1-\frac{1}{4}=\frac{3}{4}$

$$-\frac{3\pi}{16}=\int_{0}^{\infty}\left(\frac{s\cos s-\sin s}{s^3}\right)\cdot\cos\frac{s}{2}\,ds$$

or $\displaystyle\int_{0}^{\infty}\left(\frac{x\cos x-\sin x}{x^3}\right)\cos\frac{x}{2}\,dx=-\frac{3\pi}{16}$. **Ans.**

| As $\int_0^a f(x)dx=\int_0^a f(t)dt$

Example 8: Find the Fourier transform of $f(x)$

$$f(x)=\begin{cases}1, & |x|<a \text{ i.e. } -a<x<a\\ 0, & |x|>a \text{ i.e. } x>a \text{ or } x<-a\end{cases}$$

and have evaluate

$$(a)\ \int_{-\infty}^{\infty}\frac{\sin(sa)\cos(sx)}{s}ds; \qquad (b)\ \int_{0}^{\infty}\frac{\sin s}{s}ds.$$

Sol. We have the Fourier transform of $f(x)$

$$F(s)=\int_{-\infty}^{\infty}e^{isx}f(x)dx \qquad \text{...(1)}$$

$$=\int_{-\infty}^{-a}f(x)e^{isx}dx+\int_{-a}^{a}f(x)e^{isx}dx+\int_{a}^{\infty}f(x)e^{isx}dx$$

$$=\int_{-a}^{a}f(x)e^{isx}dx=\int_{-a}^{a}e^{isx}dx=\left[\frac{e^{isx}}{is}\right]_{-a}^{a}$$

$$=\frac{1}{is}\left[e^{isa}-e^{-isa}\right]=\frac{1}{is}\left[2i\sin(as)\right]$$

or $$F(x)=\frac{2\sin(as)}{s},\ s\neq 0$$

and if $s = 0$, then $$F(s)=\int_{-a}^{a}e^{0}.1dx \qquad \text{[From (1)]}$$

i.e. $$F(s)=[x]_{-a}^{a}=[a-(-a)]=2a$$

then $$f(x)=\frac{1}{2\pi}\int_{-\infty}^{\infty}F(s)e^{-isx}ds$$

$$\therefore \quad \frac{1}{2\pi}\int_{-\infty}^{\infty}\frac{2\sin(sa)}{s}e^{-isx}ds=\begin{cases}1, & |x|<a\\ 0, & |x|>a\end{cases} \quad -2$$

But $$\frac{1}{2\pi}\int_{-\infty}^{\infty}\frac{2\sin(sa)}{s}e^{-isx}\cdot ds=\frac{1}{\pi}\int_{-\infty}^{\infty}\frac{\sin(sa)\cos(sx)}{s}ds-\frac{i}{\pi}\int_{-\infty}^{\infty}\frac{\sin(sa)\sin(sx)}{s}ds$$

$$=\frac{1}{\pi}\int_{-\infty}^{\infty}\frac{\sin(sa)\cos(sx)}{s}ds-0$$

Since the integrand in the other integral is an odd function of s.

$\therefore$ From (2), we get

$$\int_{-\infty}^{\infty}\frac{\sin(sa)\cos(sx)}{s}ds=\begin{cases}\pi, & |x|<a\\ 0, & |x|>a\end{cases}$$

(*b*) If $x = 0$ and $a = 1$ in above result, then we get

$$\int_{-\infty}^{\infty} \frac{\sin s}{s} ds = \pi$$

$$\Rightarrow \quad 2\int_{0}^{\infty} \frac{\sin s}{s} ds = \pi \quad \text{or} \quad \int_{0}^{\infty} \frac{\sin s}{s} ds = \frac{\pi}{2}. \quad \textbf{Ans.}$$

Example 9: Find the Fourier complex transform of $f(x)$ if $f(x) = \begin{cases} e^{iwx}, & a < x < b \\ 0, & x < a, x > b \end{cases}$

Sol. We have $\quad F(s) = \int_{-\infty}^{\infty} e^{isx} f(x)\, dx$

$$= \left[\int_{-\infty}^{a} 0\, e^{isx} dx + \int_{a}^{b} e^{isx} e^{iwx} dx + \int_{b}^{\infty} e^{isx}.0 dx\right]$$

$$= \int_{a}^{b} e^{i(s+w)x} dx = \left[\frac{e^{i(s+w)x}}{i(s+w)}\right]_a^b$$

$$= i\left[\frac{e^{i(s+w)a} - e^{i(s+w)b}}{s+w}\right]. \quad \textbf{Ans.}$$

Example 10: Show that the Fourier transform of $f(x) = e^{-\frac{x^2}{2}}$ is $e^{-\frac{s^2}{2}}$.

Sol. We know

$$F(s) = \frac{1}{\sqrt{2\pi}} \int_{-\infty}^{\infty} f(x) e^{isx} dx = \frac{1}{\sqrt{2\pi}} \int_{-\infty}^{\infty} e^{-\frac{x^2}{2}} e^{isx} dx$$

$$= \frac{1}{\sqrt{2\pi}} \int_{-\infty}^{\infty} e^{-\frac{1}{2}x^2 + isx} dx = \frac{1}{\sqrt{2\pi}} \int_{-\infty}^{\infty} e^{-\frac{1}{2}(x^2 - 2isx)} dx$$

$$= \frac{1}{\sqrt{2\pi}} \int_{-\infty}^{\infty} e^{-\frac{1}{2}(x-is)^2 - \frac{1}{2}s^2} dx = \frac{1}{\sqrt{2\pi}} \int_{-\infty}^{\infty} e^{-\frac{1}{2}(x-is)^2} . e^{-\frac{s^2}{2}} dx,$$

$$= \frac{e^{-\frac{s^2}{2}}}{\sqrt{\pi}} . \int_{-\infty}^{\infty} e^{-y^2} dy, \text{ putting } \frac{x - is}{\sqrt{2}} = y, \text{ so that } dx = \sqrt{2} dy$$

$$= \frac{e^{-\frac{s^2}{2}}}{\sqrt{\pi}} \sqrt{\pi}, \text{ since } \int_{-\infty}^{\infty} e^{-y^2} dy = \sqrt{\pi}$$

$$= e^{-\frac{s^2}{2}}. \quad \textbf{Proved.}$$

Example 11: Find Fourier cosine transform of the function $f(x)$ if $f(x) = \begin{cases} \cos x, & 0 < x < a \\ 0, & x > a \end{cases}$

Sol. We have

$$F(s) = \int_0^\infty f(x)\cos sx\,dx$$

$$\Rightarrow \quad F(s) = \int_0^a f(x)\cos sx\,dx + \int_a^\infty f(x)\cos sx\,dx$$

$$= \int_0^a \cos x.\cos sx\,dx + \int_a^\infty 0.\cos sx\,dx$$

$$= \frac{1}{2}\int_0^a 2\cos x.\cos sx\,dx = \frac{1}{2}\int_0^a [\cos(1+s)x + \cos(1-s)x]dx$$

$$= \frac{1}{2}\left[\frac{\sin(1+s)x}{(1+s)} + \frac{\sin(1-s)x}{(1-s)}\right]_0^a$$

$$= \frac{1}{2}\left[\frac{\sin(1+s)a}{(1+s)} + \frac{\sin(1-s)a}{(1-s)}\right]. \quad \textbf{Ans.}$$

Example 12: Find the Fourier sine and cosine transform of $f(x) = e^{-ax}$.

Sol. The Fourier sine transforms of $f(x)$ is

$$F(s) = \int_0^\infty f(x)\sin(sx)dx$$

$$= \int_0^\infty e^{-ax}\sin(sx)dx = \left[\frac{e^{-ax}}{a^2+s^2}\{-a\sin(sx) - s\cos(sx)\}\right]_0^\infty$$

$$= -\left[\frac{e^{-ax}}{a^2+s^2}\{a\sin(sx) + s\cos(sx)\}\right]_0^\infty = \frac{s}{a^2+s^2}. \quad \textbf{Ans.}$$

and Fourier cosine transform is

$$F(s) = \int_0^\infty e^{-ax}.\cos(sx)dx$$

$$= \left[\frac{e^{-ax}}{a^2+s^2}\{-a\cos(sx) + s\sin(sx)\}\right]_0^\infty$$

$$= \left[0 + \frac{a}{a^2+s^2}\right] = \frac{a}{a^2+s^2}. \quad \textbf{Ans.}$$

Example 13: Find Fourier sine transform of $\frac{1}{x}$.

Sol.
$$F(s)=\int_0^\infty \frac{\sin sx}{x}dx$$

$$=\int_0^\infty \frac{\sin\theta}{\frac{\theta}{s}}\frac{d\theta}{s}$$ [Putting $sx=\theta$ so that $sdx=d\theta$]

$$=\int_0^\infty \frac{\sin\theta}{\theta}d\theta=\frac{\pi}{2}.$$ **Ans.**

Example 14: Find the Fourier cosine transform of

$$f(x)=e^{-2x}+4e^{-3x}$$

Sol. The Fourier cosine transforms of $f(x)$ is given by

$$F(s)=\int_0^\infty f(x)\cos sx\,dx$$

Putting the value of $f(x)$,we get

$$F(s)=\int_0^\infty \left(e^{-2x}+4e^{-3x}\right)\cos sx\,dx$$

$$=\int_0^\infty e^{-2x}\cos sx\,dx+\int_0^\infty 4e^{-3x}\cos sx\,dx$$

$$\left(\because \int e^{-ax}.\cos bx\,dx=\frac{e^{-ax}}{a^2+b^2}[b\sin bx-a\cos bx]\right)$$

$$=\frac{2}{s^2+4}+4.\frac{3}{s^2+9}=2\left[\frac{1}{s^2+4}+\frac{6}{s^2+9}\right].$$ **Ans.**

Example 15: Find the Fourier sine transform of $\frac{e^{-ax}}{x}$. Hence find Fourier sine transform of $1/x$.

Sol. The sine transform of the function $f(x)$ is given by *(U.T.U. 2008)*

$$F(s)=\int_0^\infty f(x)\sin(sx)dx \quad ...(1)$$

$$\therefore \quad F(s)=\int_0^\infty \frac{e^{-ax}}{x}.\sin(sx)dx.$$

Differentiating both sides w.r.t. '*s*', we get

$$\frac{d}{ds}[F(s)]=\int_0^\infty \frac{e^{-ax}}{x}(x\cos sx)dx=\int_0^\infty e^{-ax}\cos(sx)dx$$

$$=\left[\frac{e^{-ax}}{a^2+s^2}\{-a\cos(sx)+s\sin(sx)\}\right]_0^\infty$$

$$\frac{d}{ds}[F(s)] = \frac{a}{a^2+s^2}$$

Integrating w.r.t. 's', we get

$$F(s) = \int \frac{a}{a^2+s^2}\,ds = \tan^{-1}\left(\frac{s}{a}\right)+c \qquad \ldots(2)$$

But for $s = 0$, we have $F(s) = 0$ [from (1)]

Putting these values in the equation (2), we get

$$c = 0$$

$$F(s) = \tan^{-1}\left(\frac{s}{a}\right)$$

$$\therefore \quad \int_0^\infty \frac{e^{-ax}}{x}\cdot \sin(sx)dx = \tan^{-1}\left(\frac{s}{a}\right)$$

Taking $a \to 0$

$$\int_0^\infty \frac{1}{x}\sin sx\, dx = \tan^{-1}\infty = \frac{\pi}{2}. \quad \textbf{Ans.}$$

Example 16: Find the Fourier transform of $f(x)$

$$f(x) = \begin{cases} 1-\dfrac{x}{a}, & \text{if } 0<x<a \\ 1+\dfrac{x}{a}, & \text{if } -a<x<0 \\ 0, & \text{otherwise} \end{cases}$$

Sol. The Fourier transform of $f(x)$ is

$$F(s) = \int_{-\infty}^{\infty} f(x)e^{isx}dx = \int_{-a}^{0}\left(1+\frac{x}{a}\right)e^{isx}dx + \int_0^a\left(1-\frac{x}{a}\right)e^{isx}dx$$

Changing x by $-x$ in the first integral, we get

$$= \int_0^a\left(1-\frac{x}{a}\right)e^{-isx}dx + \int_0^a\left(1-\frac{x}{a}\right)e^{isx}dx$$

$$= \int_0^a 2\left(1-\frac{x}{a}\right)\frac{\left(e^{isx}+e^{-isx}\right)}{2}dx = 2\int_0^a\left(1-\frac{x}{a}\right)\cos(sx)dx$$

$$= 2\left[\left(1-\frac{x}{a}\right)\frac{\sin sx}{s} - \left(-\frac{1}{a}\right)\left(-\frac{\cos sx}{s^2}\right)\right]_0^a$$

$$= 2\left[-\frac{1}{as^2}\cos(sa)+\frac{1}{as^2}\right] = \frac{2}{as^2}[1-\cos(sa)]. \quad \textbf{Ans.}$$

Example 17: Find Fourier cosine integral of the function e^{-ax}. Hence show that

$$\int_0^{\infty} \frac{\cos \lambda x}{\lambda^2 + 1} d\lambda = \frac{\pi}{2} e^{-x}; \quad x \geq 0$$

Let $f(x) = e^{-x}$

Sol. We know that Fourier cosine integral

$$f(x) = \frac{2}{\pi} \int_0^{\infty} \cos \lambda x \, d\lambda \int_0^{\infty} f(t) \cos \lambda t \, dt$$

$$e^{-ax} = \frac{2}{\pi} \int_0^{\infty} \cos \lambda x \, d\lambda \int_0^{\infty} e^{-at} \cos \lambda t \, dt$$

$$= \frac{2}{\pi} \int_0^{\infty} \cos \lambda x \, d\lambda \left[\frac{e^{-at}}{a^2 + \lambda^2} (-a \cos \lambda t + \lambda \sin \lambda t) \right]_0^{\infty}$$

$$= \frac{2}{\pi} \int_0^{\infty} \cos \lambda x \, d\lambda \, . \, \frac{a}{a^2 + \lambda^2}$$

$$\Rightarrow \quad e^{-ax} = \frac{2a}{\pi} \int_0^{\infty} \frac{\cos \lambda x}{a^2 + \lambda^2} d\lambda \, . \quad \textbf{Ans.}$$

If $a = 1$

$$\int_0^{\infty} \frac{\cos \lambda x}{1 + \lambda^2} d\lambda = \frac{\pi}{2} e^{-x}. \quad \textbf{Proved.}$$

Example 18: Find the Fourier sine and cosine transform of

$$f(x) = \begin{cases} 1, & 0 < x < a \\ 0, & x > a \end{cases}$$

Sol. Fourier sine transform

$$F(s) = \int_0^{\infty} f(x) \sin sx \, dx = \int_0^{a} 1 \sin sx \, dx$$

$$F(s) = \left[-\frac{\cos sx}{s} \right]_0^a = \left[-\frac{\cos as}{s} + \frac{1}{s} \right]$$

Fourier cosine transform

$$F(s) = \int_0^{\infty} f(x) \cos sx \, dx = \int_0^{a} 1 \cos sx \, dx = \left[\frac{\sin sx}{s} \right]_0^a$$

$$= \frac{\sin as}{s} . \quad \textbf{Ans.}$$

Example 19: Find the Fourier cosine transform of

$$f(x) = \begin{cases} x, & \text{for} \quad 0 < x < \dfrac{1}{2} \\ 1 - x, & \text{for} \quad \dfrac{1}{2} < x < 1 \\ 0, & \text{for} \quad x > 1 \end{cases}$$

Sol. Fourier cosine transform

$$F(s) = \int_0^\infty f(x) \cos sx \, dx$$

$$= \int_0^{1/2} x \cos sx \, dx + \int_{1/2}^1 (1-x) \cos sx \, dx$$

$$= \left[x \frac{\sin sx}{s} - \left(\frac{-\cos sx}{s^2} \right) \right]_0^{1/2} + \left[(1-x) \frac{\sin sx}{s} - (-1) \frac{(-\cos sx)}{s^2} \right]_{1/2}^1$$

$$= \left[\frac{1}{2} \frac{\sin s/2}{s} + \frac{\cos s/2}{s^2} - \frac{1}{s^2} \right] + \left[-\frac{\cos s}{s^2} - \frac{1}{2} \frac{\sin s/2}{s} + \frac{\cos s/2}{s^2} \right]$$

$$= \left[-\frac{\cos s}{s^2} + \frac{2 \cos s/2}{s^2} - \frac{1}{s^2} \right]. \quad \textbf{Ans.}$$

Example 20: Obtain Fourier cosine transform of

$$f(x) = \begin{cases} x, & \text{for } 0 < x < 1 \\ 2 - x, & \text{for } 1 < x < 2 \\ 0, & \text{for } x > 2 \end{cases}$$

Sol. Fourier cosine transform

$$F(s) = \int_0^\infty f(x) \cos sx \, dx$$

$$= \left[\int_0^1 x \cos sx \, dx + \int_1^2 (2-x) \cos sx \, dx + \int_2^\infty 0 . \cos sx \, dx \right]$$

$$= \left[\left\{ x \frac{\sin sx}{s} - \left(-\frac{\cos sx}{s^2} \right) \right\}_0^1 + \left\{ (2-x) \frac{\sin sx}{s} - (-1) \left(-\frac{\cos sx}{s^2} \right) \right\}_1^2 \right]$$

$$= \left[\left(\frac{\sin s}{s} + \frac{\cos s}{s^2} \right) - \frac{1}{s^2} + \left(-\frac{\cos 2s}{s^2} \right) - \left(\frac{\sin s}{s} - \frac{\cos s}{s^2} \right) \right]$$

$$= \left[\frac{2 \cos s}{s^2} - \frac{1}{s^2} - \frac{\cos 2s}{s^2} \right]$$

$$= \left[\frac{2\cos s - 1 - (2\cos^2 s - 1)}{s^2}\right]$$

$$= \left[\frac{2\cos s - 1 - 2\cos^2 s + 1}{s^2}\right]$$

$$= \frac{2\cos s}{s^2}(1 - \cos s). \quad \textbf{Ans.}$$

Example 21: Find the complex Fourier transform of dirac-delta function $\delta(t-a)$.

Sol.
$$F\{\delta(t-a)\} = \int_{-\infty}^{\infty} e^{ist}\delta(t-a)dt$$

$$= \lim_{h\to 0} \int_a^{a+h} \frac{1}{h} e^{ist} dt$$

$$= \lim_{h\to 0} \frac{1}{h}\left(\frac{e^{ist}}{is}\right)_a^{a+h}$$

$$= \lim_{h\to 0} e^{isa}\left(\frac{e^{ish}-1}{ish}\right)$$

$$= e^{isa}. \quad \textbf{Ans.} \qquad \left[\text{since } \lim_{\theta\to 0}\frac{e^{\theta}-1}{\theta} = 1\right]$$

Note: Dirac-delta function $\delta(t-a)$ is defined as

$$\delta(t-a) = \lim_{h\to 0} I(h, t-a), \text{ where}$$

$$I(h, t-a) = \frac{1}{h} \quad \text{for} \quad a < t < a+h$$
$$= 0 \quad \text{for} \quad t < a \text{ and } t > a+h$$

Example 22: Find Fourier sine and cosine transform of (*a*) x^{n-1} (*b*) $\frac{1}{\sqrt{x}}$

Sol. (*a*)
$$F_s\left(x^{n-1}\right) = \int_0^{\infty} \sin sx . x^{n-1} dx$$

$$F_c\left(x^{n-1}\right) = \int_0^{\infty} \cos sx . x^{n-1} dx$$

$$\therefore \quad F_c\left(x^{n-1}\right) + iF_s\left(x^{n-1}\right) = \int_0^{\infty} (\cos sx + i\sin sx)x^{n-1}dx$$

$$= \int_0^{\infty} e^{isx} x^{n-1} dx$$

Putting $is\,x = -t$ so that $is\,dx = -dt$, we get

$$= \int_0^\infty e^{-t}\left(-\frac{t}{is}\right)^{n-1}\left(-\frac{dt}{is}\right)$$

$$= \frac{1}{(is)^n}(-1)^n \int_0^\infty e^{-t} t^{n-1} dt$$

$$= \frac{(i)^{2n}}{(i)^n s^n}\Gamma n = \frac{(i)^n}{s^n}\Gamma n \qquad \left|\ \text{As } \Gamma x = \int_0^\infty e^{-x}\cdot x^{n-1}\cdot de\right.$$

$$F_c(x^{n-1}) + iF_s(x^{n-1}) = \frac{\left(\cos\frac{\pi}{2} + i\sin\frac{\pi}{2}\right)^n \Gamma n}{s^n} = \frac{\left(\cos\frac{n\pi}{2} + i\sin\frac{n\pi}{2}\right)\Gamma n}{s^n}$$

Equating real and imaginary parts, we get

$$F_c\left(x^{n-1}\right) = \frac{\Gamma n}{s^n}\cos\frac{n\pi}{2}$$

$$F_s\left(x^{n-1}\right) = \frac{\Gamma n}{s^n}\sin\frac{n\pi}{2}. \quad \textbf{Ans.}$$

(*b*) $n = \frac{1}{2}$

$$F_c\left(\frac{1}{\sqrt{x}}\right) = \frac{\left\lceil\frac{1}{2}\right.}{s^{\frac{1}{2}}}\cos\frac{\pi}{4} = \frac{\sqrt{\pi}}{\sqrt{s}}\frac{1}{\sqrt{2}} = \sqrt{\frac{\pi}{2}}\frac{1}{\sqrt{s}} \qquad \left|\ \left\lceil\frac{1}{2}\right. = \sqrt{\pi}\right.$$

$$F_s\left(\frac{1}{\sqrt{x}}\right) = \frac{\left\lceil\frac{1}{2}\right.}{s^{\frac{1}{2}}}\sin\frac{\pi}{4} = \frac{\sqrt{\pi}}{\sqrt{s}}\frac{1}{\sqrt{2}} = \sqrt{\frac{\pi}{2}}\frac{1}{\sqrt{s}}. \quad \textbf{Ans.}$$

Example 23: Find the sine transform of $\dfrac{e^{ax} + e^{-ax}}{e^{\pi x} - e^{-\pi x}}$

Sol. We have
$$F(s) = \int_0^\infty \frac{e^{ax} + e^{-ax}}{e^{\pi x} - e^{-\pi x}}\sin(sx)dx$$

$$= \int_0^\infty \frac{e^{ax} + e^{-ax}}{e^{\pi x} - e^{-\pi x}}\left(\frac{e^{isx} - e^{-isx}}{2i}\right)dx$$

$$= \frac{1}{2i}\int_0^\infty \frac{e^{(a+is)x} - e^{-(a+is)x}}{e^{\pi x} - e^{-\pi x}}dx - \frac{1}{2i}\int_0^\infty \frac{e^{(a-is)x} - e^{-(a-is)x}}{e^{\pi x} - e^{-\pi x}}dx$$

$$= \frac{1}{2i}\frac{1}{2}.\tan\left(\frac{a+is}{2}\right) - \frac{1}{2i}\frac{1}{2}\tan\left(\frac{a-is}{2}\right)$$

$$\left[\because \int_0^\infty \frac{e^{ax}-e^{-ax}}{e^{\pi x}-e^{-\pi x}}dx = \frac{1}{2}\tan\left(\frac{a}{2}\right)\right]$$

$$= \frac{1}{4i}\left[\frac{\sin\left(\frac{a+is}{2}\right)}{\cos\left(\frac{a+is}{2}\right)} - \frac{\sin\left(\frac{a-is}{2}\right)}{\cos\left(\frac{a-is}{2}\right)}\right]$$

$$= \left[\frac{2\sin\left(\frac{a+is}{2}\right)\cos\left(\frac{a-is}{2}\right) - 2\sin\left(\frac{a-is}{2}\right)\cos\left(\frac{a+is}{2}\right)}{2\times 4i\cos\left(\frac{a+is}{2}\right)\cos\left(\frac{a-is}{2}\right)}\right]$$

$$= \frac{\sin a + \sin(is) - [\sin a - \sin is]}{4i[\cos(is)+\cos a]}$$

$$\left| \begin{array}{l} 2\sin A\cos B = \sin(A+B)+\sin(A-B) \\ 2\cos A\cos B = \cos(A+B)+\cos(A-B) \end{array}\right.$$

$$= \frac{2\sin(is)}{4i[\cos(is)+\cos a]} = \frac{\sinh s}{2[\cosh s+\cos a]}.$$ **Ans.**

Example 24: Find the sine and cosine transform of $x^n e^{-ax}$.

Sol. Let $f(x) = x^n e^{-ax}$

$$\therefore \quad F_s(s) = \int_0^\infty x^n e^{-ax}.\sin(sx)\,dx \qquad \text{...(1)}$$

We have $\int_0^\infty e^{-ax}.\sin(sx)\,dx = \left[\frac{e^{-ax}}{a^2+s^2}(-a\sin sx - s\cos sx)\right]_0^\infty$

$$= \frac{s}{a^2+s^2} = \frac{1}{2i}\left[\frac{1}{a-is} - \frac{1}{a+is}\right]$$

Differentiating both sides w.r.t. a, n times, we have

$$(-1)^n\int_0^\infty x^n e^{-ax}\sin sx\,dx = \frac{1}{2i}\left[\frac{d^n}{da^n}(a-is)^{-1} - \frac{d^n}{da^n}(a+is)^{-1}\right]$$

$$= \frac{1}{2i}(-1)^n n!\left[(a-is)^{-(n+1)} - (a+is)^{-(n+1)}\right]$$

Putting $a = r\cos\theta$, $s = r\sin\theta$, we get

$$= \frac{1}{2i}(-1)^n\, n!\left[2ir^{-(n+1)}\sin(n+1)\theta\right]$$

$$= (-1)^n\, n!\,\left(\frac{1}{r}\right)^{n+1}\sin(n+1)\theta$$

$$\therefore \quad \int_0^\infty x^n e^{-ax}\sin sx\, dx = n!\left[\frac{1}{\left(a^2+s^2\right)^{(n+1)/2}}\right]\sin\left\{(n+1)\tan^{-1}\left(\frac{s}{a}\right)\right\}$$

$$\left| \begin{array}{l} r = (a^2+s^2)^{1/2} \\ \theta = \tan^{-1}\left(\dfrac{s}{a}\right) \end{array}\right.$$

Hence, from (1)

$$F_s(s) = \frac{n!}{\left(a^2+s^2\right)^{\left(\frac{n+1}{2}\right)}}\sin\left\{(n+1)\tan^{-1}\left(\frac{s}{a}\right)\right\}. \quad \textbf{Ans.}$$

Also $\qquad F_s(s) = \int_0^\infty f(x)\cos(sx)dx$

$$= \int_0^\infty x^n e^{-ax}\cos(sx)dx \qquad ...(2)$$

We have $\displaystyle\int_0^\infty e^{-ax}\cos(sx)dx = \left[\frac{e^{-ax}}{a^2+s^2}(-a\cos sx + s\sin sx)\right]_0^\infty$

$$= \frac{a}{a^2+s^2} = \frac{1}{2}\left[\frac{1}{a-is}+\frac{1}{a+is}\right]$$

Differentiating both sides w.r.t. a, n times, we have

$$(-1)^n\int_0^\infty x^n e^{-ax}\cos(sx)dx = \frac{1}{2}(-1)^n\, n!\left[(a-is)^{-(n+1)}+(a+is)^{-(n+1)}\right]$$

$$\therefore \quad \int_0^\infty x^n e^{-ax}\cos(sx)dx = (-1)^n\, n!\left(\frac{1}{r}\right)^{n+1}\cos(n+1)\theta$$

Putting $a = r\cos\theta$, $s = r\sin\theta$, then we have

$$F_c(s) = \frac{n!\cos\left\{(n+1)\tan^{-1}\left(\frac{s}{a}\right)\right\}}{\left(a^2+s^2\right)^{(n+1)/2}}. \quad \textbf{Ans.}$$

Example 25: Show that

$$(a)\quad F_s[xf(x)] = -\frac{d}{ds}F_c(s) \qquad (b)\quad F_c[xf(x)] = \frac{d}{ds}F_s(s)$$

and hence find Fourier cosine and sine transform of xe^{-ax}.

Sol. (*a*) We know that $F_c(s) = \int_0^\infty f(x)\cos sx\,dx$

$$\frac{d}{ds}F_c(s) = -\int_0^\infty x\,f(x)\sin sx\,dx$$

$$= -F_s\{xf(x)\}. \quad \textbf{Proved.}$$

(*b*) And $F_s(s) = \int_0^\infty f(x)\sin sx\,dx$

$$\frac{d}{ds}F_s(s) = \int_0^\infty x\,f(x)\cos sx\,dx$$

$$= F_c\{xf(x)\}. \quad \textbf{Proved.}$$

Fourier cosine transform

$$F_c(xe^{-ax}) = \frac{d}{ds}F_s(e^{-ax})$$

$$= \frac{d}{ds}\left[\frac{s}{a^2+s^2}\right] \qquad \text{(Using example 12)}$$

$$= \frac{(a^2+s^2)-s(2s)}{(a^2+s^2)^2} = \frac{a^2-s^2}{(a^2+s^2)^2}. \quad \textbf{Ans.}$$

Fourier sine transform

$$F_s(xe^{-ax}) = -\frac{d}{ds}F_c(e^{-ax}) = -\frac{d}{ds}\left[\frac{a}{a^2+s^2}\right] \qquad \text{(Using example 12)}$$

$$= \frac{2as}{(a^2+s^2)^2}. \quad \textbf{Ans.}$$

Example 26: Find the inverse Fourier transform of $f(x) = e^{-|s|y}$, where $s \in [-\infty, \infty]$.

Sol. We have $|s| = \begin{cases} -s, \text{ if } s < 0 \\ s, \text{ if } s > 0 \end{cases}$

$$f(x) = \frac{1}{2\pi}\int_{-\infty}^{\infty} F(s)e^{-isx}\,ds$$

$$= \frac{1}{2\pi}\left[\int_{-\infty}^{0} F(s)e^{-isx}\,ds + \int_0^\infty F(s)e^{-isx}\,ds\right]$$

$$= \frac{1}{2\pi}\left[\int_{-\infty}^{0} e^{sy} e^{-isx} ds + \int_{0}^{\infty} e^{-sy} e^{-isx} ds\right]$$

$$= \frac{1}{2\pi}\left[\int_{-\infty}^{0} e^{(y-ix)s} ds + \int_{0}^{\infty} e^{-(y+ix)s} ds\right]$$

$$= \frac{1}{2\pi}\left[\left\{\frac{e^{(y-ix)s}}{(y-ix)}\right\}_{-\infty}^{0} + \left\{\frac{e^{-(y+ix)s}}{-(y+ix)}\right\}_{0}^{\infty}\right]$$

$$= \frac{1}{2\pi}\left[\frac{1}{(y-ix)}\{1-0\} - \frac{1}{(y+ix)}\{0-1\}\right]$$

$$= \frac{1}{2\pi}\left[\frac{1}{y-ix} + \frac{1}{y+ix}\right] = \frac{1}{2\pi}\left[\frac{y+ix+y-ix}{(y-ix)(y+ix)}\right]$$

$$= \frac{1}{2\pi}\left[\frac{2y}{y^2+x^2}\right] = \frac{1}{\pi}\left[\frac{y}{y^2+x^2}\right].$$ **Ans.**

Example 27: Find the Fourier cosine transform of $\frac{1}{1+x^2}$ and hence, find Fourier sine transform of $\frac{x}{1+x^2}$.

Sol. Fourier cosine transform

$$F(s) = \int_0^\infty f(x)\cos sx\, dx = I(\text{say})$$

$$\Rightarrow \quad I = \int_0^\infty \frac{1}{1+x^2}\cos sx\, dx \qquad \text{...(1)}$$

$$\Rightarrow \quad \frac{dI}{ds} = -\int_0^\infty \frac{-x\sin sx}{1+x^2} dx$$

$$= -\int_0^\infty \frac{(1+x^2-1)\sin sx}{x(1+x^2)} dx$$

$$= -\int_0^\infty \frac{\sin sx}{x} dx + \int_0^\infty \frac{\sin sx}{x(1+x^2)} dx$$

$$\frac{dI}{ds} = -\frac{\pi}{2} + \int_0^\infty \frac{\sin sx}{x(1+x^2)} dx \quad \left[\text{As} \int_0^\infty \frac{\sin n\theta}{\theta} d\theta = \frac{\pi}{2}\right] \qquad \text{...(2)}$$

$$\frac{d^2 I}{ds^2} = \int_0^\infty \frac{\cos sx}{(1+x^2)} dx$$

$$= I$$

$$\frac{d^2I}{ds^2} - I = 0 \qquad ...(3)$$

$\therefore$ Solution of (3) is $\quad I = c_1 e^s + c_2 e^{-s}$ ...(4)

$$\frac{dI}{ds} = c_1 e^s - c_2 e^{-s} \qquad ...(5)$$

When $s = 0$

$$I = \int_0^\infty \frac{1}{1+x^2} dx = \frac{\pi}{2} \quad \text{[from (1)]}$$

and
$$\frac{dI}{ds} = -\frac{\pi}{2} \quad \text{[from (2)]}$$

From equations (4) and (5)

$$c_1 + c_2 = \frac{\pi}{2}$$

and
$$c_1 - c_2 = -\frac{\pi}{2}$$

$$\therefore \quad c_1 = 0, c_2 = \frac{\pi}{2}$$

Therefore, from equation (4), we have

$$I = \frac{\pi}{2} e^{-s}$$

$$\Rightarrow \quad \int_0^\infty \frac{\cos sx}{(1+x^2)} dx = \frac{\pi}{2} e^{-s}.$$ **Ans.**

Differentiating above w.r.t. s, we get

$$\int_0^\infty \frac{-x \sin sx}{(1+x^2)} dx = -\frac{\pi}{2} e^{-s}$$

$$\int_0^\infty \frac{x}{(1+x^2)} \sin sx dx = \frac{\pi}{2} e^{-s}.$$ **Ans.**

Example 28: Find the Fourier sine transform of $f(x) = \dfrac{1}{x(a^2+x^2)}$.

Sol. We have $\quad F(s) = \int_0^\infty \frac{1}{x(a^2+x^2)} . \sin(sx) dx = I$ (say) ...(1)

Then
$$\frac{dI}{ds} = \frac{d}{ds} \int_0^\infty \frac{1}{x(a^2+x^2)} . \sin(sx) dx$$

$$\frac{dI}{ds} = \int_0^\infty \frac{\cos(sx)}{(a^2+x^2)} dx \qquad ...(2)$$

$$\therefore \quad \frac{d^2I}{ds^2} = -\int_0^\infty \frac{x\sin(sx)}{(a^2+x^2)}dx = -\int_0^\infty \frac{x^2\sin(sx)}{x(a^2+x^2)}dx$$

$$= -\int_0^\infty \frac{(x^2+a^2-a^2)\sin(sx)}{x(a^2+x^2)}dx$$

$$= -\int_0^\infty \frac{\sin(sx)}{x}dx + a^2\int_0^\infty \frac{\sin(sx)}{x(a^2+x^2)}dx$$

$$= -\frac{\pi}{2} + a^2 I$$

$$\therefore \quad \frac{d^2I}{ds^2} - a^2 I = -\frac{\pi}{2} \Rightarrow \left(D^2 - a^2\right)I = -\frac{\pi}{2}$$

where $D \equiv \frac{d}{ds}$.

Then the solution of the above differential equation is

$$I = Ae^{-as} + Be^{as} + \frac{\pi}{2a^2} \qquad \text{...(3)}$$

$$\therefore \quad \frac{dI}{ds} = -aAe^{-as} + Bae^{as} \qquad \text{...(4)}$$

Now from (1) when $s = 0$, we have $I = 0$ and from (2) when $s = 0$

$$\frac{dI}{ds} = \int_0^\infty \frac{dx}{a^2+x^2} = \frac{1}{a}\left[\tan^{-1}\left(\frac{x}{a}\right)\right]_0^\infty = \frac{\pi}{2a}$$

So putting $s = 0$ in (3) and (4), we get

$$A + B = -\frac{\pi}{2a^2} \qquad \text{...(5)}$$

and

$$-A + B = \frac{\pi}{2a^2} \qquad \text{...(6)}$$

Solving eqns. (5) and (6), we get $B = 0$, $A = -\frac{\pi}{2a^2}$

Putting the values of A and B in eqn. (3), we get

$$I = -\frac{\pi}{2a^2}e^{-as} + \frac{\pi}{2a^2} = \frac{\pi}{2a^2}(1-e^{-as}). \quad \textbf{Ans.}$$

Example 29: Solve the integral equation

$$\int_0^\infty f(x)\cos sx dx = \begin{cases} 1-s, & 0 \le s \le 1 \\ 0, & s > 1 \end{cases}$$

Hence prove that $\int_0^\infty \frac{\sin^2 t}{t^2} dt = \frac{\pi}{2}$

Sol. Let $\int_0^\infty f(x)\cos sx dx \qquad = F(s)$

then

$$f(s) = \begin{cases} 1-s, & 0 \le s \le 1 \\ 0, & s > 1 \end{cases}$$

∴ By inversion formula for Fourier consine transform, we have

$$f(x) = \frac{2}{\pi}\int_0^\infty \cos sx \cdot F(s)\, ds$$

$$= \frac{2}{\pi}\left[\int_0^1 (1-s)\cos sx\, ds + 0\right]$$

$$= \frac{2}{\pi}\left[(1-s)\cdot\frac{\sin sx}{s} + \frac{-\cos sx}{x^2}\right]_0^1 = \frac{2}{\pi}\left[-\frac{\cos x}{x^2} + \frac{1}{x^2}\right]$$

⇒

$$f(x) = \frac{2(1-\cos x)}{\pi x^2}$$

Now $\int_0^\infty f(x)\cos sx\, dx = \begin{cases} 1-s, & 0 \le s < 1 \\ 0, & s > 1 \end{cases}$

$$\Rightarrow \frac{2}{\pi}\int_0^\infty \frac{1-\cos x}{x^2}\cos sx\, dx = \begin{cases} 1-s, & 0 \le s < 1 \\ 0, & s > 1 \end{cases}$$

Putting $s = 0$, then we have

$$\frac{2}{\pi}\int_0^\infty \frac{1-\cos x}{x^2} dx = 1 \Rightarrow \int_0^\infty \frac{2\sin^2 \frac{x}{2}}{x^2} dx = \frac{\pi}{2}$$

or

$$2\int_0^\infty \frac{\sin^2(x/2)}{x^2} dx = \frac{\pi}{2}$$

Putting $\frac{x}{2} = t \Rightarrow dx = 2dt$

$$\int_0^\infty \frac{\sin^2 t}{t^2} dt = \frac{\pi}{2}.$$ **Proved.**

Example 30: Find the Fourier transform of

$$f(x)=\begin{cases}1 & \text{for } |x|\le 1\\ 0 & \text{for } |x|>1\end{cases}$$

Hence evaluate $\int_0^\infty \frac{\sin x}{x}\,dx$

Sol. The Fourier transform of $f(x)$ is defined as

$$= F(s) = \int_{-\infty}^{\infty} f(t)e^{ist}\cdot dt$$

$$= \int_{-1}^{1} e^{ist}\cdot dt = \left[\frac{e^{ist}}{is}\right]_{-1}^{1} = \frac{1}{si}\left[e^{is}-e^{-is}\right]$$

$$= \frac{2}{s}\left\{\frac{e^{is}-e^{-is}}{2i}\right\} = \frac{2\sin s}{s}$$

Now by inverse Fourier transform, we have

$$f(x) = \frac{1}{2\pi}\int_{-\infty}^{\infty} F(s)e^{-isx}\,ds$$

$$\Rightarrow \quad \frac{2}{2\pi}\int_{-\infty}^{\infty}\frac{\sin s}{s}e^{-isx}\,ds = f(x)$$

$$\Rightarrow \quad \frac{1}{\pi}\int_{-\infty}^{\infty}\frac{\sin s}{s}(\cos sx - i\sin sx)\,ds = f(x)$$

$$\Rightarrow \quad \frac{2}{\pi}\int_{0}^{\infty}\frac{\sin s}{s}\cos sx\,ds = \begin{cases}1 \text{ for } |x|\le 1\\ 0 \text{ for } |x|>1\end{cases} \quad \text{(By definite integral)}$$

Taking $x \to 0$, we get

$$\int_0^\infty \frac{\sin s}{s}\,ds = \begin{cases}\dfrac{\pi}{2} \text{ for } |x|\le 1\\ 0 \text{ for } |x|>1\end{cases}$$

$$\Rightarrow \quad \int_0^\infty \frac{\sin x}{x}\,dx = \frac{\pi}{2}. \quad \textbf{Ans.}$$

1.8 PROPERTIES OF FOURIER TRANSFORMS

(i) Linear Property

If $F_1(s)$ and $F_2(s)$ are Fourier transforms of $f_1(x)$ and $f_2(x)$ respectively, then

$$F[af_1(x)+bf_2(x)] = aF_1(s)+bF_2(s)$$

where a and b are constants.

Proof: We know that

$$F_1(s) = \int_{-\infty}^{\infty} e^{isx} \,.\, f_1(x)dx$$

and
$$F_2(s) = \int_{-\infty}^{\infty} e^{isx} \,.\, f_2(x)dx$$

Then
$$F[af_1(x) + bf_2(x)] = \int_{-\infty}^{\infty} e^{isx}[af_1(x) + bf_2dx(x)]dx$$

$$= a\int_{-\infty}^{\infty} e^{isx} \,.\, f_1(x)dx + b\int_{-\infty}^{\infty} e^{isx} \,.\, f_2(x)dx$$

$$= aF_1(s) + bF_2(s).$$ **Proved**

(ii) Change of Scale Property

If $F(s)$ is the complex Fourier transform of $f(x)$, then

$$F\{f(ax)\} = \frac{1}{a}F\left(\frac{s}{a}\right)$$

Proof: We know that
$$F(s) = \int_{-\infty}^{\infty} e^{isx} \,.\, f(x)dx$$

$$F\{f(ax)\} = \int_{-\infty}^{\infty} e^{isx} \,.\, f(ax)dx,$$

put
$$ax = t \Rightarrow dx = \frac{dt}{a}$$

$$= \int_{-\infty}^{\infty} e^{is\frac{t}{a}} f(t)\frac{dt}{a}$$

$$= \frac{1}{a}\int_{-\infty}^{\infty} e^{i\left(\frac{s}{a}\right)t} f(t)dt$$

$$= \frac{1}{a}F\left(\frac{s}{a}\right).$$ **Proved.**

(iii) Shifting Property

If $F(s)$ is the complex Fourier transform of $f(x)$, then

$$F\{f(x-a)\} = e^{isa}F(s)$$

Proof:
$$F(s) = \int_{-\infty}^{\infty} e^{isx} \,.\, f(x)dx$$

$$F\{f(x-a\} = \int_{-\infty}^{\infty} e^{isx} \,.\, f(x-a)dx, z$$

[put $x - a = t$, so that $dx = dt$]

$$= \int_{-\infty}^{\infty} e^{is(t+a)} f(t)dt = e^{isa} \int_{-\infty}^{\infty} e^{ist} f(t)dt$$

$$= e^{isa} F(s). \quad \textbf{Proved.}$$

(iv) $$F\left\{e^{iax} f(x)\right\} = F(s+a)$$

Proof: $$F\left\{e^{iax} f(x)\right\} = \int_{-\infty}^{\infty} e^{iax} f(x) e^{isx} dx = \int_{-\infty}^{\infty} e^{i(s+a)x} f(x)dx$$

$$= F(s+a). \quad \textbf{Proved.}$$

(v) Modulation Theorem

If $F(s)$ is the complex Fourier transform of $f(x)$, then

$$F\{f(x)\cos ax\} = \frac{1}{2}[F(s+a) + F(s-a)]$$

Proof: We know that $$F(s) = \int_{-\infty}^{\infty} e^{isx}. f(x)\,dx$$

$$F\{f(x)\cos ax\} = \int_{-\infty}^{\infty} e^{isx}. f(x)\cos ax\,dx$$

$$= \int_{-\infty}^{\infty} e^{isx}. f(x)\frac{e^{iax} + e^{-iax}}{2} dx$$

$$= \frac{1}{2}\int_{-\infty}^{\infty} e^{isx}. f(x)e^{iax} dx + \frac{1}{2}\int_{-\infty}^{\infty} e^{isx}.e^{-iax} dx$$

$$= \frac{1}{2}\int_{-\infty}^{\infty} e^{i(s+a)x}. f(x)dx + \frac{1}{2}\int_{-\infty}^{\infty} e^{i(s-a)x} f(x)dx$$

$$= \frac{1}{2}\left[F(s+a) + \frac{1}{2}F(s-a)\right]$$

$$= \frac{1}{2}[F(s+a) + F(s-a)]. \quad \textbf{Proved.}$$

(vi) If $F\{f(x)\} = F(s)$, then $F\left\{x^n f(x)\right\} = (-i)^n \dfrac{d^n}{ds^n} F(s).$

Proof: We know that

$$F(s) = \int_{-\infty}^{\infty} e^{isx}. f(x)dx \qquad \text{...(1)}$$

Differentiating (1) w.r.t. s both sides, n times, we get

$$\frac{d^n F(s)}{ds^n} = \int_{-\infty}^{\infty} (ix)^n e^{isx}. f(x)dx$$

$$= (i)^n \int_{-\infty}^{\infty} (x)^n e^{isx}. f(x)dx$$

$$= (i)^n F(x^n f(x))$$

$$F(x^n f(x)) = (-i)^n \frac{d^n \{F(s)\}}{ds^n}$$

(vii) $$F\{f'(x)\} = isF(s), \text{ if } f(x) \to 0 \text{ as } x \to \pm\infty$$

Proof: $$F\{f'(x)\} = \int_{-\infty}^{\infty} e^{isx} . f'(x) dx$$

$$= \int_{-\infty}^{\infty} e^{isx} . d\{f(x)\} dx = \{e^{isx} f(x)\}_{-\infty}^{\infty} - \int_{-\infty}^{\infty} e^{isx} .(is) f(x) dx$$

$$= 0 - is \int_{-\infty}^{\infty} e^{isx} . f(x) dx$$

$$= -isF(s).$$ **Proved.**

(viii) $$F\left\{\int_a^x f(x) dx\right\} = \frac{F(s)}{(-is)}$$

Proof: $$\text{Let } f_1(x) = \int_a^x f(x) dx \Rightarrow f_1'(x) = f(x)$$

$$F\{f_1'(x)\} = (-is) F_1(s) = (-is) F\{f_1(x)\}$$

$$= -isF\left\{\int_a^x f(x) dx\right\}$$

$$F\left\{\int_a^x f(x) dx\right\} = \frac{1}{(-is)} F\{f_1'(x)\}$$

$$= \frac{1}{(-is)} F\{f(x)\} = \frac{F(s)}{(-is)}.$$ **Proved.**

1.9 CONVOLUTION

The convolution of two functions $f(x)$ and $g(x)$ is defined as

$$f(x) * g(x) = \int_{-\infty}^{\infty} f(u) g(x-u) du$$

Convolution Theorem on Fourier Transforms

The Fourier transform of the convolution of $f(x)$ and $g(x)$ is the product of their Fourier transform *i.e.*,

$$F[f(x) * g(x)] = F[f(x)] . F[g(x)] = F(s) \cdot G(s)$$

Proof: We know that

$$f(x) * g(x) = \int_{-\infty}^{\infty} f(u) g(x-u) du \qquad ...(1)$$

Taking Fourier transform both sides of (1), we have

$$F[f(x)*g(x)] = F\left[\int_{-\infty}^{\infty} f(u)\cdot g(x-u)du\right]$$

$$= \int_{-\infty}^{\infty}\left[\int_{-\infty}^{\infty} f(u)\cdot g(x-u)du\right]e^{isx}dx$$

$$= \int_{-\infty}^{\infty} f(u)du\left\{\int_{-\infty}^{\infty} g(x-u)e^{isx}dx\right\}$$

$$= \int_{-\infty}^{\infty}[f(u)\cdot du\cdot F\{g(x-u)\}]$$

$$= \int_{-\infty}^{\infty} f(u)du\cdot e^{ius}G(s) \qquad \text{(Using shifting property)}$$

$$= G(s).\int_{-\infty}^{\infty} f(u)e^{ius}du$$

$$= G(s)\ .\ F(s)$$

$$= F(s)\ .\ G(s).$$ **Proved.**

By inversion

$$F^{-1}\{F(s).G(s)\} = f*g = F^{-1}\{F(s)\}*F^{-1}\{G(s)\}$$

Example 1: Find Fourier cosine transform of $e^{-a^2x^2}$ $xe^{-a^2x^2}$.

Sol. First of all we find Fourier cosine transform of e^{-x^2}

$$F_c\left\{e^{-x^2}\right\} = \int_0^{\infty} e^{-x^2}\cos sx\,dx = I \qquad ...(1)$$

Differentiating w.r.t. s, we have

$$\frac{dI}{ds} = -\int_0^{\infty} xe^{-x^2}\sin sx\,dx$$

$$= \frac{1}{2}\int_0^{\infty}(-2xe^{-x^2})\sin sx\,dx$$

$$= \frac{1}{2}\left[\left\{e^{-x^2}\sin sx\,dx\right\}_0^{\infty} - s\int_0^{\infty}\cos sx\,e^{-x^2}dx\right]$$

$$= -\frac{s}{2}\int_0^{\infty} e^{-x^2}\cos sx\,dx$$

or

$$\frac{dI}{ds} + \frac{s}{2}I = 0$$

$$\frac{dI}{I} = -\frac{s}{2}ds$$

Integrating, we have

$$\log I = -\frac{s^2}{4} + \log A$$

$$I = Ae^{-\frac{s^2}{4}} \qquad \ldots(2)$$

But when $s = 0$, from (1), we have

$$I = \int_0^\infty e^{-x^2} dx = \frac{\sqrt{\pi}}{2}$$

When $s = 0$ from (2), we have

$$I = A$$

$$\therefore \qquad A = \frac{\sqrt{\pi}}{2}$$

Hence, $$I = \int_0^\infty e^{-x^2} \cos sx\, dx = \frac{\sqrt{\pi}}{2} e^{-\frac{s^2}{4}}$$

By change of scale property

$$F_c\left\{e^{-a^2x^2}\right\} = \frac{\sqrt{\pi}}{2a} e^{\frac{-s^2}{4a^2}}. \quad \textbf{Ans.}$$

Sine Transform:

We have $\int e^{-a^2x^2} \cos sx\, dx = \frac{\sqrt{\pi}}{2a} e^{-\frac{s^2}{4a^2}}$

Differentiating w.r.t. s, we get

$$-\int_0^\infty xe^{-a^2x^2} \sin sx\, dx = \frac{\sqrt{\pi}}{2a} e^{-s^2/4a^2} \cdot \left(-\frac{2s}{4a^2}\right)$$

$$\int_0^\infty xe^{-a^2x^2} \sin sx\, dx = \frac{\sqrt{\pi}s}{4a^3} e^{-s^2/4a^2}$$

Hence $$F_s\left\{xe^{-a^2x^2}\right\} = \frac{\sqrt{\pi}s}{4a^3} e^{-s^2/4a^2}. \quad \textbf{Ans.}$$

Example 2: Find the Fourier transform of e^{-x^2}. Hence, find the Fourier transform of

(*i*) $f(x) = e^{-ax^2}$ (*ii*) $f(x) = e^{-x^2/2}$

(*iii*) $f(x) = e^{-4(x-3)^2}$ (*iv*) $f(x) = e^{-x^2} \cos 2x$

Sol. $$F(s) = \int_{-\infty}^\infty e^{isx} . f(x) dx$$

$$F\left\{e^{-x^2}\right\} = \int_{-\infty}^\infty e^{isx} . e^{-x^2} dx = \int_{-\infty}^\infty e^{-(x^2 - isx)} dx$$

$$= \int_{-\infty}^\infty e^{-\left\{\left(x - \frac{is}{2}\right)^2 + \frac{s^2}{4}\right\}} dx$$

$$= e^{-\frac{s^2}{4}} \int_{-\infty}^{\infty} e^{-\left(x-\frac{is}{2}\right)^2} dx$$

$$= e^{-\frac{s^2}{4}} \int_{-\infty}^{\infty} e^{-z^2} dz, \qquad \text{Put } x - \frac{is}{2} = z, \qquad dx = dz$$

$$= 2e^{-\frac{s^2}{4}} \int_{0}^{\infty} e^{-z^2} dz \qquad \text{(By definite integral)}$$

$$F\left\{e^{-x^2}\right\} = \sqrt{\pi}\, e^{-\frac{s^2}{4}} \qquad ...(1)$$

(*i*) By change of scale property

$$F\left\{e^{-ax^2}\right\} = \frac{1}{\sqrt{a}} \sqrt{\pi}\, e^{-\frac{1}{4}\left(\frac{s}{\sqrt{a}}\right)^2} = \sqrt{\frac{\pi}{a}}\, e^{\frac{-s^2}{4a}}$$

(*ii*) Putting $a = \frac{1}{2}$ in (*i*) part, we get

$$F\left\{e^{-\frac{x^2}{2}}\right\} = \sqrt{2\pi}\, e^{-\left(\frac{s^2}{2}\right)}$$

(*iii*) From equation (1)

$$F\left\{e^{-x^2}\right\} = \sqrt{\pi}\, e^{-\frac{s^2}{4}}$$

$$\therefore \qquad F\left\{e^{-4x^2}\right\} = \sqrt{\frac{\pi}{4}}\; e^{-\frac{s^2}{4.4}} \qquad \text{(By change of scale property)}$$

$$= \frac{\sqrt{\pi}}{2} e^{-\frac{s^2}{16}}$$

$$\therefore \qquad F\left\{e^{-4(x-3)^2}\right\} = \frac{\sqrt{\pi}}{2} .\; e^{i3s}\, e^{-\frac{s^2}{16}} \qquad \text{(By shifting property)}$$

$$= \frac{\sqrt{\pi}}{2}\, e^{\left[3is - \frac{s^2}{16}\right]}$$

(*iv*) $\because$ $$F\left\{e^{-x^2}\right\} = \sqrt{\pi}\, e^{-\frac{s^2}{4}}$$

$\therefore$ $$F(e^{-x^2}\cos 2x) = \frac{\sqrt{\pi}}{2}\left[e^{-\frac{1}{4}(s+2)^2} + e^{-\frac{1}{4}(s-2)^2}\right]$$ (By modulation Theorem)

Example 3: Find the Fourier transform of the following functions

(*i*) $f(t) = \begin{cases} t \text{ for } |t| < a \\ 0 \text{ for } |t| > a \end{cases}$ (*ii*) $f(x) = e^{-|x|}$ (*iii*) $f(x) = e^{-a|x|}, a > 0$

Sol. We have

$$F(s) = \int_{-\infty}^{\infty} e^{isx} f(x)dx = \int_{-a}^{a} e^{ist} \cdot t\, dt$$

$$= \int_{-a}^{a} (\cos st + i\sin st)t\, dt = 2i\int_{0}^{a} t\sin st\, dt$$ (By definite intergral)

$$= 2i\left[\left\{t\left(-\frac{\cos st}{s}\right)\right\}_0^a - \int_0^a\left(-\frac{\cos st}{s}\right)dt\right]$$

$$= 2i\left[-\frac{a}{s}\cos as + \left(\frac{\sin st}{s^2}\right)_0^a\right] = 2i\left[-\frac{a}{s}\cos as + \frac{1}{s^2}\sin as\right]$$

$$= \frac{2i}{s^2}(\sin as - as\cos as).$$ **Ans.**

(*ii*) $$F(s) = \int_{-\infty}^{\infty} e^{isx} f(x)dx = \int_{-\infty}^{\infty} e^{-|x|}e^{isx} \cdot dx = \int_{-\infty}^{0} e^{x}e^{isx}dx + \int_{0}^{\infty} e^{-x}e^{isx}dx$$

$$= \frac{1}{(1+is)}\left[e^{(1+is)x}\right]_{-\infty}^{0} + \frac{1}{(1-is)}\left[e^{-(1-is)x}\right]_0^{\infty} = \frac{1}{1+is} + \frac{1}{1-is}$$

$$= \frac{2}{1+s^2}.$$ **Ans.**

(*iii*) By change of scale property

$$F\{e^{-a|x|}\} = F\{f(ax)\} = \frac{1}{a}\left\{F\left(\frac{s}{a}\right)\right\} = \frac{1}{a}\left\{\frac{2}{1+\left(\frac{s}{a}\right)^2}\right\}$$ As $F(s) = \frac{2}{1+s^2}$

$$= \frac{2a}{s^2+a^2}.$$ **Ans.**

PROBLEM SET 1.1

1. Express $f(x)=\begin{bmatrix}1, & \text{for} & 0\le x\le \pi \\ 0, & \text{for} & x>\pi\end{bmatrix}$ as a Fourier sine integral and hence evaluate

$$\int_0^\infty \frac{1-\cos\pi\lambda}{\lambda}\sin\lambda x d\lambda.$$ $\left[\textbf{Ans. } \frac{\pi}{4}\right]$

2. Using Fourier integrals, show that

$$\int_0^\infty \frac{-w\sin wx}{(1+w^2)}dw = \frac{\pi}{2}e^{-x}, \quad x>0$$

3. Obtain that the Fourier cosine transform of

$$f(x)=\begin{cases}0, & \text{for } 0<x<1\\ 2-x, & \text{for } 1<x<2\\ 0, & \text{for } x>2\end{cases}$$

Also find its Fourier sine transform $\left[\textbf{Ans. } \frac{2}{s^2}\cos s(1-\cos s); \frac{2}{s^2}\sin s(1-\cos s)\right]$

4. Show that the Fourier transform of

$$f(x)=\begin{cases}a-|x|, & \text{for} \quad |x|<a\\ 0, & \text{for} \quad |x|>a>0\end{cases}$$

is $\frac{2}{s^2}(1-\cos as)$. Hence show that $\int_0^\infty \frac{\sin^2 t}{t^2}dt = \frac{\pi}{2}$.

5. Find Fourier transform of $f(x)$ if

$$f(x)=\begin{cases}x, & |x|\le a\\ 0, & |x|>a\end{cases}$$ $\left[\textbf{Ans. } \frac{2i}{s^2}(as\cos as-\sin as)\right]$

6. Show that the Fourier transform of $e^{-\frac{x^2}{2}}$ is self reciprocal.

7. Find Fourier sine and cosine transform of $\frac{1}{\sqrt{x}}$. $\left[\textbf{Ans. } \sqrt{\frac{\pi}{2s}}, \sqrt{\frac{\pi}{2s}}\right]$

8. Find Fourier transform of $e^{-a|x|}$, $a>0$. $\left[\textbf{Ans. } \frac{2a}{a^2+s^2}\right]$

9. Find Fourier sine and cosine transform of cos hx – sin hx. $\left[\textbf{Ans. } \frac{s}{1+s^2}, \frac{1}{1+s^2}\right]$

10. Find the Fourier sine and cosine transform of $ae^{-\alpha x}+be^{-\beta x}$.

$$\left[\textbf{Ans. } \frac{as}{s^2+\alpha^2}+\frac{bs}{s^2+\beta^2}, \frac{a\alpha}{s^2+\alpha^2}+\frac{b\beta}{s^2+\beta^2}\right]$$

11. Find the Fourier integral representation of following functions

(*i*) $f(x)=\begin{cases}\cos x, & \frac{-\pi}{2}<x<\frac{\pi}{2}\\ 0, & |x|>\frac{\pi}{2}\end{cases}$ $\left[\textbf{Ans. } \frac{1}{\pi}\int_{-\infty}^{\infty}\frac{\cos(\lambda\pi/2)}{1-\lambda^2}e^{i\lambda x}d\lambda\right]$

(*ii*) $f(x)\begin{cases}x, |x|<l\\ 0, |x|>l\end{cases}$ $\left[\textbf{Ans. } \frac{1}{i\pi}\int_{-\infty}^{\infty}\frac{\sin\lambda-\lambda\cos\lambda}{\lambda^2}e^{i\lambda x}d\lambda\right]$

12. Find the Fourier cosine integral of $f(x) = e^{-x}\cos x$ $\left[\textbf{Ans. } \frac{2}{\pi}\int_0^{\infty}\frac{(\lambda^2+2)\cos\lambda x\,d\lambda}{\lambda^2+4}\right]$

13. Find the Fourier sine integral of $f(x) = e^{-ax}-e^{-bx}$ $\left[\textbf{Ans. } \frac{2}{\pi}\int_0^{\infty}\frac{(b^2-a^2)\lambda\sin\lambda x\,d\lambda}{(a^2+\lambda^2)(b^2+\lambda^2)}\right]$

14. If Fourier cosine integral of a function $f(x)$ can be represented as

$$f(x)=\int_0^{\infty}A(\lambda)\cos x\lambda\,d\lambda$$

where $A(\lambda)=\frac{2}{\pi}\int_0^{\infty}f(t)\cos\lambda t\,dt$

Prove that $f(ax)=\frac{1}{a}\int_0^{\infty}A(\lambda/a)\cos x\lambda\,d\lambda,\ (a>0)$

15. Find the Fourier transform of

$$f(x)=\begin{cases}a^2-x^2, & \text{if } |x|<a\\ 0, & \text{if } |x|\geq a\end{cases} \quad \text{where } a>0$$

Deduce that $\int_0^{\infty}\frac{(\sin x-\cos x)}{x^3}dx=\frac{\pi}{4}$

16. Find the Fourier transform of $f(x)=\begin{cases}\frac{\sqrt{2\pi}}{2l}, & \text{for } -l<x<l\\ 0, & \text{Otherwise}\end{cases}$

17. Find the Fourier cosine transform of $e^{-ax}\sin ax$ $\left[\textbf{Ans. } \sqrt{\frac{2}{\pi}}\left[\frac{a(2a^2-s^2)}{s^4+4a^4}\right]\right]$

18. Find $f(x)$ of its Fourier cosine transform is e^{-as}.

19. Find the Fourier sine and cosine transforms of cos hx – sin hx

(**Hint :** $\cos hx - \sin hx = \dfrac{e^x + e^{-x}}{2} - \dfrac{e^x - e^{-x}}{2} = e^{-x}$) $\left[\textbf{Ans. } \dfrac{2}{\pi}\left(\dfrac{s}{s^2+1}\right), \dfrac{2}{\pi}\left(\dfrac{1}{s^2+1}\right)\right]$

20. Find inverse Fourier sine transform of $\dfrac{e^{-as}}{s}$. Hence also evaluate inverse Fourier sine transform of $\left(\dfrac{1}{s}\right)$. $\left[\textbf{Ans. } \dfrac{2}{\pi}\tan^{-1}\left(\dfrac{x}{a}\right);1\right]$

21. Let $f(t) = \begin{cases} e^{-xt}g(t), & \text{if } t > 0 \\ 0 & \text{if } t \le 0. \end{cases}$ Prove that the Fourier transform of $f(t)$ is the same as the Laplace transform of $g(t)$.

22. Find $f(x)$ if $\int_0^\infty f(x)\sin sx\,dx = \begin{cases} 1 & \text{if } 0 < s < 1 \\ 2 & \text{if } 1 \le s < 2 \\ 0 & \text{if } s \ge 2; \end{cases}$ $\left[\textbf{Ans. } \dfrac{2}{\pi x}(1 + \cos x - 2\cos 2x)\right]$

23. Find the Fourier cosine and sine transforms of $5e^{-2x} + 2e^{-5x}$

$\left[\textbf{Ans. } \dfrac{20}{\pi}\left\{\dfrac{1}{s^2+4} + \dfrac{1}{s^2+25}\right\}\right]$

24. Find $f(x)$ its Fourier cosine transform is $F_c(s) = \begin{cases} \dfrac{1}{\sqrt{2\pi}}\left(a - \dfrac{s}{2}\right) & \text{if } 0 < s < 2a \\ 0 & \text{if } s \ge 2a \end{cases}$

$\left[\textbf{Ans. } \dfrac{\sin^2 ax}{\pi x^2}\right]$

25. Find the Fourier sine and cosine transforms of xe^{-ax} $\left[\textbf{Ans. } \dfrac{4}{\pi}\dfrac{as}{(s^2+a^2)^2}, \dfrac{2}{\pi}\left(\dfrac{(a^2-s^2)}{(s^2+a^2)^2}\right)\right]$

26. Find the Fourier transform of $f(x) = \begin{cases} 1 - |x|, & \text{if } |x| < 1 \\ 0, & \text{if } |x| > 1 \end{cases}$ $\left[\textbf{Ans. } \sqrt{\dfrac{2}{\pi}}\left(\dfrac{1-\cos s}{s^2}\right)\right]$

27. Show that the Fourier transform of $e^{-\frac{x^2}{2}}$ is $e^{-\frac{s^2}{2}}$.

28. Using Fourier intergral representation show that

$$\int_0^\infty \frac{\cos\lambda x+\lambda\sin\lambda x}{1+\lambda^2}\,d\lambda=\begin{cases}0, & \text{if } x<0\\ \pi/2, & \text{if } x=0\\ \pi e^{-x}, & \text{if } x>0\end{cases}$$

29. Find the inverse Fourier sine transform of $s^n e^{-as}$. $\left[\textbf{Ans. } \dfrac{2\,\lfloor n \sin(x+1)x}{\pi(a^2+x^2)^{\frac{n+1}{2}}}\right]$

30. Find $f(x)$ if its Fourier cosine transform is $\dfrac{1}{2\pi}\left(a-\dfrac{s}{2}\right)$ if $s<2a$ and zero is $s\geq 2a$.

1.10 PARSEVAL'S IDENTITY FOR FOURIER TRANSFORMS

If the Fourier transform of $f(x)$ and $g(x)$ be $F(s)$ and $G(s)$ respectively, then

(*i*) $$\int_{-\infty}^{\infty} F(s)\bar{G}(s)ds=\int_{-\infty}^{\infty} f(x)\bar{g}(x)dx$$

where $\bar{G}(s)$ is the complex conjugate of $G(s)$ and $\bar{g}(x)$ is the complex conjugate of g(x).

(*ii*) $$\int_{-\infty}^{\infty} [F(s)]^2 ds=\int_{-\infty}^{\infty} |f(x)|^2 dx$$

Proof: (*i*) $\int_{-\infty}^{\infty} f(x)\bar{g}(x)dx=\int_{-\infty}^{\infty} f(x)\left[\int_{-\infty}^{\infty} \bar{G}(s)e^{isx}ds\right]dx$ Since $\bar{g}(x)=\int_{-\infty}^{\infty} \bar{G}(s)e^{isx}ds$

$$\int_{-\infty}^{\infty} f(x)\bar{g}(x)=\int_{-\infty}^{\infty} \bar{G}(s)ds\int_{-\infty}^{\infty} f(x)e^{isx}dx$$

or $$\int_{-\infty}^{\infty} f(x)\bar{g}(x)dx=\int_{-\infty}^{\infty} \bar{G}(s)F(s)ds \qquad \left[\text{since} \int_{-\infty}^{\infty} f(x)e^{isx}dx=F(s)\right] \qquad ...(1)$$

Putting $g(x)=f(x)$ in (1), we get

$$\int_{-\infty}^{\infty} F(s).\bar{F}(s)=\int_{-\infty}^{\infty} f(x)\bar{f}(x)dx$$

$$\int_{-\infty}^{\infty} [F(s)]^2 ds=\int_{-\infty}^{\infty} [f(x)]^2 dx.$$ **Proved.**

1.11 PARSEVAL'S IDENTITY FOR COSINE TRANSFORM

(*i*) $$\frac{2}{\pi}\int_0^\infty F_c(s).G_c(s)ds=\int_0^\infty f(x).g(x)dx$$

(*ii*) $$\frac{2}{\pi}\int_0^\infty |F_c(s)|^2 ds=\int_0^\infty |f(x)|^2 dx$$

1.12 PARSEVAL'S IDENTITY FOR SINE TRANSFORM

(*i*) $\frac{2}{\pi}\int_0^\infty F_s(s).G_s(s)ds = \int_0^\infty f(x).g(x)dx$

(*ii*) $\frac{2}{\pi}\int_0^\infty |F_s(s)|^2\,ds = \int_0^\infty |f(x)|^2\,dx$

Example 1: Using Parseval's identity, show that

$$\int_0^\infty \frac{dx}{\left(x^2+1\right)^2} = \frac{\pi}{4}$$

Sol. Let $f(x) = e^{-x}$ so that $F_c(s) = \frac{1}{1+s^2}$

By Parseval's identity for cosine transformation

$$\frac{2}{\pi}\int_0^\infty [F_c(s)]^2\,ds = \int_0^\infty |f(x)|^2\,dx$$

$$\frac{2}{\pi}\int_0^\infty \left|\frac{1}{(1+s^2)^2}\right|^2 ds = \int_0^\infty \left|e^{-x}\right|^2 dx = \int_0^\infty \left|e^{-2x}\right| dx = \left[\frac{e^{-2x}}{-2}\right]_0^\infty = \frac{1}{2}$$

or $$\int_0^\infty \left|\frac{1}{(1+s^2)}\right|^2 ds = \frac{1}{2}\frac{\pi}{2} = \frac{\pi}{4}. \Rightarrow \int_0^\infty \frac{1}{(x^2+1)^2}dx = \frac{\pi}{4}.$$ **Ans.**

Example 2: Using Parseval's identity, show that

$$\int_0^\infty \frac{x^2dx}{\left(x^2+1\right)^2} = \frac{\pi}{4}$$

Sol. Let $f(x) = \frac{x}{x^2+1}$ so that $F_s(s) = \frac{\pi}{2}e^{(-s)}$

By Parseval's identity for sine transformation

$$\frac{2}{\pi}\int_0^\infty |F_s(s)|^2\,ds = \int_0^\infty |f(x)|^2\,dx$$

$$\int_0^\infty \left|\frac{x}{x^2+1}\right|^2 dx = \frac{2}{\pi}\int_0^\infty \left|\frac{\pi}{2}e^{-s}\right|^2 ds$$

$$= \left(\frac{2}{\pi}\right)\left(\frac{\pi^2}{4}\right)\int_0^\infty \left|e^{-2s}\right| ds = \frac{\pi}{2}\left[\frac{e^{-2s}}{-2}\right]_0^\infty$$

$$= \frac{\pi}{2}\left[0+\frac{1}{2}\right]=\frac{\pi}{4}. \quad \textbf{Proved.}$$

Example 3: Using Parseval's identity, prove that

$$\int_0^\infty \frac{dt}{\left(a^2+t^2\right)\left(b^2+t^2\right)}=\frac{\pi}{2ab(a+b)}$$

Sol. Let $f(x)=e^{-ax}$, $g(x)=e^{-bx}$

Then $$F_c(s)=\frac{a}{a^2+s^2},\ G_c(s)=\frac{b}{b^2+s^2}$$

By Parseval's identity for Fourier cosine transformation

$$\frac{2}{\pi}\int_0^\infty F_c(s).G_c(s)ds=\int_0^\infty f(x).g(x)dx \quad \text{...(1)}$$

Putting the values of $F_o(s)$, $G_c(s)$, $f(x)$ and $g(x)$ in equation (1), we get

$$\frac{2}{\pi}\int_0^\infty\left(\frac{a}{a^2+s^2}\right)\left(\frac{b}{b^2+s^2}\right)ds=\int_0^\infty e^{-ax}.e^{-bx}dx$$

$$\Rightarrow \frac{2}{\pi}\int_0^\infty \frac{ab}{(a^2+s^2)}\frac{ds}{(b^2+s^2)}=\frac{2}{\pi}\int_0^\infty e^{-(a+b)x}dx$$

$$=\left[\frac{e^{-(a+b)x}}{-(a+b)}\right]_0^\infty=\left[0+\frac{1}{a+b}\right]$$

$$\Rightarrow \int_0^\infty \frac{ds}{\left(a^2+s^2\right)\left(b^2+s^2\right)}=\frac{\pi}{2ab}\frac{1}{(a+b)}$$

$$\int_0^\infty \frac{dt}{\left(a^2+t^2\right)\left(b^2+t^2\right)}=\frac{\pi}{2ab(a+b)}. \quad \textbf{Proved.}$$

Example 4: Using Parseval's identity, prove that

$$\int_0^\infty \frac{\sin at}{t\left(a^2+t^2\right)}dt=\frac{\pi}{2}\frac{1-e^{-a^2}}{a^2}$$

Proof: Let $$f(x)=\begin{cases}1, & 0<x<a\\ 0, & x>a\end{cases},\quad g(x)=e^{-ax}$$

$$F_c(s)=\frac{\sin as}{s},\quad G_c(s)=\frac{a}{a^2+s^2}$$

By Parseval's identity for cosine transform

$$\frac{2}{\pi}\int_0^\infty F_c(s).G_c(s)ds = \int_0^\infty f(x).g(x)dx$$

$$\frac{2}{\pi}\int_0^\infty \frac{\sin as}{s}\frac{a}{a^2+s^2}ds = \int_0^\infty f(x).e^{-ax}dx$$

$$= \int_0^a f(x)e^{-ax}dx + \int_a^\infty f(x)e^{-ax}dx$$

$$= \int_0^a 1.e^{-ax}dx + 0$$

$$= \left[\frac{e^{-ax}}{a}\right]_0^a = \frac{e^{-a^2}}{-a} - \frac{1}{-a} = \frac{1-e^{-a^2}}{a}$$

$$\Rightarrow \qquad \int_0^\infty \frac{\sin as}{s\left(a^2+s^2\right)}ds = \frac{\pi}{2}\frac{\left(1-e^{-a^2}\right)}{a^2}$$

or $$\int_0^\infty \frac{\sin at}{t\left(a^2+t^2\right)}dt = \frac{\pi}{2}\frac{\left(1-e^{-a^2}\right)}{a^2}.$$ **Proved.**

Example 5: Using Parseval's identity, prove $\int_0^\infty \left(\frac{\sin t}{t}\right)^2 dt = \frac{\pi}{2}$.

Sol. We know that

If $$f(x) = \begin{cases} 1, & \text{for } |x| < 0 \\ 0, & \text{for } |x| > a > 0 \end{cases}$$

Then $$F(s) = \sqrt{\frac{2}{\pi}}\frac{\sin as}{s}$$

Using Parseval's identity

$$\int_{-\infty}^\infty |f(t)|^2 dt = \int_{-\infty}^\infty |F(s)|^2 ds$$

$$\int_{-a}^a (1)^2 dt = \int_{-\infty}^\infty \frac{2}{\pi}\left(\frac{\sin as}{s}\right)^2 ds$$

$$2a = \frac{2}{\pi}\int_{-\infty}^\infty \left(\frac{\sin as}{s}\right)^2 ds$$

Putting $as = t$, we get $$2a = \frac{2}{\pi}\int_{-\infty}^\infty \left(\frac{\sin t}{\frac{t}{a}}\right)^2 \frac{dt}{a} \qquad \left(\because ads = dt \Rightarrow ds = \frac{dt}{a}\right)$$

$$a\pi = a\int_{-\infty}^{\infty}\left(\frac{\sin t}{t}\right)^2 dt$$

$$\int_0^{\infty}\left(\frac{\sin t}{t}\right)^2 dt = \frac{\pi}{2}.\quad \textbf{Proved.}$$

Example 6: Solve for $f(x)$ from the integral equation.

$$\int_0^{\infty} f(x)\cos sx\, dx = e^{-s}$$

Sol. $$\int_0^{\infty} f(x)\cos sx\, dx = e^{-s} \qquad ...(1)$$

$$F_c\{f(x)\} = e^{-s}$$

$$f(x) = F_c^{-1}\left[e^{-s}\right]$$

$$= \frac{2}{\pi}\left[\int_0^{\infty} e^{-s}\cos sx\, ds\right]$$

$$= \frac{2}{\pi}\int_0^{\infty} e^{-s}\cos sx\, ds$$

$$= \frac{2}{\pi}\left[\frac{e^{-s}}{1+x^2}\{-\cos sx + x\sin sx\}\right]_0^{\infty}$$

$$= \frac{2}{\pi}\cdot\frac{1}{1+x^2}.\quad \textbf{Ans.}$$

1.13 FOURIER TRANSFORM OF DERIVATIVES

We have already seen that

$$F\{f^n(x)\} = (-is)^n F(s)$$

(i) $\therefore$ $$F\left(\frac{\partial^2 u}{\partial x^2}\right) = (-is)^2 F\{u(x)\} = -s^2\bar{u},$$ where $\bar{u}$ is Fourier transform of u w.r.t. x.

(ii) $$F_c\{f'(x)\} = sF_s(s) - f(0)$$

$$F_c\{f'(x)\} = \int_0^{\infty} f'(x).\cos sx\, dx = \int_0^{\infty}\cos sx\, d\{f(x)\}$$

$$= \{f(x)\cos sx\}_0^{\infty} + s\int_0^{\infty} f(x)\sin sx\, dx$$

or $$\boxed{F_c\{f'(x) = sF_s(s) - f(0)}$$ assuming $f(x) \to 0$ as $x \to \infty$

(*iii*) $$F_s\{f'(x)\} = \int_0^\infty \sin sx d[f(x)]$$

$$= \{f(x)\sin sx\}_0^\infty - s\int_0^\infty f(x)\cos sx\, dx$$

$$\boxed{F_s\{f'(x)\} = -sF_c(s)}$$

(*iv*) $$F_c\{f''(x)\} = \int_0^\infty \cos sx d[f'(x)]$$

$$= \{f'(x)\cos sx\}_0^\infty + s\int_0^\infty f'(x)\sin sx\, dx$$

$$= -f'(0) + sF_s\{f'(x)\}$$

$$= -s^2F_c(s) - f'(0) \text{ assuming } f(x), f'(x) \to 0 \text{ as } x \to \infty$$

(*v*) $$F_s\{f''(x)\} = \int_0^\infty \sin sx d[f'(x)]$$

$$= \{f'(x)\sin sx\}_0^\infty - s\int_0^\infty f'(x)\cos sx\, dx$$

$$= -sF_c\{f'(x)\} = -s[sF_s(s) - f(0)]$$

$$\boxed{F_s\{t''(x)\} = -s^2F_s(s) + sf(0)} \text{ assuming } f(x), f'(x) \to 0 \text{ as } x \to \infty.$$

1.14 RELATIONSHIP BETWEEN FOURIER AND LAPLACE TRANSFORMS

Consider

$$f(t) = \begin{cases} e^{-xt}g(t), & \text{for } t > 0 \\ 0, & \text{for } t < 0 \end{cases} \qquad ...(1)$$

Then the Fourier transform of $f(t)$ is given by

$$F\{f(t)\} = \int_{-\infty}^\infty e^{ist}f(t)dt = \int_0^\infty e^{ist}f(t)dt$$

$$= \int_0^\infty e^{(is-x)t}g(t)dt = \int_0^\infty e^{-pt}g(t)dt \qquad [\text{where } p = x - is]$$

$$= L\{g(t)\}$$

Fourier transform of $f(t)$ = Laplace transform of $g(t)$ defined by (1).

1.15 FOURIER TRANSFORMS OF PARTIAL DERIVATIVE OF A FUNCTION

$$F\left[\frac{\partial^2 u}{\partial x^2}\right] = -s^2F(u) \text{ where } F(u) \text{ is Fourier transform of } u \text{ w.r.t. } x.$$

$$F_s\left[\frac{\partial^2 u}{\partial x^2}\right] = s(u)_{x=0} - s^2 F_s(u) \qquad \text{(sine transform)}$$

$$F_c\left[\frac{\partial^2 u}{\partial x^2}\right] = -\left[\frac{\partial u}{\partial x}\right]_{x=0} - s^2 F_c(u) \qquad \text{(cosine transform)}$$

Proof: Let $F[u(x, t)]$ be the Fourier transform of the function $u(x, t)$, *i.e.*

$$F\left[u(x,t)\right] = \int_{-\infty}^{\infty} e^{isx} u(x,t)dx$$

The Fourier transform of $\frac{\partial^2 u}{\partial x^2}$ is given by

$$F\left[\frac{\partial^2 u}{\partial x^2}\right] = \int_{-\infty}^{\infty} e^{isx} \frac{\partial^2 u}{\partial x^2} dx$$

Integrating by parts, we have

$$F\left[\frac{\partial^2 u}{\partial x^2}\right] = \left[e^{isx}\frac{\partial u}{\partial x} - \int ise^{isx}\frac{\partial u}{\partial x}dx\right]_{-\infty}^{\infty}$$

$$= \left[e^{isx}\frac{\partial u}{\partial x} - ise^{isx}u + \int (is)^2 e^{isx} udx\right]_{-\infty}^{\infty} \qquad \text{Again integration}$$

$$= \left[0 - 0 - s^2 \int_{-\infty}^{\infty} e^{isx}\,.\, udx\right] \qquad \left[\begin{matrix} u = 0, \ \dfrac{\partial u}{\partial x} = 0 \\ \text{when} \quad x \to \infty \end{matrix}\right]$$

Thus

$$\boxed{F\left[\frac{\partial^2 u}{\partial x^2}\right] = -s^2 F\left[u(x,t)\right]}$$

Similarly the Fourier sine transforms of $\frac{\partial^2 u}{\partial x^2}$ is given by

$$F_s\left[\frac{\partial^2 u}{\partial x^2}\right] = \int_0^{\infty} \frac{\partial^2 u}{\partial x^2} \sin sx\, dx$$

$$\left(\frac{\partial u}{\partial x}\sin sx\right)_0^{\infty} - s\int_0^{\infty}\frac{\partial u}{\partial x}\cos sdx = 0 - s\{u(x,t)\cdot\cos sx\}_0^{\infty} - s^2\int_0^{\infty} u \sin sx\, dx$$

or

$$\boxed{F_s\left[\frac{\partial^2 u}{\partial x^2}\right] = s\left[u(x,t)\right]_{x=0} - s^2 F_s\left[u(s,t)\right]} \qquad \text{(sine transform)}$$

Similarly

$$\boxed{F_c\left[\frac{\partial^2 u}{\partial x^2}\right] = -\left[\frac{\partial u}{\partial x}\right]_{x=0} - s^2 F_c\left[u(s,t)\right]} \qquad \text{(cosine transform)}$$

1.16 APPLICATIONS OF FOURIER TRANSFORM OF HEAT CONDUCTION (TRANSFER EQUATIONS)

In one dimensional heat transfer equations, the partial differential equation can easily be transformed into an ordinary differential equation by applying Fourier transforms. The required solution is then obtained by solving this equation and inverting by means of the complex inversion formula. This is illustrated through the following examples.

Example 1: Solve the equation

$$\frac{\partial u}{\partial t} = \frac{\partial^2 u}{\partial x^2}, \qquad x > 0,\ t > 0$$

subject to the conditions

(*i*) $u = 0$ when $x = 0,\ t > 0$

(*ii*) $u = \begin{cases} 1, & 0 < x < 1 \\ 0, & x \geq 1 \end{cases}$ when $t = 0$

(*iii*) $u(x, t)$ is bounded

(**Note:** If u, at $x = 0$ is given, take Fourier sine transform and if $\frac{\partial u}{\partial x}$ at $x = 0$ is given, use Fourier cosine transform.)

Sol. In view of the initial conditions, we apply Fourier sine transform

$$\int_0^\infty \frac{\partial u}{\partial t} \sin sx\, dx = \int_0^\infty \frac{\partial^2 u}{\partial x^2} \sin sx\, dx$$

$$\frac{d}{dt}\int_0^\infty u \sin sx\, dx = -s^2\bar{u}(s) + su(0) \qquad u = 0 \text{ when } x = 0$$

$$\frac{d\bar{u}}{dt} = -s^2\bar{u} \ \ or \ \ \frac{d\bar{u}}{dt} + s^2\bar{u} = 0$$

$$\frac{d\bar{u}}{\bar{u}} = -s^2 dt$$

$$\Rightarrow \qquad \log\bar{u} = -s^2 t + \log A$$

$$\therefore \qquad \bar{u} = Ae^{-s^2 t} \qquad \ldots(1)$$

Since $F_s[u(s, t)] = \bar{u} = \bar{u}(s,t) = \int_0^\infty u(x,t) \sin sx\, dx$

$$\Rightarrow \qquad \bar{u} = \bar{u}(s,0) = \int_0^\infty u(x,0) \sin sx\, dx$$

$$\bar{u}(s,0) = \int_0^1 1.\sin sx\, dx = \left[\frac{-\cos sx}{s}\right]_0^1 = \frac{1-\cos s}{s} \qquad \ldots(2)$$

From (2) putting the value of $\bar{u}(s,0)$ in (1), we get $\dfrac{1-\cos s}{s} = A$

$$\therefore \qquad \bar{u} = \frac{1-\cos s}{s} e^{-s^2 t} \quad \text{or} \quad u = \frac{2}{\pi}\int_0^{\infty}\left(\frac{1-\cos s}{s}\right)e^{-s^2 t}\sin sx\,dx. \quad \textbf{Ans.}$$

Example 2: Solve $\dfrac{\partial u}{\partial t} = k\dfrac{\partial^2 u}{\partial x^2}$ for $x \geq 0, t \geq 0$ under the given conditions $u = u_0$ at $x = 0, t > 0$ with initial condition $u(x, 0) = 0,\ x \geq 0$.

Sol. Taking Fourier sine transforms

$$F_s\left(\frac{\partial u}{\partial t}\right) = F_s\left(k\frac{\partial^2 u}{\partial x^2}\right)$$

$$\frac{d}{dt}\bar{u} = k\left[-s^2\bar{u} + \sqrt{\frac{2}{\pi}}\,su(0,t)\right]$$

$$= -ks^2\bar{u} + \sqrt{\frac{2}{\pi}}\,ksu_0, \text{ where } \bar{u} \text{ is the Fourier sine transform of } u.$$

$$\frac{d\bar{u}}{dt} + sk^2\bar{u} = \sqrt{\frac{2}{\pi}}\,sku_0$$

This is linear in $\bar{u}$.

$$\text{I.F.} = e^{\int sk^2 dt} = e^{sk^2 t}$$

$$\therefore \qquad \bar{u}e^{ks^2 t} = \sqrt{\frac{2}{\pi}}\,ku_0\int se^{ks^2 t}dt = \sqrt{\frac{2}{\pi}}\frac{u_0}{s}e^{ks^2 t} + c \qquad ...(1)$$

Since, $u(x, 0) = 0$, $\bar{u}(s, 0) = 0$. Using this in (1)

$$0 = \sqrt{\frac{2}{\pi}}\frac{u_0}{s} + c \quad \Rightarrow \quad c = -\sqrt{\frac{2}{\pi}}\frac{u_0}{s}$$

$$\therefore \qquad \bar{u}e^{ks^2 t} = \sqrt{\frac{2}{\pi}}\frac{u_0}{s}\left(e^{ks^2 t} - 1\right) \Rightarrow \bar{u} = \sqrt{\frac{2}{\pi}}\frac{u_0}{s}\left(1 - e^{-ks^2 t}\right)$$

By inversion theorem,

$$u(x,t) = \frac{2u_0}{\pi}\int_0^{\infty}\left(\frac{1-e^{-ks^2 t}}{s}\right)\sin sx\,ds. \quad \textbf{Ans.}$$

Example 3: Solve $\dfrac{\partial u}{\partial t} = k\dfrac{\partial^2 u}{\partial x^2}$ for $0 \leq x < \infty, t > 0$ given the conditions

(*i*) $u(x,0) = 0$ for $x \geq 0$ (*ii*) $\dfrac{\partial u}{\partial x}(0,t) = -a$ (constant)

(*iii*) $u(x,t)$ is bounded.

Sol. In this problem, $\frac{\partial u}{\partial x}$ at $x = 0$ is given. Hence, take Fourier cosine transform on both sides of the given equation.

$$F_c\left(\frac{\partial u}{\partial t}\right) = F_c\left(k\frac{\partial^2 u}{\partial x^2}\right)$$

$$\frac{d\bar{u}}{dt} = k\left(-s^2\bar{u} - \sqrt{\frac{2}{\pi}}\frac{\partial u}{\partial x}(0,t)\right)$$

$$= -ks^2\bar{u} + \sqrt{\frac{2}{\pi}}ka \qquad \text{[Using condition } (ii)]$$

$$\frac{d\bar{u}}{dt} + ks^2\bar{u} = \sqrt{\frac{2}{\pi}}ka$$

This is linear in $\bar{u}$. Therefore, solving

$$\bar{u}e^{ks^2t} = \int\sqrt{\frac{2}{\pi}}kae^{ks^2t}dt = \sqrt{\frac{2}{\pi}}ka\frac{e^{ks^2t}}{ks^2} + c$$

$$\bar{u}(s,t) = \sqrt{\frac{2}{\pi}}\frac{a}{s^2} + ce^{-ks^2t} \qquad \text{...(1)}$$

Since $\quad u(x, 0) = 0 \quad$ for $x \geq 0$

Using this in (1), we get

$$\bar{u}(s,0) = c + \sqrt{\frac{2}{\pi}}\frac{a}{s^2} = 0$$

$$\therefore \qquad c = -\sqrt{\frac{2}{\pi}}\frac{a}{s^2}$$

Substituting this in (1)

$$\bar{u}(s,t) = \sqrt{\frac{2}{\pi}}\frac{a}{s^2}\left(1 - e^{-ks^2t}\right)$$

By inversion theorem

$$u(x,t) = \frac{2}{\pi}.a\int_0^\infty \frac{1-e^{-ks^2t}}{s^2}\cos sx\,ds. \qquad \textbf{Ans.}$$

Example 4: Use Fourier sine transform to solve the equation

$$\left(\frac{\partial u}{\partial t}\right) = 2\left(\frac{\partial^2 u}{\partial x^2}\right)$$

under the conditions

(*i*) $u(0, t) = 0$ (*ii*) $u(x,0) = e^{-x}$ (*iii*) $u(x, t)$ is bounded.

Sol. The given equation is

$$\left(\frac{\partial u}{\partial t}\right) = 2\left(\frac{\partial^2 u}{\partial x^2}\right) \quad ...(1)$$

Taking Fourier sine transform on both sides of eqn. (1), we get

$$\int_0^\infty \frac{\partial u}{\partial t} \sin sx\, dx = 2\int_0^\infty \frac{\partial^2 u}{\partial x^2} \sin sx\, dx$$

$$\frac{d\bar{u}}{dt} = 2\left[s(u)_{x=0} - s^2 \bar{u}\right], \text{ where } \quad \bar{u} = \int_0^\infty u \sin sx\, dx$$

$$\Rightarrow \quad \frac{d\bar{u}}{dt} + 2s^2 \bar{u} = 0$$

Its solution is

At $\quad t = 0, \quad \bar{u} = c_1 e^{-2s^2 t}$ ...(2)

when c_1 is a constant. $(\bar{u})_{t=0} = \int_0^\infty (u)_{t=0} \sin sx\, dx$

$$= \int_0^\infty e^{-x} \sin sx\, dx$$

$$= \frac{s}{1+s^2} \quad ...(3)$$

From (2), $\quad (\bar{u})_{t=0} = c_1$ ...(4)

∴ From (3) and (4), $\quad c_1 = \dfrac{s}{1+s^2}$

From (2), $\quad \bar{u} = \dfrac{s}{1+s^2} e^{-2s^2 t}$

Taking inverse Fourier sine transform, we get

$$u(x,t) = \frac{2}{\pi}\int_0^\infty \frac{s}{1+s^2} e^{-2s^2 t} \sin sx\, ds.$$

Example 5: Use Fourier cosine transform to show that the steady temperature u in the semi-infinite solid $y > 0$ when the temperature on the surface $y = 0$ is kept at unity over the strip $|x| < a$ and at zero outside the strip is

$$\frac{1}{\pi}\left[\tan^{-1}\left(\frac{a+x}{y}\right) + \tan^{-1}\left(\frac{a-x}{y}\right)\right]$$

The results $\int_0^\infty e^{-sx} x^{-1} \sin rx\, dx = \tan^{-1}\left(\frac{r}{s}\right) (r, s > 0)$ may be assumed.

Sol. Taking Fourier cosine transform of $\left(\frac{\partial^2 u}{\partial x^2}\right)+\left(\frac{\partial^2 u}{\partial y^2}\right)=0$, we have

$$\int_0^\infty \frac{\partial^2 u}{\partial x^2}\cdot \cos sx\,dx+\int_0^\infty \frac{\partial^2 u}{\partial y^2}\cdot \cos sx\,dx=0$$

$$-\left(\frac{\partial u}{\partial x}\right)_{x=0}-s^2\bar{u}+\frac{d^2}{dy^2}(\bar{u})=0 \text{ where } \bar{u}=\int_0^\infty u\cos sx\,dx$$

$$\frac{d^2\bar{u}}{dy^2}-s^2\bar{u}=0 \quad ...(1) \quad \left(\because \frac{\partial u}{\partial x}\to 0 \text{ as } x\to\infty\right)$$

Its solution is $\bar{u}=c_1e^{sy}+c_2e^{-sy}$ ...(2)

But $\bar{u}$ is finite so $c_1 = 0$. Otherwise $\bar{u}\to\infty$ as $y\to\infty$

$\therefore$ From (2), $\bar{u}=c_2e^{-sy}$ ...(3)

But $\bar{u}=\int_0^\infty u\cos sx\,dx$

$$(\bar{u})_{y=0}=\int_0^\infty (u)_{y=0}\cos sx\,dx$$

$$=\int_0^a 1.\cos sxdx=\frac{\sin sa}{s} \quad ...(4)$$

From (3), $(\bar{u})_{y=0}=c_2$

$\therefore$ $c_2=\frac{\sin sa}{s}$

From (3), $\bar{u}=\frac{\sin sa}{s}e^{-sy}$

Applying inverse Fourier cosine transform, we get

$$u=\frac{2}{\pi}\int_0^\infty \frac{\sin sa}{s}e^{-sy}\cos sx\,ds=\frac{1}{\pi}\int_0^\infty \frac{e^{-sy}}{s}(2\sin sa\cos sx)ds$$

$$=\frac{1}{\pi}\int_0^\infty \frac{e^{-sy}}{s}\left[\sin(a+x)s+\sin(a-x)s\right]ds$$

$$=\frac{1}{\pi}\left[\tan^{-1}\left(\frac{a+x}{y}\right)+\tan^{-1}\left(\frac{a-x}{y}\right)\right]$$

Example 6: Using Fourier transform solve

$$\frac{\partial V}{\partial t} = \frac{\partial^2 V}{\partial x^2}, -\infty < x < \infty, t > 0 ; V(x,0) = f(x)$$

Sol. Taking Fourier transform of given problem, we get

$$F\left(\frac{\partial V}{dt}\right) = F\left(\frac{\partial^2 V}{\partial x^2}\right)$$

$$\Rightarrow \quad \int_{-\infty}^{\infty} \frac{\partial V}{\partial t} e^{ist} \cdot dt = -s^2 \bar{V} \qquad \Bigg| \text{ As } F\left(\frac{\partial^2 V}{\partial x^2}\right) = -s^2 \bar{V}$$

$$\Rightarrow \quad \frac{d}{dt}\int_{-\infty}^{\infty} V e^{ist} dt = -s^2 \bar{V}$$

$$\Rightarrow \quad \frac{d\bar{V}}{dt} = -s^2 \bar{V}$$

$$\Rightarrow \quad \frac{d\bar{V}}{V} = -s^2 dt$$

On integrating, we get

$$\log \bar{V} = -s^2 t + \log c$$

or

$$\bar{V} = c e^{-s^2 t} \qquad \text{...(1)}$$

Using the initial condition $V(x, 0) = f(x)$

By Fourier transform

$$\bar{V}(x,0) = \int_{-\infty}^{\infty} V(x,0) e^{isx} dx = \int_{-\infty}^{\infty} f(x) e^{isx} dx \qquad \text{...(2)}$$

Making use of relation (2) in relation (1), we get

$$\int_{-\infty}^{\infty} f(x) e^{isx} dx = c$$

$$\therefore \quad \bar{V} = e^{-s^2 t} \int_{-\infty}^{\infty} f(x) e^{isx} dx$$

Hence, By inverse fourier transform. We have

$$V = \frac{1}{2\pi}\int_{-\infty}^{\infty}\left\{e^{-s^2 t}\int_{-\infty}^{\infty} f(x) e^{isx} dx\right\} \cdot e^{-isx} ds. \quad \textbf{Ans.}$$

Example 7: Find the temperature distribution is semi-infinite bar with its end point and lateral surface insulated and with initial temperature distribution in the bar is prescribed by $f(x)$. Deduce the solution when

$$f(x) = e^{-ax}$$

Sol. The physical nature of the given problem is described by the one dimensional heat equation

$$\frac{\partial u}{\partial t} = c^2 \frac{\partial^2 u}{\partial x^2} \quad \text{for} \quad 0 < x < \infty, t > 0 \text{ with boundary}$$

condition $\frac{\partial}{\partial x} u(0,t) = 0$ (insulated)

and with initial condition $u(x, 0) = f(x)$ (given) for $0 < x < \infty$

Taking Fourier cosine transformation of heat equation, we get

$$F_c\left(\frac{\partial u}{\partial t}\right) = c^2 F_c\left(\frac{\partial^2 u}{\partial x^2}\right)$$

$$\int_0^\infty \frac{\partial u}{\partial t} \cos st dt = c^2 \left\{-\left(\frac{\partial u}{\partial x}\right)_{x=0} - s^2 F_c[u(s,t)]\right\}$$

$$\Rightarrow \quad \frac{\partial}{\partial t}\int_0^\infty u \cos st dt = c^2 \left\{0 - s^2 \bar{u}\right\} \quad \Big| F_c\{u(s,t)\} = \bar{x}$$

or $$\frac{d\bar{u}}{dt} = -s^2 c^2 \bar{u}$$

or $$\frac{d\bar{u}}{\bar{u}} = -s^2 c^2 dt$$

On integrating, we get

$$\log \bar{u} = -s^2 c^2 t + \log A \Rightarrow \bar{u} = Ae^{-s^2 c^2 t} \quad \text{...(1)}$$

At $t = 0$, $\bar{u}(s,0) = A$

Thus $A = \bar{u}(s,0) = \int_0^\infty u(x,0) \cos sx \, dx$

From the initial condition $u(x, 0) = f(x)$,

then $$A = \int_0^\infty f(x) \cos(sx) dx$$

Now, by inverse Fourier cosine transform, we get

$$u(x,t) = \frac{2}{\pi}\int_0^\infty Ae^{-c^2 s^2 t} \cdot \cos(sx) ds. \quad \textbf{Ans.}$$

Deduction: When $f(x) = e^{-ax}$, then $A = \int_0^\infty e^{-ax} \cos(sx) dx = \frac{a}{a^2 + b^2}$

Therefore $$u(x,t) = \frac{2}{\pi}\int_0^\infty \frac{a}{a^2 + b^2} e^{-c^2 s^2 t} \cdot \cos(sx) ds. \quad \textbf{Ans.}$$

PROBLEM SET 1.2

1. Apply appropriate Fourier transform to solve the partial differential equation

$$\frac{\partial v}{\partial t}=\frac{\partial^2 v}{\partial x^2};x>0,t>0$$

Subject to conditions

(*i*) $v_x(0,t)=0$ (*ii*) $v(x,0)=\begin{cases}x, & 0\le x\le 1\\ 0, & x>1\end{cases}$

(*iii*) $v(x, t)$ is bounded.

$$\left[\textbf{Ans. } v(x,t)=\frac{2}{\pi}\int_0^{\infty}\left(\frac{\sin s}{s}+\frac{\cos s-1}{s^2}\right)e^{-s^2t}\cos sx\,ds\right]$$

2. Solve the equation for high voltage semi-infinite line with the following initial and boundary conditions.

$v(x, t) = 0$ and $i(x, 0) = 0$, $v(0, t) = v_0\, u(t)$, $v(x, t)$ is finite as $x\to\infty$.

$$\left[\textbf{Ans. } v=v_0u[t-x\sqrt{LC}], \text{ for } x\le\frac{t}{\sqrt{LC}} \text{ and } v=0 \text{ for } x>\frac{t}{\sqrt{LC}}\right]$$

3. Find the temperature distribution $u(x, t)$ in a semi-infinite metal bar with the end $x = 0$ at zero temeprature and initial temeprature distribution $f(x)$. Deduce when $f(x) = e^{-x}$.

$$\left[\textbf{Ans. } u(x,t)=\frac{2}{\pi}\int_0^{\infty}\bar{u}(s,0)e^{-c^2s^2t}\cdot\sin sx ds, \text{ where } \bar{u}(s,0)=\int_0^{\infty}u(x,0)\cdot\sin sx\,dx\right.$$

$$\left.=\int_0^{\infty}f(x)\sin sx\,dx \text{ when } f(x)=e^{-x};u(x,t)=\frac{2}{\pi}\int_0^{\infty}\frac{se^{-c^2s^2t}}{1+s^2}\sin sx\,ds\right]$$

4. Find the temperature $u(x, t)$ in a semi infinite bar with $\frac{\partial u}{\partial x}=\mu$ when $x = 0$ and with initial temperature 0, Assume that $\frac{\partial u}{\partial x}\to 0$ as $x\to\infty$ and is bounded.

$$\left[\textbf{Ans. } u(x,t)=\frac{2}{\pi}\int_0^{\infty}\frac{\mu}{s^2}(1-e^{-s^2c^2t})\cos sx\,ds\right]$$

5. If the initial temperature of an infinite bar given by $\mu(x,0)=\begin{cases}1, & \text{for } -c<x<c\\ 0, & \text{otherwise}\end{cases}$, determine the temperature of an infinite bar at any point x and at any time $t > 0$.

$$\left[\textbf{Ans. } u(x,t)=\frac{1}{2\pi}\int_{-\infty}^{\infty}\frac{2}{s}\sin cs\cdot e^{-k^2s^2t}e^{-isx}\,ds\right]$$

6. Solve $\dfrac{\partial u}{\partial t} = c^2 \dfrac{\partial^2 u}{\partial x^2}, x > 0, t > 0$ with $\left(\dfrac{\partial u}{\partial x}\right)_{x=0} = 0$ and $u(x, 0) \begin{cases} x, & 0 \le x \le 1 \\ 0, & x > 1 \end{cases}$;

$$\left[\textbf{Ans. } u(x,t) = \frac{2}{\pi}\int_0^{\infty}\left[\frac{\sin s}{s} + \frac{1}{s^2}(\cos s - 1)\right] e^{-s^2c^2t} \cdot \cos sx\, ds\right]$$

7. Solve the Laplace's equation $\left(\dfrac{\partial^2 u}{\partial x^2} + \dfrac{\partial^2 u}{\partial y^2} = 0\right)$ in semi-infinite strip shown in figure

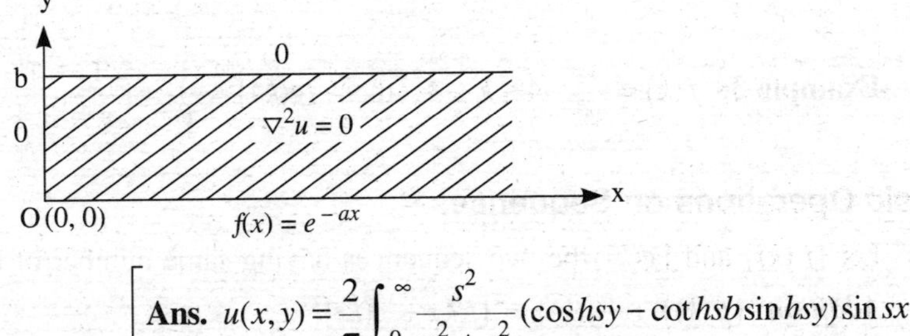

$$\left[\textbf{Ans. } u(x,y) = \frac{2}{\pi}\int_0^{\infty} \frac{s^2}{a^2 + s^2}(\cos hsy - \cot hsb \sin hsy) \sin sx\, ds\right]$$

8. If the initial temperature of an infinite bar is given by

$$\theta(x) = \begin{cases} \theta_0, & \text{for } |x| < a \\ 0, & \text{for } |x| > a \end{cases} \qquad \left[\textbf{Ans. } \theta(x,t) = \frac{\theta_0}{\pi}\int_{-\infty}^{\infty} \frac{\sin as}{s} \cdot e^{-c^2s^2t} \cos xs\, ds\right]$$

1.17 Z-TRANSFORMS

Z-transform is very useful in the discrete analysis. Difference equations are formed in discrete system and their solution and analysis are carried out by *z*-transform. Z-transform is the discrete analogue of Laplace transform. Laplace transform operates on a continuous function, whereas z-transform operates on a sequence $\{f(k)\}$ where k takes the discreate values 0, 1, 2... To every operational rule of Laplace transforms, there exists an analogus rule for *z*-transforms.

Sequence

Sequence $\{f(k)\}$ is an ordered list of real or complex numbers.

Representation of a Sequence

First method: The elementary way is to list all the members of the sequence such as

$$\{f(k)\} = \{15, 10, 7, 4, 1, -1, 0, \underset{\uparrow}{3}, 6\}$$

The symbol ↑ is used to denote the term in zero position i.e. $k = 0$, k is an index of position of a term in the sequence.

$$\{g(k)\} = \{15, 10, \underset{\uparrow}{7}, 4, 1, -1, 0, 3, 6\}$$

Two sequences $\{f(k)\}$ and $\{g(k)\}$ have same terms but these sequences are not identically the same as the zeroth term of those sequences are different.

In case the symbol ↑ is not given, then left hand end term is considered as the term corresponding to $k = 0$.

Example 1: {8, 6, 3, –1, 0, 1, 4, 5}, here

The zeroth term is 8, the left hand end term.

Second method: The second way of specifying the sequence is to define the general term of the sequence $\{f(k)\}$ as function of k.

Example 2: $f(k) = \frac{1}{3^k}$

This sequence represents $\left\{ \ldots\ldots\ldots \frac{1}{3^{-3}}, \frac{1}{3^{-2}}, \frac{1}{3^{-1}}, \underset{\uparrow}{1}, \frac{1}{3}, \frac{1}{3^2}, \frac{1}{3^3} \ldots\ldots \right\}$

Example 3: $f(k) = \frac{1}{3^k}, -4 \le k \le 3$ then $\{g(k)\} = \frac{1}{3^k} = \{\frac{1}{3^{-4}}, \frac{1}{3^{-3}}, \frac{1}{3^{-2}}, \frac{1}{3^{-1}}, 1, \frac{1}{3}, \frac{1}{3^2}, \frac{1}{3^3}\}$

Basic Operations on Sequences

Let $\{f(k)\}$ and $\{g(k)\}$ be two sequences having same number of terms.

Addition: $\{f(k)\} + \{g(k)\} = \{f(k) + g(k)\}$

Multiplication: Let a be a scalar, then $a\{f(k)\} = \{a f(k)\}$

Linearity: $a\{f(k)\} + b\{g(k)\} = \{a f(k) + b g(k)\}$

1.18 DEFINITION OF Z-TRANSFORM

The Z-transform of a sequence $\{f(k)\}$ is denoted as $Z\left[\{f(k)\}\right]$.

It is defined as $Z\left[\{f(k)\}\right] = F(z) = \sum_{k=-\infty}^{\infty} f(k) z^{-k} = \sum_{k=-\infty}^{\infty} \frac{f(k)}{z^k}$

where 1. z is a complex number.

2. Z is an operator of Z-transform.

3. $F(z)$ is the Z transform of $\{f(k)\}$.

Example 1: If $f(k) = \{15, 10, 7, \underset{\uparrow}{4}, 1, -1, 0, 3, 6\}$, then

$$Z\left[\{f(k)\}\right] = F(z) = 15z^3 + 10z^2 + 7z + 4 + \frac{1}{z} + \frac{1}{z^2} + 0 + \frac{3}{z^4} + \frac{6}{z^5}$$

Example 2: Find the z-transform of $\{a^k\}, k \ge 0$

Sol.

$$Z\left[\{a^k\}\right] = \sum_{k=0}^{\infty} a^k z^{-k}$$

$$= 1 + az^{-1} + a^2 z^{-2} + a^3 z^{-3} + \ldots\ldots\ldots\ldots\infty$$

$$= 1 + \frac{a}{z} + \frac{a^2}{z^2} + \frac{a^3}{z^3} + \ldots\ldots\ldots\ldots\infty$$

which is in G.P. whose sum $=\dfrac{a}{1-r}$

$$\therefore \qquad z\left[\left\{a^k\right\}\right]=\frac{1}{1-\dfrac{a}{z}}=\frac{z}{z-a}. \quad \textbf{Ans.}$$

1.19 PROPERTIES OF Z-TRANSFORMS

Linearity

Theorem: If $\{f(k)\}$ and $\{g(k)\}$ are such that they can be added a and b are constants, Then

$$Z\{af(k)+bg(k)\}=aZ\left[\{f(k)\}\right]+bZ\left[\{g(k\}\right]$$

Proof: $$Z\left[\{af(k)+bg(k)\}\right]=\sum_{k=-\infty}^{\infty}\left[af(k)+bg(k)\right]z^{-k} \qquad \text{[By definition]}$$

$$=\sum_{k=-\infty}^{\infty}\left[af(k)z^{-k}+bg(k)z^{-k}\right]=a\sum_{k=-\infty}^{\infty}f(k)z^{-k}+b\sum_{k=-\infty}^{\infty}g(k)z^{-k}$$

$= aZ[\{f(k)\}+bZ[\{g(k)\}]$. **Proved.**

1.20 CHANGE OF SCALE

Theorem: If $Z[\{f(k)\}] = F(z)$, then $Z\left[\left\{a^k f(k)\right\}\right]=F\left(\dfrac{z}{a}\right)$

Proof: $$F(z)=Z\left[\{f(k)\}\right]=\sum_{k=-\infty}^{\infty}f(k)z^{-k}$$

Substituting $\dfrac{z}{a}$ for z, we get

$$F\left(\frac{z}{a}\right)=\sum_{k=-\infty}^{\infty}f(k)\left(\frac{z}{a}\right)^{-k} \qquad \text{...(1)}$$

But $$Z\left[\left\{a^k f(k)\right\}\right]=\sum_{k=-\infty}^{\infty}a^k f(k)z^{-k}=\sum_{k=-\infty}^{\infty}f(k)\left(\frac{z}{a}\right)^{-k} \qquad \text{...(2)}$$

From (1) and (2)

$$Z\left[\left\{a^k f(k)\right\}\right]=F\left(\frac{z}{a}\right). \quad \textbf{Proved.}$$

1.21 SHIFTING PROPERTY

Theorem: If $\qquad Z\,[\{f\,(k)]=F\,(z),$

Then $\qquad Z\left[\{f(k\pm n\}\right]=z^{\pm n}F(z)$

Proof: $$Z\left[\{f(k \pm n\}\right] = \sum_{k=-\infty}^{\infty} f(k \pm n) z^{-k} = z^{\pm n} \sum_{k=-\infty}^{\infty} f(k \pm n) z^{-(k \pm n)}$$

$$= z^{\pm n} \sum_{k=-\infty}^{\infty} f(r) z^{-r} \quad \Big| \text{ Put } r = k \pm n$$

$$= z^{\pm n} F(z). \quad \textbf{Proved.}$$

$\Rightarrow$ **Case I.** $Z[\{f(k+n)\}] = f(k+n) z^{-k} k \geq 0$

$$= \sum_{k=0}^{\infty} f(k+n) z^{-k} = z^{n} \sum_{k=0}^{\infty} f(k+n) z^{-(k+n)}$$

Putting $r = k + n$

$$= z^{n} \sum_{r=n}^{\infty} f(r) z^{-r} = z^{n} \sum_{r=0}^{\infty} f(r) z^{-r} - z^{n} \sum_{r=0}^{n-1} f(r) z^{-r}$$

$$= z^{n} F(z) - z^{n} \sum_{r=0}^{n-1} f(r) z^{-r}$$

$\Rightarrow$ $$\boxed{Z[\{f(k+n)\}] = z^{n} F(z) - z^{n} \sum_{r=0}^{n-1} f(r) z^{-r}}$$

Case II. $$Z\left[\{f(k-n)\}\right] = \sum_{k=0}^{\infty} f(k-n) z^{-k} = z^{-n} \sum_{k=0}^{\infty} f(k-n) z^{-(k-n)}$$

Putting $r = k - n$

$$= z^{-n} \sum_{r=-n}^{\infty} f(r) z^{-r}$$

$$= z^{-n} \sum_{r=0}^{\infty} f(r) z^{-r} + z^{-n} \sum_{r=-n}^{-1} f(r) z^{-r} = z^{-n} F(z) + \sum_{r=-n}^{-1} f(r) z^{-n-r}$$

Putting $r = -m$

$$\boxed{Z[\{f(k-n)\}] = z^{-n} F(z) + \sum_{m=1}^{n} f(-m) z^{-n+m}}$$

For $n = 1$, $$Z\{f(k+1)\} = zF(z) - z \sum_{r=0}^{0} f(r) z^{-r} = zF(z) - zf(0)$$

$\Rightarrow$ $$\boxed{Z\{f(k+1)\} = zF(z) - zf(0)}$$

Similarly

For $n = 2$, $\boxed{Z\{f(k+2)\} = z^{2} F(z) - z^{2} f(0) - zf(1)}$ and so on.

1.22 MULTIPLICATION BY *K*

Theorem: If $Z\left[\{f(k)\}\right]=F(z)$, then $Z\left[\{kf(k)\}\right]=-z\dfrac{d}{dz}F(z)$

Proof:

$$Z\left[\{kf(k)\}\right]=\sum_{k=-\infty}^{\infty} kf(k)z^{-k}=-z\sum_{k=-\infty}^{\infty}-kf(k)z^{-k-1}=-z\sum_{k=-\infty}^{\infty}f(k)\left(-kz^{-k-1}\right)$$

$$=-z\sum_{k=-\infty}^{\infty}f(k)\frac{d}{dz}\left(z^{-k}\right)=-z\frac{d}{dz}\sum_{k=-\infty}^{\infty}f(k)z^{-k}$$

$$=-z\frac{d}{dz}F(z).$$ **Proved.**

In general $$Z\left[\{k^n f(k)\}\right]=\left(-z\frac{d}{dz}\right)^n F(z).$$

1.23 DIVISION BY *K*

Theorem: If $Z\left[\{f(k)\}\right]=F(z)$, then $Z\left[\left\{\dfrac{f(k)}{k}\right\}\right]=-\int^{z} z^{-1}F(z)dz$

Proof:

$$Z\left[\left\{\frac{f(k)}{k}\right\}\right]=\sum_{k=-\infty}^{\infty}\frac{f(k)}{k}z^{-k}$$

$$=\sum_{k=-\infty}^{\infty}f(k)\left(\frac{1}{k}z^{-k}\right)=-\sum_{k=-\infty}^{\infty}f(k)\int^{z}z^{-k-1}dz$$

$$=-\int^{z}\sum_{k=-\infty}^{\infty}f(k)z^{-k-1}dz=-\int^{z}z^{-1}\sum_{k=-\infty}^{\infty}f(k)z^{-k}dz=-\int^{z}z^{-1}F(z)dz$$

$$Z\left[\left\{\frac{f(k)}{k}\right\}\right]=-\int^{z}z^{-1}F(z)dz.$$ **Proved.**

1.24 INITIAL VALUE

Theorem: If $Z\,[\{f\,(k)\}]=F\,(z),\ k\geq 0$

Then $$f(0)=\lim_{z\to\infty}F(z).$$

Proof: $$Z\left[\{f(k)\}\right]=\sum_{k=0}^{\infty}f(k)z^{-k}=F(z)$$

$$f(0)+f(1)z^{-1}+f(2)z^{-2}+\ldots\ldots\ldots=F(z)$$

Taking the limit, $z\to\infty$, we get

$$f(0)=\lim_{z\to\infty}F(z).$$ **Proved.**

1.25 FINAL VALUE

Theorem: $$\lim_{k\to\infty} f(k) = \lim_{z\to 1}(z-1)F(z).$$

Proof: $$Z\left[\{f(k+1)-f(k)\}\right] = \sum_{k=0}^{\infty}\left[f(k+1)-f(k)\right]z^{-k}$$

$$zF(z)-zf(0)-F(z) = \lim_{n\to\infty}\sum_{k=0}^{n}\left[f(k+1)-f(k)\right]z^{-k}$$

$$\lim_{z\to 1}(z-1)F(z) = f(0)+\lim_{z\to 1}\ \lim_{n\to\infty}\sum\left[f(k+1)-f(k)\right]z^{-k}$$

By changing the order of limits, we get

$$\lim_{z\to 1}(z-1)F(z) = f(0)+\lim_{n\to\infty}\sum_{k=0}^{n}\lim_{z\to 1}[f(k+1)-f(k)]z^{-k}$$

$$= \lim_{n\to\infty}\left[f(0)+\sum_{k=0}^{n}\{f(k+1)-f(k)\}\right]$$

$$= \lim_{n\to\infty}[f(0)-f(0)+f(1)+f(2)-f(2)+\ldots\ldots+f(n+1)-f(n)]$$

$$= \lim_{n\to\infty} f(n+1) = \lim_{n\to\infty} f(n) = \lim_{k\to\infty} f(k).$$ **Proved.**

1.26 PARTIAL SUM

Theorem: If $Z\left[\{f(k)\}\right] = F(z)$,

Then $$Z\left[\left\{\sum_{n\to-\infty}^{k} f(n)\right\}\right] = \frac{F(z)}{1-z^{-1}}$$

Proof: Let $\{g(k)\}$ be a sequene such that

$$g(k) = \sum_{n=-\infty}^{k} f(n)$$

We are required to find $Z\{f(k)\}$,

We know that $g(k)-g(k-1) = \sum_{n=-\infty}^{k} f(n) - \sum_{n=-\infty}^{k-1} f(n) = f(k)$

$$Z\left[\{g(k)\}-\{g(k-1)\}\right] = Z\left[\{f(k)\}\right]$$

$$Z\left[\{g(k)\}-Z\{g(k-1)\}\right] = F(z)$$

$$G(z)-z^{-1}G(z) = F(z)$$

$$\sum_{n=-\infty}^{k} f(k) = G(z) = \frac{F(z)}{1-z^{-1}}.\quad \textbf{Proved.}$$

1.27 CONVOLUTION

Let two sequences be $\{f(k)\}$ and $\{g(k)\}$ and the convolution of $\{f(k)\}$ and $\{g(k)\}$ be $\{h(k)\}$ and denoted as

$$\{h(k)\} = \{f(k)\} * \{g(k)\}$$

where
$$\{h(k)\} = \sum_{n=-\infty}^{\infty} f(n)g(k-n) \quad ...(1)$$

$$= \sum_{n=-\infty}^{\infty} g(n)f(k-n) = \{g(k)\} * \{f(k)\}$$

Proof: Z-transform of (1) is

$$Z\{h(k)\} = \sum_{k=-\infty}^{\infty}\left(\sum_{n=-\infty}^{\infty}\left(f(n)g(k-n)\right)z^{-k}\right) = \sum_{n=-\infty}^{\infty}\sum_{k=-\infty}^{\infty} f(n)g(k-n)z^{-k}$$

$$= \sum_{n=-\infty}^{\infty} f(n)z^{-n}\sum_{k=-\infty}^{\infty} g(k-n)z^{-(k-n)} = \left[\sum_{n=-\infty}^{\infty} f(n)z^{-n}\right]G(z) = F(z)G(z)$$

Example 6: Find the Z-transform of $\left\{a^{|k|}\right\}$.

Sol. $Z\left[\left\{a^{|k|}\right\}\right] = \sum_{k=-\infty}^{\infty} a^{|k|}z^{-k} = \sum_{k=-\infty}^{-1} a^{-k}z^{-k} + \sum_{k=0} a^k z^{-k}$

$$= \left[.....+a^3z^3 + a^2z^2 + az\right] + [1 + az^{-1} + a^2z^{-2} + a^3z^{-3} + ...]$$

These are two G.P. and sum of G.P. $= \dfrac{a}{1-r}$

$$= \frac{az}{1-az} + \frac{1}{1-az^{-1}},\ |az| < 1 \text{ and } \left|az^{-1}\right| < 1 = \frac{az}{1-az} + \frac{z}{z-a}$$

$$= \frac{az(z-a) + z(1-az)}{(1-az)(z-a)} = \frac{az^2 - a^2z + z - az^2}{(1-az)(z-a)}$$

$$= \frac{z - a^2z}{(1-az)(z-a)}.\quad \textbf{Ans.}$$

Example 7: Find the Z-transform of $\left\{\left(\dfrac{1}{2}\right)^{|k|}\right\}$

Sol. $$Z\left[\left\{\left(\frac{1}{2}\right)^{|k|}\right\}\right]=\sum_{k=-\infty}^{\infty}\left(\frac{1}{2}\right)^{|k|}z^{-k}$$

Putting $a=\frac{1}{2}$ in example 6, we get

$$Z\left[\left\{\left(\frac{1}{2}\right)^{|k|}\right\}\right]=\frac{\left(1-\frac{1}{4}\right)}{\left(1-\frac{1}{2}z\right)\left(z-\frac{1}{2}\right)},\left|\frac{1}{2}\right|<|z|<2=\frac{\frac{3}{4}}{\left(\frac{2-z}{2}\right)\left(\frac{2z-1}{2}\right)}=\frac{3z}{(2-z)(2z-1)}$$

$$=\frac{3z}{5z-2-2z^2}.\quad \textbf{Ans.}$$

Example 8: Find the Z-transform of unit impulse

$$\delta(k)=\begin{cases}1, & k=0\\ 0, & k\neq 0\end{cases}$$

Sol. $$Z\{f(k)\}=\sum_{k=0}^{\infty}\delta(k)z^{-k}=[\ldots\ldots+0+0+1+0+0+\ldots\ldots]=1.\quad \textbf{Ans.}$$

Example 9: Find the Z-transform of discrete unit step

$$U(k)=\begin{cases}0, & k<0\\ 1, & k\geq 0\end{cases}$$

Sol. $$Z=\left[\{U(k)\}\right]=\sum_{k=0}^{\infty}U(k)z^{-k}=\left[1+z^{-1}+z^{-2}+z^{-3}+\ldots\ldots\right]$$

This is G.P. its sum is $\frac{a}{1-r}$

$$=\frac{1}{1-z^{-1}}=\frac{1}{1-\frac{1}{z}}=\frac{z}{z-1}.\quad \textbf{Ans.}$$

Example 10: Find the Z-transform of $\{f(k)\}$ where

$$f(k)=\begin{cases}5^k, & k<0\\ 3^k, & k\geq 0\end{cases}$$

Sol. $$Z\left[\{f(k)\}\right]=\sum_{k=-\infty}^{-1}5^k z^{-k}+\sum_{k=0}^{\infty}3^k z^{-k}$$

$$= \left[\ldots\ldots + 5^{-3} z^3 + 5^{-2} z^2 + 5^{-1} z^1 \right] + \left[1 + \frac{3}{z^{-1}} + \frac{9}{z^{-2}} + \frac{27}{z^{-3}} + \ldots\ldots \right]$$

These are G.P.

$$= \frac{5^{-1} z}{1 - 5^{-1} z} + \frac{1}{1 - \dfrac{3}{z^{-1}}} = \frac{z}{5 - z} + \frac{1}{1 - 3z}$$

$$= \frac{z - 3z^2 + 5z - z^2}{(5 - z)(1 - 3z)} = \frac{-4z^2 + 6z}{(5 - z)(1 - 3z)}. \quad \textbf{Ans.}$$

Example 11: Find the Z-transform of $\sin \alpha\, k, k \geq 0$.

Sol.

$$Z\left[\{\sin \alpha k\}\right] = \sum_{k=0}^{\infty} \sin \alpha k\; z^{-k} = \sum_{k=0}^{\infty} \frac{e^{i\alpha k} - e^{-i\alpha k}}{2i} z^{-k}$$

$$= \frac{1}{2i} \sum_{k=0}^{\infty} e^{i\alpha k} z^{-k} - \frac{1}{2i} \sum_{k=0}^{\infty} e^{-i\alpha k} z^{-k} = \frac{1}{2i} \sum_{k=0}^{\infty} \left(e^{i\alpha} z^{-1}\right)^k - \frac{1}{2i} \sum_{k=0}^{\infty} \left(e^{-i\alpha} z^{-1}\right)^k$$

$$= \frac{1}{2i} \left[1 + \left(e^{i\alpha} z^{-1}\right) + \left(e^{i\alpha} z^{-1}\right)^2 + \ldots\ldots \right] - \frac{1}{2i} \left[1 + \left(e^{-i\alpha} z^{-1}\right) + \left(e^{-i\alpha} z^{-1}\right)^2 + \ldots\ldots \right]$$

These are G.P. so

$$= \frac{1}{2i} \frac{1}{1 - e^{i\alpha} z^{-1}} - \frac{1}{2i} \frac{1}{1 - e^{-i\alpha} z^{-1}} = \frac{1}{2i} \frac{z}{z - e^{i\alpha}} - \frac{1}{2i} \frac{z}{z - e^{-i\alpha}}$$

$$= \frac{1}{2i} \left[\frac{z}{z - e^{i\alpha}} - \frac{z}{z - e^{-i\alpha}} \right] = \frac{1}{2i} \frac{z\left(z - e^{-i\alpha}\right) - z\left(z - e^{i\alpha}\right)}{\left(z - e^{i\alpha}\right)\left(z - e^{-i\alpha}\right)}$$

$$= \frac{1}{2i} \frac{z\left(e^{i\alpha} - e^{-i\alpha}\right)}{z^2 - z\left(e^{i\alpha} + e^{-i\alpha}\right) + 1} = \frac{z \sin \alpha}{z^2 - 2z \cos \alpha + 1}. \quad \textbf{Ans.}$$

Example 12: Find Z-transform of $\sin(3k + 5)$.

Sol.

$$F(z) = \sum_{k=0}^{\infty} \sin(3k + 5) z^{-k} = \sum_{k=0}^{\infty} \frac{e^{i(3k+5)} - e^{-i(3k+5)}}{2i} z^{-k}$$

$$= \frac{1}{2i} \sum_{k=0}^{\infty} e^{i(3k+5)} z^{-k} - \frac{1}{2i} \sum_{k=0}^{\infty} e^{-i(3k+5)} z^{-k}$$

$$= \frac{1}{2i} e^{5i} \sum_{k=0}^{\infty} \left(e^{3i} z^{-1}\right)^k - \frac{1}{2i} e^{-5i} \sum \left(e^{-3i} z^{-1}\right)^k$$

$$= \frac{1}{2i}e^{5i}\left[1+\left(e^{3i}z^{-1}\right)+\left(e^{3i}z^{-1}\right)^2+.....\right]-\frac{1}{2i}e^{-5i}\left[1+\left(e^{-3i}z^{-1}\right)+\left(e^{-3i}z^{-1}\right)^2+.....\right]$$

$$= \frac{e^{5i}}{2i}\cdot\frac{1}{1-e^{3i}z^{-1}}-\frac{1}{2i}e^{-5i}\frac{1}{1-e^{-3i}z^{-1}}\left[S=\frac{a}{1-r}\right]$$

$$= \frac{1}{2i}\frac{e^{5i}\left(1-e^{-3i}z^{-1}\right)-e^{-5i}\left(1-e^{3i}z^{-1}\right)}{\left(1-e^{3i}z^{-1}\right)\left(1-e^{-3i}z^{-1}\right)}=\frac{1}{2i}\frac{\left(e^{i5}-e^{-5i}\right)-e^{2i}z^{-1}+e^{-2i}z^{-1}}{1-e^{3i}z^{-1}-e^{-3i}z^{-1}+z^{-2}}$$

$$= \frac{\dfrac{e^{5i}-e^{-5i}}{2i}-z^{-1}\dfrac{e^{2i}-e^{-2i}}{2i}}{1-\left(e^{3i}+e^{-3i}\right)z^{-1}+z^{-2}}=\frac{\sin 5-z^{-1}\sin 2}{1-(2\cos 3)z^{-1}+z^{-2}}$$

$$= \frac{z^2\sin 5-z\sin 2}{z^2-2z\cos 3+1}, \quad |z|>1. \quad \textbf{Ans.}$$

Example 13: Find the Z-transform of $\left\{\cos\left(\frac{k\pi}{8}+\alpha\right)\right\}$

Sol. $$Z\left[\left\{\cos\left(\frac{k\pi}{8}+\alpha\right)\right\}\right]=\sum\cos\left(\frac{k\pi}{8}+\alpha\right)z^{-k}$$

$$= \sum\left\{\cos\frac{k\pi}{8}\cos\alpha-\sin\frac{k\pi}{8}\sin\alpha\right\}z^{-k}$$

$$= \sum\cos\frac{k\pi}{8}\cos\alpha\, z^{-k}-\sum\sin\frac{k\pi}{8}\sin\alpha\, z^{-k}$$

$$= \cos\alpha\sum\cos\frac{k\pi}{8}z^{-k}-\sin\alpha\sum\sin\frac{k\pi}{8}z^{-k}$$

$$= \cos\alpha\frac{z^2-z\cos\frac{\pi}{8}}{z^2-2z\cos\frac{\pi}{8}+1}-\sin\alpha\frac{z\sin\frac{\pi}{8}}{z^2-2z\cos\frac{\pi}{8}+1}$$

$$= \frac{\left(z^2-z\cos\frac{\pi}{8}\right)\cos\alpha-z\sin\frac{\pi}{8}\sin\alpha}{z^2-2z\cos\frac{\pi}{8}+1}=\frac{z^2\cos\alpha-z\left[\cos\frac{\pi}{8}\cos\alpha+\sin\frac{\pi}{8}\sin\alpha\right]}{z^2-2z\cos\frac{\pi}{8}+1}$$

$$= \frac{z^2\cos\alpha-z\cos\left(\frac{\pi}{8}-\alpha\right)}{z^2-2z\cos\frac{\pi}{8}+1}. \quad \textbf{Ans.}$$

Example 14: Find the *Z*-transform of $\cosh\left(\frac{k\pi}{2}+\alpha\right)$

Sol. $$F(z)=\sum_{k=0}^{\infty}\cosh\left(\frac{k\pi}{2}+\alpha\right)z^{-k}=\sum_{k=0}^{\infty}\frac{e^{\left(\frac{k\pi}{2}+\alpha\right)}+e^{-\left(\frac{k\pi}{2}+\alpha\right)}}{2}z^{-k}$$

$$=\frac{1}{2}\sum_{k=0}^{\infty}e^{\frac{k\pi}{2}+\alpha}z^{-k}+\frac{1}{2}\sum_{k=0}^{\infty}e^{-\frac{k\pi}{2}-\alpha}z^{-k}=\frac{1}{2}e^{\alpha}\sum_{k=0}^{\infty}\left(e^{\frac{\pi}{2}}z^{-1}\right)^{k}+\frac{1}{2}e^{-\alpha}\sum_{k=0}^{\infty}\left(e^{-\frac{\pi}{2}}z^{-1}\right)^{k}$$

$$=\frac{1}{2}e^{\alpha}\left[1+\left(e^{\frac{\pi}{2}}z^{-1}\right)+\left(e^{\frac{\pi}{2}}z^{-1}\right)^{2}+\ldots..\right]+\frac{1}{2}e^{-\alpha}\left[1+\left(e^{-\frac{\pi}{2}}z^{-1}\right)+\left(e^{-\frac{\pi}{2}}z^{-1}\right)^{2}+\ldots..\right]$$

(Geometrical series sum $=\frac{a}{1-r}$)

$$=\frac{1}{2}e^{\alpha}\frac{1}{1-e^{\frac{\pi}{2}}z^{-1}}+\frac{1}{2}e^{-\alpha}\frac{1}{1-e^{-\frac{\pi}{2}}z^{-1}}=\frac{1}{2}\frac{e^{\alpha}\left(1-e^{-\frac{\pi}{2}}z^{-1}\right)+e^{-\alpha}\left(1-e^{\frac{\pi}{2}}z^{-1}\right)}{\left(1-e^{\frac{\pi}{2}}z^{-1}\right)\left(1-e^{-\frac{\pi}{2}}z^{-1}\right)}$$

$$=\frac{\frac{e^{\alpha}+e^{-\alpha}}{2}-\frac{e^{\alpha-\frac{\pi}{2}}+e^{-\alpha+\frac{\pi}{2}}}{2}.z^{-1}}{1-e^{\frac{\pi}{2}}z^{-1}-e^{-\frac{\pi}{2}}z^{-1}+z^{-2}}=\frac{\cosh\alpha-\cosh\left(\alpha-\frac{\pi}{2}\right).z^{-1}}{1-\left(2\cosh\frac{\pi}{2}\right)z^{-1}+z^{-2}}$$

$$=\frac{z^{2}\cosh\alpha-z\cosh\left(\frac{\pi}{2}-\alpha\right)}{z^{2}-2z\cosh\frac{\pi}{2}+1}.\quad \textbf{Ans.}$$

Example 15: Find the Z-transform of a^{k}, $k\geq 0$.

Sol. We know that

$$Z[\{1\}]=\frac{z}{z-1}$$

for the given sequence, by the scale change formula the Z-transform

$$Z\left[\{a^{k}.1\}\right]=\frac{\frac{z}{a}}{\frac{z}{a}-1}=\frac{z}{z-a}.\quad \textbf{Ans.}$$

Example 16: Find the *Z*-transform of $c^k \sin \alpha k,\ k \geq 0$.

Sol. We know that

$$Z\left[\{\sin \alpha k\}\right] = \frac{z \sin \alpha}{z^2 - 2z\cos \alpha + 1} \qquad \text{(see Example 11)}$$

By applying the formula of change of scale we get

$$Z\left[\left\{c^k \sin \alpha k\right\}\right] = \frac{\frac{z}{c}\sin \alpha}{\left(\frac{z}{c}\right)^2 - 2\left(\frac{z}{c}\right)\cos \alpha + 1}$$

$$= \frac{cz \sin \alpha}{z^2 - 2cz\cos \alpha + c^2}. \quad \textbf{Ans.}$$

Example 17: Find the *Z*-transform of

(*i*) $c^k \cosh(\alpha k),\ k \geq 0$ (*ii*) $c^k \cos(\alpha k), k \geq 0$

Sol. (*i*)

$$Z\{\cosh(\alpha k)\} = \sum_{k=0}^{\infty} \left(\frac{e^{\alpha k} + e^{-\alpha k}}{2}\right) z^{-k}$$

$$= \frac{1}{2}\left[\sum_{k=0}^{\infty}\left(e^{\alpha} z^{-1}\right)^k + \sum_{k=0}^{\infty}\left(e^{-\alpha} z^{-1}\right)^k\right]$$

$$= \frac{1}{2}\left[\left(1 - e^{\alpha} z^{-1}\right)^{-1} + \left(1 - e^{-\alpha} z^{-1}\right)^{-1}\right]$$

$$= \frac{1}{2}\left[\frac{z}{z - e^{\alpha}} + \frac{z}{z - e^{-\alpha}}\right]$$

$$= \frac{z}{2}\left[\frac{2z - \left(e^{\alpha} + e^{-\alpha}\right)}{z^2 - z\left(e^{\alpha} + e^{-\alpha}\right) + 1}\right] = \frac{z(z - \cosh \alpha)}{z^2 - 2z\cosh \alpha + 1}$$

By change of scale property,

$$Z\left\{c^k \cosh(\alpha k)\right\} = \frac{\frac{z}{c}\left(\frac{z}{c} - \cosh \alpha\right)}{\left(\frac{z}{c}\right)^2 - 2\left(\frac{z}{c}\right)\cosh \alpha + 1}$$

$$= \frac{z(z - c\cosh \alpha)}{z^2 - 2cz\cosh \alpha + c^2}. \quad \textbf{Ans.}$$

(*ii*) $$Z\{\cos(\alpha k)\} = \sum_{k=0}^{\infty}\left(\frac{e^{i\alpha k}+e^{-i\alpha k}}{2}\right)z^{-k}$$

$$= \frac{1}{2}\left[\sum_{k=0}^{\infty}\left(e^{i\alpha}z^{-1}\right)^k + \sum_{k=0}^{\infty}\left(e^{-i\alpha}z^{-1}\right)^k\right]$$

$$= \frac{1}{2}\left[\left(1-e^{i\alpha}z^{-1}\right)^{-1} + \left(1-e^{-i\alpha}z^{-1}\right)^{-1}\right]$$

$$= \frac{1}{2}\left[\frac{z}{z-e^{i\alpha}} + \frac{z}{z-e^{-i\alpha}}\right]$$

$$= \frac{z}{2}\left[\frac{2z-2\cos\alpha}{z^2-z\cos\alpha+1}\right] = \frac{z(z-\cos\alpha)}{z^2-2z\cos\alpha+1}$$

By change of scale property,

$$Z\{c^k\cos(\alpha k)\} = \frac{\left(\frac{z}{c}\right)\left(\frac{z}{c}-\cos\alpha\right)}{\left(\frac{z}{c}\right)^2 - 2\left(\frac{z}{c}\right)\cos\alpha + 1}$$

$$= \frac{z(z-c\cos\alpha)}{z^2-2zc\cos\alpha+c^2}.\quad \textbf{Ans.}$$

Example 18: Find the Z-transform of $\{{}^nC_k\}$ $(0 \le k \le n)$.

Sol. $$Z\left[\{{}^nC_k\}\right] = \sum_{k=0}^{n} {}^nC_k z^{-k} = 1 + {}^nC_1 z^{-1} + {}^nC_2 z^{-2} + {}^nC_3 z^{-3} + \ldots.. + {}^nC_n z^{-n}$$

This is the expansion of Binomial theorem. $= (1+z^{-1})^n$. **Ans.**

Example 19: Find Z-transform of $\{{}^{k+n}C_n\}$.

Sol. $$Z\left[\{{}^{k+n}C_n\}\right] = \sum_{k=0}^{\infty} {}^{k+n}C_n z^{-k} \qquad \begin{pmatrix} k+n>n \\ k>0 \end{pmatrix}$$

$$= \sum_{k=0}^{\infty} {}^{k+n}C_k z^{-k} \qquad \left({}^nC_r = {}^nC_{n-r}\right)$$

$$= 1 + {}^{n+1}C_1 z^{-1} + {}^{n+2}C_2 z^{-2} + {}^{n+3}C_3 z^{-3} + \ldots\ldots\ldots$$

$$= 1 + (n+1)z^{-1} + \frac{(n+2)(n+1)}{2!}z^{-2} + \frac{(n+3)(n+2)(n+1)}{3!}z^{-3} + \ldots\ldots.$$

$$=1+(-n-1)\left(-z^{-1}\right)+\frac{(-n-1)(-n-2)}{2!}\left(-z^{-1}\right)^2+\frac{(-n-1)(-n-2(-n-3)}{3!}\left(-z^{-1}\right)^3+\ldots\ldots.$$

This is the expansion of Binomial theorem.

$$=\left(1-z^{-1}\right)^{-n-1}=\left(1-z^{-1}\right)^{-(n+1)}. \quad \textbf{Ans.}$$

Example 20: Find the Z-transform of ${}^{k+n}C_n a^k; k \geq 0$

Sol. In Example 19 $Z\left\{{}^{k+n}C_n\right\}=\left(\frac{z}{z-1}\right)^{n+1}$

Applying change of scale property we get

$$Z\left\{{}^{k+n}C_n a^k\right\}=F\left(\frac{z}{a}\right)=\left(\frac{\frac{z}{a}}{\frac{z}{a}-1}\right)^{n+1}=\left(\frac{z}{z-a}\right)^{n+1}. \quad \textbf{Ans.}$$

Example 21: Find the Z-transform of $\left\{\frac{a^k}{k!}\right\}$. $(k \geq 0)$

Sol. $$Z\left[\left\{\frac{a^k}{k!}\right\}\right]=\sum_{k=0}^{\infty}\frac{a^k}{k!}z^{-k}$$

$$=\sum_{k=0}^{\infty}\frac{\left(az^{-1}\right)^k}{k!}=1+\frac{az^{-1}}{1!}+\frac{\left(az^{-1}\right)^2}{2!}+\frac{\left(az^{-1}\right)^3}{3!}+\ldots..$$

This is exponential series.

$$=e^{az^{-1}}=e^{\frac{a}{z}}. \quad \textbf{Ans.}$$

Example 22: If $F(z)=\dfrac{2z^2+3z+12}{(z-1)^4}$ find the values of $f(2)$ and $f(3)$.

Sol. We use the initial value theorem to find $f(2)$ and $f(3)$. First of all we find $f(0)$ and $f(1)$.

$$f(0)=\lim_{z\to\infty}F(z)=\lim_{z\to\infty}\left[\frac{2z^2+3z+12}{(z-1)^4}\right]$$

$$=\lim_{z\to\infty}\left[\frac{z^2(2+3z^{-1}+12z^{-2})}{z^4(1-z^{-1})^4}\right]=\lim_{z\to\infty}\left[\frac{\left(2+\frac{3}{z}+\frac{12}{z^2}\right)}{z^2\left(1-\frac{1}{z}\right)^4}\right]=0$$

$$f(1) = \lim_{z\to\infty}[z\{F(z)-f(0)\}]$$

$$= \lim_{z\to\infty}[zF(z)]$$

As $F(z) = f(0) + \frac{1}{z}f(1) + \frac{1}{z^2}f(2) + \ldots$

or $z\{F(z)-f(0)\} = f(1) + \frac{1}{z}f(2) + \ldots$

Taking $z \to \infty$, then $f(1) = \lim_{z\to\infty}[z\{F(z)-f(0)\}]$

$$= \lim_{z\to\infty}\left[z\cdot\frac{2z^2+3z+12}{(z-1)^4}\right] = \lim_{z\to\infty}\left[\frac{\left(2+\frac{3}{z}+\frac{12}{z^2}\right)}{z\left(1-\frac{1}{z}\right)}\right] = 0$$

$\Rightarrow$ $f(1) = 0$

Similarly $f(2) = \lim_{z\to\infty}\left[z^2\left\{F(z)-f(0)-\frac{f(1)}{z}\right\}\right] = \lim_{z\to\infty}[z^2F(z)-0-0]$

$$= \lim_{z\to\infty}\left[z^2\frac{z^2\left(2+\frac{3}{z}+\frac{12}{z^2}\right)}{z^4\left(1-\frac{1}{z}\right)^4}\right] = \lim_{z\to\infty}\left[\frac{(2+3/z+12/z^2}{\left(1-\frac{1}{z}\right)^4}\right] = 2$$

$\Rightarrow$ $f(2) = 2$

$$f(3) = \lim_{z\to\infty}\left[z^3\left\{F(z)-f(0)-\frac{f(1)}{z}-\frac{f(2)}{z^2}\right\}\right] = \lim_{z\to\infty}\left[z^3\left\{F(z)-\frac{2}{z^2}\right\}\right]$$

$$= \lim_{z\to\infty}\left[z^3\left\{\frac{2z^2+3z+12}{(z-1)^4}-\frac{2}{z^2}\right\}\right]$$

$$= \lim_{z\to\infty}\left[z^3\left\{\frac{2z^4+3z^3+12z^2-2(z^4-4z^3+6z^2-4z+1)}{z^2(z-1)^4}\right\}\right]$$

$$= \lim_{z\to\infty}\frac{z^3}{z^6}\left(\frac{11z^3+8z-2}{\left(1-\frac{1}{z}\right)^4}\right)$$

$$= \lim_{z\to\infty}\left[\frac{11+\frac{8}{z^2}-\frac{2}{z^3}}{\left(1-\frac{1}{z}\right)^4}\right] = 11.$$

Hence $f(2) = 2$, $f(3) = 11$. **Ans.**

Example 23: Find the Z-transform of

(*i*) $f(k)=\frac{1}{k},\quad k\geq 1$ (*ii*) $f(k)=\frac{1}{k(k+1)},\ k\geq 0$

Sol. (*i*)
$$Z\{f(k)\}=\sum_{k=1}^{\infty}f(k)z^{-k}$$

$$Z\left\{\frac{1}{k}\right\}=\sum_{k=1}^{\infty}\frac{1}{k}z^{-k}=\frac{1}{z}+\frac{1}{2z^2}+\frac{1}{3z^3}+.....$$

$$=-\log\left(1-\frac{1}{z}\right);\quad \text{if }\left|\frac{1}{z}\right|<1$$

$$=\log\left(\frac{z}{z-1}\right);\quad \text{if }|z|>1.$$ **Ans.**

(*ii*)
$$Z\left\{\frac{1}{k(k+1)}\right\}=Z\left\{\frac{1}{k}-\frac{1}{k+1}\right\}$$

$$=Z\left\{\frac{1}{k}\right\}-Z\left\{\frac{1}{k+1}\right\}$$

$$=\log\left(\frac{z}{z-1}\right)-\sum_{k=0}^{\infty}\frac{1}{k+1}z^{-k}$$

$$=\log\left(\frac{z}{z-1}\right)-\left(1+\frac{1}{2z}+\frac{1}{3z^2}+........\right)$$

$$=\log\left(\frac{z}{z-1}\right)-z\left[\frac{1}{z}+\frac{1}{2}\left(\frac{1}{z}\right)^2+\frac{1}{3}\left(\frac{1}{z}\right)^3+......\right]$$

$$=\log\left(\frac{z}{z-1}\right)-z\left\{-\log\left(1-\frac{1}{z}\right)\right\}$$

$$=\log\left(\frac{z}{z-1}\right)-z\log\left(\frac{z}{z-1}\right)=(1-z)\log\left(\frac{z}{z-1}\right).$$ **Ans.**

Example 24: Find the Z-transform of $f * g$ where

(*i*) $f(n)=u(n),\quad g(n)=2^n u(n)$ where $u(n)$ is unit step function

(*ii*) $f(n)=3^n u(n),\quad g(n)=4^n u(n)$ using convolution theorem.

Sol. (*i*)
$$F(z)=Z\{u(n)\}=\sum_{n=0}^{\infty}u(n)z^{-n}=\frac{z}{z-1};\quad \text{if }|z|>1$$

$$u(n-a)=\begin{cases}1, & n\geq a\\ 0, & n<a\end{cases}$$

$$\therefore\quad u(n)=\begin{cases}1, & n\geq 0\\ 0, & n<0\end{cases}$$

$$G(z) = Z\{2^n u(n)\} = \sum_{n=0}^{\infty} 2^n u(n) z^{-k} = \frac{z}{z-2}; \quad \text{if } |z| > |2|$$

By convolution theorem

$$Z\{f * g\} = Z\{h(k)\} = F(z).G(z)$$

$$= \frac{z}{z-1} \cdot \frac{z}{z-2} = \frac{z^2}{(z-1)(z-2)}; \quad \text{if } |z| > |2|. \quad \textbf{Ans.}$$

(*ii*) Similarly
$$F(z) = Z\{3^n u(n)\} = \frac{z}{z-3}; \quad \text{if } |z| > |3|$$

$$G(z) = Z\{4^n u(n)\} = \frac{z}{z-4}; \quad \text{if } |z| > |4|$$

By convolution theorem

$$Z\{f * g\} = Z\{h(k)\} = F(z).G(z)$$

$$= \frac{z}{z-3} \cdot \frac{z}{z-4} = \frac{z^2}{(z-3)(z-4)}; \quad \text{if } |z| > |4|. \quad \textbf{Ans.}$$

Some Standard *Z*-Transforms

Sequence $\{f(k)\}$. for $k \geq 0$	$F(z)$: Z-transform
1. k	$z/(z-1)^2$
2. k^2	$(z^2+z)/(z-1)^3$
3. k^p	$-z\frac{d}{dz}[Z(k^{p-1})], p = +ve$
4. a^k	$z/(z-a)$
5. ku_k	$-z\frac{d}{dz}[U(z)]$
6. $\sin k\theta$	$z \sin\theta/(z^2 - 2z\cos\theta + 1)$
7. $\cos k\theta$	$\frac{z(z-\cos\theta)}{z^2-2z\cos\theta+1}$
8. $\sinh k\theta$	$\frac{z\sin h\theta}{z^2-2z\cos h\theta+1}$
9. $\cosh k\theta$	$\frac{z^2-z\cos h\theta}{z^2-2z\cos h\theta+1}$
10. $\delta(k)$	1

PROBLEM SET 1.3

Find the Z-transform of the following for ($k \geq 0$):

1. 2^k. $\left[\textbf{Ans. } \dfrac{z}{z-2}, |z| > 2\right]$

2. $\sin 2k$. $\left[\textbf{Ans. } \dfrac{z\sin 2}{z^2 - 2z\cos 2 + 1}\right]$

3. $\sin h\dfrac{k\pi}{2}$. $\left[\textbf{Ans. } \dfrac{z\sin h\dfrac{\pi}{2}}{z^2 - 2z\cos h\dfrac{\pi}{2} + 1}\right]$

4. $\sin\left(\dfrac{k\pi}{2} + \alpha\right)$. $\left[\textbf{Ans. } \dfrac{z^2\sin\alpha + z\cos\alpha}{z^2 + 1}, |z| > 1\right]$

5. $c^k \sin h(\alpha k)$. $\left[\textbf{Ans. } \dfrac{cz\sin h\alpha}{z^2 - 2cz\cos h\alpha + c^2}\right]$

6. $\cos(\alpha k)$. $\left[\textbf{Ans. } \dfrac{z(z - \cos\alpha)}{z^2 - 2z\cos\alpha + 1}\right]$

7. $\cos h(\alpha k)$. $\left[\textbf{Ans. } \dfrac{z(z - \cos h\alpha)}{z^2 - 2z\cos h\alpha + 1}\right]$

8. $\cos\left(\dfrac{k\pi}{2} + \dfrac{\pi}{4}\right)$. $\left[\textbf{Ans. } \dfrac{z^2 - z}{\sqrt{2}\left(z^2 + 1\right)}\right]$

9. $a\cos\alpha k + b\sin\alpha k$. $\left[\textbf{Ans. } \dfrac{a - z^{-1}(a\cos\alpha - b\sin\alpha)}{1 - 2z^{-1}\cos\alpha + z^{-2}}\right]$

10. $\sin h(7k)$. $\left[\textbf{Ans. } \dfrac{z\sin h7}{z^2 - 2z\cos h7 + 1}\right]$

11. $\cos h(9k)$. $\left[\textbf{Ans. } \dfrac{z(z - \cos h9)}{z^2 - 2z\cos h9 + 1}\right]$

12. $3^k \cos h(5k)$. $\left[\textbf{Ans. } \dfrac{z(z - 3\cos h5)}{z^2 - 6z\cos h5 + 9}\right]$

13. $\cos\left(\dfrac{k\pi}{3} + 5\right)$. $\left[\textbf{Ans. } \dfrac{z^2\cos 5 - z\cos\left(\dfrac{\pi}{3} - 5\right)}{z^2 - 2z\cos\dfrac{\pi}{3} + 1}\right]$

14. $\cos h\left(\dfrac{n\pi}{2} + \theta\right)$ $\left[\textbf{Ans. } \dfrac{z^2\cos h\theta - z\cos h(\pi/2 - \theta)}{z^2 - 2z\cos h\dfrac{\pi}{2} + 1}\right]$

15. Prove that $Z\left\{\dfrac{1}{(k+2)(k-1)}\right\} = (z^2 - z)\log\left(1 - \dfrac{1}{z}\right) + z$

16. If $F(z) = \dfrac{2z^2 + 3z + 4}{(z-3)^2}$, find $f(1)$ and $f(2)$. $[\textbf{Ans. } 2, 21]$

17. Using $Z\{k^2\} = \dfrac{z^2+2}{(z-1)^3}$, show that $Z\{(k+1)^2\} = \dfrac{z^3+z^2}{(z-1)^3}$

18. If $Z\{f(k)\} = \dfrac{z}{z-1} + \dfrac{z}{z^2+1} + 1$, find $Z[\{f(k+2)\}]$.

1.28 INVERSE Z-TRANSFORM

Inverse Z-transform is a process for determining the sequence which generates given Z-transform.

If $f(z)$ is the Z-transform of the sequence $\{f(k)\}$, then $\{f(k)\}$ is called the inverse Z-transform of $F(z)$.

The operator for inverse Z-transform is Z^{-1}.

If $$Z\{f(k)\} = F(z), \text{ then}$$
$$Z^{-1}[f(z)] = \{f(k)\}.$$

1.29 METHOD OF FINDING INVERSE Z-TRANSFORMS

We have the following method of finding inverse Z-transforms:

1. Convolution method.
2. Long division method.
3. Binomial expansion method.
4. Partial Fraction method.
5. Residue method.

1. Convolution Method

We know that

$$Z\{f * g\} = F(z)G(z)$$

$$\therefore \quad Z^{-1}\{F(z)G(z)\} = f * g = \sum_{m=0}^{k} f(m)g(k-m)$$

Example 1: Using convolution theorem evaluate

$$Z^{-1}\left\{\frac{z^2}{(z-1)(z-3)}\right\}.$$

Sol. We know that

$$Z^{-1}\{F(z).G(z)\} = f * g$$

Let $$F(z) = \frac{z}{z-1} \quad \therefore\ f(k) = (1)^k$$

and $$G(z) = \frac{z}{z-3} \quad \therefore\ g(k) = (3)^k$$

Now, $$Z^{-1}\{F(z).G(z)\} = (1)^k * (3)^k = \sum_{m=0}^{k} 1^m (3)^{k-m}$$

$$= 3^k \sum_{m=0}^{k} \left(\frac{1}{3}\right)^m \qquad \text{[which is G.P.]}$$

$$= 3^k \left\{ \frac{\left(\frac{1}{3}\right)^{k+1} - 1}{\frac{1}{3} - 1} \right\} = \frac{(1)^{k+1} - (3)^{k+1}}{1-3}$$

$$= \frac{1}{2}\left(3^{k+1} - 1\right). \quad \textbf{Ans.}$$

2. Long Division Method

Example 2: Find $Z^{-1}\left[\dfrac{1}{z-2}\right]$.

Sol. Case I. $|z| > 2$

$$\begin{array}{r|l} & \dfrac{1}{z} + \dfrac{2}{z^2} + \dfrac{4}{z^3} \\ \hline z-2 & 1 \\ & 1 - \dfrac{2}{z} \\ \hline & \dfrac{2}{z} \\ & \dfrac{2}{z} - \dfrac{4}{z^2} \\ \hline & \dfrac{4}{z^2} \\ & \dfrac{4}{z^2} - \dfrac{8}{z^3} \\ \hline & \dfrac{8}{z^3} \end{array}$$

$$\frac{1}{z-2} = \frac{1}{z} + \frac{2}{z^2} + \frac{4}{z^3} + \frac{8}{z^4} + \ldots\ldots$$

$$= z^{-1} + 2z^{-2} + 4z^{-3} + 8z^{-4} + \ldots\ldots + 2^{k-1}z^{-k} + \ldots..$$

$$= \left\{2^{k-1}\right\} z^{-k}$$

$$Z^{-1}\left[\frac{1}{z-2}\right]=\left\{2^{k-1}\right\}. \quad \textbf{Ans.}$$

Case II. $|z|<2$

$$\begin{array}{r|l}
 & -\frac{1}{2}-\frac{z}{4}-\frac{z^2}{8}-\frac{z^3}{16} \\
\hline
-2+z & 1 \\
 & 1-\frac{z}{2} \\
\hline
 & \frac{z}{2} \\
 & \frac{z}{2}-\frac{z^2}{4} \\
\hline
 & \frac{z^2}{4} \\
 & \frac{z^2}{4}-\frac{z^3}{8} \\
\hline
 & \frac{z^3}{8} \\
 & \frac{z^3}{8}-\frac{z^4}{16} \\
\hline
 & \frac{z^4}{16}
\end{array}$$

$$\frac{1}{z-2}=-\frac{1}{2}-\frac{z}{4}-\frac{z^2}{8}-\frac{z^3}{16}-\ldots\ldots..$$

$$=-\frac{1}{2}\left[1+\frac{z}{2}+\frac{z^2}{2^2}+\frac{z^3}{2^3}+\ldots\ldots..+\frac{z^k}{2^k}+\ldots..\right]$$

$$=-\frac{1}{2}-\frac{z}{2^2}-\frac{z^2}{2^3}-\ldots\ldots..-\frac{z^k}{2^{k+1}}+\ldots..=\left\{2^{-k-1}\right\}z^k$$

$$Z^{-1}\left[\frac{1}{z-2}\right]=\left\{2^{-k-1}\right\}. \quad \textbf{Ans.}$$

3. Binomial Expansion and Partial Fraction

Example 3: Find the inverse Z-transform of $\frac{4z}{z-a}$.

(*i*) $|z|>|a|$ (*ii*) $|z|<|a|$

Sol. Case I. $|z| > |a|$

$$\frac{4z}{z-a} = \frac{4z}{z}\frac{1}{1-\frac{a}{z}} = 4\left(1-\frac{a}{z}\right)^{-1} = 4\left[1+\frac{a}{z}+\frac{a^2}{z^2}+\frac{a^3}{z^3}+.....\right]$$

$$= 4 + 4az^{-1} + 4a^2z^{-2} + 4a^3z^{-3} + ... + 4a^kz^{-k} + ... = \left\{4a^k\right\}z^{-k}$$

$$Z^{-1}\left(\frac{4z}{z-a}\right) = \left\{4a^k\right\}. \quad \textbf{Ans.}$$

Case II. $|z| < |a|$

$$\frac{4z}{z-a} = -\frac{4}{a}\frac{z}{(1-\frac{z}{a})} = \frac{-4z}{a}\left(1-\frac{z}{a}\right)^{-1}$$

$$= -\frac{4z}{a}\left[1+\frac{z}{a}+\frac{z^2}{a^2}+\frac{z^3}{a^3}+.....\right],$$

$$= -\frac{4z}{a} - \frac{4z^2}{a^2} - \frac{4z^3}{a^3} - .....$$

$$Z^{-1}\left\{\frac{4z}{z-a}\right\} = \{f(k)\}$$

where, $$\{f(k)\} = \left\{......-\frac{4}{a^4},-\frac{4}{a^3},-\frac{4}{a^2},-\frac{4}{a}\right\}. \quad \textbf{Ans.}$$

4. Partial Fraction Method

Here we split the given $F(z)$ into partial fractions whose inverse transform can be written directly.

Example 4: Find the Z-transform of

(*a*) $\dfrac{z}{z^2+7z+10}$, (*b*) $\dfrac{z^3-20z}{(z-2)^2(z-4)}$

Sol. (*a*) $$\frac{F(z)}{z} = \frac{1}{z^2+7z+10} = \frac{1}{(z+2)(z+5)} = \frac{A}{(z+2)} + \frac{B}{(z+5)} = \frac{1}{3(z+2)} - \frac{1}{3(z+5)}$$

$\therefore$ $$F(z) = \frac{z}{3(z+2)} - \frac{z}{3(z+5)}$$

$$f(k) = Z^{-1}\{F(z)\} = \frac{1}{3}Z^{-1}\left\{\frac{z}{z+2}\right\} - \frac{1}{3}Z^{-1}\left\{\frac{z}{z+5}\right\}$$

$$= \left\{ \frac{1}{3}(-2)^k - \frac{1}{3}(-5)^k \right\}$$

(*b*) $$F(z) = \frac{z^3 - 20z}{(z-2)^3(z-4)} \quad \text{or} \quad \frac{F(z)}{z} = \frac{z^2 - 20}{(z-2)^3(z-4)}$$

Now $$\frac{F(z)}{z} = \frac{z^2 - 20}{(z-2)^3(z-4)} = \frac{A + Bz + Cz^2}{(z-2)^3} + \frac{D}{z-4}$$

$\Rightarrow$ $D = -1/2$, A = 6, $B = 0$ and $C = 1/2$

$\therefore$ $$\frac{F(z)}{z} = \frac{6 + \frac{1}{2}z^2}{(z-2)^3} + \frac{-\frac{1}{2}}{z-4}$$

$$F(z) = \frac{1}{2} \times \frac{12z + z^3}{(z-2)^3} - \frac{1}{2} \times \frac{z}{z-4} = \frac{1}{2}\left\{ \frac{z(z-2)^2 + 4z^2 + 8z}{(z-2)^3} - \frac{z}{z-4} \right\}$$

$$= \frac{1}{2}\left\{ \frac{z}{z-2} + 2.\frac{2z^2 + 4z}{(z-2)^3} - \frac{z}{z-4} \right\}$$

Now $$f(k) = Z^{-1}\{F(z)\} = \frac{1}{2}\{2^k + 2k^2 2^k - 4^4\} \qquad \left[\because Z^{-1}\left\{ \frac{az^2 + a^2 z}{(z-a)^3} \right\} = k^2 a^k \right]$$

$$= \frac{1}{2}\{2^{k-1} + k^2 2^k - 2^{2k-1}\}. \quad \textbf{Ans.}$$

Example 5: Find the inverse *Z*-transform of $\frac{1}{(z-3)(z-2)}$.

(*a*) $|z| < 2$, (*b*) $2 < |z| < 3$ and (*c*) $|z| > 3$

Sol. (*a*) $|z| < 2$

$$F(z) = \frac{1}{(z-3)(z-2)} = \frac{1}{(z-3)} - \frac{1}{(z-2)}$$

$$F(z) = -\frac{1}{3}\frac{1}{1-\frac{z}{3}} + \frac{1}{2}\frac{1}{1-\frac{z}{2}} = -\frac{1}{3}\left(1 - \frac{z}{3}\right)^{-1} + \frac{1}{2}\left(1 - \frac{z}{2}\right)^{-1}$$

$$= -\frac{1}{3}\left(1 + \frac{z}{3} + \frac{z^2}{3^2} + \frac{z^3}{3^3} + \ldots\ldots.\right) + \frac{1}{2}\left(1 + \frac{z}{2} + \frac{z^2}{2^2} + \frac{z^3}{2^3} + \ldots.\right)$$

$$= -\frac{1}{3} - \frac{z}{3^2} - \frac{z^2}{3^3} - \frac{z^3}{3^4} - \ldots.. + \frac{1}{2} + \frac{z}{2^2} + \frac{z^2}{2^3} + \frac{z^3}{2^4} + \ldots\ldots$$

$$= = \sum_{k=-\infty}^{0} -3^{k-1} z^{-k} + \sum_{k=-\infty}^{0} 2^{k-1} \cdot z^{-k}$$

Therefore coefficient of z^{-k}, when $k > 0 = 0$ and

coefficient of z^{-k} when $k < 0 = -3^{k-1} + 2^{k-1}$

$\Rightarrow \quad \{f(k)\} = -3^{k-1} + 2^{k-1}, k < 0.$ **Ans.**

(*b*) $2 < |z| < 3$, $\quad F(z) = -\frac{1}{3}\frac{1}{1-\frac{z}{3}} - \frac{1}{z}\frac{1}{1-\frac{2}{z}} = -\frac{1}{3}\left(1-\frac{z}{3}\right)^{-1} - \frac{1}{z}(1-\frac{2}{z})^{-1}$

$$= -\frac{1}{3}\left(1+\frac{z}{3}+\frac{z^2}{3^2}+\frac{z^3}{3^3}+\ldots\ldots\right) - \frac{1}{z}\left(1+\frac{2}{z}+\frac{2^2}{z^2}+\frac{2^3}{z^3}+\ldots\right)$$

$$= -\frac{1}{3} - \frac{z}{3^2} - \frac{z^2}{3^3} - \frac{z^3}{3^4} - \ldots\ldots - \frac{1}{z} - \frac{2}{z^2} - \frac{2^2}{z^3} - \frac{2^3}{z^4} - \ldots\ldots$$

$$= \ldots\ldots - \frac{z^3}{3^4} - \frac{z^2}{3^3} - \frac{z}{3^2} - \frac{1}{3} - \frac{1}{z} - \frac{2}{z^2} - \frac{2^2}{z^3} - \frac{2^3}{z^4} - \ldots\ldots$$

$$= \ldots\ldots\left(-\frac{z^3}{3^4} - \frac{z^2}{3^3} - \frac{z}{3^2} - \frac{1}{3}\right) - (z^{-1} + 2z^{-2} + 2^2 z^{-3} + 2^3 z^{-4} + \ldots\ldots)$$

$$= \sum_{k=-\infty}^{0} -3^{k-1} \cdot z^{-k} + \sum_{k=-\infty}^{0} -2^{k-1} \cdot z^{-k}$$

$\Rightarrow \quad f(k) = \left\{-2^{k-1}\right\}, k > 0$

$f(k) = \left\{-3^{k-1}\right\}, k \le 0.$ **Ans.**

(*c*) $|z| > 3$

$$F(z) = \frac{1}{z}\frac{1}{1-\frac{3}{z}} - \frac{1}{z}\frac{1}{1-\frac{2}{z}} = \frac{1}{z}\left(1-\frac{3}{z}\right)^{-1} - \frac{1}{z}(1-\frac{2}{z})^{-1}$$

$$= \frac{1}{z}\left(1+\frac{3}{z}+\frac{3^2}{z^2}+\frac{3^3}{z^3}+\ldots\ldots\right) - \frac{1}{z}\left(1+\frac{2}{z}+\frac{2^2}{z^2}+\frac{2^3}{z^3}+\ldots\right)$$

$$= \frac{1}{z} + \frac{3}{z^2} + \frac{3^2}{z^3} + \frac{3^3}{z^4} + \ldots\ldots - \frac{1}{z} - \frac{2}{z^2} - \frac{2^2}{z^3} - \frac{2^3}{z^4} - \ldots\ldots$$

$$= \left\{3^{k-1} - 2^{k-1}\right\} z^{-k}, \quad k \ge 1$$

$= 0, k \le 0.$ **Ans.**

Example 6: Find the inverse of Z-transform of $\dfrac{1}{(z-5)^3}, |z| > 5.$

Sol.
$$F(z) = \frac{1}{(z-5)^3}$$
$$= \frac{1}{z^3}\frac{1}{\left(1-\frac{5}{z}\right)^3} = \frac{1}{z^3}(1-5z^{-1})^{-3}$$
$$= z^{-3}\left[1+15z^{-1}+6(5z^{-1})^2+10(5z^{-1})^3+......+\frac{(n+1)(n+2)}{2}(5z^{-1})^n+......\right]$$
$$= z^{-3}\left\{\frac{(k+1)(k+2)}{2}5^k\right\}z^{-k} = \left\{\frac{(k+1)(k+2)}{2}5^k\right\}z^{-k-3}$$

Replacing k by k–3, we get
$$F(z) = \left\{\frac{(k-3+1)(k-3+2)}{2}5^{k-3}\right\}z^{-k} = \left\{\frac{(k-2)(k-1)}{2}5^{k-3}\right\}z^{-k}$$
$$f(k) = Z^{-1}\{F(z)\} = \begin{cases}\dfrac{(k-2)(k-1)}{2}5^{k-3}, & k \geq 3\\ 0, & k < 3\end{cases}$$ **Ans.**

Example 7: Obtain $Z^{-1}\dfrac{1}{(z-\frac{1}{2})(z-\frac{1}{3})}$, when (a) $\frac{1}{3} < |z| < \frac{1}{2}$ $(b)\frac{1}{2} < |z|$.

Sol. (a) $\frac{1}{3} < |z| < \frac{1}{2}$
$$\frac{1}{(z-\frac{1}{2})(z-\frac{1}{3})} = 6\left[\frac{1}{\left(z-\frac{1}{2}\right)} - \frac{1}{\left(z-\frac{1}{3}\right)}\right] = \frac{6}{z-\frac{1}{2}} - \frac{6}{z-\frac{1}{3}}$$
$$= \frac{6}{-\frac{1}{2}(1-2z)} - \frac{1}{z}\frac{6}{(1-\frac{1}{3}z^{-1})}$$
$$= -12[1+2z+(2z)^2+(2z)^3+.....] - \frac{6}{z}\left(1+\frac{1}{3z}+\frac{1}{(3z)^2}+.....\right)$$
$$= -12\sum_{k=-\infty}^{0} 2^k z^{-k} - 6\sum_{k=1}^{\infty}\frac{z^{-k}}{3^{k-1}}$$

Hence $f(k) = -12.2^{-k};$ if $k \leq 0$

$f(k) = -\dfrac{6}{3^{k-1}};$ if $k > 0.$ **Ans.**

(*b*) $\frac{1}{2} < |z|$

$$F(z) = -\frac{6}{z-\frac{1}{2}} - \frac{6}{z-\frac{1}{3}} = \frac{1}{z}\frac{6}{\left(1-\frac{1}{2z}\right)} - \frac{1}{z}\frac{6}{\left(1-\frac{1}{3z}\right)}$$

$$= 6z^{-1}\left(1-\tfrac{1}{2}z^{-1}\right)^{-1} - 6z^{-1}\left(1-\tfrac{1}{3}z^{-1}\right)^{-1}$$

$$= 6z^{-1}\left[1+\tfrac{1}{2}z^{-1}+\left(\tfrac{1}{2}\right)^2 z^{-2}+....\right] - 6z^{-1}\left[1+\tfrac{1}{3}z^{-1}+\left(\tfrac{1}{3}\right)^2 z^{-2}+....\right]$$

$$= 6\left[z^{-1}+\tfrac{1}{2}z^{-2}+\left(\tfrac{1}{2}\right)^2 z^{-3}+....\right] - 6\left[z^{-1}+\tfrac{1}{3}z^{-2}+\left(\tfrac{1}{3}\right)^2 z^{-3}+....\right]$$

$$= 6\sum_{k=1}^{\infty}\left\{\frac{1}{2^{k+1}}\right\}z^{-k} - 6\sum_{k=1}^{\infty}\left\{\frac{1}{3^{k+1}}\right\}z^{-k}$$

$\therefore$ $$f(k) = 6\left[\frac{1}{2^{k-1}} - \frac{1}{3^{k-1}}\right], k \geq 1. \quad \textbf{Ans.}$$

Example 8: Obtain $Z^{-1}\dfrac{2z^2-10z+13}{(z-3)^2(z-2)}$, when $2<|z|<3$.

Sol. Let $$\frac{2z^2-10z+13}{(z-3)^2(z-2)} = \frac{A}{(z-3)^2} + \frac{B}{(z-3)} + \frac{C}{(z-2)}$$

Converting into partial fraction, we get

$$\frac{2z^2-10z+13}{(z-3)^2(z-2)} = \frac{1}{(z-3)^2} + \frac{1}{(z-3)} + \frac{1}{(z-2)}$$

$$= \frac{1}{9}\frac{1}{\left(1-\frac{z}{3}\right)^2} - \frac{1}{3}\frac{1}{1-\frac{z}{3}} + \frac{1}{z}\frac{1}{1-\frac{2}{z}}$$

$$= \tfrac{1}{9}\left(1-\tfrac{z}{3}\right)^{-2} - \tfrac{1}{3}\left(1-\tfrac{z}{3}\right)^{-1} + \tfrac{1}{z}\left(1-\tfrac{2}{z}\right)^{-1}$$

$$= \frac{1}{9}\left[1+\frac{2z}{3}+\frac{3z^2}{9}+\frac{4z^3}{27}+........\right] - \frac{1}{3}\left[1+\frac{z}{3}+\frac{z^2}{9}+\frac{z^3}{27}+.......\right] + \frac{1}{z}\left[1+\frac{2}{z}+\frac{4}{z^2}+\frac{8}{z^3}+......\right]$$

$$= \frac{1}{3^2}+\frac{2z}{3^3}+\frac{3z^2}{3^4}+\frac{4z^3}{3^5}+......-\frac{1}{3}-\frac{z}{3^2}-\frac{z^2}{3^3}-\frac{z^3}{3^4}-.......+\frac{1}{z}+\frac{2}{z^2}+\frac{4}{z^3}+\frac{8}{z^4}+.......$$

$$= \sum_{k=-\infty}^{0}\frac{k+1}{3^{k+2}}z^{-k} - \sum_{k=-\infty}^{0}\frac{1}{3^{k+1}}z^{-k} + \sum_{k=1}^{\infty}2^{k-1}\cdot z^{-k}$$

$$\therefore \qquad f(k) = \begin{cases} \dfrac{k+1}{3^{k+2}} - \dfrac{1}{3^{k+1}} \; ; k \le 0 \\ 2^{k-1} \; ; k \ge 1 \end{cases}$$

Hence $Z^{-1}[F(z)] = 2^{k-1}$ if $k \ge 1$, and $Z^{-1}[F(z)] = -(k+2)3^{k-2}, k \le 0$. **Ans.**

$$\text{As } \frac{k+1}{3^{k+2}} - \frac{1}{3^{k+1}} = -(k+2)3^{k-2}$$

5. Residue Method

Take the contour C such that all the poles of the function $f(z)$ lie within the contour. Then by residue method

$f(k)$ = sum of the residues of $z^{k-1}F(z)$ at its poles.

where residue for simple pole $z = z_i$ is $= \left[(z - z_i) z^{k-1} F(z)\right]_{z=z_i}$

Residue of order n at the pole $z = z_i$ $= \dfrac{1}{\lfloor (n-1)} \left[\dfrac{d^{n-1}}{dz^{n-1}} (z - z_i)^n z^{k-1} F(z) \right]_{z=z_i}$

Example 9: Evaluate $Z^{-1}\left[\dfrac{z}{z^2 + 7z + 10}\right]$.

Sol. $$F(z) = \frac{z}{z^2 + 7z + 10} = \frac{z}{(z+2)(z+5)}$$

Poles are given by, $(z + 2)(z + 5) = 0$ or $z = -2, -5$

There are two simple poles. Consider a contour $|z| > 5$

Residue at $(z = -2)$ $= \left[(z+2) z^{k-1} \dfrac{z}{(z+2)(z+5)}\right]_{z=-2} = \dfrac{(-2)^k}{3}$

Residue at $(z = -5)$ $= \left[(z+5) z^{k-1} \dfrac{z}{(z+2)(z+5)}\right]_{z=-5} = \dfrac{(-5)^k}{-3}$

$\therefore f(k)$ = Sum of the residues $= \dfrac{(-2)^k}{3} + [\dfrac{(-5)^k}{-3}] = \dfrac{1}{3}\left\{(-2)^k - (-5)^k\right\}$. **Ans.**

Example 10: Evaluate $Z^{-1}\left[\dfrac{z}{(z-2)(z-3)}\right]$.

Sol. The poles are given by, $(z - 2)(z - 3) = 0 \Rightarrow z = 2$ and $z = 3$

There are two simple pole. Let us consider the contour $|z| > 3$.

Residue at $(z = 2) = \left[(z-2)z^{k-1}\dfrac{z}{(z-2)(z-3)}\right]_{z=2} = \left[\dfrac{z^k}{z-3}\right]_{z=2} = \dfrac{2^k}{-1}$

Residue at $(z = 3) = \left[(z-3)z^{k-1}\dfrac{z}{(z-2)(z-3)}\right]_{z=3} = \left[\dfrac{z^k}{z-2}\right]_{z=3} = \dfrac{3^k}{1}$

Hence, $f(k)$ = Sum of the residues = $3^k - 2^k$. **Ans.**

Example 11: Evaluate $Z^{-1}\left[\dfrac{3z^2-18z+26}{(z-2)(z-3)(z-4)}\right]$.

Sol. The poles are given by $(z-2)(z-3)(z-4) = 0 \quad \Rightarrow \quad z = 2, 3, 4$

There are three poles. Let us consider the contour $|z| > 4$.

Residue at $(z = 2) = \left[\dfrac{(z-2)z^{k-1}(3z^2-18z+26)}{(z-2)(z-3)(z-4)}\right]_{z=2}$.

$$= \left[\frac{3z^{k+1}-18z^k+26z^{k-1}]}{(z-3)(z-4)}\right]_{z=2} = \frac{3.2^{k+1}-18.2^k+26.2^{k-1}}{(-1)(-2)}$$

$$= 3.2^k - 9.2^k + 13.2^{k-1}$$

$$= -6.2^k + 13.2^{k-1} = -12.2^{k-1} + 13.2^{k-1} = 2^{k-1}$$

Residue at $(z = 3) = \left[\dfrac{(z-3)z^{k-1}(3z^2-18z+26)}{(z-2)(z-3)(z-4)}\right]_{z=3} = \left[\dfrac{3z^{k+1}-18z^k+26z^{k-1}]}{(z-2)(z-4)}\right]_{z=3}$.

$$= \left[\frac{3.3^{k+1}-18.3^k+26.3^{k-1}]}{(-1)}\right] = -3.3^{k+1} + 18.3^k - 26.3^{k-1}$$

$$= -27.3^{k-1} + 54.3^{k-1} - 26.3^{k-1} = 3^{k-1}$$

Residue at $(z = 4) = \left[\dfrac{(z-4)z^{k-1}(3z^2-18z+26)}{(z-2)(z-3)(z-4)}\right]_{z=4} = \left[\dfrac{3z^{k+1}-18z^k+26z^{k-1}]}{(z-2)(z-3)}\right]_{z=4}$

$$= \left[\frac{3.4^{k+1}-18.4^k+26.4^{k-1}]}{2}\right] = \frac{3}{2}.4^{k+1} - 9.4^k + 13.4^{k-1}$$

$$= 24.4^{k-1} - 36.4^{k-1} + 13.4^{k-1} = 4^{k-1}$$

Hence, $f(k)$ = Sum of the residues $= 2^{k-1} + 3^{k-1} + 4^{k-1}, k > 0$. **Ans.**

Example 12: Evaluate $Z^{-1}\left[\dfrac{9z^3}{(3z-1)^2(z-2)}\right]$.

Sol. The poles are given by $(3z-1)^2(z-2) = 0 \Rightarrow z = 1/3$ (order 2), 2.

There are two poles *i.e.*, one is simple pole at $z = 2$ and other is pole of order 2 at $z = 1/3$. Let us consider the contour $|z| > 2$.

$$\text{Residue at } (z=2) = \left[\frac{(z-2)z^{k-1}9z^3}{(3z-1)^2(z-2)}\right]_{z=2} = \left[\frac{9z^{k+2}}{(3z-1)^2}\right]_{z=2} = \frac{9.2^{k+2.}}{25}$$

$$\text{Residue at } (z=1/3) = \frac{d}{dz}\left[\frac{\left(z-\frac{1}{3}\right)^2 .z^{k-1}.9z^3}{(3z-1)^2.(z-2)}\right]_{z=\frac{1}{3}} = \frac{d}{dz}\left[\frac{9z^{k+2}}{9(z-2)}\right]_{z=\frac{1}{3}} = \frac{d}{dz}\left[\frac{z^{k+2}}{(z-2)}\right]_{z=\frac{1}{3}}$$

$$= \left[\frac{(k+2)z^{k+1}}{(z-2)} - \frac{z^{k+2}}{(z-2)^2}\right]_{z=\frac{1}{3}} = -\frac{3}{5}(k+2)\left(\frac{1}{3}\right)^{k+1} - \frac{9}{25}\left(\frac{1}{3}\right)^{k+2} = -\left(\frac{1}{3}\right)^k\left(\frac{5k+11}{25}\right)$$

$$\text{Hence, } f(k) = \text{Sum of the residues} = \frac{9}{25}.2^{k+2} - \left(\frac{1}{3}\right)^k\left(\frac{5k+11}{25}\right). \quad \textbf{Ans.}$$

Example 13: Using residue method, evaluate $Z^{-1}\left\{\dfrac{8z-z^3}{(4-z)^3}\right\}$.

Sol. Here $F(z) = \left\{\dfrac{8z-z^3}{(4-z)^3}\right\}$.

Poles are given by $z = 4$ (order 3). Consider a contour $|z| = 5$.

Residue of $F(z)$ at $(z = 4)$

$$= \frac{1}{(3-1)!}\left[\frac{d^2}{dz^2}\left\{(z-4)^3 z^{k-1}.\frac{8z-z^3}{(4-z)^3}\right\}\right]_{z=4}$$

$$= \frac{1}{2}\left[\frac{d^2}{dz^2}\left\{(z^3-8z)z^{k-1}\right\}\right]_{z=4} = \frac{1}{2}\left[\frac{d^2}{dz^2}(z^{k+2}-8z^k)\right]_{z=4}$$

$$= \frac{1}{2}\left[\frac{d}{dz}\left\{(k+2)z^{k+1} - 8kz^{k-1}\right\}\right]_{z=4} = \frac{1}{2}\left[(k+2)(k+1)z^k - 8k(k-1)z^{k-2}\right]_{z=4}$$

$$= \frac{1}{2}\left[(k+2)(k+1)(4)^k - 8k(k-1)(4)^{k-2}\right] = \frac{1}{2}\left[(k+2)(k+1)(4)^k - \tfrac{1}{2}k(k-1)(4)^k\right]$$

$$= \frac{4^k}{2}\left[k^2+3k+2-\tfrac{1}{2}(k^2-k)\right] = \frac{4^k}{2}\left(\frac{k^2}{2}+\frac{7k}{2}+2\right) = (k^2+7k+4)(4)^{k-1}$$

$$\therefore\ Z^{-1}\left\{\frac{8z-z^3}{(4-z)^3}\right\} = \text{sum of the residues} = \left\{\left(k^2+7k+4\right)(4)^{k-1}\right\}. \quad \textbf{Ans.}$$

Example 14: Using residue's method, show that

$$Z^{-1}\left\{\frac{3z^2+2}{(5z-1)(5z+2)}\right\} = \frac{53}{75}(0.2)^k + \frac{31}{75}(-0.4)^k.$$

Sol. Here, $$F(z) = \frac{3z^2+2}{(5z-1)(5z+2)}$$

Poles are given by $(5z-1)(5z+2)=0$ *i.e.*, $z = 1/5, -2/5$ which are simple poles.

Consider a contour $|z| = 1$

Residue at ($z = 1/5$) $$= \lim_{z\to\frac{1}{5}}\left[\left(z-\frac{1}{5}\right)z^{k-1}\cdot\frac{3z^2+2}{(5z-1)(5z+2)}\right]$$

$$= \lim_{z\to\frac{1}{5}}\left[\frac{1}{5}\cdot\frac{1}{(5z+2)}z^{k-1}(3z^2+2)\right]$$

$$= \frac{1}{15}\cdot\left(\frac{1}{5}\right)^{k-1}\left(\frac{3}{25}+2\right) = \frac{53}{3}\cdot\frac{1}{(5)^{k+2}} = \frac{53}{75}(0.2)^k$$

Residues at ($z = -2/5$) $$= \lim_{z\to-\frac{2}{5}}\left(z+\frac{2}{5}\right)z^{k-1}\cdot\frac{3z^2+2}{(5z-1)(5z+2)} = \lim_{z\to-\frac{2}{5}}\left[\frac{1}{5}\cdot\frac{1}{(5z-1)}z^{k-1}.(3z^2+2)\right]$$

$$= \frac{1}{5}\cdot\left(-\frac{1}{3}\right)\left(\frac{-2}{5}\right)^{k-1}\left(\frac{12}{25}+2\right) = \frac{(-1)^k.2^{k-1}}{(5)^{k+2}}\cdot\frac{62}{3}$$

$$Z^{-1}\{F(z)\} = \frac{62}{75}(-1)^k.2^{k-1}(0.2)^k = \frac{31}{75}(-0.4)^k$$

Hence, $f(k)$ = Sum of the residues

$$= \frac{53}{75}(0.2)^k + \frac{31}{75}(-0.4)^k. \quad \textbf{Ans.}$$

Example 15: Find $Z^{-1}\left\{\frac{z^2}{z^2+4}\right\}. |z| > 2.$

Sol. The poles are determined by, $z^2 + 4 = 0$ *i.e.*, $z = \pm 2i$

There are two poles at $z = 2i$ and $z = -2i$.

Let us consider a contour $|z| > 2$

Residue at ($z = 2i$) $$= \left[\frac{(z-2i)z^{k-1}.z^2}{z^2+4}\right]_{z=2i} = \left[\frac{z^{k+1}}{z+2i}\right]_{z=2i} = \frac{(2i)^{k+1}}{4i} = 2^{k-1}(i)^k$$

$$= 2^{k-1}\left[\cos\frac{\pi}{2} + i\sin\frac{\pi}{2}\right]^k = 2^{k-1}\left[\cos\frac{k\pi}{2} + i\sin\frac{k\pi}{2}\right]$$

Residue at ($z = -2i$) $$= \left[\frac{(z+2i)z^{k-1}.z^2}{z^2+4}\right]_{z=-2i} = \left[\frac{z^{k+1}}{z-2i}\right]_{z=-2i} = \frac{(-2i)^{k+1}}{-4i} = 2^{k-1}(-i)^k$$

$$= 2^{k-1}\left[\cos\frac{\pi}{2} - i\sin\frac{\pi}{2}\right]^k = 2^{k-1}\left[\cos\frac{k\pi}{2} - i\sin\frac{k\pi}{2}\right]$$

$f(k)$ = sum of the residues $$= 2^{k-1}\left[\cos\frac{k\pi}{2} + i\sin\frac{k\pi}{2}\right] + 2^{k-1}\left[\cos\frac{k\pi}{2} - i\sin\frac{k\pi}{2}\right]$$

$$= 2^{k-1}\left[2\cos\frac{k\pi}{2}\right] = 2^k\cos\frac{k\pi}{2}.$$ **Ans.**

PROBLEM SET 1.4

1. Find $Z^{-1}\left\{\frac{z^2}{(z-a)(z-b)}\right\}$ using convolution theorem. $\left[\textbf{Ans. } \frac{a^{k+1}-b^{k+1}}{a-b}\right]$

2. Find the inverse of the Z-transform of the following:

   (*a*) $\left\{\frac{z^2}{(z-\frac{1}{4})(z-\frac{1}{5})}\right\}$, $|z| < \frac{1}{5}$ $\left[\textbf{Ans. } 4\left(\frac{1}{5}\right)^k - 5\left(\frac{1}{4}\right)^k\right]$

   (*b*) $\left\{\frac{z^3}{(z-\frac{1}{2})^2(z-1)}\right\}$, $|z| < \frac{1}{2}$ $\left[\textbf{Ans. } -4(1)^k + (k+3)(\tfrac{1}{2})^k, (k<0)\right]$

3. Find the inverse Z-transform of $\frac{z}{z-a}$.

   (*a*) $|z| > |a|$, (*b*) $|z| < |a|$. $\left[\textbf{Ans. } (a)\ \{a^k\},\ (b)\ \left\{-\tfrac{1}{a^k}\right\}\right]$

4. Evaluate $Z^{-1}\left[\frac{2z^2-5z}{(z-3)(z-2)}\right]$, $|z| > 3$. $\left[\textbf{Ans. } \left\{2^k + 3^k\right\}, k \geq 0\right]$

5. Evaluate inverse *Z*-transform of the following:

(*a*) $\dfrac{z}{z-e^{\alpha}}$ $\left[\textbf{Ans. } \{e^{k\alpha}\}\right]$ (*b*) $\dfrac{2z}{(z-2)^2},\ |z|>2$ $\left[\textbf{Ans. } \{k2^k\}, k \ge 0\right]$

(*c*) $\dfrac{z^2+z\cos\alpha}{z^2-2z\cos\alpha+2},\ |z|>1$ $\left[\textbf{Ans. } \{\cos\alpha k\}, k \ge 0\right]$

(*d*) $-\log(1-2z^{-1}),\ |z|>2$ $\left[\textbf{Ans. } \dfrac{2^k}{k}, k \ge 1\right]$ (*e*) $\dfrac{ze^{-a}}{(z-e^{-a})^2},\ |z|>|e^{-a}|$ $\left[\textbf{Ans. } \{ke^{-ak}\}\right]$

6. Find the inverse *Z*-transform of the following functions by residue method:

(*a*) $\dfrac{z}{(z-1)(z-2)}$ $\left[\textbf{Ans. } 2^k-1, k \ge 0\right]$ (*b*) $\dfrac{z^2}{z^2+1}$ $\left[\textbf{Ans. } \cos\dfrac{k\pi}{2}\right]$

(*c*) $\dfrac{z(z^2-1)}{(z^2+1)^2}$ $\left[\textbf{Ans. } \left\{\frac{k}{2}\left[(i)^{k-2}+(-i)^{k-1}\right]\right\}, k=0,1,2\right]$

7. Evaluate $Z^{-1}\left\{\dfrac{z^3-20z}{(z-2)^3(z-4)}\right\}$ $\left[\textbf{Ans. } 2^{k-1}+k^2 2^k-4^k\right]$

8. Prove that $Z^{-1}\left[\left(\dfrac{z}{z-a}\right)^2\right]=(k+1)a^k$

9. Find $Z^{-1}\left[(z-5)^{-3}\right]$ when $|z|>5$. $\left[\textbf{Ans. } \begin{array}{ll}\frac{1}{2}(k-1)(k-2)5^{k-3} & \text{for } k \ge 3\\ 0, & \text{for } k<3\end{array}\right]$

10. Find the inverse *Z*-transform of $\dfrac{4z^2-2z}{z^3-5z^2+8z-4}$. $\left[\textbf{Ans. } 2-2^{k+1}+3k\cdot 2^k\right]$

11. Evaluate $Z^{-1}\left\{\dfrac{z}{z^2+11z+24}\right\}$ $\left[\textbf{Ans. } \dfrac{1}{5}\{(-3)^k-(-8)^k\}\right]$

12. Evaluate $Z^{-1}\left\{\dfrac{z}{z^3-z^2+z-1}\right\}$ $\left[\textbf{Ans. } \dfrac{1}{2}\left\{1-\cos\left(\dfrac{k\pi}{2}\right)-\sin\left(\dfrac{k\pi}{2}\right)\right\}\right]$

13. Evaluate $Z^{-1}\left\{\dfrac{z^2+z}{(z-1)(z^2+1)}\right\}$ $\left[\textbf{Ans. } 1-\cos\dfrac{k\pi}{2}\right]$

14. Evaluate $Z^{-1}\left\{\dfrac{z}{(z+1)^2(z-1)}\right\}$ $\left[\textbf{Ans. } \dfrac{1}{4} - \dfrac{k(-1)^k}{2} - \dfrac{(-1)^k}{4}\right]$

15. Evaluate $Z^{-1}\left\{\dfrac{8z^2}{(2z-1)(4z-1)}\right\}$ $\left[\textbf{Ans. } 2\left(\dfrac{1}{2^k} - \dfrac{1}{4^k}\right)\right]$

1.30 DIFFERENCE EQUATION

A difference equation is a relation between the differences of an unknown function at one or more general values of the arguments

Thus $$\Delta y_{(n+1)} + y_{(n)} = 2 \quad \text{...(1)}$$

and $$\Delta y_{(n+1)} + \Delta^2 y_{(n-1)} = 1 \quad \text{...(2)}$$

are difference equations.

Equation (1) may be written as

$$y_{n+2} - y_{n+1} + y_n = 2 \quad \text{...(3)}$$

$$[\because \Delta y_{n+1} = y_{n+2} - y_{n+1}],\ [\because \Delta^2 y_{n+1} = y_{n+1} - 2y_n + y_{n-1}]$$

From equation (2) $$y_{n+2} - 2y_n + y_{n-1} = 1 \quad \text{...(4)}$$

1.31 ORDER OF A DIFFERENCE EQUATION

The order of a difference equation is the difference between the largest and the smallest argument occurring in the difference equation divided by the unit of increment.

$\therefore$ From equation (3)

$$\text{Order} = \frac{\text{Largest argument} - \text{Smallest argument}}{\text{Unit of increment}}$$

$$= \frac{(n+2)-n}{1} = 2$$

And order of equation (4) is $\dfrac{(n+2)-(n-1)}{1} = 3$

Theorem: $Z(y_{k+n}) = z^n\left(\bar{y} - y_0 - \dfrac{y_1}{z} - \ldots\ldots - \dfrac{y_{n-1}}{z^{n-1}}\right)$, where $Z(y_k) = \bar{y}$

Proof: L.H.S. $$Z(y_{k+n}) = \sum_{k=0}^{\infty} y_{k+n} z^{-k} = z^n \sum_{k=0}^{\infty} y_{k+n} z^{-(k+n)}$$

Putting $k + n = m$

$$Z(y_{k+n}) = z^n \sum_{m=n}^{\infty} y_m z^{-m}$$

$$Z(y_{k+n}) = z^n \left[\sum_{m=0}^{\infty} y_m z^{-m} - \sum_{m=0}^{n-1} y_m z^{-m} \right]$$

$$= z^n \left[\bar{y} - y_0 - \frac{y_1}{z} - \frac{y_2}{z^2} - \ldots\ldots\ldots - \frac{y_{n-1}}{z^{n-1}} \right]$$

Notes: (1) For $n = 1, 2, 3, \ldots\ldots$, we have

$$\boxed{Z(y_{k+1}) = z(\bar{y} - y_0)}$$

$$\boxed{Z(y_{k+2}) = z^2(\bar{y} - y_0 - \frac{y_1}{z})}$$

$$\boxed{Z(y_{k+3}) = z^3(\bar{y} - y_0 - \frac{y_1}{z} - \frac{y_2}{z^2})} \text{ and so on.}$$

(2) If $Z(y_k) = \bar{y}$, then

$$Z(y_{k-n}) = z^{-n}\bar{y}.$$

Example 1: Solve by Z-transform: $y_{k+2} - 3y_{k+1} + 2y_k = 0; y_0 = 0, y_1 = 1$.

Sol. Take Z-transform on both sides

$$Z(y_{k+2}) - 3Z(y_{k+1}) + 2Z(y_k) = Z(0)$$

$$z^2(\bar{y} - y_0 - \frac{y_1}{z}) - 3z(\bar{y} - y_0) + 2\bar{y} = 0$$

$$\left(z^2 - 3z + 2\right)\bar{y} = z$$

$$\bar{y} = \frac{z}{z^2 - 3z + 2} = \frac{z}{(z-1)(z-2)}$$

$$\bar{y} = \frac{z}{z-2} - \frac{z}{z-1}$$

Now, taking inverse Z-transform, we get

$$y_k = Z^{-1}\left(\frac{z}{z-2} - \frac{z}{z-1} \right)$$

$$= \left\{2^k - 1\right\}, \text{ where } k = 0, 1, 2. \quad \textbf{Ans.}$$

Example 2: Solve by Z-transform: $6y_{k+2} - y_{k+1} - y_k = 0; y_0 = 0, y_1 = 1$.

Sol.

$$6y_{k+2} - y_{k+1} - y_k = 0$$

Taking Z-transform on both sides, we get

$$Z[6y_{k+2} - y_{k+1} - y_k] = 0$$

$$Z(6y_{k+2}) - Z(y_{k+1}) - Z(y_k) = Z(0)$$

$$6(z^2\bar{y} - z^2 y_0 - zy_1) - (z\bar{y} - zy_0) - \bar{y} = 0$$

On putting the value of y_0 and y_1, we get

$$6z^2\bar{y} - 6z - z\bar{y} - \bar{y} = 0$$

$$(6z^2 - z - 1)\bar{y} = 6z$$

$$\bar{y} = \frac{6z}{6z^2 - z - 1} = \frac{6z}{(3z+1)(2z-1)} = \frac{z^{-1}}{\left(1+\frac{z^{-1}}{3}\right)\left(1-\frac{z^{-1}}{2}\right)} = \frac{\frac{6}{5}}{1-\frac{z^{-1}}{2}} - \frac{\frac{6}{5}}{1+\frac{z^{-1}}{3}}$$

$$y_k = Z^{-1}\left[\frac{\frac{6}{5}}{1-\frac{z^{-1}}{2}}\right] - Z^{-1}\left[\frac{\frac{6}{5}}{1+\frac{z^{-1}}{3}}\right] = \frac{6}{5}\left(\frac{1}{2}\right)^k - \frac{6}{5}\left(-\frac{1}{3}\right)^k$$

$$= \frac{6}{5}\left[\left(\frac{1}{2}\right)^k - \left(-\frac{1}{3}\right)^k\right].\qquad \textbf{Ans.}$$

Example 3: Solve the difference equation $y_{k+3} - 3y_{k+2} + 3y_{k+1} - y_k = U(k)$,

$$y_0 = y_1 = y_2 = 0 \qquad \text{(By Z-transforms)}$$

Sol. $\qquad y_{k+3} - 3y_{k+2} + 3y_{k+1} - y_k = U(k)$

Taking Z-transform on both sides, we get

$$Z[y_{k+3} - 3y_{k+2} + 3y_{k+1} - y_k] = ZU(k)$$

$$Z[y_{k+3}] - 3Z[y_{k+2}] + 3Z[y_{k+1}] - Z[y_k] = ZU(k)$$

$$[z^3\bar{y} - z^3 y_0 - z^2 y_1 - zy_2] - 3[z^2\bar{y} - z^2 y_0 - zy_1] + 3[z\bar{y} - zy_0] - \bar{y} = ZU(k)$$

Putting the values of $y_0 = y_1 = y_2 = 0$ in the above equation

$$z^3\bar{y} - 3z^2\bar{y} + 3z\bar{y} - \bar{y} = \frac{1}{1-z^{-1}}$$

$$(z^3 - 3z^2 + 3z - 1)\bar{y} = \frac{1}{1-z^{-1}}$$

$$(z-1)^3\bar{y} = \frac{1}{1-z^{-1}}$$

$$\bar{y} = \frac{1}{(z-1)^3(1-z^{-1})} = \frac{1}{z^3(1-z^{-1})^3(1-z^{-1})} = z^{-3}(1-z^{-1})^{-4}$$

$$y_k = \text{Coefficient of } z^{-k} \text{ in } z^{-3}(1-z^{-1})^{-4}$$

$$= \text{Coefficient of } z^{-k-3} \text{ in } (1-z^{-1})^{-4} = \frac{(k-2)(k-1)k}{6}, k \geq 3. \quad \textbf{Ans.}$$

Example 4: Solve the following difference equations using Z-transform:

$$y_{k+1} - 2y_{k-1} = 0; \quad k \geq 1, \quad y_0 = 1$$

Sol. Taking Z-transform of given equation

$$Z(y_{k+1}) - 2Z(y_{k-1}) = 0$$

$$z(\bar{y} - y_0) - 2\bar{y}z^{-1} = 0 \Rightarrow \left(\frac{z^2-2}{z}\right)\bar{y} = z \qquad | \text{ As } y_0 = 1$$

$$\Rightarrow \quad \bar{y} = \frac{z^2}{z^2-2} \quad \text{or} \quad \frac{\bar{y}}{z} = \frac{z}{(z-\sqrt{2})(z+\sqrt{2})} = \frac{1}{2}\left(\frac{1}{z-\sqrt{2}} + \frac{1}{z+\sqrt{2}}\right)$$

$$\Rightarrow \quad \bar{y} = \frac{1}{2}\left(\frac{z}{z-\sqrt{2}} + \frac{z}{z+\sqrt{2}}\right) \Rightarrow Z\{y_k\} = \frac{1}{2}\left(\frac{z}{z-\sqrt{2}} + \frac{z}{z+\sqrt{2}}\right)$$

$$\therefore \quad y_k = \frac{1}{2}\left[Z^{-1}\left(\frac{z}{z-\sqrt{2}}\right) + Z^{-1}\left(\frac{z}{z+\sqrt{2}}\right)\right] = \frac{1}{2}\left[\left(\sqrt{2}\right)^k + \left(-\sqrt{2}\right)^k\right]. \quad \textbf{Ans.}$$

Example 5: Solve by Z-transform $y_{n+2} - 4y_n = 0$ given that $y_0 = 0$, $y_1 = 2$ *(U.T.U. 2008)*

Sol. Taking Z-transform of given equation, we get

$$Z(y_{n+2}) - 4Z(y_n) = 0$$

$$\text{or} \quad z^2\left(\bar{y} - y_0 - \frac{y_1}{z}\right) - 4\bar{y} = 0 \Rightarrow z^2\left(\bar{y} - 0 - \frac{2}{z}\right) - 4\bar{y} = 0$$

$$\text{or} \quad \bar{y} = \frac{2z}{(z-2)(z+2)} \quad \text{or} \quad y_k = 2Z^{-1}\left[\frac{z}{(z-2)(z+2)}\right]$$

$$\text{or} \quad y_k = \frac{1}{2}\left[Z^{-1}\left(\frac{z}{z-2}\right) - Z^{-1}\left(\frac{z}{z+2}\right)\right] = \frac{1}{2}\left[(2)^k - (-2)^k\right]. \quad \textbf{Ans.}$$

Example 6: Use Z-transform to solve the difference equation.

$$y_{n+2} - 2y_{n+1} + y_n = 3n + 5$$

Sol.

$$y_{n+2} - 2y_{n+1} + y_n = 3n + 5 \qquad \ldots(1)$$

Taking Z-Transform in (1), on both sides, we get

$$Z\{y_{n+2}\} - 2Z\{y_{n+1}\} + Z\{y_n\} = 3Z\{n\} + 5Z\{1\}$$

$$z^2\left[\bar{y} - y_0 - y_1 z^{-1}\right] - 2z\left[\bar{y} - y_0\right] + \bar{y} = 3.\frac{z}{(z-1)^2} + 5.\frac{z}{z-1}$$

$$(z^2-2z+1)\bar{y}=\frac{5z^2-2z}{(z-1)^2}+y_0(z^2-2z)+y_1z$$

$$\bar{y}=\frac{5z^2-2z}{(z-1)^4}+y_0(z^2-2z)/(z-1)^2+y_1z/(z-1)^2$$

On inversion, we obtain

$$y_n=Z^{-1}\left[\frac{5z^2-2z}{(z-1)^4}\right]+y_0Z^{-1}\left[\frac{z^2-2z}{(z-1)^2}\right]+y_1Z^{-1}\left[\frac{z}{(z-1)^2}\right] \quad ...(2)$$

Now for $Z^{-1}\left\{\frac{5z^2-2z}{(z-1)^4}\right\}$ $\quad z = 1$ (order 4)

$$\text{Res} =\frac{1}{\lfloor 3}\left[\frac{d^3}{dz^3}\cdot(z-1)^4\cdot\frac{z^{n-1}(5z^2-2z)}{(z-1)^4}\right]_{z=1}$$

$$=\frac{1}{6}\left[\frac{d^3}{dz^3}\{5z^{n+1}-2z^n\}\right]_{z=1}$$

$$=\frac{1}{6}[5(n+1)n(n-1)z^{n-2}-2n(n-1)(n-2)z^{n-3}]_{z=1}$$

$$=\frac{1}{6}[5(n+1)n(n-1)-2n(n-1)(n-2)]$$

$$=\frac{n(n-1)}{6}[5n+5-2n+4]$$

Res $(z = 1)$

$$=\frac{3(n+3)\cdot n(n-1)}{6}=\frac{1}{2}n(n-1)(n+3)$$

$$\therefore \qquad z^{-1}\left[\frac{5z^2-2z}{(z-1)^4}\right]=\frac{1}{2}n(n-1)(n+3)$$

Next $z = 1$ (order 2)

$$\text{Res} =\left[\frac{d}{dz}(z-1)^2\cdot z^{n-1}\cdot\frac{(z^2-2z)}{(z-1)^2}\right]_{z=1}$$

$$=\left[\frac{d}{dz}(z^{n+1}-2z^n)\right]_{z=1}$$

$$=[(n+1)z^n-2nz^{n-1}]_{z=1}=1-n$$

$\therefore \qquad z^{-1}\left\{\dfrac{z^2-2z}{(z-1)^2}\right\} = 1-z$

Similarly $\qquad z^{-1}\left[\dfrac{z}{(z-1)^2}\right] = n$

Putting these values in equation (2)

$$y_n = \frac{1}{2}n(n-1)(n+3) + y_0(1-n) + y_1 n$$

$$y_n = \frac{1}{2}n(n-1)(n+3) + y_0 + (y_1 - y_0)n$$

$$y_n = \frac{1}{2}n(n-1)(n+3) + c_0 + c_1 n, \text{ where } y_0 = c_0 \text{ and } y_1 - y_0 = c_1$$

Example 7: Using Z-transform, solve the following difference equation.

$$y_{k+2} + 4y_{k+1} + 3y_k = 3^k, \text{ given that } y_0 = 0 \text{ and } y_1 = 1$$

Sol. $\qquad y_{k+2} + 4y_{k+1} + 3y_k = 3^k \qquad \ldots(1)$

Let $\qquad Z(y_k) = \bar{y}$

Taking Z-transform in (1), on both sides, we get

$$Z(y_{k+2}) + 4Z(y_{k+1}) + 3Z(y_k) = Z(3^k)$$

$$\left\{z^2\bar{y} - z^2 y_0 - zy_1\right\} + 4\left\{z\bar{y} - zy_0\right\} + 3\bar{y} = \frac{z}{z-3} \qquad \ldots(2)$$

Putting the value of y_0 and y_1 in equation (2), we get

$$\left\{z^2\bar{y} - z\right\} + 4\left\{z\bar{y}\right\} + 3\bar{y} = \frac{z}{z-3}$$

$$\left(z^2 + 4z + 3\right)\bar{y} = \frac{z}{z-3} + z \;\Rightarrow\; \left(z^2 + 4z + 3\right)\bar{y} = \frac{z^2 - 2z}{z-3}$$

$$\Rightarrow \qquad \bar{y} = \frac{z^2 - 2z}{(z-3)(z^2 + 4z + 3)} = \frac{z^2 - 2z}{(z+1)(z+3)(z-3)} = \frac{z(z-2)}{(z+1)(z+3)(z-3)}$$

$$y_k = z^{-1}\left\{\frac{z(z-2)}{(z+1)(z+3)(z-3)}\right\}$$

The poles are $z = -1, -3, 3$

$$\therefore \qquad \text{Res } (z=-1) = \left[\frac{(z+1)z^{k-1}\cdot z(z-2)}{(z+1)(z+3)(z-3)}\right]_{z=-1} = \frac{-3(-1)^k}{2\cdot(-4)} = \frac{3}{8}(-1)^k$$

Similarly $\qquad \text{Res } (z=-3) = -\dfrac{5}{12}(-3)^k$

and $\text{Res}\,(z = 3) = \dfrac{3^k}{24}$

Therefore $y_k = \dfrac{3}{8}(-1)^k - \dfrac{5}{12}(-3)^k + \dfrac{1}{24}(3)^k$. **Ans.**

Example 8: Solve by Z-transform $y_{k+1} = 7y_k + 10x_k$

$$x_{k+1} = y_k + 4x_k; \qquad y_0 = 3, x_0 = 2.$$

Sol. $y_{k+1} = 7y_k + 10x_k$

Taking Z-transform, we get

$$z(\bar{y} - y_0) = 7\bar{y} + 10\bar{x}$$

$$\Rightarrow \quad (7 - z)\bar{y} + 10\bar{x} = -3z \qquad ...(1)$$

Applying Z-transform to $x_{k+1} = y_k + 4x_k$, we get

$$z(\bar{x} - x_0) = 4\bar{x} + \bar{y}$$

$$\Rightarrow \quad \bar{y} - (z - 4)\bar{x} = -2z \qquad ...(2)$$

Eliminating $\bar{y}$ from (1) and (2), we get

$$(z^2 - 11z + 18)\bar{x} = 2z^2 - 11z$$

$$\bar{x} = \frac{2z^2 - 11z}{(z-2)(z-9)} = \frac{z}{(z-9)} + \frac{z}{(z-2)}$$

$$\therefore \quad x_k = 9^k + 2^k \qquad \text{(Take inverse } Z\text{-transform)}$$

The equation is

$$x_{k+1} = y_k + 4x_k$$

$$\therefore \quad y_k = x_{k+1} - 4x_k$$

$$= 9^{k+1} + 2^{k+1} - 4(9^k + 2^k) = 5.9^k - 2.2^k$$

Hence, the solution is $x_k = 9^k + 2^k$

$$y_k = \{5.9^k - 2.2^k\}. \quad \textbf{Ans.}$$

Example 9: Solve using Z-transform: $y_{x+2} - 2\cos\alpha y_{x+1} + y_x = 0$, $y_0 = 0$, $y_1 = 1$

Sol. Take Z-transform on both sides,

$$z^2\left(\bar{y} - y_0 - \frac{y_1}{z}\right) - 2z\cos\alpha.(\bar{y} - y_0) + \bar{y} = 0$$

$$\Rightarrow \quad z^2\left(\bar{y} - \frac{1}{z}\right) - 2z\cos\alpha.\bar{y} + \bar{y} = 0$$

$$(z^2 - 2z\cos\alpha + 1)\bar{y} = z$$

$$\bar{y} = \frac{z}{z^2 - 2z\cos\alpha + 1} = \frac{z}{(z - e^{i\alpha})(z - e^{-i\alpha})} \qquad \left| \text{ As } \cos\alpha = \frac{e^{i\alpha} + e^{i\alpha}}{2} \right.$$

$$y_x = z^{-1}\left\{\frac{z}{(z - e^{i\alpha})(z - e^{-i\alpha})}\right\}$$

The poles are $2 = e^{i\alpha}, e^{-i\alpha}$

$$\text{Res}(z = e^{i\alpha}) = \left[\frac{(z \cdot e^{i\alpha})z^{x-1} \cdot z}{(z - e^{i\alpha})(z - e^{-i\alpha})}\right]_{z = e^{i\alpha}}$$

$$= \frac{e^{x\alpha}}{e^{i\alpha} - e^{-i\alpha}}$$

Similarly $\qquad \text{Res}(z = e^{-i\alpha}) = \dfrac{e^{-x\alpha}}{e^{-i\alpha} - e^{i\alpha}}$

Hence $\qquad y_x = \dfrac{e^{-ix\alpha}}{e^{-i\alpha} - e^{i\alpha}} - \dfrac{e^{-ix\alpha}}{e^{i\alpha} - e^{i\alpha}}$

$$y_x = \frac{\dfrac{e^{-ix\alpha} - e^{-ix\alpha}}{2c}}{\dfrac{e^{-i\alpha} - e^{i\alpha}}{2i}}$$

or $\qquad y_x = \left\{\dfrac{\sin\alpha x}{\sin\alpha}\right\}.$ **Ans.**

Example 10: Solve by *Z*-transform.

$$y_{k+1} + \frac{1}{4}y_k = \left(\frac{1}{4}\right)^k; \qquad (k \geq 0), y_0 = 0$$

Sol. We have $\qquad y_{k+1} + \dfrac{1}{4}y_k = \left(\dfrac{1}{4}\right)^k \qquad$ ...(1)

Take Z-transform on both sides of (1), we get

$$Z\left[y_{k+1} + \frac{1}{4}y_k\right] = Z\left[\left(\frac{1}{4}\right)^k\right]$$

$$Z[y_{k+1}] + Z\left[\frac{1}{4}y_k\right] = Z\left[\left(\frac{1}{4}\right)^k\right]$$

$$z\bar{y} - zy_0 + \frac{1}{4}\bar{y} = \frac{1}{1-\frac{1}{4}z^{-1}}; \qquad |z| < \frac{1}{4}$$

$$z\bar{y} - 0 + \frac{1}{4}\bar{y} = \frac{1}{1-\frac{1}{4}z^{-1}}$$

$$\left(z+\frac{1}{4}\right)\bar{y} = \frac{1}{\left(1-\frac{1}{4z}\right)}$$

$$\bar{y} = \frac{1}{\left(z+\frac{1}{4}\right)\left(1-\frac{1}{4z}\right)} = \frac{z}{\left(z+\frac{1}{4}\right)\left(z-\frac{1}{4}\right)}$$

The poles are $\quad z = \frac{1}{4}, -\frac{1}{4}$

$$\text{Res}\left(z=\frac{1}{4}\right) = \left[\frac{\left(z-\frac{1}{4}\right)z^{k-1}\cdot z}{\left(z+\frac{1}{4}\right)\left(z-\frac{1}{4}\right)}\right]_{z=\frac{1}{4}}$$

$$= \frac{\left(\frac{1}{4}\right)^k}{\frac{1}{2}} = 2\cdot\left(\frac{1}{4}\right)^k$$

Similarly $\quad \text{Res}\left(z=-\frac{1}{4}\right) = -2\left(-\frac{1}{4}\right)^k$

Hence $\quad y_k = -2\left(-\frac{1}{4}\right)^k + 2\left(\frac{1}{4}\right)^k.$ **Ans.**

Example 11: Solve:

$y_k + \frac{1}{4}y_{k-1} = U_{(k)} + \left(\frac{1}{3}\right)U_{(k-1)}$, where $U_{(k)}$ is the unit impulse function.

Sol.
$$y_k + \frac{1}{4}y_{k-1} = U_{(k)} + \left(\frac{1}{3}\right)U_{(k-1)} \qquad \text{...(1)}$$

Taking Z-transform on both sides of (1), we get

$$Z[\{y_k\}]+\frac{1}{4}Z[\{y_{k-1}\}]=Z[U_{(k)}]+\left(\frac{1}{3}\right)Z[\{U_{(k-1)}\}]$$

$$\bar{y}+\frac{1}{4}z^{-1}\bar{y}=\left[1+\frac{1}{3}z^{-1}\right]$$

$$\left[1+\frac{1}{4}z^{-1}\right]\bar{y}=\left[1+\frac{1}{3}z^{-1}\right]U_{(k)}$$

$$\bar{y}=\frac{1+\frac{1}{3}z^{-1}}{1+\frac{1}{4}z^{-1}}=\frac{z+\frac{1}{3}}{z+\frac{1}{4}}$$

There is only one simple pole at $z=-1/4$.

Let us consider the contour $|z|>\frac{1}{4}$.

Residue at $\left(z=-\frac{1}{4}\right)=\left[\left(z+\frac{1}{4}\right)z^{k-1}\frac{z+\frac{1}{3}}{z+\frac{1}{4}}\right]_{z=-\frac{1}{4}}=\left[z^k+\frac{z^{k-1}}{3}\right]_{z=-\frac{1}{4}}$

$$=\left(-\frac{1}{4}\right)^k+\frac{1}{3}\left(-\frac{1}{4}\right)^{k-1}=-\frac{1}{4}\left(-\frac{1}{4}\right)^{k-1}+\frac{1}{3}\left(-\frac{1}{4}\right)^{k-1}=\frac{1}{12}\left(-\frac{1}{4}\right)^{k-1}$$

Hence, y_k = Residue $=\frac{1}{12}\left(-\frac{1}{4}\right)^{k-1}$. **Ans.**

Example 12: Solve:

$$y_k+\frac{1}{25}y_{k-2}=\left(\frac{1}{5}\right)^k\cos\frac{k\pi}{2};(k\geq 0);\text{ by residue method.}$$

Sol. Take Z-transform on both sides, we get

$$Z\left(y_k+\frac{1}{25}y_{k-2}\right)=Z\left[\left(\frac{1}{5}\right)^k\cos\frac{k\pi}{2}\right]$$

$$\Rightarrow \qquad \bar{y}+\frac{1}{25}z^{-2}\bar{y}=\frac{z^2}{z^2+\frac{1}{25}}$$

$\left[\begin{array}{l}\therefore Z(c^k\cos\alpha k)=\dfrac{z(z-c\cos\alpha)}{z^2-2cz\cos\alpha+c^2}\\ \text{Here }\alpha=\dfrac{\pi}{2},c=\dfrac{1}{5}\end{array}\right]$

$$\therefore \qquad \bar{y} = \frac{z^2}{\left(z^2+\frac{1}{25}\right)\left(1+\frac{1}{25}z^{-2}\right)} = \frac{z^4}{\left(z^2+\frac{1}{25}\right)^2}$$

There are two poles of second order at $z = i/5$ and $z = -i/5$.

Consider a contour $|z| = \frac{1}{5}$.

$$\text{Residue at} \quad (z = i/5) = \left[\frac{d}{dz}\left\{\left(z-\frac{i}{5}\right)^2 .z^{k-1}.\frac{z^4}{\left(z^2+\frac{1}{25}\right)^2}\right\}\right]_{z=\frac{i}{5}}$$

$$= \left[\frac{d}{dz}.\left\{\frac{z^{k+3}}{\left(z+\frac{i}{5}\right)^2}\right\}\right]_{z=\frac{i}{5}} = \left[\frac{\left(z+\frac{i}{5}\right)^2 (k+3)z^{k+2} - z^{k+3}.z\left(z+\frac{i}{5}\right)}{\left(z+\frac{i}{5}\right)^4}\right]_{z=\frac{i}{5}}$$

$$= \left[\frac{\left(z+\frac{i}{5}\right)(k+3)z^{k+2} - 2z^{k+3}}{\left(z+\frac{i}{5}\right)^3}\right]_{z=\frac{i}{5}}$$

$$= \frac{(2i/5)(k+3)(i/5)^{k+2} - 2(i/5)^{k+3}}{(2i/5)^3}$$

$$= \left(\frac{5}{2i}\right)^3\left[(2k+6)\left(\frac{i}{5}\right)^{k+3} - 2\left(\frac{i}{5}\right)^{k+3}\right]$$

$$= \left(\frac{k+2}{4}\right).\left(\frac{1}{5}\right)^k .i^k = \left(\frac{k+2}{4}\right)\left(\frac{1}{5}\right)^k\left(\cos\frac{\pi}{2}+i\sin\frac{\pi}{2}\right)^k$$

$$= \frac{(k+2)}{4.(5)^k}.\left(\cos\frac{k\pi}{2}+i\sin\frac{k\pi}{2}\right)$$

$$\text{Similarly, residue at } (z = -i/5) \quad = \frac{(k+2)}{4.(5)^k}\left(\cos\frac{k\pi}{2}-i\sin\frac{k\pi}{2}\right)$$

$$y_k = \text{Sum of residues} = \frac{(k+2)}{2.(5)^k}\cos\frac{k\pi}{2}.$$ **Ans.**

Example 13: Solve by Z-transform the difference equation

$$y_{k+2} + 6y_{k+1} + 9y_k = 2^k \text{, given that } y_0 = 0 \text{ and } y_1 = 0$$

Sol. We have $\quad y_{k+2} + 6y_{k+1} + 9y_k = 2^k \qquad \ldots(1)$

Taking Z-transform in (1) on both sides, we get

$$Z(y_{k+2}) + 6Z(y_{k+1}) + 9Z(y_k) = Z(2^k)$$

$$\left\{z^2(\bar{y} - y_0 - y_1/z)\right\} + 6z\{\bar{y} - y_0\} + 9\bar{y} = \frac{z}{z-2} \qquad \ldots(2)$$

Putting the value of y_0 and y_1 in equation (2), we get

$$\left(z^2 + 6z + 9\right)\bar{y} = \frac{z}{z-2}$$

$$\Rightarrow \qquad \bar{y} = \frac{z}{(z-2)(z+3)^2}$$

Poles are given by $(z-2)(z+3)^2 = 0 \quad \Rightarrow \quad z = 2, -3, -3$

There are two poles out of which one is simple and other is of order two.

Consider the contour $|z| = 4$

Residue at $(z = 2)$

$$= \left[(z-2)z^{k-1}.\frac{z}{(z-2)(z+3)^2}\right]_{z=2}$$

$$= \left[\frac{z^k}{(z+3)^2}\right]_{z=2} = \frac{2^k}{25}$$

Residue at $(z = -3)$

$$= \frac{1}{(2-1)!}\left[\frac{d^{2-1}}{dz^{2-1}}\left\{(z+3)^2 .z^{k-1}\frac{z}{(z-2)(z+3)^2}\right\}\right]_{z=-3}$$

$$= \left[\frac{d}{dz}\left\{\frac{z^k}{(z-2)}\right\}\right]_{z=-3} = \left[\frac{(z-2).kz^{k-1} - z^k}{(z-2)^2}\right]_{z=-3}$$

$$= \frac{-5k(-3)^{k-1} - (-3)^k}{25} = -\frac{1}{5}k(-3)^{k-1} - \frac{(-3)^k}{25}$$

Here, $f(k)$ = Sum of the residues

$$= \left\{\frac{2^k}{25} - \frac{k}{5}(-3)^{k-1} - \frac{(-3)^k}{25}\right\}. \qquad \textbf{Ans.}$$

Example 14: Using Z-transform solve the following difference equation:

$$y_{k+2} + 4y_{k+1} + 3y_k = 3^k,$$

Given $y_0 = 1$ and $y_1 = 1$

Sol. Taking Z-transform on both sides, we have

$$Z(y_{k+2})+4Z(y_k)+3Z(y_k)=Z(3^k)$$

$$\Rightarrow \quad z^2\left(\bar{y}-y_0-\frac{y_1}{z}\right)+4(z\bar{y}-zy_0)+3\bar{y}=\frac{z}{z-3}$$

$$\Rightarrow \quad (z^2+4z+3)\,\bar{y}-z=\frac{z}{z-3}$$

$$\Rightarrow \quad \bar{y}=\frac{z}{(z-3)(z+1)(z+3)}+\frac{z}{(z+1)(z+3)}$$

Taking inverse Z-transform,

$$y_k=Z^{-1}\left\{\frac{z}{(z-3)(z+1)(z+3)}\right\}+\left\{\frac{z}{(z+1)(z+3)}\right\}$$

$$Z^{-1}\{A(z)\}+Z^{-1}\{B(z)\}$$

Consider the region $\quad |z|=4$

Residue of A (z) at $(z = 3)$ $=\lim\limits_{z\to 3}(z-3).z^{k-1}.\dfrac{z}{(z-3)(z+1)(z+3)}=\dfrac{3^k}{24}$

Residue of A (z) at $(z = -1)$ $=\lim\limits_{z\to -1}(z+1).z^{k-1}.\dfrac{z}{(z-3)(z+1)(z+3)}=\dfrac{(-1)^k}{(-8)}$

Residue of A (z) at $(z = -3)$ $=\lim\limits_{z\to -3}(z+3).z^{k-1}.\dfrac{z}{(z-3)(z+1)(z+3)}=\dfrac{(-3)^k}{12}$

Residue of B (z) at $(z = -1)$ $=\lim\limits_{z\to -1}(z+1).z^{k-1}.\dfrac{z}{(z+1)(z+3)}=\dfrac{(-1)^k}{2}$

Next

Residue of B (z) at $(z = -3)$ $=\lim\limits_{z\to -3}(z+3).z^{k-1}.\dfrac{z}{(z+1)(z+3)}=\dfrac{(-3)^k}{-2}$

$\therefore$ y_k = Sum of the residues $=\left\{\dfrac{3^k}{24}-\dfrac{(-1)^k}{8}+\dfrac{(-3)^k}{12}\right\}+\left\{\dfrac{(-1)^k}{2}-\dfrac{(-3)^k}{2}\right\}$

$$=\left\{\frac{3^k}{24}+\frac{3}{8}(-1)^k-\frac{5}{12}(-3)^k\right\}. \qquad \textbf{Ans.}$$

Example 15: Using Z-transforms solve the system of difference equations

$$x_{k+1}-7x_k-10y_k=0$$

$$y_{k+1}-x_k-4y_k=0, \quad \text{given } x_0=3,\ y_0=2$$

Sol. We have

$$x_{k+1} - 7x_k - 10y_k = 0 \quad ...(1)$$

$$y_{k+1} - x_k - 4y_k = 0 \quad ...(2)$$

Taking Z-transforms in equation (1), we get

$$Z(x_{k+1}) - 7Z(x_k) - 10Z(y_k) = 0$$

$$z[\bar{x} - x_0] - 7\bar{x} - 10\bar{y} = 0$$

$$z(\bar{x} - 3) - 7\bar{x} - 10\bar{y} = 0$$

or $$(z-7)\bar{x} - 10\bar{y} = 3z \quad ...(3)$$

Again taking Z-transforms in equation (2) and using $y_0 = 2$, we get

$$z(\bar{y} - 2) - \bar{x} - 4\bar{y} = 0$$

or $$(z-4)\bar{y} - \bar{x} = 2z \quad ...(4)$$

Eliminating $\bar{x}$ from (3) and (4), we get

$$(z-4)(z-7)\bar{y} - 10\bar{y} = 3z + 2z(z-7)$$

or $$(z^2 - 11z + 18)\bar{y} = z(2z - 11)$$

$$\therefore \quad \frac{\bar{y}}{z} = \frac{2z-11}{z^2 - 11z + 18} = \frac{2z-11}{(z-9)(z-2)} = \frac{1}{z-9} + \frac{1}{z-2}$$

$$\Rightarrow \quad \bar{y} = \frac{z}{z-9} + \frac{z}{z-2}$$

$$\Rightarrow \quad y_k = Z^{-1}\left[\frac{z}{z-9}\right] + Z^{-1}\left[\frac{z}{z-2}\right] = 9^k + 2^k$$

From (2), we get

$$x_k = y_{k+1} - 4y_k$$

$$\therefore \quad x_k = 9^{k+1} + 2^{k+1} - 4(9^k + 2^k)$$

$$= 9(9)^k + 2(2)^k - 4(9^k + 2^k)$$

$$= 5.9^k - 2.2^k$$

Hence $x_k = 5.9^k - 2.2^k$ and $y_k = 9^k + 2^k$ **Ans.**

PROBLEM SET 1.5

Solve the following difference equation by Z-transform:

1. $y_{k+1} + y_k = 1$ if $y_0 = 0$. $\left[\textbf{Ans. } \frac{1}{2}\{1 - (-1)^k\}\right]$

2. $y_k - \frac{5}{6}y_{k-1} + \frac{1}{6}y_{k-2} = U_k$. $\left[\textbf{Ans. } y_k = \left[3 - 3\left(\frac{1}{2}\right)^k + \left(\frac{1}{3}\right)^k\right]U(k)\right]$

3. $y_{k+1} - y_{k-1} = U_k, \quad y_0 = 0.$ $\left[\textbf{Ans.} \left[\dfrac{k}{2} + \dfrac{1}{4}\right] U(k)\right]$

4. $y_k - \dfrac{1}{3} y_{k-1} = \left\{\left(\dfrac{1}{3}\right)^k\right\}, \quad k \geq 0, \ y_0 = 0.$ $\left[\textbf{Ans.}\ y_k = \left\{\dfrac{k+1}{3^k}\right\}\right]$

5. $y_k + \left(\dfrac{1}{16}\right) y_{k-2} = \left\{\left(\dfrac{1}{4}\right)^k \cos \dfrac{k\pi}{2}\right\}, \quad k \geq 0.$ $\left[\textbf{Ans.} \left\{\dfrac{(k+2)}{(4)^{k+1}} .2 \cos \dfrac{k\pi}{2}\right\}\right]$

6. $y_{k+2} - 4y_{k+1} + 3y_k = 0$ if $y_0 = 2, y_1 = 4.$ $[\textbf{Ans.}\ 1 + 3k]$

7. $y_{k+2} - 4y_{k+1} + 4y_k = 0$ if $y_0 = 0, y_1 = 0.$ $[\textbf{Ans.}\ 2^k (1-k)]$

8. $y_{k+2} - 5y_{k+1} + 6y_k = 5^k$ if $y_0 = 0, y_1 = 0.$ $\left[\textbf{Ans.}\ \dfrac{2k}{3} - \dfrac{3^k}{2} + \dfrac{5^k}{6}\right]$

9. $y_{k+2} - 5y_{k+1} + 6y_k = 4^k$ if $y_0 = 0, y_1 = 1.$ $\left[\textbf{Ans.}\ \dfrac{1}{2}\left(4^k - 2^k\right)\right]$

10. $y_k = y_{k-1} + y_{k-2},$ if $y(0) = 1, y(1) = 1.$ $\left[\textbf{Ans.}\ \dfrac{1}{\sqrt{5}} \left(\dfrac{1}{2}\right)^{k+1} \left\{(1+\sqrt{5})^{k+1} - (1-\sqrt{5})^{k+1}\right\}\right]$

11. $y_{k+1} - 2y_k = 1,$ where $y_0 = 0.$ $[\textbf{Ans.}\ 2^k - 1]$

12. $y_{k+2} - 4y_k = k - 1.$ $\left[\textbf{Ans.}\ y_k = c_1 (2)^k + c_2 (-2)^k - \dfrac{k}{3} + \dfrac{1}{9}\right]$

13. $x_{k+1} - y_k = 1, y_{k+1} - x_k = 1$ given $x_0 = 0, y_0 = 1.$

Ans. $y_k = \dfrac{1}{2}\left\{\sin \dfrac{k\pi}{2} - \cos \dfrac{k\pi}{2} - 1\right\}$

$x_k = \dfrac{1}{2}\left\{1 - \cos \dfrac{k\pi}{2} - \sin \dfrac{k\pi}{2}\right\}$

UNIT II

Functions of a Complex Variable I

2.1 COMPLEX NUMBERS

Any ordered pair (x, y) which can be expressed in the form of $z = x + iy$, is called a complex number. The real numbers x and y are called as the real and imaginary parts of z. It can be written as

$$R(z) = x, \ I(z) = y$$

The term $\sqrt{x^2 + y^2}$ taken with the positive sign is defined to be modulus of z and written as $|z|$

Therefore, $|z| = \sqrt{(x^2 + y^2)}$.

If $z = x + iy$, then the complex numbers $x - iy = \bar{z}$, is called the conjugate of the complex number $z = x + iy$

Hence $\qquad |\bar{z}| = |z|, \ |z|^2 = z\bar{z}$

And it is easy to verify that

$$R(z) = \text{Real part of } z = \frac{z + \bar{z}}{2}$$

$$I(z) = \text{Imaginary part of } z = \frac{z - \bar{z}}{2i}$$

2.2 FUNCTION OF A COMPLEX VARIABLE

A complex variable w is said to be function of the complex variable z, if for each value of $z = x + iy$ in a certain region R, there correspond one or more definite values of $w = u + iv$.

The function of a complex variable is written as:

$$w = f(z) = u + iv$$

where u and v both are the function of x and y

i.e. $\qquad u = u(x, y) \quad \text{and} \quad v = v(x, y)$

If w takes only one value for each value of z in the region R, then w is said to be uniform or single valued function of z. If there correspond two or more values of w for some or all value of z in the region R, then w is called a many-alued or multi-valued function of z.

2.3 CONTINUITY

A single valued function $w = f(z) = u + iv$ is said to be continuous at $z = z_0$ if $\lim_{z \to z_0} f(z) = f(z_0)$ and further $f(z)$ is said to be continuous in any region R of the z-plane, if it is continuous at every point of the region R.

2.4 DERIVATIVE OF *f*(z)

If a function $f(z)$ is single valued in a region R then the derivative of $f(z)$ is denoted by $f'(z)$ and is defined as $\dfrac{dw}{dz} = f'(z) = \lim_{\delta z \to 0} \dfrac{f(z+\delta z) - f(z)}{\delta z}$.

Provided the limit exists and is independent of the path along which $\delta z \to 0$.

Remark: For particular point $z = a$, the derivative of $f(z)$ is defined as $f'(a) = \lim_{h \to 0} \dfrac{f(a+h) - f(a)}{h}$, and we can say that $f(z)$ is differentiable at $z = 0$.

2.5 ANALYTIC FUNCTION

A function $f(z)$ which is single valued and differentiable at any point $z = a$ is said to be analytic at $z = a$. If the function is analytic at every point of a certain region R it said to be analytic in that region. A function $f(z)$ may be differentiable in a region except possibly for a finite number of points. These points are called singular points or singulaties of $f(z)$ in the region.

2.6 NECESSARY AND SUFFICIENT CONDITIONS FOR *f*(z) TO BE ANALYTIC

(a) The necessary conditions for *f*(z) to be analytic or Cauchy's-Reimann equations

Theorem 1: The necessary conditions for a function $f(z) = u + iv$ to be analytic at all the points in the region R are

(*i*) $\dfrac{\partial u}{\partial x} = \dfrac{\partial v}{\partial y}$ (*ii*) $\dfrac{\partial u}{\partial y} = -\dfrac{\partial v}{\partial x}$ provided $\dfrac{\partial u}{\partial x}, \dfrac{\partial u}{\partial y}, \dfrac{\partial v}{\partial x}, \dfrac{\partial v}{\partial y}$ exist.

Proof: Let $f(z)$ be an analytic function in a region R,

$\Rightarrow$ $$f(z) = u + iv,$$

where u and v are the functions of x and y.

Let δu and δv be the increments of u and v respectively and δz be the corresponding increments of z.

then $$f(z+\delta z) = (u+\delta u) + i(v+\delta v)$$

Now $\dfrac{f(z+\delta z)-f(z)}{\delta z}=\dfrac{(u+\delta u)+i(v+\delta v)-(u+iv)}{\delta z}=\dfrac{\delta u+i\delta v}{\delta z}=\dfrac{\delta u}{\delta z}+i\dfrac{\delta v}{\delta z}$

Taking $\delta z \to 0$

$$\lim_{\delta z\to 0}\frac{f(z+\delta z)-f(z)}{\delta z}=\lim_{\delta z\to 0}\left(\frac{\delta u}{\delta z}+i\frac{\delta v}{\delta z}\right) \quad \text{or} \quad f'(z)=\lim_{\delta z\to 0}\left(\frac{\delta u}{\delta z}+i\frac{\delta v}{\delta z}\right) \qquad ...(1)$$

(i) Along x-axis

$z = x + iy$ but on x axis, $y = 0$

$\therefore$ $z = x,$ $\delta z = \delta x,\ \delta y = 0$

Putting these values in (1), we have

$$f'(z)=\lim_{\delta x\to 0}\left(\frac{\delta u}{\delta x}+i\frac{\delta v}{\delta x}\right)=\frac{\partial u}{\partial x}+i\frac{\partial v}{\partial x} \qquad ...(2)$$

(ii) Along y-axis

$z = x + iy$ but on y axis, $x = 0$

$z = iy$ $\delta x = 0$ and $\delta z = i\delta y$

Putting these values in (1), we get

$$f'(z)=\lim_{\delta y\to 0}\left(\frac{\delta u}{i\delta y}+\frac{i\delta v}{i\delta y}\right)=\lim_{\delta y\to 0}\left(-i\frac{\delta u}{\delta y}+\frac{\delta v}{\delta y}\right)=-i\frac{\partial u}{\partial y}+\frac{\partial v}{\partial y} \qquad ...(3)$$

If $f(z)$ is differentiable, then above two values of $f'(z)$ must be same

From (2) and (3), we get

$$\frac{\partial u}{\partial x}+i\frac{\partial v}{\partial x}=-i\frac{\partial u}{\partial y}+\frac{\partial v}{\partial y}$$

Equating real and imaginary parts, we have

$$\frac{\partial u}{\partial x}=\frac{\partial v}{\partial y},\quad \frac{\partial v}{\partial x}=-\frac{\partial u}{\partial y}$$

$$\Rightarrow \quad \boxed{\frac{\partial u}{\partial x}=\frac{\partial v}{\partial y}},\quad \boxed{\frac{\partial u}{\partial y}=-\frac{\partial v}{\partial x}}$$

The above equations are known as **Cauchy-Reimann** equations.

(b) Sufficient condition for $f(z)$ to be analytic

Theorem 2: The function $w = f(z)$ is analytic in a region R if

(i) $\dfrac{\partial u}{\partial x}=\dfrac{\partial v}{\partial y},\quad \dfrac{\partial u}{\partial y}=-\dfrac{\partial v}{\partial x}$ *(ii)* $\dfrac{\partial u}{\partial x},\dfrac{\partial u}{\partial y},\dfrac{\partial v}{\partial x},\dfrac{\partial v}{\partial y}$ continuous function of x and y in region R.

Proof: Let $f(z)$ be a single valued function having

$$\frac{\partial u}{\partial x},\frac{\partial u}{\partial y},\frac{\partial v}{\partial x},\frac{\partial v}{\partial y}$$

At each point in the region R then satisfying C-R equations

i.e.,
$$\frac{\partial u}{\partial x}=\frac{\partial v}{\partial y} \text{ and } \frac{\partial u}{\partial y}=-\frac{\partial v}{\partial x}$$

Since,

$$f(z+\delta z)=u(x+\delta x, y+\delta y)+iv(x+\delta x, y+\delta y)$$

Expand by Taylor's theorem and neglecting the higher power in terms of δx and δy

$$f(z+\delta z)=\left[u(x,y)+\left(\frac{\partial u}{\partial x}\delta x+\frac{\partial u}{\partial y}\delta y\right)\right]+i\left[v(x,y)+\left(\frac{\partial v}{\partial x}\delta x+\frac{\partial v}{\partial y}\delta y\right)\right]$$

$$=\left[u(x,y)+iv(x,y)\right]+\left(\frac{\partial u}{\partial x}+i\frac{\partial v}{\partial x}\right)\delta x+\left(\frac{\partial u}{\partial y}+i\frac{\partial v}{\partial y}\right)\delta y$$

$$=f(z)+\left(\frac{\partial u}{\partial x}+i\frac{\partial v}{\partial x}\right)\delta x+\left(\frac{\partial u}{\partial y}+i\frac{\partial v}{\partial y}\right)\delta y \quad \text{...(1)}$$

But by C-R equations $\frac{\partial u}{\partial x}=\frac{\partial v}{\partial y}$

and $\frac{\partial u}{\partial y}=-\frac{\partial v}{\partial x}$

From (1), we get

$$f(z+\delta z)=f(z)+\left(\frac{\partial u}{\partial x}+i\frac{\partial v}{\partial x}\right)\delta x+\left(-\frac{\partial v}{\partial x}+i\frac{\partial u}{\partial x}\right)\delta y$$

$$=f(z)+\left(\frac{\partial u}{\partial x}+i\frac{\partial v}{\partial x}\right)\delta x+\left(\frac{\partial u}{\partial x}+i\frac{\partial v}{\partial x}\right)i\delta y$$

$$=f(z)+\left(\frac{\partial u}{\partial x}+i\frac{\partial v}{\partial x}\right)(\delta x+i\delta y)$$

$$=f(z)+\left(\frac{\partial u}{\partial x}+i\frac{\partial v}{\partial x}\right)\delta z$$

$$\Rightarrow \frac{f(z+\delta z)-f(z)}{\delta z}=\frac{\partial u}{\partial x}+i\frac{\partial v}{\partial x}$$

$$f'(z)=\lim_{\delta z\to 0}\frac{f(z+\delta z)-f(z)}{\delta z}=\frac{\partial u}{\partial x}+i\frac{\partial v}{\partial x}$$

But $\frac{\partial u}{\partial x}$ and $\frac{\partial v}{\partial x}$ are exist so $f'(z)$is also exist. Hence $f(z)$ is analytic.

2.7 CAUCHY-REIMANNIAN (C-R) EQUATIONS IN POLAR COORDINATE

We know that the relation between polar and cartesian coordinate

$$x = r\cos\theta$$

$$y = r\sin\theta$$

$$z = x + iy = r(\cos\theta + i\sin\theta) = re^{i\theta}$$

$$\therefore \quad w = f(z) = f(re^{i\theta}) \ \text{i.e.}\ u + iv = f(re^{i\theta}) \qquad ...(1)$$

Differentiating (1) partially w.r.t. r, we have

$$\frac{\partial u}{\partial r} + i\frac{\partial v}{\partial r} = f'(re^{i\theta}).e^{i\theta} \qquad ...(2)$$

Again differentiating (1) partially w.r.t. θ, we have

$$\frac{\partial u}{\partial \theta} + i\frac{\partial v}{\partial \theta} = f'(re^{i\theta}).ire^{i\theta} = ir\left\{f'(re^{i\theta}).e^{i\theta}\right\}, \qquad ...(3)$$

Using eqn. (2) in (3), then we get

$$\frac{\partial u}{\partial \theta} + i\frac{\partial v}{\partial \theta} = ir\left\{\frac{\partial u}{\partial r} + i\frac{\partial v}{\partial r}\right\} = ir\frac{\partial u}{\partial r} - r\frac{\partial v}{\partial r}, \text{ as } i^2 = -1$$

Equating real and imaginary part, we obtain

$$\frac{\partial u}{\partial \theta} = -r\frac{\partial v}{\partial r} \Rightarrow \boxed{\frac{\partial v}{\partial r} = -\frac{1}{r}\frac{\partial u}{\partial \theta}}$$

and

$$\frac{\partial v}{\partial \theta} = r\frac{\partial u}{\partial r} \Rightarrow \boxed{\frac{\partial u}{\partial r} = \frac{1}{r}\frac{\partial v}{\partial \theta}}$$

These are C-R equations in polar form.

2.8 DERIVATIVE OF W IN POLAR FORM

Theorem 3: To prove that $\dfrac{dw}{dz} = e^{-i\theta}\dfrac{\partial w}{\partial r} = -\dfrac{i}{r}e^{-i\theta}\dfrac{\partial w}{\partial \theta}$

Proof: Since $w = f(z)$

$$\frac{dw}{dz} = f'(z) = \frac{\partial u}{\partial x} + i\frac{\partial v}{\partial x} = \frac{\partial}{\partial x}(u + iv) = \frac{\partial w}{\partial x}$$

$$\Rightarrow \quad \frac{dw}{dz} = \frac{\partial w}{\partial x}$$

$$\Rightarrow \quad \frac{dw}{dz} = \frac{\partial w}{\partial r}\frac{\partial r}{\partial x} + \frac{\partial w}{\partial \theta}\frac{\partial \theta}{\partial x} \qquad ...(1)$$

But $r^2 = x^2 + y^2$

$$2r\frac{\partial r}{\partial x} = 2x \Rightarrow \frac{\partial r}{\partial x} = \frac{x}{r}$$

$$\frac{\partial r}{\partial x}=\frac{r\cos\theta}{r}=\cos\theta,\ \theta=\tan^{-1}\frac{y}{x}\quad\Rightarrow\quad\frac{\partial\theta}{\partial x}=\frac{1\times x^2}{x^2+y^2}\left(-\frac{y}{x^2}\right)$$

$$\Rightarrow\quad\frac{\partial r}{\partial x}=\cos\theta,\quad\frac{\partial\theta}{\partial x}=-\frac{y}{r^2}=-\frac{r\sin\theta}{r^2}=-\frac{\sin\theta}{r}$$

Using these values in equation (1)

$$\frac{dw}{dz}=\frac{\partial w}{\partial r}\cos\theta+\frac{\partial w}{\partial\theta}\left(-\frac{\sin\theta}{r}\right)$$

$$=\cos\theta\frac{\partial w}{\partial r}+\frac{\partial}{\partial\theta}(u+iv)\left(-\frac{\sin\theta}{r}\right)$$

$$=\cos\theta\frac{\partial w}{\partial r}-\frac{\sin\theta}{r}\left\{\frac{\partial u}{\partial\theta}+i\frac{\partial v}{\partial\theta}\right\}$$

but
$$\left.\begin{aligned}\frac{\partial v}{\partial r}&=-\frac{1}{r}\frac{\partial u}{\partial\theta}\Rightarrow\frac{\partial u}{\partial\theta}=-r\frac{\partial v}{\partial r}\\ \frac{\partial u}{\partial r}&=\frac{1}{r}\frac{\partial v}{\partial\theta}\Rightarrow\frac{\partial v}{\partial\theta}=r\frac{\partial u}{\partial r}\end{aligned}\right\}\quad\ldots(2)$$

So
$$\frac{dw}{dz}=\cos\theta\frac{\partial w}{\partial r}-\frac{\sin\theta}{r}\left\{-r\frac{\partial v}{\partial r}+ir\frac{\partial u}{\partial r}\right\}$$

$$=\cos\theta\frac{\partial w}{\partial r}-i\sin\theta\left\{\frac{\partial u}{\partial r}+i\frac{\partial v}{\partial r}\right\}$$

$$=(\cos\theta-i\sin\theta)\frac{\partial w}{\partial r},\quad\Bigg|\ \text{As }\frac{\partial w}{\partial r}=\frac{\partial u}{\partial r}+i\frac{\partial v}{\partial r}$$

$$\Rightarrow\quad\boxed{\frac{dw}{dz}=e^{-i\theta}\frac{\partial w}{\partial r}}$$

Again

$$\frac{dw}{dz}=\frac{\partial w}{\partial r}\cos\theta-\frac{\sin\theta}{r}\frac{\partial w}{\partial\theta}=\left(\frac{\partial u}{\partial r}+i\frac{\partial v}{\partial r}\right)\cos\theta-\frac{\sin\theta}{r}\frac{\partial w}{\partial\theta}$$

Using equation (2), then we get

$$\frac{dw}{dz}=\left(\frac{1}{r}\frac{\partial v}{\partial\theta}-i\frac{1}{r}\frac{\partial u}{\partial\theta}\right)\cos\theta-\frac{\sin\theta}{r}\frac{\partial w}{\partial\theta}$$

$$=-\frac{i}{r}\left\{\frac{\partial u}{\partial\theta}+i\frac{\partial v}{\partial\theta}\right\}\cos\theta-\frac{\sin\theta}{r}\frac{\partial w}{\partial\theta}$$

$$\frac{dw}{dz}=-\frac{i}{r}\cos\theta\frac{\partial w}{\partial\theta}-\frac{\sin\theta}{r}\frac{\partial w}{\partial\theta}\quad\Bigg|\ \frac{\partial w}{\partial\theta}=\frac{\partial u}{\partial\theta}+i\frac{\partial u}{\partial\theta}$$

$$= -\frac{i}{r}(\cos\theta - i\sin\theta)\frac{\partial w}{\partial \theta}$$

$$\Rightarrow \quad \boxed{\frac{dw}{dz} = -\frac{i}{r}e^{-i\theta}\frac{\partial w}{\partial \theta}}$$

2.9 HARMONIC FUNCTION

Any function of x, y say $u(x, y)$, having continuous partial derivatives of first and second order and also satisfying Laplace's equation is call Harmonic function Laplace's equation is $\frac{\partial^2 u}{\partial x^2} + \frac{\partial^2 u}{\partial y^2} = 0,$

Theorem 4: If $w = f(z)$ be an analytic function then u and v both are harmonic functions.

Proof: By C-R equations.

$$\frac{\partial u}{\partial x} = \frac{\partial v}{\partial y}, \quad \frac{\partial u}{\partial y} = -\frac{\partial v}{\partial x}$$

$$\Rightarrow \quad \frac{\partial^2 u}{\partial x^2} = \frac{\partial^2 v}{\partial x \partial y}, \quad \frac{\partial^2 u}{\partial y^2} = -\frac{\partial^2 v}{\partial y \partial x}$$

$$\Rightarrow \quad \frac{\partial^2 u}{\partial x^2} + \frac{\partial^2 u}{\partial y^2} = 0$$

and for the function v

$$\Rightarrow \quad \frac{\partial^2 v}{\partial y^2} = \frac{\partial^2 u}{\partial x \partial y}, \quad \frac{\partial^2 v}{\partial x^2} = -\frac{\partial^2 u}{\partial x \partial y}$$

$$\Rightarrow \quad \frac{\partial^2 v}{\partial x^2} + \frac{\partial^2 v}{\partial y^2} = 0$$

Hence, u, v are harmonic functions.

2.10 ORTHOGONAL SYSTEM

Two curves are said to be orthogonal to each other when they intersect at right angle at each of their points of intersection.

Theorem 5: If $f(z) = u + iv$ is an analytic function of $z = x + iy$, then the family of curves $u(x, y)$ = constant; $v(x, y)$ = constant. Form an orthogonal system.

Proof: For the curve $u(x, y) = c_1$

We have on differentiating

$$du = \frac{\partial u}{\partial x}dx + \frac{\partial u}{\partial y}dy = 0$$

$$\Rightarrow \qquad \frac{dy}{dx} = -\frac{\partial u}{\partial x}\bigg/\frac{\partial u}{\partial y} = m_1 \text{(say)}$$

and for the curve $v(x, y) = c_2$

$$dv = \frac{\partial v}{\partial x}dx + \frac{\partial v}{\partial y}dy = 0$$

$$\Rightarrow \qquad \frac{dy}{dx} = -\frac{\partial v}{\partial x}\bigg/\frac{\partial v}{\partial y} = m_2 \text{(say)}$$

$$m_1 m_2 = \left(\frac{\frac{\partial u}{\partial x}}{\frac{\partial u}{\partial y}}\right) \times \left(\frac{\frac{\partial v}{\partial x}}{\frac{\partial v}{\partial y}}\right) = -\frac{\frac{\partial v}{\partial y}}{\frac{\partial v}{\partial x}} \times \frac{\frac{\partial v}{\partial x}}{\frac{\partial v}{\partial y}}$$

$\because$ C-R equation $\frac{\partial u}{\partial x} = \frac{\partial v}{\partial y}$, $\frac{\partial u}{\partial y} = -\frac{\partial v}{\partial x}$

$$\Rightarrow \qquad m_1 m_2 = -1. \quad \textbf{Proved.}$$

Theorem 6: An analytic function of a constant modulus is constant.

Proof: $f(z) = u + iv$

But $|f(z)| = |u + iv| = \text{constant}$

$$\Rightarrow \qquad \sqrt{u^2 + v^2} = \text{constant} = k$$

Squaring both sides, we get

$$u^2 + v^2 = k^2 \qquad \text{...(1)}$$

Diffentiating equation (1) w.r.t. 'x'

$$2u\frac{\partial u}{\partial x} + 2v\frac{\partial v}{\partial x} = 0$$

$$u\frac{\partial u}{\partial x} + v\frac{\partial v}{\partial x} = 0 \qquad \text{...(2)}$$

Again from (1) $u\frac{\partial u}{\partial y} + v\frac{\partial v}{\partial y} = 0$

$$\Rightarrow \qquad u\left(-\frac{\partial v}{\partial x}\right) + v\left(\frac{\partial u}{\partial x}\right) = 0 \qquad \text{....(3)}$$

$\frac{\partial u}{\partial x} = \frac{\partial v}{\partial y}$, $\frac{\partial u}{\partial y} = -\frac{\partial v}{\partial x}$

Adding equations (2) and (3) with squaring, we get

$$(u^2 + v^2)\left\{\left(\frac{\partial u}{\partial x}\right)^2 + \left(\frac{\partial v}{\partial x}\right)^2\right\} = 0$$

Since $$u^2 + v^2 \neq 0$$

So $$\left(\frac{\partial u}{\partial x}\right)^2 + \left(\frac{\partial v}{\partial x}\right)^2 = 0$$

or $$\left[\sqrt{\left(\frac{\partial u}{\partial x}\right)^2 + \left(\frac{\partial v}{\partial x}\right)^2}\right]^2 = |f'(z)|^2 = 0$$

or $$|f'(z)| = 0 \quad \Rightarrow \quad f(z) = \text{constant.}$$ **Proved.**

2.11 DETERMINATION OF THE CONJUGATE FUNCTION

If $f(z) = u + iv$ is an analytic function where both $u(x, y)$ and $v(x, y)$ are conjugate function. Being given one of these say u, determine the other v.

Since v is a function of x, y therefore,

$$dv = \frac{\partial v}{\partial x}dx + \frac{\partial v}{\partial y}dy$$

$$dv = -\frac{\partial u}{\partial y}dx + \frac{\partial u}{\partial x}dy \quad \ldots(1)$$

$\left| \frac{\partial u}{\partial x} = \frac{\partial v}{\partial y} \right.$, C-R equation

$\left| \frac{\partial u}{\partial y} = -\frac{\partial v}{\partial x} \right.$

The R.H.S. of (1) is of the form

$$Mdx + Ndy,$$

where $$M = -\frac{\partial u}{\partial y}, N = \frac{\partial u}{\partial x}$$

so that $$\frac{\partial M}{\partial y} = \frac{\partial}{\partial y}\left(-\frac{\partial u}{\partial y}\right) = -\frac{\partial^2 u}{\partial y^2}$$

and $$\frac{\partial N}{\partial x} = \frac{\partial}{\partial x}\left(\frac{\partial u}{\partial x}\right) = \frac{\partial^2 u}{\partial x^2} = -\frac{\partial^2 u}{\partial y^2}$$

$$\Rightarrow \quad \frac{\partial M}{\partial y} = \frac{\partial N}{\partial x}$$

$\Big[$ u is harmonic

$$\frac{\partial^2 u}{\partial x^2} + \frac{\partial^2 u}{\partial y^2} = 0$$

$$\Rightarrow \frac{\partial^2 u}{\partial x^2} = -\frac{\partial^2 u}{\partial y^2}\Big]$$

Hence, equation (1) satisfies the condition of exact differential equation.

So equation (1) can be integrated and thus v determined.

2.12 MILNE'S THOMSON METHOD

With the help of this method $f(z)$ can be directly constructed without finding u or v.

Since $$z = x + iy \text{ and } \bar{z} = x - iy$$

so that $$x = \frac{z + \bar{z}}{2}, y = \frac{z - \bar{z}}{2i}$$

$$w = f(z) = u + iv = u(x, y) + iv(x, y)$$

or $$f(z) = u\left(\frac{z+\bar{z}}{2}, \frac{z-\bar{z}}{2i}\right) + iv\left(\frac{z+\bar{z}}{2}, \frac{z-\bar{z}}{2i}\right) \quad ...(1)$$

This relation is an identity in two independent variable $z, \bar{z}$ putting $\bar{z} = z$

We have $x = z$ and $y = 0$

From equation (1)

$$\Rightarrow \quad f(z) = u(z, 0) + iv(z, 0) \quad ...(2)$$

We know that

$$f'(z) = \frac{dw}{dz} = \frac{\partial w}{\partial x} = \frac{\partial}{\partial x}(u + iv)$$

$$\Rightarrow \quad f'(z) = \frac{\partial u}{\partial x} + i\frac{\partial v}{\partial x}$$

$$f'(z) = \frac{\partial u}{\partial x} - i\frac{\partial u}{\partial y} \quad ...(3) \qquad \left| \frac{\partial u}{\partial y} = \frac{-\partial v}{\partial x} \right.$$

Let $$\frac{\partial u}{\partial x} = \phi_1(x, y) = \phi_1(z, 0) \qquad \left| \text{As } x = z \text{ and } y = 0 \right.$$

and $$\frac{\partial u}{\partial y} = \phi_2(x, y) = \phi_2(z, 0)$$

From (3) $$f'(z) = \phi_1(z, 0) - i\phi_2(z, 0)$$

$$\Rightarrow \quad f(z) = \int\{\phi_1(z, 0) - i\phi_2(z, 0)\}dz + C$$

Similarly if $v\ (x, y)$ is given then we can obtained

$$f(z) = \int\{\psi_1(z, 0) + i\psi_2(z, 0)\}dz + C$$

where $$\left(\frac{\partial v}{\partial x}\right)_{(z,0)} = \psi_2(z, 0)$$

$$\left(\frac{\partial v}{\partial y}\right)_{(z,0)} = \psi_1(z, 0)$$

Example 1: If ϕ and ψ are functions of x and y satisfying Laplace's equation, show that $s+it$ is analytic where

$$s = \frac{\partial \phi}{\partial y} - \frac{\partial \psi}{\partial x} \text{ and } t = \frac{\partial \phi}{\partial x} + \frac{\partial \psi}{\partial y}$$

Sol. For the function $st + it$ to be analytic,

$$\frac{\partial s}{\partial x} = \frac{\partial t}{\partial y} \text{ and } \frac{\partial s}{\partial y} = -\frac{\partial t}{\partial x} \qquad ...(A)$$

Now
$$\frac{\partial s}{\partial x} = \frac{\partial}{\partial x}\left(\frac{\partial \phi}{\partial y} - \frac{\partial \psi}{\partial x}\right) = \frac{\partial^2 \phi}{\partial x \partial y} - \frac{\partial^2 \psi}{\partial x^2} \qquad ...(1)$$

$$\frac{\partial t}{\partial y} = \frac{\partial}{\partial y}\left(\frac{\partial \phi}{\partial x} + \frac{\partial \psi}{\partial y}\right) = \frac{\partial^2 \phi}{\partial y \partial x} + \frac{\partial^2 \psi}{\partial y^2}$$

or
$$\frac{\partial t}{\partial y} = \frac{\partial^2 \phi}{\partial y \partial x} - \frac{\partial^2 \psi}{\partial x^2} \qquad ...(2) \qquad \left| \text{ As } \frac{\partial^2 \psi}{\partial x^2} + \frac{\partial^2 \psi}{\partial y^2} = 0 \right.$$

and
$$\frac{\partial s}{\partial y} = \frac{\partial^2 \phi}{\partial y^2} - \frac{\partial^2 \psi}{\partial y \partial x} \qquad ...(3)$$

$$\frac{\partial t}{\partial x} = \frac{\partial^2 \phi}{\partial x^2} + \frac{\partial^2 \Psi}{\partial y \partial x}$$

$$= -\frac{\partial^2 \phi}{\partial y^2} + \frac{\partial^2 \psi}{\partial y \partial x} \qquad \left| \text{ As } \frac{\partial^2 \phi}{\partial x^2} + \frac{\partial^2 \phi}{\partial y^2} = 0 \right.$$

$$\Rightarrow \qquad \frac{\partial t}{\partial x} = -\left(\frac{\partial^2 \phi}{\partial y^2} - \frac{\partial^2 \psi}{\partial y \partial x}\right) \qquad ...(4)$$

Making use of equations (1), (2), (3) and (4)

The C-R equations (*A*) are satisfied

Hence the function $s + it$ is analytic.

Example 2: Determine an analytic function $f(z)$, in terms of z whose real part is $e^{-x}(x \sin y - y \cos y)$.

Sol. Here
$$u = e^{-x}(x \sin y - y \cos y)$$

$$\frac{\partial u}{\partial x} = -e^{-x}(x \sin y - y \cos y) + e^{-x} \sin y = e^{-x}(y \cos y - x \sin y + \sin y)$$

$$\frac{\partial u}{\partial y} = e^{-x}(x \cos y + y \sin y - \cos y)$$

$$\therefore \qquad \phi_1(z, 0) = \left(\frac{\partial u}{\partial x}\right)_{(z,\,0)} = e^{-z}(0 - z \cdot 0 + 0) = 0$$

and
$$\phi_2(z, 0) = \left(\frac{\partial u}{\partial y}\right)_{(z,\,0)} = e^{-z}(z + 0 - 1) = e^{-z}(z - 1)$$

By Milne's Thomson method, we have

$$f(z) = \int \{\phi_1(z, 0) - i\phi_2(z, 0)\}dz + C$$

$$= \int \{0 - ie^{-z}(z-1)\}dz + C = i\int e^{-z}(1-z)dz + C$$

$$= i\{-e^{-z}(1-z) - \int e^{-z}dz\} + C$$

$$= i\{-e^{-z} + ze^{-z} + e^{-z}\} + C = ize^{-z} + C$$

$\Rightarrow$ $f(z) = ize^{-z} + C$. **Ans.**

Example 3: Find the analytic function of which the real part is $u = e^x(x \cos y - y \sin y)$, given $f(1) = e$.

Sol. $u = e^x(x \cos y - y \sin y)$

$$\frac{\partial u}{\partial x} = e^x(x \cos y - y \sin y) + e^x \cos y = e^x(x\cos y - y\sin y + \cos y)$$

and $$\frac{\partial u}{\partial y} = e^x(-x\sin y - \sin y - y\cos y)$$

$$\frac{\partial u}{\partial x} = \frac{\partial v}{\partial y}, \quad \frac{\partial u}{\partial y} = -\frac{\partial v}{\partial x}$$

We know that $$dv = \frac{\partial v}{\partial x}dx + \frac{\partial v}{\partial y}dy$$

$$dv = -\frac{\partial u}{\partial y}dx + \frac{\partial u}{\partial x}dy$$

$$dv = -e^x(-x\sin y - \sin y - y\cos y)dx + e^x(x\cos y - y\sin y + \cos y)dy$$

$$\int dv = \int e^x(x\sin y + \sin y + y\cos y)dx + \int(\text{those terms which free from } x)dy + C$$

$$v = \sin y\int xe^x dx + \int(y\cos y + \sin y)e^x dx + \int 0 dy + C$$

$$= \sin y \cdot e^x(x-1) + (y\cos y + \sin y)e^x + C$$

$$= [\sin y \cdot (x-1) + y\cos y + \sin y]e^x + C$$

$$I = \int xe^x dx = xe^x - \int e^x dx = e^x(x-1)$$

Therefore $$f(z) = u + iv = e^x(x\cos y - y\sin y) + i[e^x\{\sin y \cdot (x-1) + y\cos y + \sin y\} + C]$$

$$= xe^x(\cos y + i\sin y) + iye^x(\cos y + i\sin y) + iC$$

$$= (xe^x + iye^x)(\cos y + i\sin y) + iC$$

$$= e^x(x + iy)(\cos y + i\sin y) + iC$$

$$= ze^x e^{iy} + C_1, \text{ where } C_1 = iC$$

$$= ze^{x+iy} + C_1 \Rightarrow f(z) = ze^z + C_1 \Rightarrow f(1) = e + C_1 \Rightarrow e = e + C_1$$

$\Rightarrow$ $C_1 = 0$

Hence $f(z) = ze^z$. **Ans.**

Examples 4: Show that the function defined by $f(z)=\sqrt{|xy|}$ is not regular at the origin although Cauchy-Riemannian equations are satisfied.

Sol. We have $f(z)=\sqrt{|xy|}$

$$u(x,y)+iv(x,y) = \sqrt{|xy|}$$

$\therefore$ $$u(x,y) = \sqrt{|xy|} \quad \text{and} \quad v(x,y)=0$$

At the origin, we have

$$\frac{\partial u}{\partial x} = \lim_{x\to 0}\frac{u(x,0)-u(0,0)}{x} = \lim_{x\to 0}\frac{0-0}{x} = 0$$

$$\frac{\partial u}{\partial y} = \lim_{y\to 0}\frac{u(0,y)-u(0,0)}{y} = \lim_{y\to 0}\frac{0-0}{y} = 0$$

$$\frac{\partial v}{\partial x} = \lim_{x\to 0}\frac{v(x,0)-v(0,0)}{x} = \lim_{x\to 0}\frac{0-0}{x} = 0$$

$$\frac{\partial v}{\partial y} = \lim_{y\to 0}\frac{v(0,y)-v(0,0)}{y} = \lim_{y\to 0}\frac{0-0}{y} = 0$$

From these values, we have

$$\frac{\partial u}{\partial x} = \frac{\partial v}{\partial y} \text{ and } \frac{\partial u}{\partial y} = -\frac{\partial v}{\partial x}$$

Next $$f'(z) = \lim_{z\to 0}\frac{f(z)-f(0)}{z} = \lim_{z\to 0}\frac{f(z)}{z} = \lim_{z\to 0}\frac{\sqrt{|xy|}}{z}$$

We find $f'(z)$ along the line $y = mx$, which passes through the origin
i.e., $z \to 0$ along the line $y = mx$

$\therefore$ $$f'(z) = \lim_{x\to 0}\frac{\sqrt{|mx^2|}}{x+imx} = \lim_{x\to 0}\frac{\sqrt{|m|}}{(1+im)} = \frac{\sqrt{|m|}}{(1+im)}$$

The $f'(z)$ depends on m. Therefore $f'(z)$ does not exist at the origin. Hence the function $f(z)$ is not regular at the origin.

Example 5: (*i*) Prove that the function sinh z is analytic and find its derivative. (*ii*) Show that $f(z)$ =log z is analytic everywhere in the complex plane except at the origin and that its derivative is $\left(\frac{1}{z}\right)$.

Sol. (*i*) $f(z)=\sinh z=\sinh(x+iy)=\frac{1}{i}\sin i(x+iy)$

$\sin i\theta = i\sin h\theta$
$\Rightarrow \sin h\theta = \frac{\sin i\theta}{i}$
$\cos i\theta = \cosh\theta$

$\Rightarrow$ $$u+iv=\frac{1}{i}\sin(xi-y)$$

$$u+iv=\frac{1}{i}\{\sin xi\cos y-\cos xi\sin y\}$$

$$= \frac{1}{i} i \sinh x \cos y + \frac{i^2}{i} \cosh x \sin y$$

$$u + iv = \sinh x \cos y + i \cosh x \sin y$$

$$\Rightarrow \qquad u = \sinh x \cos y, \ v = \cosh x \sin y$$

$$\frac{\partial u}{\partial x} = \cosh x \cos y, \ \frac{\partial v}{\partial x} = \sinh x \sin y$$

$$\frac{\partial u}{\partial y} = -\sinh x \sin y, \ \frac{\partial v}{\partial y} = \cosh x \cos y$$

From these relations $\quad \frac{\partial u}{\partial x} = \frac{\partial v}{\partial y}$ and $\frac{\partial u}{\partial y} = -\frac{\partial v}{\partial x}$

Thus the four partial derivatives being continuous everywhere and C-R equations being satisfied, the function $f(z)$ is analytic everywhere.

Derivative of $f(z) = f'(z) = \frac{\partial u}{\partial x} + i\frac{\partial v}{\partial x} = \cosh x \cos y + i \sinh x \sin y$

$$f'(z) = \cos ix \cos y + \sin ix \sin y$$

$$f'(z) = \cos(ix - y) = \cos(ix + i^2 y)$$

$$f'(z) = \cos i(x + iy) = \cosh(x + iy)$$

$$\Rightarrow \qquad f'(z) = \cosh z. \quad \textbf{Ans.}$$

(*ii*) $f(z) = \log z = \log(x + iy) = \frac{1}{2}\log(x^2 + y^2) + i \tan^{-1}\frac{y}{x}$

$$\Rightarrow \qquad u + iv = f(z) = \frac{1}{2}\log(x^2 + y^2) + i \tan^{-1}\frac{y}{x}$$

$$\log(\alpha + i\beta) = \frac{1}{2}\log(\alpha^2 + \beta^2) + i \tan^{-1}\frac{\beta}{\alpha}$$

Equating real and imaginary part

$$u = \frac{1}{2}\log(x^2 + y^2), v = \tan^{-1}\frac{y}{x}$$

$$\frac{\partial u}{\partial x} = \frac{1}{2(x^2 + y^2)} \times 2x = \frac{x}{(x^2 + y^2)}, \frac{\partial u}{\partial y} = \frac{y}{(x^2 + y^2)}$$

$$\frac{\partial v}{\partial x} = \frac{1}{1 + \frac{y^2}{x^2}}\left(-\frac{y}{x^2}\right) = -\frac{y}{x^2 + y^2}, \frac{\partial v}{\partial y} = \frac{1}{1 + \frac{y^2}{x^2}}\left(\frac{1}{x}\right) = \frac{x}{x^2 + y^2}$$

From these relations $\frac{\partial u}{\partial x} = \frac{\partial y}{\partial y}$ and $\frac{\partial u}{\partial y} = -\frac{\partial v}{\partial x}$

If $z = 0$ *i.e.* $x^2 + y^2 = 0$ *i.e.* $x = 0, y = 0$

At $z = 0$, $\qquad \frac{\partial u}{\partial x} = \lim_{x \to 0} \frac{u(x,0) - u(0,0)}{x} = \lim_{x \to 0} \frac{\log x - \log 0}{x} = \infty$

The function $f(z)$ does not exist at origin.

Hence, the function $f(z)$ is analytic everywhere in the complex plane except at the origin.

Now $$f'(z) = \frac{dw}{dz} = \frac{\partial w}{\partial x} = \frac{\partial u}{\partial x} + i\frac{\partial v}{\partial x}$$

$$f'(z) = \frac{x - iy}{x^2 + y^2} = \frac{x - iy}{(x + iy)(x - iy)} = \frac{1}{(x + iy)} = \frac{1}{z}$$

$$f'(z) = \frac{1}{z}.$$ **Ans.**

Example 6: (*i*) If $u(x, y) = x^2 - y^2$ and $v(x, y) = \dfrac{y}{x^2 + y^2}$, prove that both u and v satisfy Laplace's equation but are not harmonic conjugates.

(*ii*) Given that $u(x, y) = x^2 - y^2$ and $v(x, y) = -\dfrac{y}{x^2 + y^2}$ prove that both u and v are harmonic functions but $u + iv$ is not an analytic function of z.

Sol. (*i*) $u = x^2 - y^2,\ v = y / x^2+y^2$

$$\frac{\partial u}{\partial x} = 2x, \frac{\partial u}{\partial y} = -2y$$

$$\frac{\partial^2 u}{\partial x^2} = 2, \frac{\partial^2 u}{\partial y^2} = -2$$

Since $$\frac{\partial^2 u}{\partial x^2} + \frac{\partial^2 u}{\partial y^2} = 2 - 2 = 0$$

Hence, u satisfies Laplace's equation.

$$v = y/x^2+y^2$$

$\therefore$ $$\frac{\partial v}{\partial x} = \frac{-2xy}{(x^2 + y^2)^2} \Rightarrow \frac{\partial^2 v}{\partial x^2} = \frac{6x^2 y - 2y^3}{(x^2 + y^2)^3}$$

$$\frac{\partial v}{\partial y} = \frac{x^2 - y^2}{(x^2 + y^2)^2} \Rightarrow \frac{\partial^2 v}{\partial y^2} = \frac{-6x^2 y + 2y^3}{(x^2 + y^2)^3}$$

$\Rightarrow$ $$\frac{\partial^2 v}{\partial x^2} + \frac{\partial^2 v}{\partial y^2} = 0$$

Hence, $v(x, y)$ satisfies Laplace's equation.

But $$\frac{\partial u}{\partial x} \neq \frac{\partial v}{\partial y} \text{ and } \frac{\partial v}{\partial x} \neq -\frac{\partial u}{\partial y}$$

Thus u and v are not harmonic conjugates.

(ii) $u = x^2 - y^2$

$$\frac{\partial u}{\partial x} = 2x, \quad \frac{\partial u}{\partial y} = -2y$$

$$\frac{\partial^2 u}{\partial x^2} = 2, \quad \frac{\partial^2 u}{\partial y^2} = -2$$

$$\Rightarrow \quad \frac{\partial^2 u}{\partial x^2} + \frac{\partial^2 u}{\partial y^2} = 0$$

$u(x, y)$ is harmonic function.

and
$$v = -\frac{y}{x^2 + y^2}$$

$$\frac{\partial v}{\partial x} = \frac{2xy}{(x^2+y^2)^2}, \quad \frac{\partial v}{\partial y} = \frac{y^2 - x^2}{(x^2+y^2)^2}$$

$$\frac{\partial^2 v}{\partial x^2} = \frac{2y^3 - 6x^2 y}{(x^2+y^2)^3}, \quad \frac{\partial^2 v}{\partial y^2} = \frac{6x^2 y - 2y^3}{(x^2+y^2)^3}$$

$$\Rightarrow \quad \frac{\partial^2 v}{\partial x^2} + \frac{\partial^2 v}{\partial y^2} = 0$$

Hence, $v(x, y)$ is harmonic.

But
$$\frac{\partial u}{\partial x} \neq \frac{\partial v}{\partial y} \text{ and } \frac{\partial u}{\partial y} \neq -\frac{\partial v}{\partial x}$$

Therefore, $u + iv$ is not analytic.

Example 7: Show that the function $f(z) = e^{-z^{-4}}$, $z \neq 0$ and $f(0) = 0$ is not analytic at $z = 0$ although Cauchy-Reimann equations are satisfied at the point. How would you explain this?

Sol.
$$w = f(z) = u(x, y) + iv(x, y)$$

If
$$z = 0$$

$$f(0) = 0 \quad \Rightarrow \quad u(0,0) = 0 = v(0,0)$$

Now
$$u + iv = e^{-z^{-4}} = e^{-(x+iy)^{-4}}$$

$\Rightarrow$ At
$$y = 0$$

$$u(x,0) + iv(x,0) = e^{-x^{-4}}$$

Equating real and imaginary part.

$$u(x,0) = e^{-x^{-4}}, \quad v(x,0) = 0$$

And at
$$x = 0$$

$$u(0,y) + iv(0,y) = e^{-y^{-4}}$$

Equating real and imaginary part.

Now $$u(0, y) = e^{-y^{-4}}, v(0, y) = 0$$

$$\frac{\partial u}{\partial x} = \lim_{h \to 0} \frac{u(x,0) - u(0,0)}{x} = \lim_{x \to 0} \frac{e^{-x^{-4}} - 0}{x}$$

$$= \lim_{x \to 0} \frac{1}{x}\left[\frac{1}{e^{x^{-4}}}\right] = \lim_{x \to 0} \frac{1}{x}\left[\frac{1}{e^{\frac{1}{x^4}}}\right]$$

$$= \lim_{x \to 0} \frac{1}{x}\left[\frac{1}{1 + \frac{1}{x^4} + \frac{1}{2!x^8} + \ldots\ldots}\right] = \lim_{x \to 0}\left[\frac{1}{x + \frac{1}{x^3} + \ldots\ldots}\right]$$

$$\frac{\partial u}{\partial x} = \frac{1}{\infty} = 0$$

Similarly $\frac{\partial u}{\partial y} = 0$

Again

$$\frac{\partial v}{\partial x} = \lim_{x \to 0} \frac{v(x,0) - v(0,0)}{x} = \lim_{x \to 0} \frac{0 - 0}{x} = 0$$

Similarly $\frac{\partial v}{\partial y} = 0$. Thus $\frac{\partial u}{\partial x} = \frac{\partial v}{\partial y}, \frac{\partial u}{\partial y} = -\frac{\partial v}{\partial x}$

i.e., C-R equations are satisfied

Now derivative of $f(z)$ at $z = 0$

$$f'(0) = \lim_{z \to 0} \frac{f(z) - f(0)}{z} = \lim_{z \to 0} \frac{e^{-z^{-4}} - 0}{z}$$

Let $z = 0$ along the path $z = re^{i\pi/4}$

$\therefore$ $$Z^{-4} = r^{-4}e^{-i\pi} = r^{-4}(\cos\pi - i\sin\pi) = r^{-4}(-1-0) = -r^{-4}$$

$\therefore$ $$f'(0) = \lim_{r \to 0} \frac{e^{r-4}}{re^{i\pi/4}} = \lim_{r \to 0} \frac{1}{re^{i\pi/4}}\left[\frac{1}{e^{r4}}\right]$$

or $$f'(0) = \lim_{r \to 0} \frac{1}{re^{i\pi/4}}\left[1 + \frac{1}{r^4} + \frac{1}{\lfloor 2} \cdot \frac{1}{r^8} + \ldots\right] = \frac{1}{0} = \infty$$

This shows that $f'(z)$ does not exist at $z = 0$ *i.e.*, the function $f(z)$ is not differentiable at $z = 0$. Hence the given function is not analytic at $z = 0$.

Example 8: Find the values of a and b such that the function

$f(z)=x^2+ay^2-2xy+i(bx^2-y^2+2xy)$ is analytic. Also find $f'(z)$.

Sol.
$$f(z)=u+iv=x^2+ay^2-2xy+i(bx^2-y^2+2xy)$$
$$u=x^2+ay^2-2xy,\quad v=bx^2-y^2+2xy$$
$$\frac{\partial u}{\partial x}=2x-2y,\ \frac{\partial u}{\partial y}=2ay-2x,\ \frac{\partial v}{\partial x}=2bx+2y,\ \frac{\partial v}{\partial y}=-2y+2x$$

But $f(z)$ is analytic, so
$$\frac{\partial u}{\partial x}=\frac{\partial v}{\partial y}\quad\Rightarrow\quad 2x-2y=2x-2y\ \text{(which is satisfied)}$$

and
$$\frac{\partial u}{\partial y}=-\frac{\partial v}{\partial x}\quad\Rightarrow\quad 2ay-2x=-2bx-2y \qquad \text{... (1)}$$

Equating the coefficient of x and y in equation (1)

$a=-1$ and $b=1$. **Ans.**

Now
$$f'(z)=\frac{dw}{dz}=\frac{\partial w}{\partial x}=\frac{\partial u}{\partial x}+i\frac{\partial v}{\partial x}=2x-2y+i(2bx+2y)$$
$$f'(z)=2x-2y+i(2x+2y)$$
$$=2(x+iy)+2(xi-y)$$
$$=2z+2(xi+i^2y)$$
$$=2z+2i(x+iy)=2z+2iz$$
$$f'(z)=2z(1+i).$$ **Ans.**

Example 9: Discuss the analyticity of the function
$$f(z)=z\bar{z}$$
or

Prove that the function $f(z)=|z|^2$ is continuous everywhere but nowhere differentiable except at the origin.

Sol.
$$f(z)=u+iv=(x+iy)(x-iy)=x^2+y^2$$
$\Rightarrow$
$$u=x^2+y^2, v\neq 0$$
$$\frac{\partial u}{\partial x}=2x,\ \frac{\partial u}{\partial y}=2y,\ \frac{\partial v}{\partial x}=0,\ \frac{\partial v}{\partial y}=0$$

At origin *i.e.* $z=0\ \Rightarrow\ x=0, y=0$

So
$$\frac{\partial u}{\partial x}=0,\ \frac{\partial u}{\partial y}=0$$

$$\Rightarrow \qquad \frac{\partial u}{\partial x} = \frac{\partial v}{\partial y} \text{ and } \frac{\partial u}{\partial y} = -\frac{\partial v}{\partial x}$$

Hence, C-R equation are satisfied

Again $$f'(0) = \lim_{z \to 0} \frac{f(z) - f(0)}{z}$$

$$= \lim_{z \to 0} \frac{z\bar{z} - 0}{z} = \lim_{z \to 0} (\bar{z})$$

$$= \lim_{z \to 0} (x - iy)$$

Let z $\to$ 0 along the line $y = mx$ then

$$f'(0) = \lim_{x \to 0} (x - imx) = 0$$

Therefore $f'(0)$ is unique *i.e.*, $f(z)$ is differentiable at the origin but if be take any other point except the origin example $z = 1$, *i.e.*, $x \to 1$ then $f'(z)$ depends on the values of m. So the given function is not differentiable except the origin. Hence it is analytic only at the origin.

Example 10: Show that the function $e^x(\cos y + i \sin y)$ is holomorphic (analytic) and find $f'(z)$.

Sol. We have $$f(z) = e^x(\cos y + i \sin y) = u + iv$$

Equating real and imaginary part

$$u = e^x \cos y, v = e^x \sin y$$

$$\frac{\partial u}{\partial x} = e^x \cos y, \frac{\partial u}{\partial y} = -e^x \sin y \text{ and } \frac{\partial v}{\partial x} = e^x \sin y, \frac{\partial v}{\partial y} = e^x \cos y$$

From these $$\frac{\partial u}{\partial x} = \frac{\partial v}{\partial y} \text{ and } \frac{\partial u}{\partial y} = -\frac{\partial v}{\partial x}$$

Hence, C-R, equations are satisfied. Therefore $f(z)$ is analytic.

Now $$f'(z) = \frac{\partial u}{\partial x} + i\frac{\partial v}{\partial x} = e^x \cos y + ie^x \sin y$$

$$= e^x(\cos y + i \sin y)$$

$$= e^x e^{iy} = e^{x+iy}$$

$$\Rightarrow \qquad f'(z) = e^z. \textbf{ Ans.}$$

Example 11: Show that the function $f(z) = z|z|$ is not analytic anywhere.

Sol. $$f(z) = z|z| = (x + iy)\sqrt{x^2 + y^2} = u + iv$$

$$u = x\sqrt{x^2 + y^2} \text{ and } v = y\sqrt{x^2 + y^2}$$

$$\frac{\partial u}{\partial x} = \sqrt{x^2+y^2} + \frac{x \times 2x}{2\sqrt{x^2+y^2}} = \sqrt{x^2+y^2} + \frac{x^2}{\sqrt{x^2+y^2}}$$

$$\frac{\partial u}{\partial y} = \frac{2xy}{2\sqrt{x^2+y^2}} = \frac{xy}{\sqrt{x^2+y^2}}$$

$$\frac{\partial v}{\partial x} = \frac{xy}{\sqrt{x^2+y^2}} \text{ and } \frac{\partial v}{\partial y} = \sqrt{x^2+y^2} + \frac{y^2}{\sqrt{x^2+y^2}}$$

$$\Rightarrow \quad \frac{\partial u}{\partial x} \neq \frac{\partial v}{\partial x} \text{ and } \frac{\partial u}{\partial x} \neq -\frac{\partial v}{\partial x}$$

Therefore, C-R equations are not satisfied. Hence, the given function is not analytic.

Example 12: In two dimensional fluid flow, the stream function is $\psi = -\frac{y}{x^2+y^2}$ find the velocity potential ϕ.

Sol. $\psi = -\frac{y}{x^2+y^2}, \quad \frac{\partial \psi}{\partial x} = \frac{2xy}{(x^2+y^2)^2}, \quad \frac{\partial \psi}{\partial y} = \frac{y^2-x^2}{(x^2+y^2)^2}$

$$d\phi = \frac{\partial \phi}{\partial x}dx + \frac{\partial \phi}{\partial y}dy = \frac{\partial \psi}{\partial y}dx - \frac{\partial \psi}{\partial x}dy$$

$u = \phi, v = \psi$

$$\frac{\partial \phi}{\partial x} = \frac{\partial \psi}{\partial y}$$

$$\frac{\partial \phi}{\partial y} = -\frac{\partial \psi}{\partial x}$$

$$d\phi = \frac{y^2-x^2}{(x^2+y^2)^2}dx - \frac{2xy}{(x^2+y^2)^2}dy$$

$$= \frac{(y^2-x^2)dx - 2xydy}{(x^2+y^2)^2} = \frac{(x^2+y^2)dx - 2x^2dx - 2xydy}{(x^2+y^2)^2}$$

$$= \frac{(x^2+y^2)dx - x(2xdx+2ydy)}{(x^2+y^2)^2} = \frac{(x^2+y^2)dx - xd(x^2+y^2)}{(x^2+y^2)^2}$$

$\Rightarrow \quad d\phi = d\left\{\frac{x}{x^2+y^2}\right\}$, On integrating, we get $\phi = \frac{x}{x^2+y^2} + C$, where C is any constant. **Ans.**

Example 13: Find the analytic function $w = u + iv$ if $v = \log(x^2+y^2) + x - 2y$

Sol.

$$v = \log(x^2+y^2) + x - 2y$$

$$\frac{\partial v}{\partial y} = \frac{2y}{x^2+y^2} - 2 = \psi_1(x,y), \quad \frac{\partial v}{\partial x} = \frac{2x}{x^2+y^2} + 1 = \psi_2(x,y)$$

$$\therefore \quad \psi_1(z,0) = -2 \text{ and } \psi_2(z,0) = \frac{2}{z} + 1$$

By Milne's Thomson Method

$$f(z) = \int [\psi_1(z,0) + i\psi_2(z,0)]\, dz + C \qquad \left| \begin{array}{l} y = 0 \\ x = z \end{array} \right.$$

$$f(z) = \int \left[-2 + i\left(\frac{2}{z} + 1 \right) \right] dz + C$$

$$f(z) = -2z + 2i \log z + iz + C$$

$$f(z) = z(i-2) + 2i \log z + C. \quad \textbf{Ans.}$$

Example 14: Determine the function $w = f(z)$ if $u = e^{2x}(x\cos 2y - y\sin 2y)$

Sol.

$$u = e^{2x}(x\cos 2y - y\sin 2y)$$

$$\frac{\partial u}{\partial x} = 2e^{2x}(x\cos 2y - y\sin 2y) + e^{2x}(\cos 2y)$$

$$= e^{2x}(2x\cos 2y - 2y\sin 2y + \cos 2y) = \phi_1(x, y)$$

$$\frac{\partial u}{\partial y} = e^{2x}(-2x\sin 2y - 2y\cos 2y - \sin 2y) = \phi_2(x, y)$$

Now $\quad x = z,\ y = 0$

$\therefore \quad \phi_1(z,0) = e^{2z}(2z+1)$

and $\quad \phi_2(z,0) = e^{2z}(0) = 0$

By Milne's Thomson method, we have

$$f(z) = \int [\phi_1(z,0) - i\phi_2(z,0)]\, dz + C$$

$$= \int \left[e^{2z}(2z+1) - 0 \right] dz + C$$

$$f(z) = (2z+1)\frac{e^{2z}}{2} - \int 2\frac{e^{2z}}{2}\, dz + C$$

$$f(z) = ze^{2z} + C. \quad \textbf{Ans.}$$

where C is integral constant.

Example 15: Show that the function $u(x, y) = e^{-2xy}\sin(x^2 - y^2)$ is harmonic. Also find analytic function $w = f(z)$.

Sol. $u(x, y) = e^{-2xy}\sin(x^2 - y^2)$

$$\frac{\partial u}{\partial x} = -2ye^{-2xy}\sin(x^2 - y^2) + 2xe^{-2xy}\cos(x^2 - y^2)$$

$$\frac{\partial^2 u}{\partial x^2} = 4y^2 e^{-2xy} \sin(x^2 - y^2) - 4xye^{-2xy} \cos(x^2 - y^2)$$
$$+ 2e^{-2xy} \cos(x^2 - y^2) - 4xye^{-2xy} \cos(x^2 - y^2) - 4x^2 e^{-2xy} \sin(x^2 - y^2)$$

$$\frac{\partial u}{\partial y} = -2xe^{-2xy} \sin(x^2 - y^2) - 2ye^{-2xy} \cos(x^2 - y^2)$$

$$\frac{\partial^2 u}{\partial y^2} = 4x^2 e^{-2xy} \sin(x^2 - y^2) + 4xye^{-2xy} \cos(x^2 - y^2)$$
$$- 2e^{-2xy} \cos(x^2 - y^2) + 4xye^{-2xy} \cos(x^2 - y^2) - 4y^2 e^{-2xy} \sin(x^2 - y^2)$$

From these equations $\frac{\partial^2 u}{\partial x^2} + \frac{\partial^2 u}{\partial y^2} = 0.$

$\Rightarrow$ The function $u(x, y)$ is harmonic.

Now, $\phi_1(z, 0) = 2z\cos z^2$ and $\phi_2(z, 0) = -2z\sin z^2$

By Milne's Thomson method

$$w = f(z) = \int \{\phi_1(z,0) - i\phi_2(z,0)\} dz + C$$
$$= 2\int \left\{ z\cos z^2 + iz\sin z^2 \right\} dz + C$$
$$= 2\int ze^{iz^2} dz + C$$

$$f(z) = -ie^{iz^2} + C. \quad \textbf{Ans.}$$

Example 16: Show that the function $u = \frac{1}{2}\log(x^2 + y^2)$ is harmonic and find its harmonic conjugate.

Sol. Here $u = \frac{1}{2}\log(x^2 + y^2)$

$$\frac{\partial u}{\partial x} = \frac{1}{2} \cdot \frac{1}{(x^2 + y^2)} \cdot 2x = \frac{x}{x^2 + y^2}$$

$$\frac{\partial^2 u}{\partial x^2} = \frac{(x^2 + y^2) - 2x^2}{(x^2 + y^2)^2} = \frac{y^2 - x^2}{(x^2 + y^2)^2}$$

and
$$\frac{\partial u}{\partial y} = \frac{1}{2(x^2 + y^2)} \cdot 2y = \frac{y}{x^2 + y^2}$$

$$\frac{\partial^2 u}{\partial y^2} = \frac{(x^2 + y^2) - 2y^2}{(x^2 + y^2)^2} = \frac{x^2 - y^2}{(x^2 + y^2)^2}$$

$$\therefore \qquad \frac{\partial^2 u}{\partial x^2}+\frac{\partial^2 u}{\partial y^2}=0$$

Hence u is a harmonic function.

Next
$$dv=\frac{\partial v}{\partial x}dx+\frac{\partial v}{\partial y}dy=\left(-\frac{\partial u}{\partial y}\right)dx+\left(\frac{\partial u}{\partial x}\right)dy \qquad \text{(By C-R equations)}$$

$$=\left(\frac{-y}{x^2+y^2}\right)dx+\left(\frac{x}{x^2+y^2}\right)dy$$

$$dv=\frac{xdy-ydx}{x^2+y^2}=d\left(\tan^{-1}\frac{y}{x}\right)$$

On integrating, we get

$$v=\tan^{-1}\frac{y}{x}+C. \quad \textbf{Ans.}$$

Example 17: If $u+v=\dfrac{2\sin 2x}{e^{2y}+e^{-2y}-2\cos 2x}$ and $f(z)=u+iv$ is an analytic function of $z=x+iy$, find $f(z)$ in terms of z.

Sol. Here $\qquad f(z)=u+iv$

$\therefore \qquad if(z)=iu-v$

Adding, $\qquad (1+i)f(z)=(u-v)+i(u+v)$

$\Rightarrow \qquad F(z)=U+iV$

where $F(z)=(1+i)f(z)$, $U=u-v$ and $V=u+v$

It means that we have been given

$$V=\frac{2\sin 2x}{e^{2y}+e^{-2y}-2\cos 2x}$$

or
$$V=\frac{2\sin 2x}{2\cosh 2y-2\cos 2x}$$

or
$$V=\frac{\sin 2x}{\cosh 2y-\cos 2x}$$

Now
$$\frac{\partial V}{\partial y}=\frac{-2\sin 2x\sinh 2y}{(\cosh 2y-\cos 2x)^2}=\psi_1(x,y) \quad \text{(say)}$$

and
$$\frac{\partial V}{\partial x}=\frac{2\cos 2x(\cosh 2y-\cos 2x)-2\sin^2 2x}{(\cosh 2y-\cos 2x)^2}$$

$$\frac{\partial V}{\partial x}=\frac{2\cos 2x\cosh 2y-2}{(\cosh 2y-\cos 2x)^2}=\psi_2(x,y) \quad \text{(say)}$$

$\therefore \qquad \psi_1(z,0)=0$

and $$\psi_2(z,0)=\frac{2(\cos 2z-1)}{(1-\cos 2z)^2}=\frac{-2}{1-\cos 2z}=\frac{-2}{1-1+2\sin^2 z}=-\text{cosec}^2 z$$

By Milne's Thomson method, we have

$$F(z)=\int\{\psi_1(z,0)+i\psi_2(z,0)\}dz+c$$

$$=\int -i\,\text{cosec}^2 z\,dz+c=i\cot z+c$$

Replacing $F(z)$ by $(1+i)f(z)$, we get

$$(1+i)f(z)=i\cot z+c$$

$$\Rightarrow \qquad f(z)=\frac{i}{1+i}\cot z+\frac{c}{1+i}$$

$$\therefore \qquad f(z)=\frac{1}{2}(1+i)\cot z+c_1 \qquad \text{where} \quad c_1=\frac{c}{1+i}$$

Example 18: Let $f(z)=u(r,\theta)+iv(r,\theta)$ be an analytic function. If $u=-r^3\sin 3\theta$, then construct the corresponding conjugate v and find $f(z)$ in terms of z.

Sol. $$u=-r^3\sin 3\theta$$

$$\frac{\partial u}{\partial r}=-3r^2\sin 3\theta$$

$$\frac{\partial u}{\partial \theta}=-3r^3\cos 3\theta$$

C-R Equations in polar

$$\frac{\partial v}{\partial r}=-\frac{1}{r}\frac{\partial u}{\partial \theta}$$

$$\frac{\partial u}{\partial r}=\frac{1}{r}\frac{\partial v}{\partial \theta}$$

So $$dv=\frac{\partial v}{\partial r}dr+\frac{\partial v}{\partial \theta}d\theta$$

$$dv=\left(-\frac{1}{r}\frac{\partial u}{\partial \theta}\right)dr+\left(r\frac{\partial u}{\partial r}\right)d\theta$$

$$=\left\{\frac{1}{r}\times(3r^3\cos 3\theta)\right\}dr+r(-3r^2\sin 3\theta)d\theta$$

$$dv=3r^2\cos 3\theta\,dr-3r^3\sin 3\theta\,d\theta$$

$$dv=d(r^3\cos 3\theta)$$

$$\int dv=\int d(r^3\cos 3\theta)+C \qquad \Rightarrow \qquad v \qquad \Rightarrow r^3\cos 3\theta+C$$

$$f(z)=u+iv=-r^3\sin 3\theta+ir^3\cos 3\theta+C$$

$$=i^2r^3\sin 3\theta+ir^3\cos 3\theta+C$$

$$= r^3 i\{\cos 3\theta + i\sin 3\theta\} + C$$

$$= r^3 i e^{3i\theta} + C = i(re^{i\theta})^3 + C$$

$$f(z) = iz^3 + C. \qquad \textbf{Ans.} \qquad \Big| \text{ As } z = re^{i\theta}$$

Example 19: If $f(z) = \dfrac{x^3 y(y - ix)}{x^6 + y^2}, z \neq 0$ and $f(0) = 0$, show that $\dfrac{f(z) - f(0)}{z} \to 0$ as $z \to 0$ along any radius vector but not as $z \to 0$ in any manner, *i.e.* $f(z)$ is not differentiable at $z = 0$.

Sol.
$$\frac{f(z) - f(0)}{z} = \frac{f(z) - 0}{z} = \frac{f(z)}{z}$$

$$= \frac{x^3 y(y - ix)}{(x^6 + y^2)z} = \frac{-ix^3 y(x + iy)}{(x^6 + y^2)z} = \frac{ix^3 yz}{(x^6 + y^2)z}$$

$$\frac{f(z) - f(0)}{z} = \frac{-ix^3 y}{(x^6 + y^2)}$$

Along the path $y = mx$, which is radius vector

$$\lim_{z\to 0} \frac{f(z) - f(0)}{z} = \lim_{z\to 0} \frac{-ix^3 y}{(x^6 + y^2)} = \lim_{x\to 0} \frac{-ix^3 mx}{x^2(x^4 + m^2)} = \lim_{x\to 0} \frac{-ix^2 m}{(x^4 + m^2)}$$

$$\lim_{z\to 0} \frac{f(z) - f(0)}{z} = 0 \quad \text{along any radius vector, as } z \to 0.$$

Now along the curve $y = x^3$

$$\lim_{z\to 0} \frac{f(z) - f(0)}{z} = \lim_{x\to 0} \frac{-ix^6}{(x^6 + x^6)} = -\frac{i}{2} \neq 0$$

Hence
$$\lim_{z\to 0} \frac{f(z) - f(0)}{z} \neq 0$$

along any path except radius vector. Therefore $f'(z)$ does not exit at $z = 0$, so $f(z)$ is not differentiable at $z = 0$.

Example 20: Prove that the function $f(z)$ defined by

$$f(z) = \frac{x^3(1 + i) - y^3(1 - i)}{x^2 + y^2}, \; z \neq 0 \text{ and } f(0) = 0$$

is continuous and Cauchy-Riemannian equations are satisfied at the origin, yet $f'(0)$ does not exist.

Sol. Here
$$u + iv = \frac{x^3(1 + i) - y^3(1 - i)}{x^2 + y^2} \qquad \Big| \text{As } w = u + iv = f(z)$$

$$u + iv = \frac{(x^3 - y^3) + i(x^3 + y^3)}{(x^2 + y^2)}$$

$$\Rightarrow \qquad u = \frac{x^3 - y^3}{x^2 + y^2}, \; v = \frac{x^3 + y^3}{x^2 + y^2} \quad (z \neq 0)$$

Here we see that both u and v are rational and finite for all values of $z \neq 0$, so u and v are continuous at all those point for which $z \neq 0$. Hence $f(z)$ is continuous where $z \neq 0$.

Next, at the origin

$$\frac{\partial u}{\partial x} = \lim_{x \to 0} \frac{u(x, 0) - u(0, 0)}{x} = \lim_{x \to 0} \frac{x - 0}{x} = 1$$

$$\frac{\partial u}{\partial y} = \lim_{y \to 0} \frac{u(0, y) - x(0, 0)}{y} = \lim_{y \to 0} \frac{-y - 0}{y} = -1$$

$$\frac{\partial v}{\partial x} = \lim_{x \to 0} \frac{v(x, 0) - v(0, 0)}{x} = \lim_{y \to 0} \frac{x - 0}{x} = 1$$

$$\frac{\partial v}{\partial y} = \lim_{y \to 0} \frac{v(0, y) - v(0, 0)}{y} = \lim_{y \to 0} \frac{y - 0}{y} = 1$$

From these values,

$$\frac{\partial u}{\partial x} = \frac{\partial v}{\partial y} \text{ and } \frac{\partial u}{\partial y} = -\frac{\partial v}{\partial x}$$

Hence C-R equations are satisfied at origin.

Again $$f'(0) = \lim_{z \to 0} \frac{f(z) - f(0)}{z} = \lim_{z \to 0} \frac{(x^3 - y^3) + i(x^3 + y^3)}{(x^2 + y^2)(x + iy)} \qquad |\text{As } f(0) = 0$$

Let $z \to 0$ along the line $y = mx$

$$\therefore \qquad f'(0) = \lim_{x \to 0} \frac{(x^3 - m^3 x^3) + i(x^3 + m^3 x^3)}{(x^2 + m^2 x^2)(x + imx)} = \lim_{x \to 0} \frac{(1 - m^3) + i(1 + m^3)}{(1 + m^2)(1 + im)}$$

$$\Rightarrow \qquad f'(0) = \frac{(1 - m^3) + i(1 + m^3)}{(1 + m^2)(1 + im)}$$

Since $f'(0)$ depends on m *i.e.,* $f'(0)$ is not unique. Therefore the function $f(z)$ is not analytic at origin *i.e.,* $(z = 0)$.

Example 21: If $f(z)$ is a regular function of z, prove that

$$\left(\frac{\partial^2}{\partial x^2} + \frac{\partial^2}{\partial y^2} \right) |f(z)|^2 = 4 |f'(z)|^2. \qquad \text{(U.T.U. 2008)}$$

Sol. We know that

$$f(z) = u + iv$$

$$\therefore \qquad |f(z)| = \sqrt{u^2 + v^2}$$

$$\Rightarrow \qquad |f(z)|^2 = u^2 + v^2$$

Differentiate partially w.r.t. x and y respectively, we get

$$\frac{\partial}{\partial x}\left\{|f(z)|^2\right\} = 2u\frac{\partial u}{\partial x} + 2v\frac{\partial v}{\partial x} 0$$

and
$$\frac{\partial}{\partial y}\left\{|f(z)|^2\right\} = 2u\frac{\partial u}{\partial y} + 2v\frac{\partial v}{\partial y}$$

$\therefore$
$$\frac{\partial^2}{\partial x^2}\left\{|f(z)|^2\right\} = \frac{\partial}{\partial x}\left\{2u\frac{\partial u}{\partial x} + 2v\frac{\partial v}{\partial x}\right\}$$

or
$$\frac{\partial^2}{\partial x^2}\left\{|f(z)|^2\right\} = 2\left\{u\frac{\partial^2 u}{\partial x^2} + \left(\frac{\partial u}{\partial x}\right)^2 + v\frac{\partial^2 v}{\partial x^2} + \left(\frac{\partial v}{\partial x}\right)^2\right\} \quad ...(1)$$

Similary
$$\frac{\partial^2}{\partial y^2}\left\{|f(z)|^2\right\} = 2\left\{u\frac{\partial^2 u}{\partial y^2} + \left(\frac{\partial u}{\partial y}\right)^2 + v\frac{\partial^2 v}{\partial y^2} + \left(\frac{\partial v}{\partial y}\right)^2\right\} \quad ...(2)$$

Adding (1) and (2), we get

$$\left(\frac{\partial^2}{\partial x^2} + \frac{\partial^2}{\partial y^2}\right)|f(z)|^2 = 2\left[u\left(\frac{\partial^2 u}{\partial x^2} + \frac{\partial^2 u}{\partial y^2}\right) + \left(\frac{\partial u}{\partial x}\right)^2 + \left(\frac{\partial u}{\partial y}\right)^2\right.$$

$$\left. + v\left(\frac{\partial^2 v}{\partial x^2} + \frac{\partial^2 v}{\partial y^2}\right) + \left(\frac{\partial v}{\partial x}\right)^2 + \left(\frac{\partial v}{\partial y}\right)^2\right] \quad ...(3)$$

Since $f(z)$ is analytic (regular) function, so we have

$$\frac{\partial u}{\partial x} = \frac{\partial v}{\partial y}, \ \frac{\partial u}{\partial y} = -\frac{\partial v}{\partial x} \text{ and } \frac{\partial^2 u}{\partial x^2} + \frac{\partial^2 u}{\partial y^2} = 0, \ \frac{\partial^2 v}{\partial x^2} + \frac{\partial^2 v}{\partial y^2} = 0 \quad (u \text{ and } v \text{ are harmonic})$$

$\therefore$ From (3), we get

$$\left(\frac{\partial^2}{\partial x^2} + \frac{\partial^2}{\partial y^2}\right)|f(z)|^2 = 2\left[0 + \left(\frac{\partial u}{\partial x}\right)^2 + \left(\frac{\partial v}{\partial x}\right)^2 + \left(\frac{\partial v}{\partial x}\right)^2 + \left(\frac{\partial u}{\partial x}\right)^2\right]$$

$$= 4\left[\left(\frac{\partial u}{\partial x}\right)^2 + \left(\frac{\partial v}{\partial x}\right)^2\right] = 4|f'(z)|^2 \qquad \Big| \text{ As } f'(z) = \frac{\partial u}{\partial x} + i\frac{\partial v}{\partial x}$$

$\Rightarrow$
$$\left(\frac{\partial^2}{\partial x^2} + \frac{\partial^2}{\partial y^2}\right)|f(z)|^2 = 4|f'(z)|^2.$$ **Proved.**

Example 22: Prove that $\left(\frac{\partial^2}{\partial x^2} + \frac{\partial^2}{\partial y^2}\right) = 4\frac{\partial^2}{\partial z \partial \bar{z}}$

Sol. We have $x + iy = z$ and $x - iy = \bar{z}$

so that $x = \frac{1}{2}(z + \bar{z}),\ y = -\frac{i}{2}(z - \bar{z})$

$$\therefore \quad \frac{\partial x}{\partial z} = \frac{1}{2},\ \frac{\partial x}{\partial \bar{z}} = \frac{1}{2},\ \frac{\partial y}{\partial z} = -\frac{i}{2} \text{ and } \frac{\partial y}{\partial \bar{z}} = \frac{i}{2}$$

Now
$$\frac{\partial}{\partial z} = \frac{\partial}{\partial x}\frac{\partial x}{\partial z} + \frac{\partial}{\partial y}\frac{\partial y}{\partial z} = \frac{1}{2}\left(\frac{\partial}{\partial x} - i\frac{\partial}{\partial y}\right)$$

and
$$\frac{\partial}{\partial \bar{z}} = \frac{\partial}{\partial x}\cdot\frac{\partial x}{\partial \bar{z}} + \frac{\partial}{\partial y}\cdot\frac{\partial y}{\partial \bar{z}} = \frac{1}{2}\left(\frac{\partial}{\partial x} + i\frac{\partial}{\partial y}\right)$$

Hence
$$\frac{\partial^2}{\partial z \partial \bar{z}} = \frac{1}{4}\left(\frac{\partial^2}{\partial x^2} + \frac{\partial^2}{\partial y^2}\right)$$

$$\Rightarrow \quad \left(\frac{\partial^2}{\partial x^2} + \frac{\partial^2}{\partial y^2}\right) = 4\frac{\partial^2}{\partial y \partial \bar{z}}.$$ **Proved**

Example 23: If $f(z)$ is analytic function of z, prove that

$$\left(\frac{\partial^2}{\partial x^2} + \frac{\partial^2}{\partial y^2}\right)|Rf(z)|^2 = 3|f'(z)|^2.$$

Sol. Since $f(z) = u + iv \quad \therefore \quad Rf(z) = u$ or $|Rf(z)|^2 = u^2$

Now
$$\frac{\partial}{\partial x}|Rf(z)|^2 = \frac{\partial}{\partial x}u^2 = 2u\frac{\partial u}{\partial x}$$

and
$$\frac{\partial^2}{\partial x^2}|Rf(z)|^2 = \frac{\partial}{\partial x}\left(2u\frac{\partial u}{\partial x}\right) = 2\left(\frac{\partial u}{\partial x}\right)^2 + 2u\frac{\partial^2 u}{\partial x^2} \qquad ...(1)$$

Similarly
$$\frac{\partial^2}{\partial y^2}|Rf(z)|^2 = \frac{\partial}{\partial y}\left(2u\frac{\partial u}{\partial y}\right) = 2\left(\frac{\partial u}{\partial y}\right)^2 + 2u\frac{\partial^2 u}{\partial y^2} \qquad ...(2)$$

Adding (1) and (2), we get

$$\left(\frac{\partial^2}{\partial x^2} + \frac{\partial^2}{\partial y^2}\right)|Rf(z)|^2 = 2\left[\left(\frac{\partial u}{\partial x}\right)^2 + \left(\frac{\partial u}{\partial y}\right)^2\right] + 2u\left[\frac{\partial^2 u}{\partial x^2} + \frac{\partial^2 u}{\partial y^2}\right]$$

$$= 2\left[\left(\frac{\partial u}{\partial x}\right)^2 + \left(\frac{\partial u}{\partial y}\right)^2\right] \qquad \text{(Since } u \text{ is harmonic)}$$

$$= 2\left[\left(\frac{\partial u}{\partial x}\right)^2 + \left(\frac{\partial u}{\partial x}\right)^2\right] \qquad \left(\text{As } \frac{\partial u}{\partial y} = -\frac{\partial u}{\partial x}\right)$$

$$\Rightarrow \quad \left(\frac{\partial^2}{\partial x^2}+\frac{\partial^2}{\partial y^2}\right)|Rf(z)|^2 = 2|f'(z)|^2 \qquad \left(\text{As } f'(z)=\frac{\partial u}{\partial x}+i\frac{\partial u}{\partial x}\right). \quad \textbf{Proved.}$$

Example 24: If $w = f(z)$ is a regular function of z, prove that

$$\left(\frac{\partial^2}{\partial x^2}+\frac{\partial^2}{\partial y^2}\right)\log|f'(z)| = 0$$

Sol. We know that $\left(\frac{\partial^2}{\partial x^2}+\frac{\partial^2}{\partial y^2}\right) = 4\frac{\partial^2}{\partial z\,\partial \bar{z}}$

$$\text{Here}\left(\frac{\partial^2}{\partial x^2}+\frac{\partial^2}{\partial y^2}\right)\{\log|f'(z)|\} = 4\frac{\partial^2}{\partial z\,\partial \bar{z}}\{\log|f'(z)|\}$$

$$= 4\frac{\partial^2}{\partial z\partial \bar{z}}\left\{\frac{1}{2}\log|f'(z)|^2\right\}$$

$$= 2\frac{\partial^2}{\partial z\partial \bar{z}}[\log\{f'(z)f'(\bar{z})\}]$$

$$= 2\frac{\partial^2}{\partial z\partial \bar{z}}[\log f'(z)+\log f'(\bar{z})]$$

$$= 2\frac{\partial}{\partial z}\left[\frac{f''(\bar{z})}{f'(\bar{z})}\right] = 0 \quad \Big| \text{ Since } f'(\bar{z}) \text{ and } f''(\bar{z}) \text{ are independent of } z.$$

Proved.

PROBLEM SET 2.1

1. An electrostatic field in the xy plane is given by the potential function $\phi = 3x^2y - y^3$, find the stream function ψ.

   $\left[\textbf{Ans. } -x^3+3xy^2+C\right]$

2. If $f(z) = u+iv$ is an analytic function of z and $u-v = \dfrac{\cos x+\sin x-e^{-y}}{2\cos x-2\cosh y}$. Prove that

   $f(z) = \frac{1}{2}\left[1-\cot\frac{z}{2}\right]$ where $f\left(\frac{\pi}{2}\right) = 0.$

3. If $u - v = (x - y)(x^2 + 4xy + y^2)$ and $f(z) = u+iv$ is an analytic function of $z = x + iy$, find $f(z)$ in terms of z.

   $\left[\textbf{Ans. } f(z) = -iz^3 + C\right]$

4. If $u = x^2 - y^2$, $v = \dfrac{-y}{x^2+y^2}$, then both satisfy Laplace's equation but $u + iv$ does not satisfy Laplace's equations.

5. Derive the Cauchy's Reimannian equations for an analytic function.

6. Prove that $\psi = \log[(x-1)^2 + (y-2)^2]$ is harmonic in every region which does not include the point (1, 2). Find a function ϕ such that $(\phi + i\psi)$ is an analytic function of the complex variable $z = x + iy$. Express $(\phi + i\psi)$ as a function of z.

$$\left[\textbf{Ans. } \phi = -2\tan^{-1}\left(\frac{y-2}{x-1}\right); 2i(z-1-2i)\right]$$

7. Derive the Cauchy-Reimannian equation for an analytic function in polar form and show that $\dfrac{\partial^2 u}{\partial r^2} + \dfrac{1}{r}\dfrac{\partial u}{\partial r} + \dfrac{1}{r^2}\dfrac{\partial^2 u}{\partial \theta^2} = 0.$

8. Determine k such that the function $f(z) = \dfrac{1}{2}\log(x^2+y^2) + i\tan^{-1}\dfrac{kx}{y}$ be an analytic function.

[**Ans.** $k = -1$]

9. Show that $u + iv = \dfrac{x - iy}{x - iy + a}$ where $a \neq 0$ is not an analytic function of $z = x + iy$ whereas $u - iv$ is such a function.

10. If the potential function is $\log(x^2 + y^2)$, find the flux function and the complex potential function.

$$\left[\textbf{Ans. } 2\tan^{-1}\frac{y}{x}, 2\log z + C\right]$$

11. Find the constants a, b and c so that the following are differentiable at every point.

(*i*) $f(z) = x + ay - i(bx + cy)$. [**Ans.** $a = b; c = -1$]

(*ii*) $f(z) = \cos x(\cosh y + a\sinh y) + i\sin x(\cosh y + b\sinh y)$. [**Ans.** $a = b = -1$]

12. Prove that an analytic function whose real part is constant is itself a constant.

13. Prove that an analytic function whose imaginary part is constant is itself a constant.

14. Given $v(x, y) = x^4 - 6x^2y^2 + y^4$ find $f(z)$. [**Ans.** $f(z) = iz^4 + c$]

15. Find $f(z)$ if $u + v = \dfrac{x}{x^2+y^2}$ given that $f(1) = 1$. $\left[\textbf{Ans. } \dfrac{1+i}{2z} + \dfrac{1-i}{2}\right]$

16. If $f(z) = u + iv$ is a regular function of z in any domain prove that

(*i*) $\left(\dfrac{\partial^2}{\partial x^2} + \dfrac{\partial^2}{\partial y^2}\right)|f(z)|^p = p^2\,|f(z)|^{p-2}\,|f'(z)|^2$

(*ii*) $\left(\dfrac{\partial^2}{\partial x^2} + \dfrac{\partial^2}{\partial y^2}\right)|u|^p = p(p-1)|u|^{p-2}\,|f'(z)|^2$

17. Find $f(z)$, when $u = \dfrac{2\cos x \cosh y}{\cos 2x + \cosh 2y}$. $\left[\textbf{Ans. } f(z) = \sec z\right]$

18. Find $f(z)$, when $v = x^3 - 3xy^2 + 3x^2 - 3y^2 + 1$. $\left[\textbf{Ans. } i(z^3 + 3z)\right]$

19. Find $f(z)$, when $u - v = (x - y)(x^2 + 4xy + y^2)$. $\left[\textbf{Ans. } f(z) = \cot z\right]$

20. Prove that $u(z)$ and $u(\bar{z})$ are simultaneously harmonic.

21. Prove that function $u(x, y)$ and $u(x^2 - y^2, 2xy)$ are simultaneously harmonic.

2.13 LINE INTEGRAL IN THE COMPLEX PLANE

Let a function $f(z) = u + iv$ be continuous in region R and ab, be two points in this region, then the integral $f(z)$ from a to b is defined as.

Let $f(z)$ is continuous at every point of the closed curve C let there be any partition ($a = z_0, z_1, z_2, \ldots z_{r-1}, z_r, \ldots z_n, = b$) of the curve C.

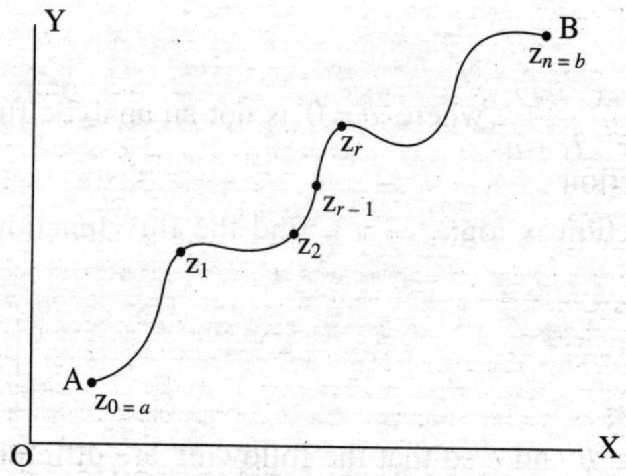

From the sum

$$S_n = \sum_{r=1}^{n} (z_r - z_{r-1}) f(\xi_r)$$

where $z_{r-1} \le \xi_r \le z_r$ if it exists uniquely, for the path and for every partition is called the integral of $f(z)$ over the curve C from a to b and defined as $\int_a^b f(z)dz$ or $\int_C f(z)dz$.

Thus
$$\int_a^b f(z)dz = \lim_{n\to\infty} \sum_{r=1}^{\infty} (z_r - z_{r-1}) f(\xi_r) \text{ where } z_{r-1} \le \xi_r \le z_r.$$

Example 1: Evaluate $\int_C |dz|$

Sol. We know that $\int_C f(z)dz = \lim_{n\to\infty} \sum_{r=1}^{n} (z_r - z_{r-1})\, f(\xi_r)$

$$\int_C |dz| = \lim_{n\to\infty}\left[\sum_{r=1}^{n} |z_r - z_{r-1}|\right] \qquad |f(z)| = 1, \therefore\ |f(\xi_r)| = 1$$

$$= \lim_{n\to\infty}\left[|z_1 - z_0| + |z_2 - z_1| + \ldots + |z_n - z_{n-1}|\right]$$

$$= \lim_{n\to\infty}\left[chord z_1 z_0 + chord z_2 z_1 + ... + chord z_n z_{n-1}\right] = \text{arc AB}. \quad \textbf{Ans.}$$

Example 2: Find $\int_C z\,dz$

Sol. We know that $\int_C f(z)dz = \lim_{n\to\infty}\sum_{r=1}^{n}(z_r - z_{r-1})\, f(\xi_r)$

Since $f(z) = z$, so $f(\xi_r) = z_r$

$$\therefore \qquad \int_C z\,dz = \lim_{n\to\infty}\left[\sum_{r=1}^{n}\{z_r(z_r - z_{r-1})\}\right] \qquad ...(1)$$

and

$$\int_C z\,dz = \lim_{n\to\infty}\left[\sum_{r=1}^{n}\{z_{r-1}(z_r - z_{r-1})\}\right] \qquad ...(2)$$

Adding (1) and (2), then we get

$$2\int_C z\,dz = \lim_{n\to\infty}\left[\sum_{r=1}^{n}\{(z_r^2 - z_{r-1}^2)\}\right]$$

$$= \lim_{n\to\infty}\left[\left(z_1^2 - z_2^2\right) + \left(z_2^2 - z_1^2\right) + \left(z_3^2 - z_2^2\right) + ... + \left(z_n^2 - z_{n-1}^2\right)\right]$$

$$= \lim_{n\to\infty}\left(z_n^2 - z_0^2\right) = b^2 - a^2 \qquad \left[\because \text{At } z_n = b,\ z_0 = a\right]$$

$$\Rightarrow \qquad \int_C z\,dz = \frac{1}{2}(b^2 - a^2). \quad \textbf{Ans.}$$

Example 3: Find the value of integral $\int_0^{1+i}(x - y + ix^2)dz$

(*i*) along the straight line from $z = 0$ to $z = 1 + i$.

(*ii*) along the real axis from $z = 0$ to $z = 1$ and then along a line parallel to the imaginary axis, from $z = 1$ to $z = 1 + i$.

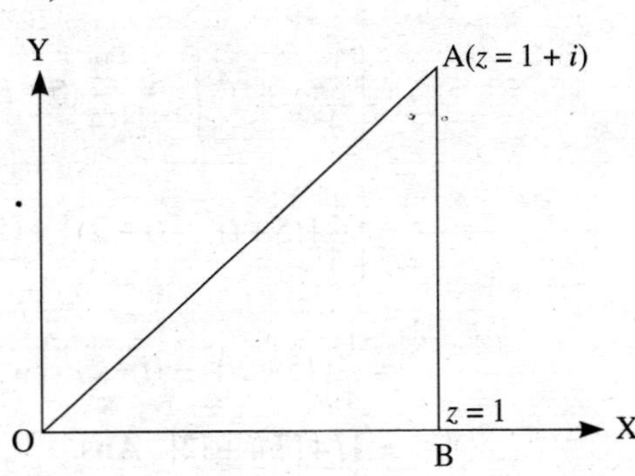

Sol. (*i*) Along the straight line OA

$$\int_{OA} f(z)dz = \int_{OA}(x - y + ix^2)(dx + idy)$$

The equation of line OA is $y = x$

$$\int_0^1 (x - x + ix^2)(dx + idx) = i(1+i)\int_0^1 x^2 dx = \frac{i(1+i)}{3} = \frac{i-1}{3}. \quad \textbf{Ans.}$$

(*ii*) $$\int_{OB} f(z)dz = \int_{OB} (x - y + ix^2)(dx + idy) = \int_0^1 (x - 0 + ix^2)\cdot dx \quad \Big| \text{ Along real line, } y = 0$$

$$\int_0^1 (x + ix^2)\,dx = \frac{1}{2} + \frac{i}{3}. \quad \textbf{Ans.}$$

and parallel to imaginary axis *i.e.* along to the line BA

$$\int_{BA} f(z)dz = \int_{BA} (x - y + ix^2)(dx + idy)$$

$$= \int_0^1 (1 - y + i)\,idy \quad \Big| \text{ As } x = 1 \text{ so } dx = 0$$

$$= \left[iy - \frac{iy^2}{2} - y\right]_0^1 = \left[i - \frac{i}{2} - 1\right] = -1 + \frac{i}{2}. \quad \textbf{Ans.}$$

Therefore the integral along OB and then along BA $= \frac{1}{2} + \frac{i}{3} - 1 + \frac{i}{2} = -\frac{1}{2} + \frac{5}{6}i.$ **Ans.**

Example 4: Evaluate $\int_{-2+i}^{5+3i} z^3 dz$.

Sol. Let the path of integration joining the points (–2, 1) and (5, 3) be along the curve made up of

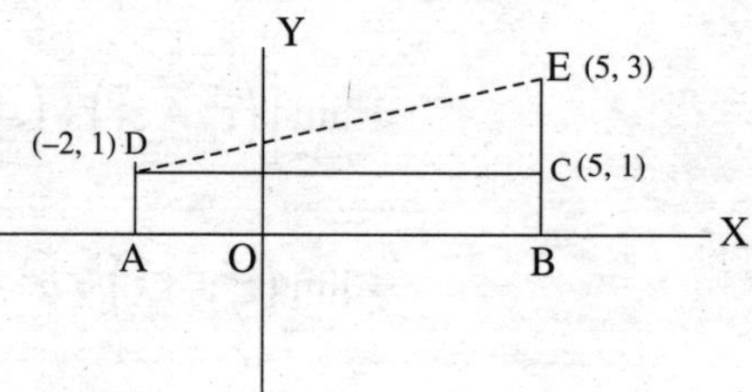

(*i*) a line parallel to real axis *i.e.* line DC parallel to x-axis from the point (–2, 1) to (5, 1) thus $z = x + iy \Rightarrow z = x + i(y = 1)$ and x-varies from –2 to 5.

(*ii*) A line parallel to imaginary axis from points (5,1) to (5,3) thus

$z = x + iy = 5 + iy$, $(x = 5)$ and y-varies from 1 to 3.

Hence $$\int_{-2+i}^{5+3i} z^3 dz = \int_{DC} z^3 dz + \int_{CE} z^3 dz = \int_{-2}^{5} (x+i)^3 dx + i\int_1^3 (5+iy)^3 dy$$

$$= \left[\frac{1}{4}(x+i)^4\right]_{-2}^{5} + \left[\frac{1}{4}(5+iy)^4\right]_1^3 = \left[\frac{1}{4}\{(5+i)^4 - (-2+i)^4\}\right] + \left[\frac{1}{4}\{(5+3i)^4 - (5+i)^4\}\right]$$

$$= \left[\frac{1}{4}\{(5+i)^4 - (i-2)^4 + (5+3i)^4 - (5+i)^4\}\right]$$

$$= \frac{1}{4}\left[(5+3i)^4 - (i-2)^4\right]$$

$= 1/4\,[34i + 13]$. **Ans.**

Example 5: Evaluate $\int_0^{2+i} (\bar{z})^2\, dz$, along the real axis from $z = 0$ to $z = 2$ and then along a line parallel to y-axis from $z = 2$ to $z = 2 + i$.

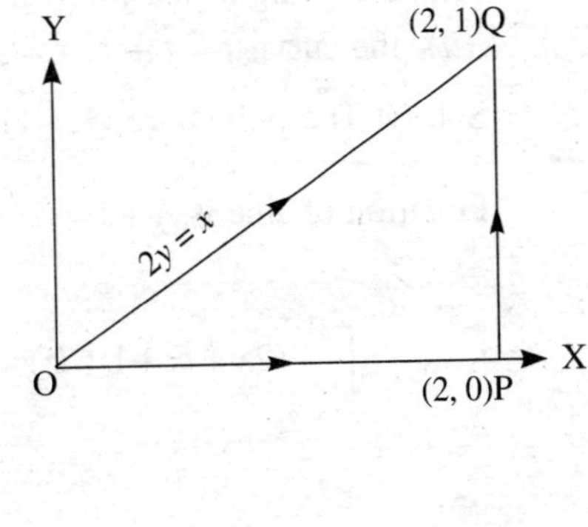

Sol. $\int_0^{2+i} (\bar{z})^2 dz = \int_{OP} (\bar{z})^2\, dz + \int_{PQ} (\bar{z})^2\, dz$

$$= \int_{OP} (x - iy)^2 (dx + idy) + \int_{PQ} (x - iy)^2 (dx + idy)$$

$$= \int_0^2 x^2 dx + \int_0^1 (2 - iy)^2 idy$$

$$= \left[\frac{x^3}{3}\right]_0^2 + \left[\frac{-(2-iy)^3}{3}\right]_0^1$$

$$= \frac{8}{3} + \frac{11}{3}i + 2 = \frac{11}{3}i + \frac{14}{3}.$$ **Ans.**

Example 6: Evaluate $\int_{1-i}^{2+3i} (z^2 + z)dz$ along the line joining the points (1, –1) and (2, 3).

Sol. The points are (1,–1) and (2,3) equation of line is

$$y + 1 = \frac{3+1}{2-1}(x - 1) \qquad \left|\; y - y_1 = \frac{y_2 - y_1}{x_2 - x_1}(x - x_1)\right.$$

$$y + 1 = 4x - 4 \;\Rightarrow y = 4x - 5$$

Along the line $y = 4x - 5$

$$\int_{1-i}^{2+3i} (z^2 + z)\, dz = \int_1^2 \{(x + iy)^2 + (x + iy)\}(dx + idy)$$

$$= \int_1^2 \left[\{x + i(4x - 5)\}^2 + \{x + i(4x - 5)\}\right](dx + i4dx)$$

$$= \int_1^2 \{(-15x^2 + 40x - 25) + i(8x^2 - 10x) + x + i(4x - 5)\}(1 + 4i)\, dx$$

$$= (1 + 4i)\int_1^2 (-15x^2 + 41x - 25) + i(8x^2 - 6x - 5)\, dx$$

$$= (1 + 4i)\int_1^2 (8i - 15)x^2 + (-6i + 41)x - (5i + 25)\, dx$$

$$= (1 + 4i)\left[(8i - 15)\frac{x^3}{3} + (-6i + 41)\frac{x^2}{2} - (5i + 25)x\right]_1^2$$

$$= (1 + 4i)\left[\frac{3}{2} + \frac{14i}{3}\right] = -\frac{103}{6} + \frac{32i}{3} = \frac{-103 + 64i}{6}.$$ **Ans.**

Example 7: Evaluate $\int_{1-i}^{2+i}(2x+iy+1)\,dz$, along

(*i*) the straight line joining $(1 - i)$ to $(2 + i)$

(*ii*) the curve $x = t + 1$, $y = 2t^2 - 1$.

Sol. (*i*) The points are (1, –1) and (2, 1)

Equation of line is $y+1=\frac{2}{1}(x-1) \Rightarrow y+1=2x-2 \Rightarrow y=2x-3$

Now $\int_{1-i}^{2+i}(2x+iy+1)(dz)=\int_{1}^{2}(2x+i2x-3i+1)(dx+2idx)$

$$=(1+2i)\int_{1}^{2}\{2x(i+1)+(1-3i)\}dx$$

$$=(1+2i)\Big[x^2(i+1)+(1-3i)x\Big]_1^2$$

$$=(1+2i)\big[4(i+1)+2(1-3i)-(i+1)-(1-3i)\big]$$

$= 8i + 4$. **Ans.**

(*ii*) $\int_{1-i}^{2+i}(2x+iy+1)dz=\int_{0}^{1}\Big[2(t+1)+i(2t^2-1)+1\Big](dt+4itdt)$

$$=\int_0^1(2t+2+2t^2i-i+1)(1+4it)\,dt$$

$$=\int_0^1(2t^2i+2t+3-i)dt+4i\int_0^1\{2t^3i+2t^2+(3-i)t\}\,dt$$

$$=\left[\frac{2}{3}t^3i+t^2+(3-i)t\right]_0^1+4i\left[\frac{t^4}{2}i+\frac{2}{3}t^3+(3-i)\frac{t^2}{2}\right]_0^1$$

$$=\left[\frac{2i}{3}+1+(3-i)\right]+4i\left[\frac{i}{2}+\frac{2}{3}+\frac{3}{2}-\frac{i}{2}\right]$$

$=4+\frac{25}{3}i.$ **Ans.**

Lower limit

$1=t+1 \Rightarrow t=0$

and $-1=2t^2-1 \Rightarrow t=0$

Upper limit

$2=t+1 \Rightarrow t=1$

and $-1=2t^2-1 \Rightarrow t^2=1 \Rightarrow t=1$

Example 8: Evaluate the integral $\int_C \log z\,dz$ where C is the unit circle $|z|=1$.

Sol. Since C be a circle of radius unity

So $z=e^{i\theta}$

$dz=ie^{i\theta}d\theta$

$\therefore$ $\int_C \log_e z\,dz=\int_0^{2\pi}\log_e e^{i\theta}\times ie^{i\theta}.d\theta=\int_0^{2\pi}i^2\theta e^{i\theta}d\theta$

$\log_e e^{i\theta}=i\theta$

$i^2=-1$

$$= -\int_0^{2\pi} \theta e^{i\theta} d\theta = -\left[\left\{\frac{\theta e^{i\theta}}{i}\right\}_0^{2\pi} - \frac{1}{i}\int_0^{2\pi} e^{i\theta} d\theta\right]$$

$$= -\left[\left\{\frac{2\pi e^{2\pi i}}{i} - 0\right\} + i\left\{\frac{e^{i\theta}}{i}\right\}_0^{2\pi}\right]$$

$$= -\left[\frac{2\pi}{i} + (e^{2\pi i} - e^0)\right]$$

$$= 2\pi i + (1-1) = 2\pi i. \quad \textbf{Ans.}$$

Example 9: Evaluate $\int_C |z|^2\, dz$, around the square with vertices at (0,0), (1,0), (1,1) and (0,1).

Sol. $$\int_C |z|^2\, dz = \int_{OA} (x^2+y^2)(dx+idy) + \int_{AB} (x^2+y^2)(dx+idy)$$

$$+\int_{BC} (x^2+y^2)(dx+idy) + \int_{CO} (x^2+y^2)(dx+idy)$$

$$\int_C |z|^2\, dz = \int_0^1 (x^2+0)dx + \int_0^1 (1+y^2)idy + \int_1^0 (x^2+1)dx + \int_1^0 y^2 idy$$

$$= \int_0^1 x^2 dx + i\int_0^1 dy + i\int_0^1 y^2 dy - \int_0^1 x^2 dx - \int_0^1 dx - i\int_0^1 y^2 dy$$

$$= -\int_0^1 dx + i\int_0^1 dy = -[x]_0^1 + i[y]_0^1 = (-1) + i. \quad \textbf{Ans.}$$

2.14 SIMPLE CLOSED CURVE

A curve is called simple closed curve if it does not cross itself and the region within is called simply connected region.

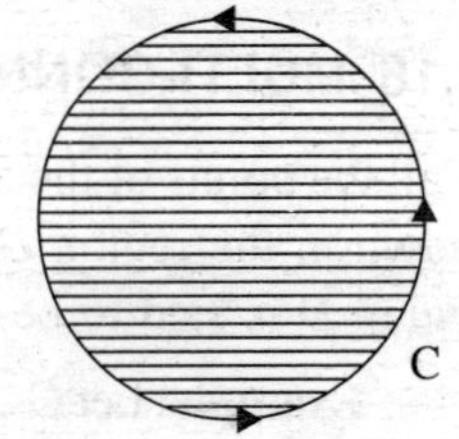

Domain (or region)

The word domain means connex (or connected). A set of points in the Argand diagram is said to be connex if any two of points can be joined by a continuous arc which consists only of the points of the set.

Thus an open connex set of points mean an open domain or open region. If the boundary points of the set are also added to an open domain, it is then called a closed domain.

Jordon arc

A Jordan arc is defined as a continuous arc without multiple point.

Contour

A continuous chain of a finite number of regular arc is called contour.

Closed contour

If the contour is closed and does not intersect itself, then it is called closed contour.

Example: Boundaries of circles, triangles and rectangles.

2.15 CAUCHY'S THEOREM (OR CAUCHY'S INTEGRAL THEOREM)

If $f(z)$ is a analytic function of z and $f'(z)$ is continuous at each point within and on a closed contour C, then

$$\int_C f(z)dz = 0 \quad \text{or} \quad \oint_C f(z)dz = 0$$

Proof: Let D be the region which consists of all points within and on the contour C.

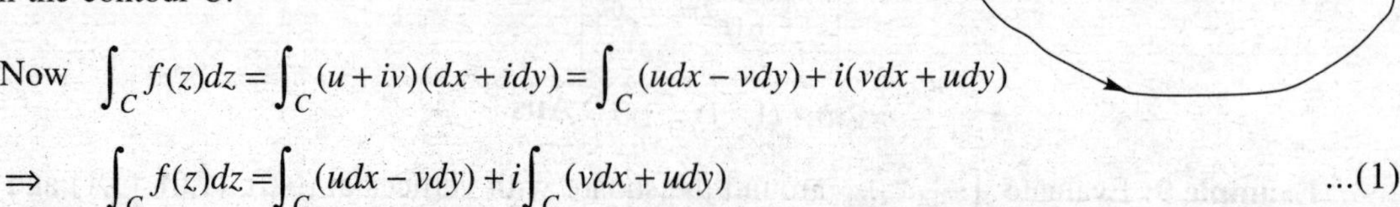

Now $\displaystyle\int_C f(z)dz = \int_C (u+iv)(dx+idy) = \int_C (udx - vdy) + i(vdx + udy)$

$$\Rightarrow \quad \int_C f(z)dz = \int_C (udx - vdy) + i\int_C (vdx + udy) \qquad \ldots(1)$$

Since $f'(z)$ is continuous in the region D, therefore u_x, u_y, v_x, v_y will be continuous in D. This is the condition for satisfied of green's theorem. So by Green's theorem from equation (1)

$$\int_C f(z)dz = \iint_D \left(-\frac{\partial v}{\partial x} - \frac{\partial u}{\partial y}\right)dx\,dy + i\iint_D \left(\frac{\partial u}{\partial x} - \frac{\partial v}{\partial y}\right)dx\,dy$$

$$\int_C f(z)dz = \iint_D \left(\frac{\partial u}{\partial y} - \frac{\partial u}{\partial y}\right)dx\,dy + i\iint_D \left(\frac{\partial v}{\partial y} - \frac{\partial v}{\partial y}\right)dx\,dy$$

$$\int_C f(z)dz = 0$$

Green's Theorem

$$\int_C (Pdx + Qdy) = \iint_D \left(\frac{\partial Q}{\partial x} - \frac{\partial P}{\partial y}\right)dx\,dy$$

C-R Equations

$$\frac{\partial u}{\partial x} = \frac{\partial v}{\partial y} \quad \text{and} \quad \frac{\partial u}{\partial y} = -\frac{\partial v}{\partial x}$$

Finally $\boxed{\int_C f(z)dz = 0}$.

2.16 MULTI-CONNECTED REGION

If all the points of the area bounded by two or more closed curves drawn in the region D, are the points of the region D, then the region D is said to be multi-connected region..

Example: Let $C_1, C_2, \ldots\ldots\ldots$.. are the closed curves are draw in a region D.

If all the points, of the area lying between the closed curves $C, C_1, C_2, \ldots\ldots\ldots$..the area which is interior to C and exterior the the other curves $C_1, C_2, \ldots\ldots\ldots$..are points of the region D, the region D is said multiconnected region.

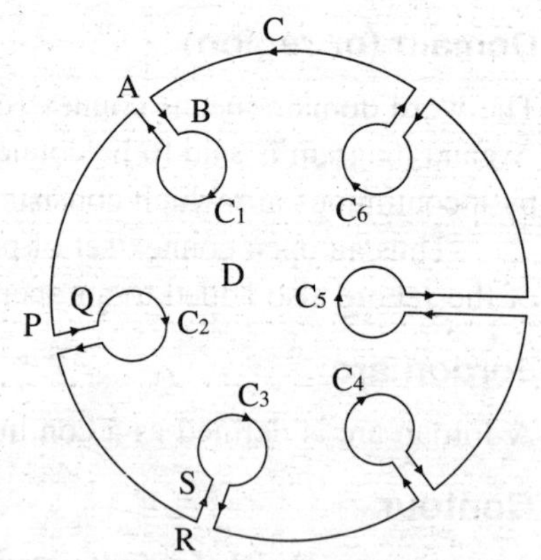

Cross Cut: The lines drawn in a multiconnected region, without intersecting any one of the curves, which make a multiconnected region in a simply-connected are called cross cut or cuts, the lines AB, PQ.,,,,, etc. are called cuts or cross cut.

2.17 EXTENSION OF CAUCHY'S THEOREM TO MULTI-CONNECTED REGION

If C is a closed curve and C_1, C_2,..........etc. are the other closed curves which lie inside C, and if a function $f(z)$ is analytic in the region but between these curves, and continuous on C, then

$$\int_C f(z)dz = \int_{C_1} f(z)dz + \int_{C_2} f(z)dz + \int_{C_3} f(z)dz + \ldots\ldots$$

where integral along each curve is taken anticlockwise direction.

Proof: From figure (Art. 2.18), we see that the rounding over C is anticlockwise whenever the rounding over each of the curves C_1, C_2, C_3,..........etc. is clockwise therefore by Cauchy's theorem.

$$\int_C f(z)dz + \int_{AB} f(z)dz - \int_{C_1} f(z)dz + \int_{BA} f(z)dz + \int_{PQ} f(z)dz - \int_{C_2} f(z)dz + \cdot \int_{QP} f(z)dz + \ldots\ldots = 0$$

But $$\int_{AB} f(z)dz = -\int_{BA} f(z)dz$$

$$\int_{PQ} f(z)dz = -\int_{PQ} f(z)dz \text{....etc.}$$

$$\int_C f(z)dz - \int_{C_1} f(z) - \int_{C_2} f(z) - \int_{C_3} f(z) \ldots\ldots = 0$$

or $$\int_C f(z)dz = \int_{C_1} f(z) + \int_{C_2} f(z) + \int_{C_3} f(z) + \ldots\ldots$$

where all integrals are taken in anticlockwise direction.

Remark: If there is only curves C_1 with in C.

Then $$\int_C f(z)dz + \int_{AB} f(z)dz + \int_{BA} f(z)dz - \int_{C_1} f(z) \cdot dz = 0$$

$$\Rightarrow \quad \int_C f(z)dz - \int_{C_1} f(z) \cdot dz = 0$$

$$\int_C f(z)dz = \int_{C_1} f(z) \cdot dz.$$

2.18 CAUCHY'S INTEGRAL FORMULA

If $f(z)$ is analytic within and on a closed conour C and a is any point within C, then

$$f(a) = \frac{1}{2\pi i} \int_C \frac{f(z)dz}{z-a}$$

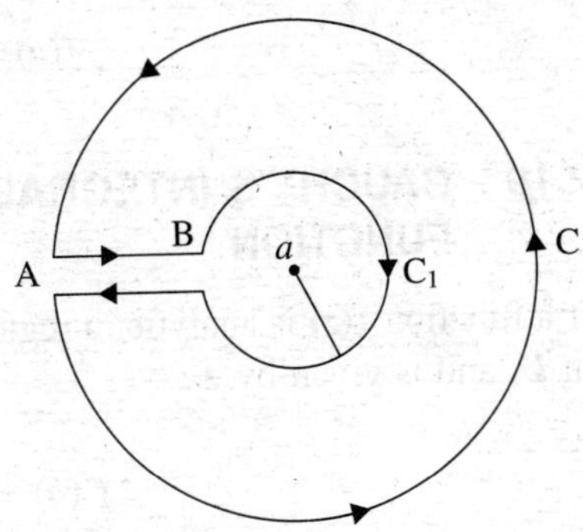

Proof: We describe a small circle about the point 'a' of a radius r lying entirely within C.

Let us consider a function $\dfrac{f(z)}{z-a}$, which is analytic except $z = a$

Now by Cauchy's theorem for multiconnected region.

$$\int_C f(z)dz - \int_{C_1} f(z) \cdot dz = 0$$

$$\Rightarrow \qquad \int_C f(z)dz = \int_{C_1} f(z)\cdot dz$$

or
$$\int_C \frac{f(z)}{(z-a)}dz = \int_{C_1} \frac{f(a)}{(z-a)}dz$$

$$\Rightarrow \qquad \int_C \frac{f(z)}{(z-a)}dz - \int_{C_1} \frac{f(a)}{(z-a)}dz = \int_{C_1} \frac{f(z)-f(a)}{(z-a)}dz$$

$$\Rightarrow \qquad \int_C \frac{f(z)}{(z-a)}dz - f(a)\times 2\pi i = \int_{C_1} \frac{f(z)-f(a)}{(z-a)}dz \qquad \ldots(1)$$

$$z-a = re^{i\theta}$$
$$dz = re^{i\theta}d\theta$$
$$\int_C \frac{1}{(z-a)}dz = \int_0^{2\pi} \frac{re^{i\theta}id\theta}{re^{i\theta}}$$
$$= 2\pi i$$

and
$$\int_{C_1} \frac{f(z)-f(a)}{z-a}dz$$

$$= \int_0^{2x} \frac{f(a+re^{i\theta})-f(a)}{re^{i\theta}} ire^{i\theta}d\theta$$

$$= i\int_0^{2x} \{f(a+re^{i\theta} - f(a)\}\, d\theta$$

Let $r \to 0$ (at $z = a$)

$$\therefore \qquad \int_{C_1} \frac{f(z)-f(a)}{z-a}dz = i\int_0^{2\pi} \{f(a)-f(a)\}\, d\theta = 0 \qquad \ldots(2)$$

From (1) and (2), we get

$$\int_C \frac{f(z)}{(z-a)}dz - 2\pi i f(a) = 0$$

or
$$\boxed{f(a) = \frac{1}{2\pi i}\int_C \frac{f(z)}{(z-a)}dz}$$

Remark: For multiconnected region we can prove the integral from

$$f(a) = \frac{1}{2\pi i}\int_C \frac{f(z)dz}{(z-a)} - \frac{1}{2\pi i}\int_{C_1} \frac{f(z)dz}{(z-a)} - \frac{1}{2\pi i}\int_{C_2} \frac{f(z)dz}{(z-a)} - \ldots\text{ etc}$$

2.19 CAUCHY'S INTEGRAL FORMULA FOR THE DERIVATIVE OF AN ANALYTIC FUNCTION

If a function $f(z)$ is analytic in region D, then its derivative at any any point $z = a$ of D is also analytic in D, and is given by

$$f'(a) = \frac{1}{2\pi i}\int_C \frac{f(z)}{(z-a)^2}dz$$

where C is any closed contour in D surrounding the point $z = a$.

By Cauchy's integral formula, we have

$$f(a) = \frac{1}{2\pi i}\int_C \frac{f(z)}{(z-a)}dz$$

Differentiating w.r.to, a, we get

$$f'(a) = \frac{1}{2\pi i}\int_C \frac{\partial}{\partial a}\frac{f(z)}{(z-a)}dz$$

$\frac{d}{da}\int f(ax)\,dx = \int \frac{\partial}{\partial a} f(ax)\,dx$
Leibnitz' rule of differentiation under sign of integration

$\Rightarrow$
$$\boxed{f'(a) = \frac{1}{2\pi i}\int_C \frac{f(z)dz}{(z-a)^2}}$$

2.20 HIGHER ORDER DERIVATIVES

If a function $f(z)$ is analytic within and on a closed contour C and a is any point within C then derivatives of all orders are analytic and are given by $f^n(a) = \frac{n!}{2\pi i}\int_C \frac{f(z)dz}{(z-a)^{n+1}}$

Proof: We know that

$$f'(a) = \frac{1}{2\pi i}\int_C \frac{f(z)dz}{(z-a)^2} \quad ...(1)$$

Differentiating equation (1), w.r.to 'a', we get

$$f''(a) = \frac{1}{2\pi i}\int_C \frac{\partial}{\partial a}\left\{\frac{f(z)}{(z-a)^2}\right\}dz = \frac{1\cdot 2}{2\pi i}\int_C \frac{f(z)}{(z-a)^3}dz$$

or
$$f''(a) = \frac{\underline{|2}}{2\pi i}\int_C \frac{f(z)}{(z-a)^3}dz$$

....

....

$$\boxed{f^n(a) = \frac{n!}{2\pi i}\int_C \frac{f(z)dz}{(z-a)^{n+1}}}$$

This implies that derivatives of $f(z)$ of all orders are analytic if $f(z)$ is analytic.

2.21 CAUCHY'S INEQUALITY

If $f(z)$ is analytic within and on a circle C given by

$$|z-a| = R \text{ and if } |f(z)| \le M \text{ for every } z \text{ on } C, \text{ then } |f^n(a)| \le \frac{Mn!}{R^n}$$

Proof: We know that $f^n(a) = \frac{n!}{2\pi i}\int_C \frac{f(z)dz}{(z-a)^{n+1}}$

$$\therefore \quad \left|f^n(a)\right| = \left|\frac{n!}{2\pi i}\int_C \frac{f(z)dz}{(z-a)^{n+1}}\right|$$

$$\leq \frac{n!}{|2\pi i|}\int_C \frac{|f(z)||dz|}{|z-a|^{n+1}}$$

$$\leq \frac{Mn!}{2\pi}\int_C \frac{|dz|}{R^{n+1}}$$

$$\leq \frac{Mn!}{2\pi}\int_0^{2\pi} \frac{Rd\theta}{R^n R}$$

$$\leq \frac{Mn!}{2\pi}\times\frac{2\pi}{R^n}$$

As $\left|\int_C f(z)dz\right| \leq \int_C |f(z)||dz|$

$\because \; z-a = Re^{i\theta}$

$\therefore \quad dz = Rie^{i\theta}d\theta$

and $|dz| = Rd\theta$

$$\Rightarrow \quad \boxed{\left|f^n(a)\right| \leq \frac{Mn!}{R^n}}$$

2.22 LIOUVILLE'S THEOREM

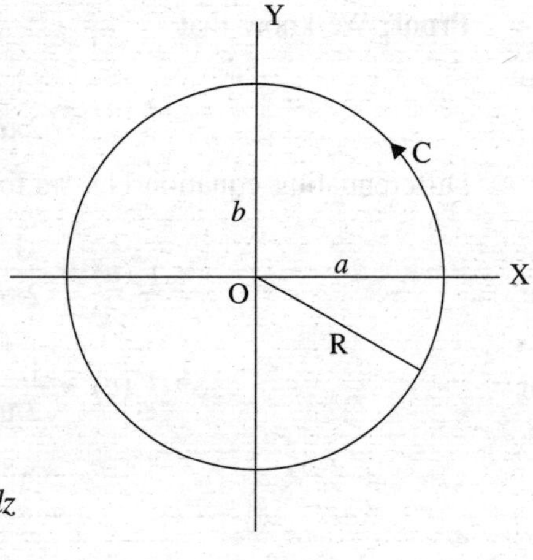

If a function $f(z)$ is analytic for all finite values of z, and is bounded, is constant.

Or

If an entire function is bounded for all values of z, then it is constant.

Proof: Let a, b be any two points of z-plane.

We take contour C to be a large circle, with its center at origin and radius R, enclosing the points a, b. Now by Cauchy's integral formula

$$f(a) = \frac{1}{2\pi i}\int_C \frac{f(z)}{(z-a)}dz, f(b) = \frac{1}{2\pi i}\int_C \frac{f(z)}{(z-b)}dz$$

$$\therefore \quad f(a) - f(b) = \frac{1}{2\pi i}\int_C \left\{\frac{1}{z-a} - \frac{1}{z-b}\right\} f(z)dz = \frac{1}{2\pi i}\int_C \frac{(a-b)}{(z-a)(z-b)} f(z)dz$$

$$\text{Now} \quad |f(a) - f(b)| = \left|\frac{(a-b)}{2\pi i}\int_C \frac{f(z)dz}{(z-a)(z-b)}\right|$$

$$\leq \frac{(a-b)}{2\pi}\int_C \frac{|f(z)||dz|}{|(z-a)(z-b)|}$$

$$\leq \frac{M|(a-b)|}{2\pi}\int_0^{2\pi} \frac{Rd\theta}{(|z|+|a|)(|z|+|b|)}$$

Since $f(z)$ is bounded so $|f(z)| \leq M$ and $z = Re^{i\theta} \Rightarrow |dz| = Rd\theta$

$$\leq \frac{M|(a-b)|2\pi R}{2\pi(R+|a|)(R+|b|)} \leq \frac{M|a-b|}{(R+|a|)(R+|b|)} \to 0 \text{ as } R \to \infty$$

$f(a) - f(b) = 0$ or $f(a) = f(b)$ showing thereby $f(z)$ is constant.

Aliter: By Cauchy's inequality, we have

$$|f^n(a)| \leq \frac{M\lfloor n}{R^n}$$

or $$|f^n(z)| \leq \frac{M\lfloor n}{R^n} \quad \text{(For any finite value of } z\text{)}$$

for $$n = 1$$

$$|f'(z)| \leq \frac{M}{R^n}$$

As $R \to \infty, f'(z) = 0 \Rightarrow \boxed{f(z) = \text{constant}}$

Rouche's Theorem (without proof) : If $f(z)$ and $g(z)$ are two analytic function with in and on closed contour C and $|g(z)| < |f(z)|$ *i.e.* $\frac{|g(z)|}{|f(z)|} < 1$, then $f(z)$ and $f(z) + g(z)$ have same number of zeros inside C.

2.23 THE FUNDAMENTAL THEOREM OF ALGEBRA

Every polynomial of degree n in the complex plane has at least one zero.

Or

Every polynomial of degree n has exactly n zeros.

Proof: Let us consider a polynomial of degree n

$$F(z) = a_0 + a_1 z + a_2 z^2 + \ldots + a_n z^n \text{ where } a_n \neq 0$$

Suppose that $F(z)$ has no zeros, then $F(z) \neq 0$ for any value of z.

Now $f(z) = \frac{1}{F(z)}$ is analytic everywhere. Also $f(z) \to 0$ as $|z| \to \infty$ so that $|f(z)|$ is bounded for all z. Liouville's theorem states that a function $f(z)$ which is analytic everywhere and bounded is constant. This is a contradiction since $F(z)$ is not a constant. Hence $F(z)$ is zero for at least one value of z.

Aliter: Let us consider the polynomial of degree n

$$F(z) = a_0 + a_1 z + a_2 z^2 + \ldots + a_n z^n, \quad \text{where } a_n \neq 0$$

Let $$f(z) = a_n z^n, g(z) = a_0 + a_1 z + a_2 z^2 + \ldots + a_{n-1} z^{n-1}$$

Let C be a circle of radius R with center origin *i.e.* $|z| = R$

Now $$\frac{|g(z)|}{|f(z)|} = \frac{|a_0 + a_1 z + a_2 z^2 + \ldots + a_n z^n|}{|a_n z^n|}$$

$$\leq \frac{|a_0| + |a_1||z| + |a_2||z|^2 + ... + |a_n||z|^n}{|a_n||z|^n}$$

$$\leq \frac{|a_0| + |a_1| R + |a_2| R^2 + ... + |a_n| R^n}{|a_n| R^n}$$

$$\frac{|g(z)|}{|f(z)|} \leq \frac{|a_0| R^{n-1} + |a_1| R^{n-1} + |a_2| R^{n-1} + ... + |a_n| R^{n-1}}{|a_n| R^n}$$

Since R is very large so we can take R^{n-1} in numerator for simplicity

$$= \frac{|a_0| + |a_1| + |a_2| + ... + |a_n|}{|a_n| R} < 1 \text{ if } R \text{ is very large}$$

$$\Rightarrow \quad \frac{|g(z)|}{|f(z)|} < 1 \quad \text{on } C$$

Hence, by Rouche's theorem the given polynomial $g(z) + f(z)$ has the same number of zeros as $f(z)$. But $f(z)$, *i.e.* $a_n z^n$ has clearly n zeros located at $z = 0$, therefore, $g(z) + f(z)$ also has n zeros.

Example 1: Prove that all the roots of $z^7 - 5z^3 + 12 = 0$ lie between the circles $|z| = 1$ and $|z| = 2$.

Sol. Let C_1 be the circle $|z| = 1$ and C_2 be the circle $|z| = 2$. The polynomial is $z^7 - 5z^3 + 12 = 0$

For the circle $|z| = 1$

Let $f(z) = 12$

and $$g(z) = z^7 - 5z^3$$

$$\left|\frac{g(z)}{f(z)}\right| = \frac{|z^7 - 5z^3|}{|12|} \leq \frac{|z|^7 + 5|z|^3}{12} = \frac{6}{12} = \frac{1}{2}$$

$$\left|\frac{g(z)}{f(z)}\right| \leq 1 \quad \text{on } C_1$$

Hence, by Rouche's theorem $f(z) + g(z) = z^7 - 5z^3 + 12$ has the same number of zeros inside C_1 as $f(z) = 12$. But $f(z)$ has no zeros inside C_1. It means that $z^7 - 5z^3 + 12$ has no zeros inside C_1.

Now on the circle $|z| = 2$

Let $g(z) = -5z^3 + 12$ and $f(z) = z^7$.

$$\left|\frac{g(z)}{f(z)}\right| = \frac{|-5z^3 + 12|}{|z|^7} \leq \frac{5|z|^3 + 12}{z^7} = \frac{5 \times 8 + 12}{128} = \frac{52}{128} < 1$$

$$\left|\frac{g(z)}{f(z)}\right| < 1$$

Hence, by Rouche's theorem $f(z) + g(z) = z^7 - 5z^3 + 12$ has the same number of zeros as $f(z) = z^7$. But all the seven zeros of z^7 lie inside $|z| = 2$. Hence, all the seven zeros of $z^7 - 5z^3 + 12$ lie inside $|z| = 2$.

Hence, all the roots of the equation $z^7 - 5z^3 + 12 = 0$ lie inside $|z| = 2$ but outside $|z| = 1$ *i.e.*, all the roots of the given polynomial lie between $|z| = 1$ and $|z| = 2$.

Example 2: If $a > e$, use Rouche's theorem (fundamental theorem of algebra) to prove that the equation $e^z = az^n$ has n roots inside the circle $|z| = 1$.

Sol. The given equation is $e^z = az^n \Rightarrow e^z - az^n = 0$

Let $f(z) = -az^n, g(z) = e^z$

$$\frac{|g(z)|}{|f(z)|} = \frac{|e^z|}{|az^n|} = \frac{|e^z|}{a|z^n|}$$

$$= \frac{\left|1 + z + \frac{z^2}{2!} + \ldots\right|}{a|z^n|}$$

$$\leq \frac{1 + |z| + \left|\frac{z^2}{2!}\right| + \ldots}{a|z^n|}$$

$$= \frac{1 + 1 + \frac{1}{2!} + \ldots}{a} = \frac{e}{a}$$

But $a > e$ so $\frac{e}{a} < 1$

$$\Rightarrow \quad \left|\frac{g(z)}{f(z)}\right| < 1 \text{ on } C \text{ is satisfied.}$$

Hence $f(z) + g(z) = e^z - az^n$ has the same number of zeros as $f(z) = -az^n$. But az^n has n zeros all locate at $z = 0$ inside $|z| = 1$. Hence, $e^z - az^n$ has n zeros inside $|z| = 1$.

Example 3: Use Cauchy's integral formula to evaluate $\int_C \frac{\sin \pi z^2 + \cos \pi z^2}{(z-1)(z-2)} dz$, where C is the circle $|z| = 3$. *(U.T.U. 2008)*

Sol: The points are $z - 1 = 0$ and $z - 2 = 0$

$$z = 1, 2$$

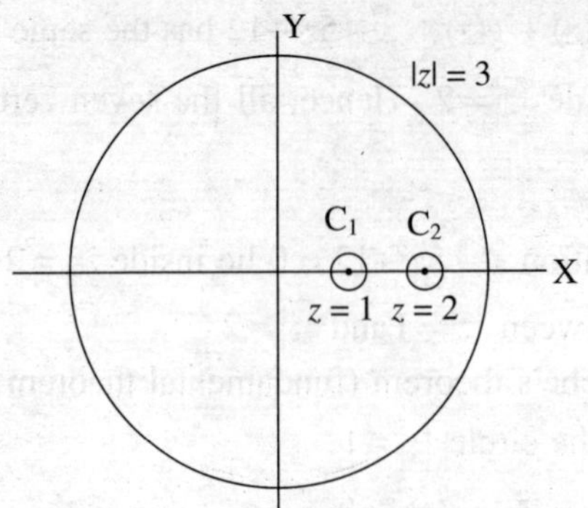

Both the points $z = 1, 2$ lies within the circle $|z| = 3$

$$\therefore \quad \int_C \frac{\sin \pi z^2 + \cos \pi z^2}{(z-1)(z-2)}\, dz$$

$$= \int_{C_1} \frac{\left(\frac{\sin \pi z^2 + \cos \pi z^2}{z-2}\right)}{z-1}\, dz + \int_{C_2} \frac{\left(\frac{\sin \pi z^2 + \cos \pi z^2}{z-1}\right)}{z-2}\, dz$$

$$\int_C \frac{\sin \pi z^2 + \cos \pi z^2}{(z-1)(z-2)}\, dz = 2\pi i \left[\frac{\sin \pi z^2 + \cos \pi z^2}{z-2}\right]_{z=1} + 2\pi i \left[\frac{\sin \pi z^2 + \cos \pi z^2}{z-1}\right]_{z=2}$$

$$= 2\pi i \left[\frac{\sin \pi + \cos \pi}{-1}\right] + 2\pi i \left[\frac{\sin 4\pi + \cos 4\pi}{1}\right]$$

$$= 2\pi i + 2\pi i$$

$$= 4\pi i. \quad \textbf{Ans.}$$

Example 4: (*i*) Find the value of $\int_C \frac{2z^2 + z}{z^2 - 1}\, dz$, where C is the circle of unit radius with center at $z = 1$.

(*ii*) Evaluate by Cauchy's integral formula $\int_C \frac{dz}{z(z + \pi i)}$, where C is $|z + 3i| = 1$.

Sol. (*i*) For the poles or points $(z^2 - 1) = 0 \Rightarrow (z-1)(z+1) = 0$

We see that only $z = 1$ lies

Within the circle so we shall consider only one pole $z = 1$,

By Cauchy's integral formula

$$\int_C \frac{2z^2 + z}{z^2 - 1}\, dz = \int_C \frac{2z^2 + z}{(z+1)(z-1)}\, dz = \int_C \frac{\left(\frac{2z^2 + z}{z+1}\right)}{z-1}\, dz$$

$$= 2\pi i \left[\frac{2z^2 + z}{z+1}\right]_{z=1} = 2\pi i \left[\frac{2+1}{2}\right] = 3\pi i. \quad \textbf{Ans.}$$

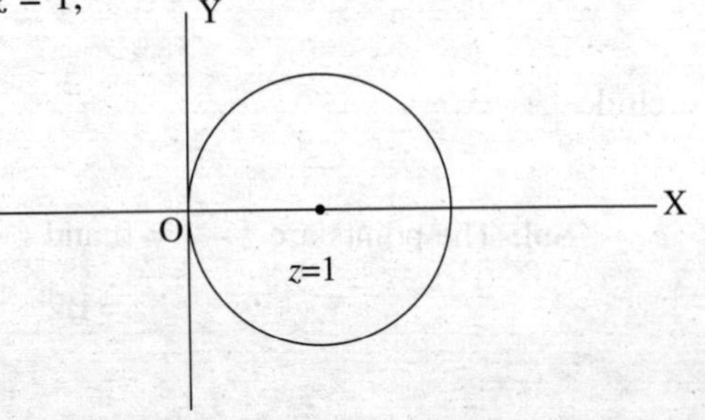

(*ii*) For the poles $z(z+\pi i)=0 \;\Rightarrow z=0, -\pi i$

$\Rightarrow\; z=-\pi i$, lie within the circle C therefore,

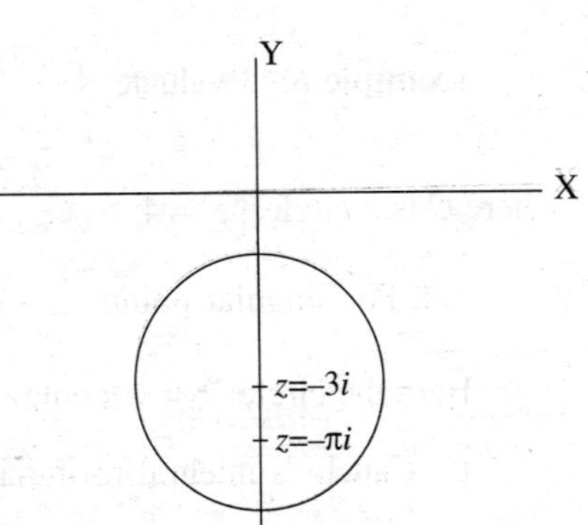

$$\int_C \frac{dz}{z(z+\pi i)} = \int_C \frac{\dfrac{dz}{z}}{(z+\pi i)} = 2\pi i\left[\frac{1}{z}\right]_{z=-\pi i}$$

$$= 2\pi i\left(-\frac{1}{\pi i}\right) = -2. \quad \textbf{Ans.}$$

Example 5: Evaluate $\displaystyle\int_C \frac{z^2+1}{z^2-1}\,dz$, where C is circle,

(*i*) $|z| = \dfrac{3}{2}$ (*ii*) $|z-1| = 1$ (*iii*) $|z| = \dfrac{1}{2}$.

Sol. (*i*) Here both the pole $z = 1$ and $z = -1$ lie within the circle $|z| = \dfrac{3}{2}$

$\therefore$ by Cauchy's integral formula, we have

$$\int_C \frac{z^2+1}{z^2-1}\,dz = \int_C \frac{\left(\dfrac{z^2+1}{z-1}\right)}{z+1}\,dz + \int_C \frac{\left(\dfrac{z^2+1}{z+1}\right)}{z-1}\,dz = 2\pi i\left\{\frac{z^2+1}{z-1}\right\}_{z=-1} + 2\pi i\left\{\frac{z^2+1}{z+1}\right\}_{z=1}$$

$$= 2\pi i\left\{\frac{2}{-2}\right\} + 2\pi i\left\{\frac{2}{2}\right\} = -2\pi i + 2\pi i = 0. \quad \textbf{Ans.}$$

(*ii*) Here only the pole $z = 1$, lies with in the circle $|z-1| = 1$

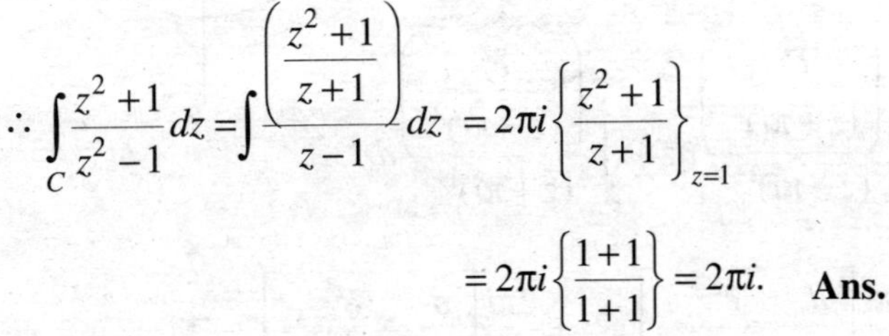

$$\therefore \int_C \frac{z^2+1}{z^2-1}\,dz = \int \frac{\left(\dfrac{z^2+1}{z+1}\right)}{z-1}\,dz = 2\pi i\left\{\frac{z^2+1}{z+1}\right\}_{z=1}$$

$$= 2\pi i\left\{\frac{1+1}{1+1}\right\} = 2\pi i. \quad \textbf{Ans.}$$

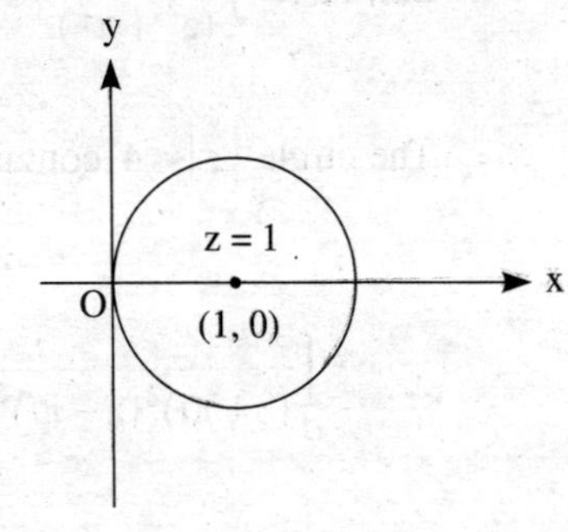

(*iii*) In this case both poles $z = 2, -2$ lie outside the circle $|z| = \dfrac{1}{2}$. Hence by Cauchy's Integral Theorem, we have

$$\int_C \frac{z^2+1}{z^2-1}\,dz = 0. \quad \textbf{Ans.}$$

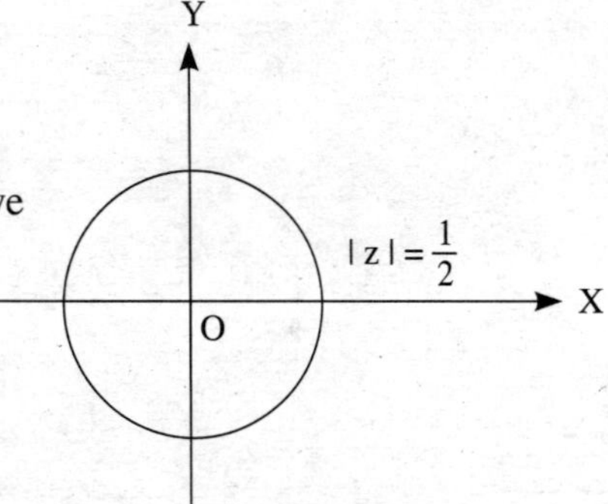

Example 6: Evaluate $\int_C \frac{\sin(\pi z)+\cos(\pi z)}{(z-1)(z-2)}dz$

where C is a circle $|z| = 4$.

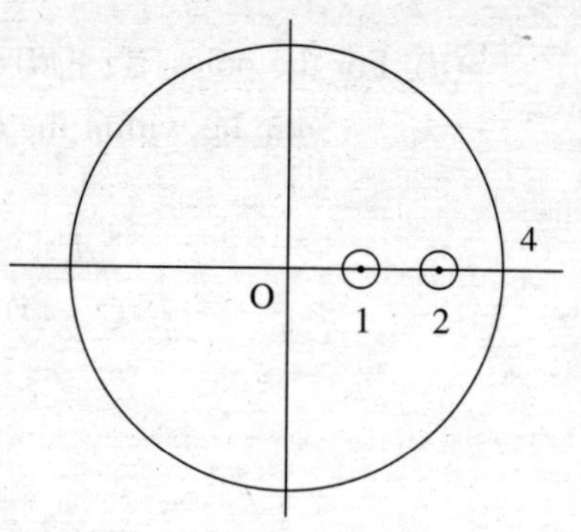

Sol. For singular points $(z-1)(z-2)=0 \Rightarrow z = 1, 2$

Here the circle $|z| = 4$ contain both singular point $z = 1, 2$

By Cauchy's integral formula, we have

$$\int_C \frac{\sin \pi z+\cos \pi z}{(z-1)(z-2)}dz = \int_{C_1} \frac{\left(\dfrac{\sin \pi z+\cos \pi z}{z-2}\right)}{(z-1)}dz + \int_{C_2} \frac{\left(\dfrac{\sin \pi z+\cos \pi z}{z-1}\right)}{(z-2)}dz$$

$$= 2\pi i\left[\frac{\sin \pi z+\cos \pi z}{z-2}\right]_{z=1} + 2\pi i\left[\frac{\sin \pi z+\cos \pi z}{z-1}\right]_{z=2}$$

$$= 2\pi i\left[\frac{0-1}{-1}\right] + 2\pi i\left[\frac{0+1}{2-1}\right] = 2\pi i + 2\pi i = 4\pi i. \quad \textbf{Ans.}$$

Example 7: Evaluate $\int_C \frac{e^z}{(z^2+\pi^2)^2}dz$, where C is $|z| = 4$

Sol. Here $\int_C \frac{e^z}{(z^2+\pi^2)}dz = \int_C \frac{e^z}{(z+\pi i)^2(z-\pi i)^2}dz$

The circle $|z| = 4$ contain both singular points $z = \pi i$ and $z = -\pi i$

$$\therefore \quad \int_C \frac{e^z}{(z+\pi i)^2(z-\pi i)^2}dz = \int_{C_1} \frac{\left(\dfrac{e^z}{(z+\pi i)^2}\right)}{(z-\pi i)^{1+1}}dz + \int_{C_2} \frac{\left(\dfrac{e^z}{(z-\pi i)^2}\right)}{(z+\pi i)^{1+1}}dz$$

$$= \frac{2\pi i}{\lfloor 1}\left[\frac{d}{dz}\frac{e^z}{(z+\pi i)^2}\right]_{z=\pi i} + \frac{2\pi i}{\lfloor 1}\left[\frac{d}{dz}\frac{e^z}{(z-\pi i)^2}\right]_{z=-\pi i}$$

$$= 2\pi i\left[\left\{\frac{e^z}{(z+\pi i)^2} - \frac{2e^z}{(z+\pi i)^3}\right\}_{z=\pi i} + \left\{\frac{e^z}{(z-\pi i)^2} - \frac{2e^z}{(z-\pi i)^3}\right\}_{z=-\pi i}\right]$$

$$= 2\pi i\left[-\frac{e^{\pi i}}{4\pi^2} + \frac{2e^{\pi i}}{8\pi^3 i} - \frac{e^{-\pi i}}{4\pi^2} - \frac{2e^{-\pi i}}{8\pi^3 i}\right]$$

$$= 2\pi i\left[\frac{1}{4\pi^2} + \frac{i}{4\pi^3} + \frac{1}{4\pi^2} - \frac{i}{4\pi^3}\right] = 2\pi i \times \frac{2}{4\pi^2} = \frac{i}{\pi}. \quad \textbf{Ans.}$$

Example 8: Evaluate

(*i*) $\int_C \frac{2z+1}{z^2+z} dz$ where C is $|z| = \frac{1}{2}$ (*ii*) $\int_C \left(\frac{e^{-z}}{z+1}\right) dz$, where C is the circle.

(*a*) $|z| = 2$ (*b*) $|z| = \frac{1}{2}$

Sol. (*i*) For the poles $z^2 + z = 0 \Rightarrow z(z+1) = 0 \Rightarrow z = 0, z = -1$

only the pole $z = 0$ lies within the circle, by Cauchy's integral formula

$$\int_C \frac{f(z)dz}{(z-a)} = 2\pi i\left[f(z)\right]_{z=a}$$

$$\therefore \qquad \int_C \frac{2z+1}{z^2+z} dz = \int_C \frac{\left\{\frac{2z+1}{(z+1)}\right\}}{z} dz$$

$$= 2\pi i \left\{\frac{2z+1}{z+1}\right\}_{z=0} = 1 \times 2\pi i = 2\pi i$$

(*ii*) For the pole $z + 1 = 0 \Rightarrow z = -1$

(*a*) $|z| = 2$, the pole $z = -1$ lies within the circle $|z| = 2$, So by Cauchy's integral formula

$$\int_C \frac{e^{-z}}{z+1} dz = 2\pi i\left[e^{-z}\right]_{z=-1} = 2\pi ie. \quad \textbf{Ans.}$$

(*b*) $|z| = \frac{1}{2}$, in this case the pole $z = -1$ lies outside the circle so by Cauchy's integral theorem, we have

$$\int_C f(z)dz = 0$$

$$\therefore \qquad \int_C \frac{e^{-z}}{z+1} dz = 0. \quad \textbf{Ans.}$$

Example 9: If $f(a) = \int_C \frac{3z^2+7z+1}{z-a} dz$, where C is the circle $x^2 + y^2 = 4$, find the values of $f(3), f'(1-i)$ and $f''(1-i)$.

Sol. The center of circle is at origin and the radius is 2.

At $a = 3$, $f(3) = \int_C \frac{3z^2+7z+1}{z-3} dz,$

for the poles $z - 3 = 0 \quad \Rightarrow z = 3$.

The pole $z = 3$ lies outside the circle by Cauchy's theorem $\int_C f(z)dz = 0$

$$\therefore \quad f(3) = \int_C \left(\frac{3z^2 + 7z + 1}{z - 3}\right) dz = 0 \quad \Rightarrow f(3) = 0. \quad \textbf{Ans.}$$

Now by Cauchy's integral formula

$$f(a) = (3z^2 + 7z + 1)_{z=a} 2\pi i$$

$$f(a) = (3a^2 + 7a + 1) 2\pi i$$

$$f'(a) = (6a + 7) 2\pi i$$

$$f''(a) = 12\pi i$$

At $a = 1 - i$ $\quad f'(1-i) = \{6 - 6i + 7\} 2\pi i = (13 - 6i) 2\pi i = 2\pi(13i + 6)$. **Ans.**

and $\quad f''(1-i) = 12\pi i$. **Ans.**

Example 10: Evaluate $\int_C \frac{z-1}{(z+1)^2 (z-2)} dz$ where C is $|z - i| = 2$.

Sol. For the poles $(z+1)^2 (z-2) = 0 \Rightarrow z = -1$ (order 2), $z = 2$. The pole $z = -1$ lies within the circle and the pole $z = 2$ lies outside the circle.

By Cauchy's integral formula for derivative

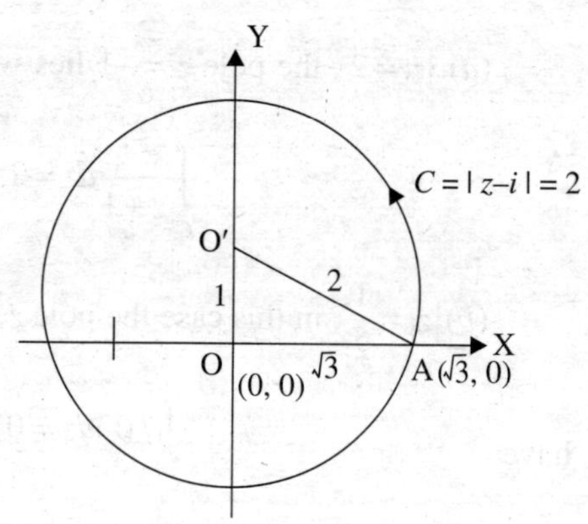

$$f'(a) = \frac{1}{2\pi i} \int_C \frac{f(z)}{(z-a)^2} dz$$

$$\therefore \quad \int_C \frac{\left(\frac{z-1}{z-2}\right)}{(z+1)^2} dz = 2\pi i \left[\frac{d}{dz}\left\{\frac{z-1}{z-2}\right\}\right]_{z=-1}$$

$$= 2\pi i \left[\frac{-(z-1)}{(z-2)^2} + \frac{1}{z-2}\right]_{z=-1} = 2\pi i \left[\frac{2}{9} - \frac{1}{3}\right] = -\frac{2\pi i}{9}. \quad \textbf{Ans.}$$

Example 11: Find the value of $\int_C \frac{z}{z^2 + 1} dz$, where C is a circle s.t.

(*a*) $\left|z + \frac{1}{z}\right| = 2 \qquad$ (*b*) $|z + i| = 1$

Sol. (*a*) For the poles $z^2 + 1 = 0 \quad \Rightarrow z^2 = i^2 \quad \Rightarrow z = \pm i$

$$\left|z + \frac{1}{z}\right| = 2 \quad \Rightarrow \left|z^2 + 1\right| = 2|z| \quad \Rightarrow \left|(x + iy)^2 + 1\right| = 2\sqrt{x^2 + y^2}$$

$$\Rightarrow \left|x^2 - y^2 + 1 + i2xy\right| = 2\sqrt{x^2 + y^2}$$

$$\Rightarrow \sqrt{(x^2 - y^2 + 1)^2 + 4x^2y^2} = 2\sqrt{x^2 + y^2}$$

Squaring on both sides, we get

$$x^4 + y^4 + 1 - 2x^2y^2 + 2x^2 - 2y^2 + 4x^2y^2 = 4x^2 + 4y^2$$

$$\Rightarrow x^4 + y^4 + 1 - 2y^2 + 2x^2y^2 - 2x^2 = 4y^2$$

$$\Rightarrow (x^2 + y^2 - 1)^2 = 4y^2$$

$$\Rightarrow x^2 + y^2 - 1 = \pm 2y$$

$$\Rightarrow x^2 + y^2 \mp 2y = 1$$

$$\Rightarrow x^2 + y^2 \mp 2y + 1 = 2$$

$$\Rightarrow x^2 + (y \mp 1)^2 = 2$$

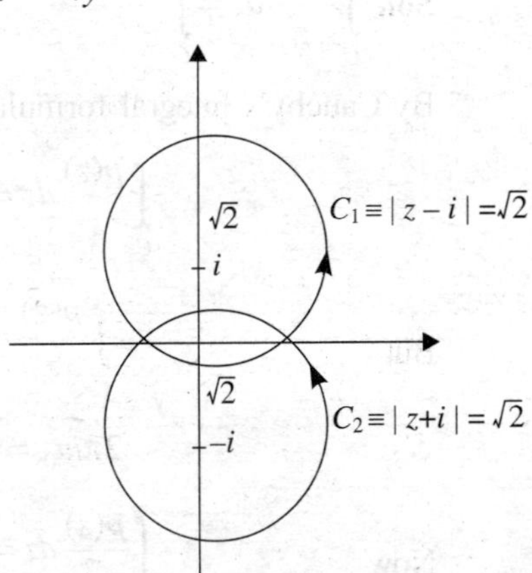

or $\quad x^2 + (y-1)^2 = 2 \quad$ and $\quad x^2 + (y+1)^2 = 2$

These are the equation of two circle having center (0,1) and (0, –1) by of same radius $\sqrt{2}$.

Hence, the both pole $z = i, -i$ lies within the circle.

By Cauchy's integral formula, we have

$$\int_C \frac{z}{z^2 + 1} dz = \int_{C_1} \frac{\left(\frac{z}{z+i}\right)}{(z-i)} dz + \int_{C_2} \frac{\left(\frac{z}{z-i}\right)}{(z+i)} dz$$

$$= 2\pi i \left[\frac{z}{z+i}\right]_{z=i} + 2\pi i \left[\frac{z}{z-i}\right]_{z=-i}$$

$$= 2\pi i \times \frac{1}{2} + 2\pi i \times \frac{1}{2} = 2\pi i. \quad \textbf{Ans.}$$

(*b*) $|z + i| = 1$

In this region only the pole $z = -i$, lies within the circle so

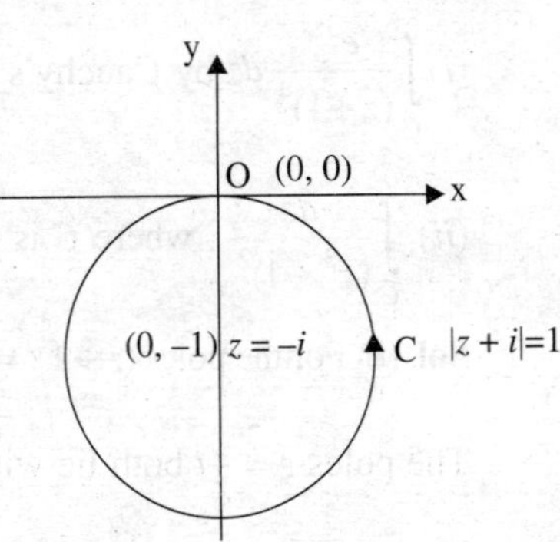

$$\int_C \frac{z}{z^2 + 1} dz = \int_C \frac{\left(\frac{z}{z-i}\right)}{(z+i)} dz = 2\pi i \left[\frac{z}{z-i}\right]_{z=-i}$$

$$= 2\pi i \times \frac{1}{2} = \pi i. \quad \textbf{Ans.}$$

Example 12: Suppose $p(z) = a_0 + a_1 z + a_2 z^2$ and

$$\int_C \frac{p(z)}{z} dz = \int_C \frac{p(z)}{z^2} dz = \int_C \frac{p(z)}{z^3} dz = 2\pi i$$

where C is the circle $|z| = 1$. Find the values of a_0, a_1 and a_2.

Sol. $\int_C \frac{p(z)}{z} dz = \int_C \frac{a_0 + a_1 z + a_2 z^2}{z} dz$, the pole $z = 0$ lies within the circle $|z| = 1$

By Cauchy's integral formula.

$$\int_C \frac{p(z)}{z} dz = 2\pi i \left[a_0 + a_1 z + a_2 z^2 \right]_{z=0} = 2\pi i a_0$$

But $$\int_C \frac{p(z)}{z} = 2\pi i$$

$\therefore$ $$2\pi i a_0 = 2\pi i \quad \Rightarrow a_0 = 1. \quad \textbf{Ans.}$$

Now $$\int_C \frac{p(z)}{z^2} dz = 2\pi i \left\{ \frac{d}{dz}(a_0 + a_1 z + a_2 z^2) \right\}_{z=0} = 2\pi i (a_1 + 2a_2 z)_{z=0} = 2\pi i a_1$$

$\Rightarrow$ $$2\pi i a_1 = 2\pi i \quad \Rightarrow \quad a_1 = 1. \quad \textbf{Ans.}$$

and $$\int_C \frac{p(z)}{z^3} dz = \frac{2\pi i}{2!} \left[\frac{d^2}{dz^2}(a_0 + a_1 z + a_2 z^2) \right]_{z=0}$$

$$= \frac{2\pi i}{2!}(2a_2) = 2\pi i a_2$$

$\Rightarrow$ $$2\pi i a_2 = 2\pi i \quad \Rightarrow \quad a_2 = 1. \quad \textbf{Ans.}$$

Example 13: Evaluate

(*i*) $\int_C \frac{e^{3z}}{(z+1)^4} dz$ by Cauchy's integral formula where $|z| = 2$.

(*ii*) $\int_C \frac{dz}{(z^3 - 1)^2}$, where C is the circle, | 2–1| = 1

Sol. (*i*) For the poles $(z + 1)^4 = 0 \;\Rightarrow\; z = i^2 \;\Rightarrow\; z = \pm i$ (order 4)

The poles $z = \pm i$ both lie within the circle so by Cauchy's

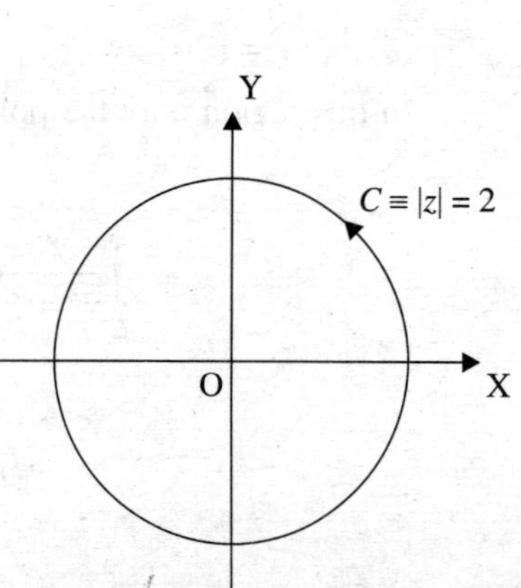

integral formula for derivative

$$\int_C \frac{e^{3z}}{(z+1)^4}dz = \frac{2\pi i}{3!}\left[\frac{d^3}{dz^3}\{e^{3z}\}\right]_{z=-1}$$

$$= \frac{2\pi i}{6}\left[27e^{3z}\right]_{z=-1} = \frac{9\pi i}{e^3}. \textbf{ Ans.}$$

$$f^n(a) = \frac{n!}{2\pi i}\int_C \frac{f(z)}{(z-a)^{n+1}}dz$$

$$\Rightarrow \int_C \frac{f(z)}{(z-a)^{n+1}}dz = \frac{2\pi i}{n!}\frac{d^n}{dz^n}\left[f(z)\right]_{z=a}$$

(*ii*) $(z^3-1)^2 = 0$

$$(z-1)^2(z^2+z+1)^2 = 0$$

$$z = 1, \quad \frac{-1 \pm i\sqrt{3}}{2}\{\text{of order } 2\}$$

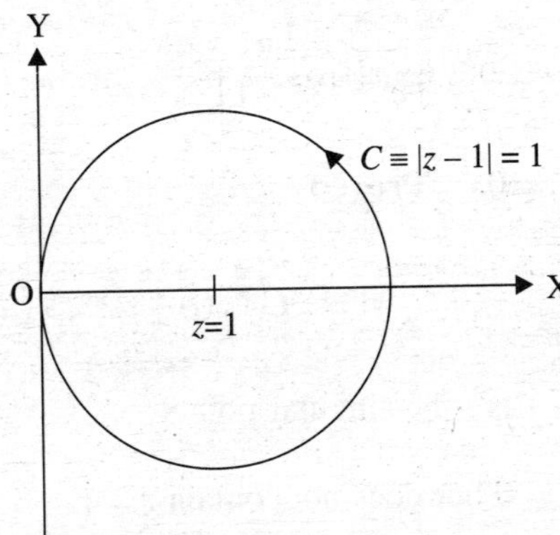

Only the pole $z = 1$ lies within the circle so by Cauchy's integral formula for derivative.

$$\int_C \frac{dz}{(z^3-1)^2} = \int_C \frac{\frac{1}{(z^2+z+1)^2}}{(z-1)^2}dz$$

$$= 2\pi i\left\{\frac{d}{dz}\left(\frac{1}{(z^2+z+1)^2}\right)\right\}_{z=1}$$

$$= 2\pi i\left\{\frac{-2(2z+1)}{(z^2+z+1)^3}\right\}_{z=1}$$

$$= -4\pi i \times \frac{3}{27} = -\frac{4\pi i}{9}. \textbf{ Ans.}$$

Example 14: (*i*) Prove that $\int_C \frac{dz}{(z-a)} = 2\pi i$ (*ii*) $\int_C (z-a)^n dz = 0, |z-a| = r.$

Sol. (*i*) The equation of the circle is $z - a = re^{i\theta}, dz = re^{i\theta} i d\theta$

$$\therefore \qquad \int_C \frac{dz}{z-a} = \int_0^{2\pi} \frac{rie^{i\theta}}{re^{i\theta}} d\theta = i\int_0^{2\pi} d\theta = 2\pi i. \quad \textbf{Ans.}$$

$$(ii) \int_C (z-a)^n dz = \int_0^{2\pi} (re^{i\theta})^n . rie^{i\theta} d\theta = i\int_0^{2\pi} r^{n+1} e^{in\theta} e^{i\theta}\, d\theta$$

$$= r^{n+1} i \int_0^{2\pi} e^{i(n+1)\theta} d\theta$$

$$= ir^{n+1}\left[\frac{e^{i(n+1)\theta}}{i(n+1)}\right]_0^{2\pi}$$

$$= \frac{r^{n+1}}{n+1}\left[e^{i(n+1)2\pi} - 1\right]$$

$$= 0. \quad \textbf{Proved.}$$

$$\begin{aligned} e^{i2\pi(n+1)} &= \cos(2n+2)\pi \\ &\quad + i\sin(2n+2)\pi \\ &= 1 + 0 = 1 \end{aligned}$$

Example 15: Evaluate $\int_C \frac{e^z}{z(1-z)^3} dz$, where C is (*i*) $|z| = \frac{1}{2}$ (*ii*) $|z-1| = \frac{1}{2}$ (*iii*) $|z| = 2$.

Sol. $z(1 - z)^3 = 0 \Rightarrow z = 0, 1$ are the singular points

(*i*) $C \equiv |z| = \frac{1}{2}$ contains $z = 0$ but does not contain $z = 1$

Using integral formula

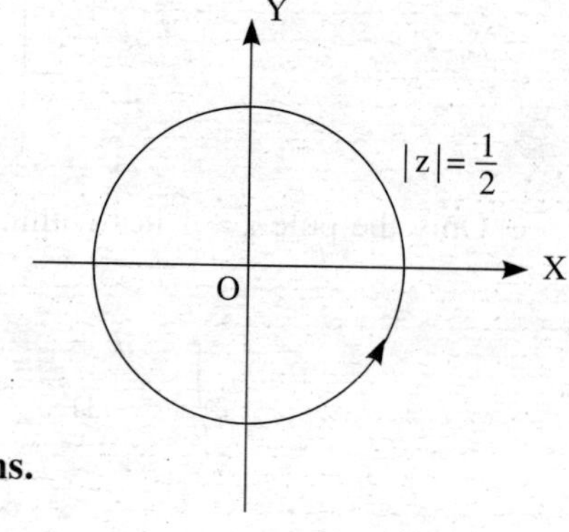

$$\int_C \frac{e^z}{z(1-z)^3} dz = \int_C \frac{\{e^z/(1-z)^3\}}{(z-0)} dz$$

$$= 2\pi i \left\{\frac{e^z}{(1-z)^3}\right\}_{z=0} = 2\pi i. \quad \textbf{Ans.}$$

(*ii*) The circle $|z| = \frac{1}{2}$ contains 1 but not 0

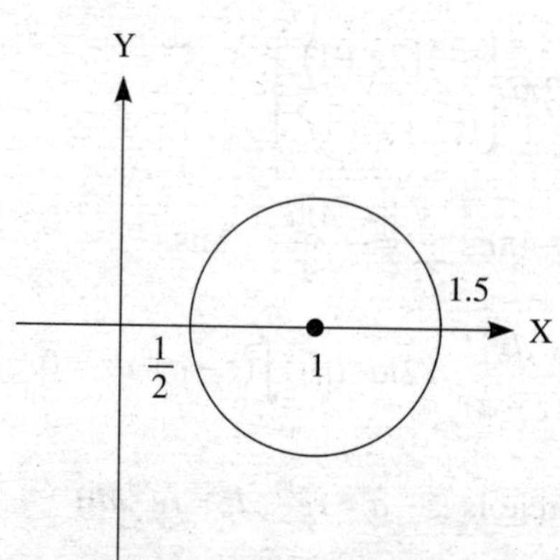

$$\therefore \quad \int_C \frac{e^z}{z(1-z)^3}dz = \int_C \frac{\left(\frac{e^z}{z}\right)}{(1-z)^{2+1}}dz = -\int_C \frac{(e^z/z)}{(z-1)^{2+1}}dz$$

$$= -\frac{2\pi i}{\underline{|2}}\left[\frac{d^2}{dz^2}\left(\frac{e^z}{z}\right)\right]_{z=1} \qquad \left| \; f^n(a) = \frac{\underline{|n}}{2\pi i}\int_C \frac{f(z)}{(z-a)^{n+1}}dz \right.$$

$$= -\pi i\left[\frac{(2-z)e^z}{z^3}\right]_{z=1} = -\pi ei. \quad \textbf{Ans.}$$

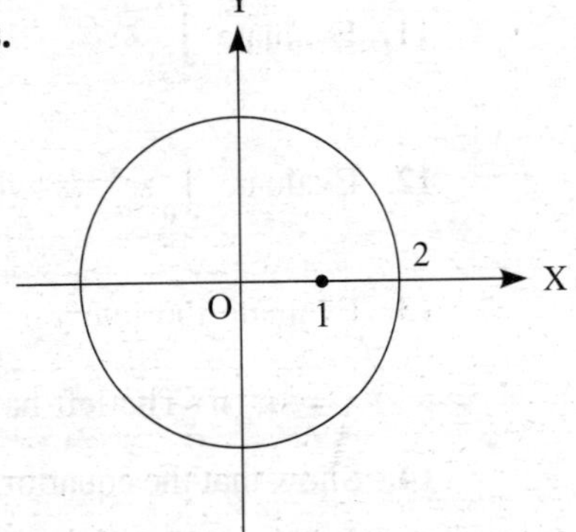

(*iii*) Here the circle $|z| = 2$ contains both the sigular points

$\therefore$ By Cauchy's theorem for multi-connected region,

$$\int_C \frac{e^z}{z(1-z)^3}dz = \int_{C_1} \frac{e^z/(1-2)^3}{z}dz + \int_{C_2} \frac{(e^z/z)}{(1-z)^3}dz$$

$$= 2\pi i + (-\pi ei) = \pi i(2-e). \quad \textbf{Ans.}$$

PROBLEM SET 2.2

1. Evaluate $\int_C dz.$ **[Ans.** $(b - a)$ = chord AB]

2. Evaluate $\int_0^{1+i} z^2 dz.$ $\left[\textbf{Ans.}\ \frac{1}{3}(1+i)^3\right]$

3. Evaluate $\int_C (z^2 + 3z + 2)dz,$ where C is the arc of the cuboid $x = a(\theta + \sin\theta),$ $y = a(1-\cos\theta)$ between the points $(0,0)$ and $(\pi a, 2a)$.

$\left[\textbf{Ans.}\ 2\pi a + 1/3(\pi a + 12a)^4 + 3/2\pi a\right]$

4. Evaluate $\int_C (x^2 - iy)dz$ along the path (*i*) $y = x$, (*ii*) $y = x^2$.

[Ans. (*i*) $5/6 - i/6$ (*ii*) $5/6 + i/6$]

5. Integrate $f(z) = x^2 + ixy$ from A (1, 1), B (2,4) along the curve $x = t, y = t^2$.

$\left[\textbf{Ans.}\ \frac{-151}{5} + \frac{45}{4}i\right]$

6. Evaluate $\int_{(0,0)}^{(4,1)} \left[(3x^2+4xy+3y^2)dx+2(x^2+3xy+4y^2)dy\right]$, along (*i*) $y = x^2$ and (*ii*) $y = x$. Does the value of integral depend on the path. [**Ans.** (*i*) 26/3 (*ii*) 26/3]

7. Evaluate $\int_0^{2\pi} z^2 \sin 4z dz$. [**Ans.** $-\pi^2$] **8.** Evaluate $\int_0^1 ze^{2z} dz$. $\left[\textbf{Ans. } \frac{1}{4}(e^2+1)\right]$

9. Evaluate $\int_{2i}^{3} \sin z dz$. $[\textbf{Ans. } \cos h2 - \cos 3]$ **10.** Prove that $\int_3^{-3} \sqrt{z}\, dz = -2\sqrt{3}(1+i)$.

11. Evaluate $\int_{-1}^{1} z^i dz$. $\left[\textbf{Ans. } \frac{1+e^{-n}}{2}(1-i)\right]$

12. Evaluate $\int_0^i ze^{z^2} dz$ where $t = z^2$. $\left[\textbf{Ans. } \frac{1}{2}(e^{-1}-1)\right]$

13. Evaluate the integral $\int_C f(z)dz$, where C is the contour. (*i*) the straight line from $z = i$ to $z = -i$, (*ii*) The left half of the unit circle $|z| = 1$ from $z = -i$ to i. [**Ans.** (*i*) 2, (*ii*) $2i$]

14. Show that the equation $z^5 + 15z + 1 = 0$ has one root in the disc $|z| < 3/2$ and four roots in the annulus $3/2 < |z| < 2$.

15. (*i*) Evaluate the following integral by using Cauchy's integral formula $\int_C \frac{4-3z}{z(z-1)(z-2)} dz$, where C is the circle $|z| = \frac{1}{2}$. [**Ans.** $2\pi i$]

(*ii*) Evaluate $\int_C \frac{3z^2+z}{z^2-1} dz$, where C is the circle $|z-1| = 1$. [**Ans.** $4\pi i$]

16. (*i*) Evaluate the following integral by using Cauchy's integral formula $\int_C \frac{3z^2+z+1}{(z^2-1)(z+3)} dz$, where C is the circle $|z| = 2$. $\left[\textbf{Ans. } -\frac{\pi i}{4}\right]$

17. Use Cauchy's integral formula to evaluate $\int_C \frac{e^{2z}}{(z+1)^4} dz$, where C is the circle $|z| = 2$.

$\left[\textbf{Ans. } 8\pi i/3e^2\right]$

18. Evaluate $\int_C \frac{3z^2+7z+1}{z+1} dz$, where C is the circle

(*a*) $|z| = 1.5$ (*b*) $|z+i| = 1$. [**Ans.** (*a*) $-6\pi i$, (*b*) 0]

19. Evaluate $\int_C \frac{e^{2z}}{(z-1)(z-2)}\,dz$, where C is a circle $|z| = 3$ $\left[\textbf{Ans. } 2\pi i(e^4 - e^2)\right]$

20. Evaluate $\int_C \frac{\cos \pi z}{(z^2-1)}\,dz$, where C is a rectangle with vertices $\pm$ 2, $\pm$ i. $\left[\textbf{Ans. } 0\right]$

21. Evaluate $\int_C \frac{\cos 2\pi z}{(2z-1)(z-3)}\,dz, C: |z| = 1.$ $\left[\textbf{Ans. } \frac{2\pi i}{5}\right]$

22. Determine $f(4), f(i), f'(-1), f''(-i)$, if $f(\alpha) = \int_C \frac{4z^3+z+5}{z-\alpha}\,dz$, where C : ellipse $\frac{x^2}{4} + \frac{y^2}{9} = 1.$

$\left[\textbf{Ans. } 0, 2\pi(i-1), -14\pi i, 16\pi i\right]$

23. Evaluate $\int_C \frac{e^z}{(z-1)^2(z^2+4)}\,dz, C: |z-1| = \frac{1}{2}$. $\left[\textbf{Ans. } \frac{6e\pi}{25}\right]$

24. Evaluate $\int_C \frac{e^z}{z^3}\,dz, C: |z| = 1.$ $\left[\textbf{Ans. } \pi i\right]$

25. Evaluate $\int_C \frac{(z^4 - 3z^2 + 6)}{(z+i)^3}\,dz,\ C: |z| = 2.$ $\left[\textbf{Ans. } -18\pi i\right]$

UNIT III

Functions of a Complex Variable II

3.1 REPRESENTATION OF A FUNCTION BY POWER SERIES

Power Series: A series of the form $\sum_{n=0}^{\infty} a_n z^n$ or $\sum_{n=0}^{\infty} a_n (z-a)^n$, whose terms are variables, is called a power series, where z is a complex variable and a_n are complex constants.

The second form can be reduced to the first form by changing origin. If

$$z = \xi + a \text{ then } \sum_{n=0}^{\infty} a_n (z-a)^n = \sum_{n=0}^{\infty} a_n \xi^n .$$

The form $\sum_{n=0}^{\infty} a_n z^n$ being simpler. Hence, we shall consider only the first form.

3.1.1 Absolute Convergence of Power Series $\sum_{n=0}^{\infty} a_n z^n$

The power series $\sum_{n=0}^{\infty} a_n z^n$ is said to be absolutely convergent if series $\sum_{n=0}^{\infty} \left| a_n z^n \right|$ is convergent.

If the series $\sum_{n=0}^{\infty} a_n z^n$ convergent but the series $\sum_{n=0}^{\infty} |a_n||z^n|$ not converge then the series $\sum_{n=0}^{\infty} a_n z^n$ is called conditionally convergent.

3.1.2 Radius of Convergence of the Power Series

Consider the power series $\sum_{n=0}^{\infty} a_n z^n$.

Let $\sum u_n = \sum_{n=0}^{\infty} a_n z^n$, where u_n is the function of z.

If $\lim_{n\to\infty} |u_n|^{1/n} < 1$, then $\sum u_n$ is convergent.

$\Rightarrow \quad \lim_{n\to\infty} |a_n z^n|^{1/n} < 1 \quad \Rightarrow \quad \lim_{n\to\infty} \left\{|a_n|^{1/n} . |z|\right\} < 1$

Let $\lim_{n\to\infty} |a_n|^{1/n} = \frac{1}{R}$, we get

$$\frac{|z|}{R} < 1 \text{ or } |z| < R.$$

$\Rightarrow \sum a_n z^n$ is convergent or divergent according as $|z| < R$ or $|z| > R$ where R is the radius of convergence.

By Cauchy's theorem on limit

$$\frac{1}{R} = \lim_{n\to\infty} |a_n|^{1/n} = \lim_{n\to\infty} \left|\frac{a_{n+1}}{a_n}\right|$$

Example 1: Find the radius of the following power series.

(*i*) $\sum \frac{n!}{n^n} z^n$ (*ii*) $\sum \frac{2^{-n} z^n}{1+in^2}$ (*iii*) $\sum \frac{(n!)^2}{(2n)!} z^n$

Sol. (*i*) $a_n = \frac{n!}{n^n}, \quad a_{n+1} = \frac{(n+1)!}{(n+1)^{n+1}}$

$$\frac{a_{n+1}}{a_n} = \frac{(n+1)!n^n}{(n+1)^{n+1}(n)!} = \frac{(n+1)n!n^n}{(n+1)^n(n+1)(n)!} = \frac{1}{\left(1+\frac{1}{n}\right)^n}$$

$\therefore \quad \frac{1}{R} = \lim_{n\to\infty} \left|\frac{a_{n+1}}{a_n}\right| = \lim_{n\to\infty} \frac{1}{\left(1+\frac{1}{n}\right)^n} = \frac{1}{e}. \qquad \left| \lim_{x\to\infty} \left(1+\frac{1}{x}\right)^x = e \right.$

$\Rightarrow \quad R = e$ **Ans.**

(*ii*) $a_n = \frac{2^{-n}}{1+in^2}, \qquad |a_n| = \frac{2^{-n}}{\sqrt{1+n^4}}$

$$\frac{1}{R} = \lim_{n\to\infty} \left[\frac{2^{-n}}{\sqrt{1+n^4}}\right]^{1/n} = \lim_{n\to\infty} \frac{1}{2(1+n^4)^{1/2n}} \qquad \therefore \frac{1}{R} = \lim_{n\to\infty} [a_n]^{1/n}$$

$$= \lim_{n\to\infty} \frac{1}{2}[1+n^4]^{-1/2n} = \lim_{n\to\infty} \frac{1}{2(n^2)^{1/n}} \left[1+\frac{1}{n^4}\right]^{-1/2n}$$

$$= \lim_{n\to\infty} \frac{1}{2(n^{1/n})^2} \left[1-\frac{1}{2n^5}+\ldots\right]$$

$$= \frac{1}{2}\left(\lim_{n\to\infty} \frac{1}{n^{1/n}}\right)^2 \times 1 \qquad \left| \lim_{n\to\infty} \frac{1}{n^{1/n}} = 1 \right.$$

$$\frac{1}{R} = \frac{1}{2} \times 1 \times 1 = \frac{1}{2}$$

$$\Rightarrow \qquad R = 2. \quad \textbf{Ans.}$$

(*iii*)
$$a_n = \frac{(n!)^2}{(2n)!}, \quad a_{n+1} = \frac{\{(n+1)!\}^2}{\{2(n+1)\}!} = \frac{[(n+1)n!]^2}{(2n+2)(2n+1)(2n)!}$$

$$\frac{a_{n+1}}{a_n} = \frac{(n+1)^2}{(2n+2)(2n+1)} = \frac{(n+1)}{2(2n+1)} = \frac{1+\frac{1}{n}}{4\left(1+\frac{1}{2n}\right)}$$

$$\therefore \qquad \frac{1}{R} = \lim_{n\to\infty}\left|\frac{a_{n+1}}{a_n}\right| = \frac{1+0}{4(1+0)} = \frac{1}{4}$$

$$\Rightarrow \qquad R = 4. \quad \textbf{Ans.}$$

3.2 TAYLOR'S SERIES OR THEOREM

If a function $f(z)$ is analytic at every point inside a circle C, with its center at $z = a$ and radius R, then at each point z inside C.

$$f(z) = f(a) + (z-a)f'(a) + \frac{(z-a)^2}{2!} f''(a) + ... + \frac{(z-a)^n}{n!} f^n(a) + ...$$

or
$$f(z) = \sum_{n=0}^{\infty} a_n (z-a)^n \qquad \text{where, } a_n = \frac{f^n(a)}{n!}$$

Proof: Lef $f(z)$ is analytic within the circle C whose equation is $|w - a| = R$. Let z be any point within C s.t. $|z - a| = r$.

Now by Cauchy's integral formula

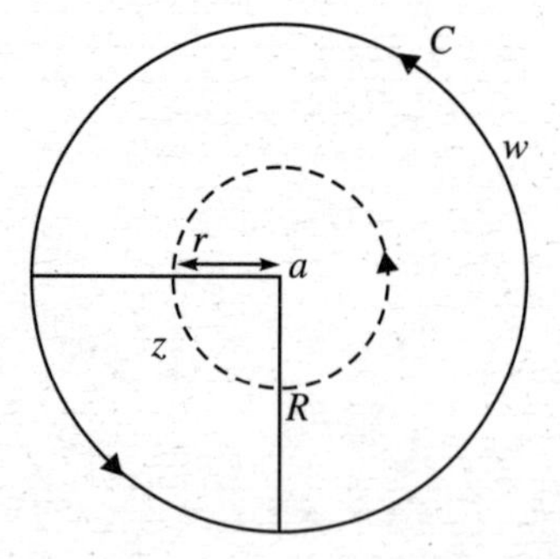

$$f(z) = \frac{1}{2\pi i}\int_C \frac{f(w)dw}{w-z}$$

$$\text{or } \quad f(z) = \frac{1}{2\pi i}\int_C \frac{f(w)dw}{(w-a)-(z-a)}$$

$$= \frac{1}{2\pi i}\int_C \frac{f(w)}{(w-a)}\left[1 - \frac{z-a}{w-a}\right]^{-1} dw$$

$$= \frac{1}{2\pi i}\int_C \frac{f(w)}{(w-a)}\left[1 + \left(\frac{z-a}{w-a}\right) + \left(\frac{z-a}{w-a}\right)^2 + ... + \left(\frac{z-a}{w-a}\right)^n\right] dw$$

$$= \frac{1}{2\pi i}\int_C \frac{f(w)}{(w-a)}dw + (z-a)\frac{1}{2\pi i}\int_C \frac{f(w)}{(w-a)^2}dw + (z-a)^2\frac{1}{2\pi i}\int_C \frac{f(w)}{(w-a)^3}dw$$

$$+ .... + (z-a)^n \frac{1}{2\pi i}\int_C \frac{f(w)}{(w-a)^{(n+1)}}dw + ...$$

$$\Rightarrow \qquad \boxed{f(z) = f(a) + (z-a)f'(a) + \frac{(z-a)^2}{2!}f''(a) + ... + \frac{(z-a)^n}{n!}f^n(a) + ...}$$

or

$$f(z) = \sum_{n=0}^{\infty} \frac{(z-a)^n}{n!} f^n(a)$$

or

$$f(z) = \sum_{n=0}^{\infty} a_n (z-a)^n, \qquad \text{where } a_n = \frac{f^n(a)}{n!}$$

Deduction: (*i*) If $z = a + h$, then

$$f(a+h) = f(a) + hf'(a) + \frac{h^2}{2}f''(a) + ...$$

(*ii*) If $a = 0$, then

$$f(z) = \sum_{n=0}^{\infty} a_n z^n \quad \text{or} \quad f(z) = \sum_{n=0}^{\infty} \frac{f^n(0) z^n}{n!}.$$

3.3 LAURENT'S SERIES OR THEOREM

It is a series of positive and negative integer powers of $(z - a)$ by which we can represent a given function $f(z)$ in an annulus (*a* circular ring with centre *a*) in which $f(z)$ is analytic. $f(z)$ may not analytic outside the ring as well as in its "hole".

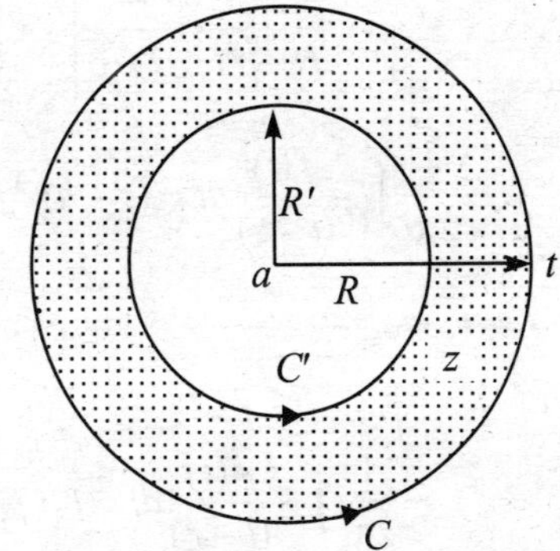

Statement: If a function $f(z)$ is analytic in the closed ring bounded by two concentric circles C and C' of center a and radius R and R', $(R' < R)$. If z is any point of the annulus, then

$$f(z) = \sum_{n=0}^{\infty} a_n (z-a)^n + \sum_{n=1}^{\infty} b_n (z-a)^{-n}$$

where

$$a_n = \frac{1}{2\pi i}\int_C \frac{f(t)dt}{(t-a)^{n+1}}, \quad b_n = \frac{1}{2\pi i}\int_{C'} \frac{f(t)dt}{(t-a)^{-n+1}}$$

Proof: Let the function $f(z)$ is analytic in the closed ring bounded by two concentric circles C and C' of center a and radius R and R'; $(R' < R)$. Let z be any point in shaded region then by Cauchy's integral formula for connected region.

$$f(z) = \frac{1}{2\pi i}\int_C \frac{f(t)}{(t-z)}dt - \frac{1}{2\pi i}\int_{C'} \frac{f(t)}{(t-z)}dt \qquad ...(1)$$

for the first integral t lies on C

$$\therefore \qquad |t-a| > |z-a| \quad i.e., \quad \left|\frac{z-a}{t-a}\right| < 1$$

Now,
$$\frac{1}{(t-z)} = \frac{1}{(t-a)-(z-a)} = \frac{1}{t-a}\left[1-\frac{z-a}{t-a}\right]^{-1}$$

$$\frac{1}{(t-z)} = \frac{1}{t-a}\left[1+\left(\frac{z-a}{t-a}\right)+\left(\frac{z-a}{t-a}\right)^2+....\right]$$

$$\Rightarrow \frac{1}{2\pi i}\int_C \frac{f(t)}{(t-z)}dt = \frac{1}{2\pi i}\int_C \frac{f(t)}{(t-a)}dt + (z-a)\int_C \frac{f(t)}{(t-a)^2}dt + ...$$

$$= a_0 + a_1(z-a) + a_2(z-a)^2 + ...$$

$$a_n = \frac{1}{2\pi i}\int_C \frac{f(t)}{(t-a)^{n+1}}dt$$

Putting $n = 0, 1, 2, 3$

$$\Rightarrow \frac{1}{2\pi i}\int_C \frac{f(t)}{(t-z)}dt = \sum_{n=0}^{\infty} a_n(z-a)^n \qquad ...(2)$$

Now for the second integral

$$\because \qquad |t-a| < |z-a| \quad \text{so that} \quad \frac{|t-a|}{|z-a|} < 1$$

Now,
$$\frac{1}{(t-z)} = \frac{1}{(t-a)-(z-a)} = \frac{-1}{z-a}\left[1-\frac{t-a}{z-a}\right]^{-1}$$

$$\frac{1}{(t-z)} = \frac{-1}{z-a}\left[1+\left(\frac{t-a}{z-a}\right)+\left(\frac{t-a}{z-a}\right)^2+\left(\frac{t-a}{z-a}\right)^3+.....\right]$$

$$\therefore -\frac{1}{2\pi i}\int_{C'} \frac{f(t)}{(t-z)}dt = (z-a)^{-1}\frac{1}{2\pi i}\int_{C'} f(t)dt + (z-a)^{-2}\frac{1}{2\pi i}\int_{C'} \frac{f(t)}{(t-a)^{-1}}dt$$
$$+ (z-a)^{-3}\frac{1}{2\pi i}\int_{C'} \frac{f(t)}{(t-a)^{-2}}dt + ...$$

$$\Rightarrow \qquad -\frac{1}{2\pi i}\int_{C'} \frac{f(t)}{(t-z)}dt = b_1(z-a)^{-1} + b_2(z-a)^{-2} + ...$$

or
$$-\frac{1}{2\pi i}\int_{C'} \frac{f(t)}{(t-z)}dt = \sum_{n=1}^{\infty} b_n(z-a)^{-n}$$

$$\Rightarrow \qquad \frac{1}{2\pi i}\int_{C'} \frac{f(t)}{(t-z)}dt = -\sum_{n=1}^{\infty} b_n(z-a)^{-n} \qquad ...(3)$$

$$b_n = \frac{1}{2\pi i}\int_{C'} \frac{f(t)}{(t-a)^{-n+1}}dt$$

Putting $n = 1, 2, 3$

$$b_1 = \frac{1}{2\pi i}\int_{C'} f(t)dt$$

$$b_2 = \frac{1}{2\pi i}\int_{C'} \frac{f(t)dt}{(t-a)^{-1}}$$

and so on

Using equation (2) and (3) in equation (1), then we get

$$f(z) = \sum_{n=0}^{\infty} a_n(z-a)^n + \sum_{n=1}^{\infty} b_n(z-a)^{-n}$$

where
$$a_n = \frac{1}{2\pi i}\int_C \frac{f(t)}{(t-a)^{n+1}}dt, \; b_n = \frac{1}{2\pi i}\int_{C'} \frac{f(t)}{(t-a)^{-n+1}}dt,$$

Deduction

Since $a_n = \frac{1}{2\pi i}\int_{C_0} \frac{f(t)}{(t-a)^{n+1}}\,dt$ and $b_n = \frac{1}{2\pi i}\int_{C_0} \frac{f(t)}{(t-a)^{-n+1}}\,dt$

where C_0 is any path between C and C'
replace $n = -n$, then

$$a_{-n} = \frac{1}{2\pi i}\int_{C_0} \frac{f(t)}{(t-a)^{-n+1}}\,dt = b_n$$

$\therefore$ From above $f(z) = \sum_{n=-\infty}^{\infty} a_n (z-a)^n$ with $a_n = \frac{1}{2\pi i}\int_{C} \frac{f(t)}{(t-a)^{n+1}}\,dt$

Example 2: Obtain the Taylor's and Laurent's series which represent the function $\frac{(z^2-1)}{(z+2)(z+3)}$ in the regions.

(*i*) $|z| < 2$ (*ii*) $2 < |z| < 3$ (*iii*) $|z| > 3$

Sol. Let
$$f(z) = \frac{(z^2-1)}{(z+2)(z+3)} = 1 - \frac{5z+7}{(z+2)(z+3)}$$

$\Rightarrow$
$$f(z) = 1 + \frac{3}{(z+2)} - \frac{8}{(z+3)}. \quad \text{(By partial fraction)}$$

(*i*) $|z| < 2 \Rightarrow \frac{|z|}{2} < 1$

So
$$f(z) = 1 + \frac{3}{2\left(1+\frac{z}{2}\right)} - \frac{8}{3\left(1+\frac{z}{3}\right)}$$

$$= 1 + \frac{3}{2}\left(1+\frac{z}{2}\right)^{-1} - \frac{8}{3}\left(1+\frac{z}{3}\right)^{-1}$$

$$= 1 + \frac{3}{2}\left\{1 - \frac{z}{2} + \left(\frac{z}{2}\right)^2 - \left(\frac{z}{2}\right)^3 + \ldots\right\} - \frac{8}{3}\left\{1 - \frac{z}{3} + \left(\frac{z}{3}\right)^2 - \left(\frac{z}{3}\right)^3 + \ldots\right\}$$

$$= 1 + \frac{3}{2}\sum_{n=0}^{\infty}(-1)^n\left(\frac{z}{2}\right)^n - \frac{8}{3}\sum_{n=0}^{\infty}(-1)^n\left(\frac{z}{3}\right)^n$$

$$= 1 + \sum_{n=0}^{\infty}(-1)^n\left[\frac{3}{2^{n+1}} - \frac{8}{3^{n+1}}\right]z^n. \quad \textbf{Ans.}$$

This is Taylor's series valid for $|z| < 2$.

(*ii*) $2<|z|<3$

$$\Rightarrow \qquad \frac{2}{|z|}<1, \frac{|z|}{3}<1$$

$$f(z)=1-\frac{5z+7}{(z+2)(z+3)}=1+\frac{3}{(z+2)}-\frac{8}{(z+3)}$$

$$=1+\frac{3}{z\left(1+\frac{2}{z}\right)}-\frac{8}{3\left(1+\frac{z}{3}\right)}=1+\frac{3}{z}\left(1+\frac{2}{z}\right)^{-1}-\frac{8}{3}\left(1+\frac{z}{3}\right)^{-1}$$

$$=1+\frac{3}{z}\left[1-\frac{2}{z}+\left(\frac{2}{z}\right)^2-\left(\frac{2}{z}\right)^3+\ldots\right]-\frac{8}{3}\left(1-\frac{z}{3}+\left(\frac{z}{3}\right)^2-\left(\frac{z}{3}\right)^3+\ldots\right)$$

$$=1+\frac{3}{z}\sum_{n=0}^{\infty}(-1)^n\left(\frac{2}{z}\right)^n-\frac{8}{3}\sum_{n=0}^{\infty}(-1)^n\left(\frac{z}{3}\right)^n$$

$$=1+3\sum_{n=0}^{\infty}\frac{(-1)^n 2^n}{z^{n+1}}-8\sum_{n=0}^{\infty}\frac{(-1)^n z^n}{3^{n+1}}$$

This is a Laurent's series in region $2<|z|<3$.

(*iii*) $|z|>3 \quad\Rightarrow\quad \frac{3}{|z|}<1$

$$\text{Now} \quad f(z)=1+\frac{3}{z+2}-\frac{8}{z+3}$$

$$f(z)=1+\frac{3}{z\left(1+\frac{2}{z}\right)}-\frac{8}{z\left(1+\frac{3}{z}\right)}$$

$$f(z)=1+3\sum_{n=0}^{\infty}\frac{(-1)^n 2^n}{z^{n+1}}-\frac{8}{z}\sum_{n=0}^{\infty}\frac{(-1)^n 3^n}{z^n}$$

$$f(z)=1+3\sum_{n=0}^{\infty}\frac{(-1)^n 2^n}{z^{n+1}}-8\sum_{n=0}^{\infty}\frac{(-1)^n 3^n}{z^{n+1}}. \qquad \textbf{Ans.}$$

It is Laurent's series.

Example 3: Expand $f(z)=1/\{(z-1)(z-2)\}$ in the region

(*i*) $|z|<1$ (*ii*) $1<|z|<2$ (*iii*) $|z|>2$. (*U.T.U. 2008*)

Sol. We have $f(z)=\frac{1}{(z-1)(z-2)}=\frac{1}{z-2}-\frac{1}{z-1}$ (By P.F.)

(*i*) $|z| < 1$

$$f(z) = \frac{1}{-2\left(1-\frac{z}{2}\right)} + \frac{1}{(1-z)} = -\frac{1}{2}\left(1-\frac{z}{2}\right)^{-1} + (1-z)^{-1}$$

$$= -\frac{1}{2}\left[1+\frac{z}{2}+\frac{z^2}{2^2}+\ldots+\frac{z^n}{2^n}\right] + \left[1+z+z^2+\ldots+z^n\right]$$

$$= -\frac{1}{2}\sum_{n=0}^{\infty}\frac{z^n}{2^n} + \sum_{n=0}^{\infty} z^n = \sum_{n=0}^{\infty}\left(1-\frac{1}{2^{n+1}}\right)z^n \quad \textbf{Ans.}$$

which is Taylor's series expansion

(*ii*) $1 < |z| < 2$, here $\frac{1}{|z|} < 1$ and $\frac{|z|}{2} < 1$

$$\therefore \quad f(z) = -\frac{1}{2\left(1-\frac{z}{2}\right)} - \frac{1}{z\left(1-\frac{1}{z}\right)} = -\sum_{n=0}^{\infty}\frac{z^n}{2^{n+1}} - \frac{1}{z}\left(1-\frac{1}{z}\right)^{-1}$$

$$= -\sum_{n=0}^{\infty}\frac{z^n}{2^{n+1}} - \frac{1}{z}\left(1+\frac{1}{z}+\frac{1}{z^2}+\ldots+\frac{1}{z^n}+\ldots\right) = -\sum_{n=0}^{\infty}\frac{z^n}{2^{n+1}} - \sum_{n=0}^{\infty}\frac{1}{z^{n+1}}$$

$$= -\left[\sum_{n=0}^{\infty}\frac{z^n}{2^{n+1}} + \sum_{n=0}^{\infty}\frac{1}{z^{n+1}}\right]. \quad \textbf{Ans.}$$

(*iii*) $|z| > 2$, here $\frac{2}{|z|} < 1$

$$\therefore \quad f(z) = \frac{1}{z\left(1-\frac{2}{z}\right)} - \sum_{n=0}^{\infty}\frac{1}{z^{n+1}} = \frac{1}{z}\left(1-\frac{2}{z}\right)^{-1} - \sum_{n=0}^{\infty}\frac{1}{z^{n+1}}$$

$$= \frac{1}{z}\left(1+\frac{2}{z}+\frac{2^2}{z^2}+\cdots+\frac{2^n}{z^n}+\cdots\right) - \sum_{n=0}^{\infty}\frac{1}{z^{n+1}}$$

$$= \sum_{n=0}^{\infty}\left(\frac{2^n}{z^{n+1}} - \frac{1}{z^{n+1}}\right) = \sum_{n=0}^{\infty}(2^n-1)z^{-n-1}$$

Example 4: Expand the following function in a Laurent's series:

$$f(z) = \frac{1}{z(z-1)(z-2)} \text{ for } |z-1| < 1$$

Sol.

$$f(z) = \frac{1}{z(z-1)(z-2)} = \frac{1}{2z} - \frac{1}{z-1} + \frac{1}{2(z-2)} \qquad \text{(By Partial Fraction)}$$

$$= \frac{1}{2(z-1+1)} - \frac{1}{(z-1)} + \frac{1}{2(z-1-1)}$$

$$= \frac{1}{2}\{1+(z-1)\}^{-1} - \frac{1}{z-1} - \frac{1}{2}\{1-(z-1)\}^{-1} \qquad \Big|\text{ As } |z-1|<1$$

$$= \frac{1}{2}\sum_{n=0}^{\infty}(-1)^n (z-1)^n - \frac{1}{z-1} - \frac{1}{2}\sum_{n=0}^{\infty}(z-1)^n \quad \textbf{Ans.}$$

which represents Laurent's sereis expansion

Example 5: If $0 < |z-1| < 2$, then express $f(z) = \dfrac{z}{(z-1)(z-3)}$ in a series of positive and negative powers of $(z - 1)$.

Sol. $0 < |z-1| < 2,$

$\therefore$ $\dfrac{|z-1|}{2} < 1$

$$f(z) = \frac{z}{(z-1)(z-3)} = -\frac{1}{2(z-1)} + \frac{3}{2(z-3)}$$

$$= \frac{-1}{2(z-1)} + \frac{3}{2(z-1-2)} = -\frac{1}{2(z-1)} + \frac{3}{-4\left(1-\frac{z-1}{2}\right)}$$

$$= -\frac{1}{2(z-1)} - \frac{3}{4}\left\{1-\left(\frac{z-1}{2}\right)\right\}^{-1}$$

$$= -\frac{1}{2(z-1)} - \frac{3}{4}\left\{1+\frac{z-1}{2}+\left(\frac{z-1}{2}\right)^2+\ldots\right\}$$

$$= -\frac{1}{2(z-1)} - \frac{3}{4}\sum_{n=0}^{\infty}\left(\frac{z-1}{2}\right)^n. \quad \textbf{Ans.}$$

Example 6: Prove that $\log z = (z-1) - \dfrac{(z-1)^2}{2} \ldots +$

Sol. By Taylor's Theorem

$$f(z) = f(a) + (z-a)f'(a) + \frac{(z-a)^2}{2!}f''(a) + \frac{(z-a)^3}{3!}f'''(a) + \ldots \qquad \ldots(1)$$

Taking, $a = 1$

Let $f(z) = \log z \Rightarrow f(1) = \log 1 = 0$

$$f'(z) = \frac{1}{z} \quad \Rightarrow f'(1) = 1$$

$$f''(z) = -\frac{1}{z^2} \quad \Rightarrow f''(1) = -1$$

$$f'''(z) = \frac{2}{z^3} \quad \Rightarrow f'''(1) = 2 \text{ etc.}$$

Using these values in equation (1), then we get

$$f(z) = \log z = 0 + (z-1)\times 1 + \frac{(z-1)^2}{2}(-1) + \ldots$$

$$\Rightarrow \qquad \log z = (z-1) - \frac{(z-1)^2}{2} + \frac{(z-1)^3}{3} - \ldots \qquad \textbf{Proved.}$$

Example 7: Write all possible Laurent series for the function $f(z) = \dfrac{1}{z(z+2)^3}$ about the pole $z = -2$.

Sol. The given pole is $z = -2$ *i.e.*, $z + 2 = 0$, we put $z + 2 = t$ for expansion

$$f(z) = \frac{1}{(t-2)t^3} = \frac{1}{-2t^3\left(1-\frac{t}{2}\right)}$$

$$= -\frac{1}{2t^3}\left(1-\frac{t}{2}\right)^{-1}$$

$$f(z) = -\frac{1}{2t^3}\left\{1+\frac{t}{2}+\frac{t^2}{4}+\frac{t^3}{8}+\frac{t^4}{16}+\ldots\right\}$$

$$= -\frac{1}{2t^3} - \frac{1}{4t^2} - \frac{1}{8t} - \frac{1}{16} - \frac{t}{32} \ldots$$

$$\Rightarrow \qquad f(z) = -\frac{1}{2(z+2)^3} - \frac{1}{4(z+2)^2} - \frac{1}{8(z+2)} - \frac{1}{16} - \frac{(z+2)}{32} \ldots \quad \textbf{Ans.}$$

Example 8: For the function $f(z) = \dfrac{4z-1}{z^4-1} + \dfrac{1}{z-1}$, find all Taylor's or Laurent series about the center zero.

Sol.

$$f(z) = \frac{4z-1}{z^4-1} + \frac{1}{z-1} = \frac{4z-1}{(z-1)(z+1)(z^2+1)} + \frac{1}{z-1}$$

$$f(z) = \frac{3}{4(z-1)} + \frac{5}{4(z+1)} + \frac{-2z+\frac{1}{2}}{(z^2+1)} + \frac{1}{z-1}$$

Let $|z|<1 \quad \Rightarrow \quad \dfrac{|z|}{1}<1$

$$\therefore \qquad f(z)=\frac{-3}{4(1-z)}+\frac{5}{4(1+z)}+\left(-2z+\frac{1}{2}\right)\frac{1}{(1+z^2)}-\frac{1}{1-z}$$

$$=-\frac{7}{4(1-z)}+\frac{5}{4(1+z)}+\left(-2z+\frac{1}{2}\right)\frac{1}{(1+z^2)}=-\frac{7}{4}(1-z)^{-1}+\frac{5}{4}(1+z)^{-1}+\left(-2z+\frac{1}{2}\right)(1+z^2)^{-1}$$

$$=-\frac{7}{4}\{1+z+z^2+....\}+\frac{5}{4}\{1-z+z^2-z^3+....\}+\left(-2z+\frac{1}{2}\right)\{1-z^2+z^4....\}$$

$$=\left\{-\frac{7}{4}-\frac{7}{4}z-\frac{7}{4}z^2-\frac{7}{4}z^3.....\right\}+\left\{\frac{5}{4}-\frac{5}{4}z+\frac{5}{4}z^2.....\right\}+$$

$$\{-2z+2z^3-2z^5+....\}+\left\{\frac{1}{2}-\frac{1}{2}z^2+\frac{1}{2}z^4......\right\}$$

$f(z)=-5z-z^2-z^3.....$ **Ans.**

Example 9: Obtain the expression for $\dfrac{(z-2)(z+2)}{(z+1)(z+4)}$ which are valid when

(*i*) $|z|<1$ (*ii*) $1<|z|<4$ (*iii*) $|z|>4$

Sol.
$$f(z)=\frac{(z-2)(z+2)}{(z+1)(z+4)}=1-\frac{5z+8}{(z+1)(z+4)}$$

$$\Rightarrow \qquad f(z)=1-\frac{1}{(1+z)}-\frac{4}{(z+4)}$$

(*i*) $|z|<1$

$$f(z)=1-(1+z)^{-1}-\left(1+\frac{z}{4}\right)^{-1}$$

$$\Rightarrow \qquad f(z)=1-\{1-z+z^2-...+(-1)^n z^n+...\}-\left\{1-\frac{z}{4}+\left(\frac{z}{4}\right)^2......+(-1)^n\left(\frac{z}{4}\right)^n\right\}$$

$$f(z)=-1+\left\{z-z^2+...(-1)^{n+1}z^n+...\right\}+\left\{\frac{z}{4}-\left(\frac{z}{4}\right)^2+...(-1)^{n+1}\left(\frac{z}{4}\right)^n+...\right\}$$

$$f(z)=-1+\sum_{n=0}^{\infty}(-1)^{n+1}\left[1+4^{-n}\right]z^n. \qquad \textbf{Ans.}$$

(*ii*) $1 < |z| < 4 \Rightarrow \dfrac{1}{|z|} < 1, \; \dfrac{|z|}{4} < 1$

$$f(z) = 1 - \frac{1}{z}\left(1 + \frac{1}{z}\right)^{-1} - \left(1 + \frac{z}{4}\right)^{-1}$$

$$= 1 - \frac{1}{z}\left\{1 - \frac{1}{z} + \frac{1}{z^2} \ldots + \ldots\right\} - \left\{1 - \frac{z}{4} + \left(\frac{z}{4}\right)^2 \ldots + \ldots\right\}$$

$$= \sum_{n=0}^{\infty} (-1)^n \left[\frac{1}{z^n} - \left(\frac{z}{4}\right)^n\right]. \quad \textbf{Ans.}$$

(*iii*) $|z| > 4 \quad \Rightarrow \quad \dfrac{4}{|z|} < 1$

$$f(z) = 1 - \frac{1}{z}\left(1 + \frac{1}{z}\right)^{-1} - \frac{4}{z}\left(1 + \frac{4}{z}\right)^{-1}$$

$$= 1 - \frac{1}{z}\left\{1 - \frac{1}{z} + \frac{1}{z^2} \ldots + \ldots\right\} - \frac{4}{z}\left\{1 - \frac{4}{z} + \frac{4}{z^2} \ldots + \ldots\right\}$$

$$= 1 - \frac{1}{z}\sum_{n=0}^{\infty} (-1)^n . \frac{1}{z^n} - \frac{4}{z}\sum_{n=0}^{\infty} (-1)^n . \left(\frac{4}{z}\right)^n$$

$$= 1 - \frac{1}{z}\sum_{n=0}^{\infty} (-1)^n \left[\frac{1}{z^{n+1}} + \left(\frac{4}{z}\right)^{n+1}\right] = 1 + \sum_{n=0}^{\infty} \frac{(-1)^{n+1}}{z^{n+1}}(1 + 4^{n+1}). \quad \textbf{Ans.}$$

Example 10: Expand $\dfrac{1}{(z+1)(z+3)}$ in the region

(*a*) $|z| < 1$ (*b*) $1 < |z| < 3$

(*c*) $|z| > 3$

Sol. We have $\quad f(z) = \dfrac{1}{(z+1)(z+3)} = \dfrac{1}{2}\left[\dfrac{1}{z+1} - \dfrac{1}{z+3}\right]$

(*a*) $|z| < 1$

$$f(z) = \frac{1}{2}\left[(1+z)^{-1} - \frac{1}{3}\left(1 + \frac{z}{3}\right)^{-1}\right] = \frac{1}{2}\left[\sum_{n=0}^{\infty} (-1)^n z^n - \frac{1}{3}\sum_{n=0}^{\infty} (-1)^n \left(\frac{z}{3}\right)^3\right]. \quad \textbf{Ans.}$$

It is Taylor's series expansion.

(*b*) $1<|z|<3$ here $\frac{1}{|z|}<1$ and $\frac{|z|}{3}<1$

$$f(z)=\frac{1}{2}\left[\frac{1}{z}\left(1+\frac{1}{z}\right)^{-1}-\frac{1}{3}\left(1+\frac{z}{3}\right)^{-1}\right]=\frac{1}{2}\left[\frac{1}{z}\sum_{n=0}^{\infty}(-1)^n\left(\frac{1}{z}\right)^n-\frac{1}{3}\sum_{n=0}^{\infty}(-1)^n\left(\frac{z}{3}\right)^n\right].\ \textbf{Ans.}$$

which represent at Laurent's series expansion.

(*c*) $|z|>3$ here $\left|\frac{3}{z}\right|<1$

$$f(z)=\frac{1}{2}\left[\frac{1}{2}\left(1+\frac{1}{z}\right)^{-1}-\frac{1}{z}\left(1+\frac{3}{z}\right)^{-1}\right]$$

$$=\frac{1}{2}\left[\frac{1}{z}\sum_{n=0}^{\infty}(-1)^n\left(\frac{1}{z}\right)^n-\frac{1}{z}\sum_{n=0}^{\infty}(-1)^n\left(\frac{3}{2}\right)^n\right]\quad \textbf{Ans.}$$

which represents at Laurent's series expansion.

Example 11: Find the Taylor's or Laurent's series which represent the function $\frac{1}{(1+z^2)(z+2)}$.

(*i*) When $|z|<1$ (*ii*) When $1<|z|<2$ (*iii*) When $|z|<2$.

Sol. $f(z)=\frac{1}{(1+z^2)(z+2)}=\frac{A}{(z+2)}+\frac{Bz+C}{1+z^2}=\frac{1}{5}\left[\frac{1}{z+2}-\frac{z-2}{(1+z^2)}\right]$ (By Partial Fraction)

(*i*) When $|z|<1$

Then $$f(z)=\frac{1}{5}\left[\frac{1}{z+2}-\frac{z-2}{(1+z^2)}\right]$$

$$f(z)=\frac{1}{5}\left[\frac{1}{2}\left(1+\frac{z}{2}\right)^{-1}-(z-2)(1+z^2)^{-1}\right]$$

$$=\frac{1}{10}\left[1-\frac{z}{2}+\left(\frac{z}{2}\right)^2....+..\right]-\frac{1}{5}\cdot(z-2)\left[1-z^2+z^4....\right]$$

$$f(z)=\frac{1}{10}\sum_{n=0}^{\infty}(-1)^n\left(\frac{z}{2}\right)^n+\frac{(2-z)}{5}\sum_{n=0}^{\infty}(-1)^n z^{2n}$$

which is a Taylor's series. **Ans.**

(*ii*) When $1<|z|<2$, then $\frac{1}{|z|}<1$ and $\frac{|z|}{2}<1$

So
$$f(z)=\frac{1}{5}\left[\frac{1}{2\left(1+\frac{z}{2}\right)}-\frac{z-2}{\left(1+z^2\right)}\right]$$

$$=\frac{1}{5}\cdot\frac{1}{2}\left(\frac{1}{\left(1+\frac{z}{2}\right)}\right)-\frac{(z-2)}{5z^2}\left(1+\frac{1}{z^2}\right)^{-1}$$

$$=\frac{1}{10}\sum_{n=0}^{\infty}(-1)^n\left(\frac{z}{2}\right)^n+\frac{(2-z)}{5z^2}\left(1-\frac{1}{z^2}+\frac{1}{z^4}....+\right)$$

$$=\frac{1}{10}\sum_{n=0}^{\infty}(-1)^n\left(\frac{z}{2}\right)^n+\frac{(2-z)}{5z^2}\sum_{n=0}^{\infty}(-1)^n\frac{1}{z^{2n}}$$

There are negative and positive powers of z in above expression so it is Laurent's expansion of $f(z)$ within the annulus. $1<|z|<2$.

(*iii*) When $|z|<2$ or $\frac{2}{|z|}<1$

Now
$$f(z)=\frac{1}{5}\left[\frac{1}{z+2}-\frac{z-2}{\left(1+z^2\right)}\right]$$

$$=\frac{1}{5z}\left(1+\frac{2}{z}\right)^{-1}-\frac{(z-2)}{5z^2}\left(1+\frac{1}{z^2}\right)^{-1}$$

$$=\frac{1}{5z}\left[1-\frac{2}{z}+\left(\frac{2}{z}\right)^2....+..\right]-\frac{(z-2)}{5z^2}\left[1-\frac{1}{z^2}+\frac{1}{z^4}....\right]$$

$$f(z)=\frac{1}{5z}\sum_{n=0}^{\infty}(-1)^n\left(\frac{2}{z}\right)^n-\frac{(z-2)}{5z^2}\sum_{n=0}^{\infty}(-1)^n\frac{1}{z^{2n}}$$

This is Laurent's expansion within the annulus $|z|<2$.. **Ans.**

Example 12: Expand cos z about the point $z=\frac{\pi}{2}$.

Sol. By Taylor's theorem

$$f(z)=\sum_{n=0}^{\infty}a_n\left(z-\frac{\pi}{2}\right)^n \quad \text{where } a_n=\frac{f^n(\pi/2)}{n!}$$

$$f(z) = \cos z \qquad \Rightarrow \quad f(\pi/2) = 0$$
$$f'(z) = -\sin z = \cos(z + \pi/2) \qquad \Rightarrow \quad f'(\pi/2) = -1$$
$$f''(z) = -\cos z = \cos(z + 2\pi/2) \qquad \Rightarrow \quad f''(\pi/2) = 0$$
$$f'''(z) = \sin z = \cos(z + 3\pi/2) \qquad \Rightarrow \quad f'''(\pi/2) = 1$$

....

$$f^n(z) = \cos\left(z + \frac{n\pi}{2}\right) \qquad \Rightarrow \quad f^n(\pi/2) = \cos\left(\frac{\pi}{2} + \frac{n\pi}{2}\right) = -\sin\frac{n\pi}{2}$$

$$a_n = \frac{-\sin\frac{n\pi}{2}}{n!}$$

$a_0 = 0$, but $a_2 = \dfrac{f''(\pi/2)}{2!} = 0 = a_4 = a_6$ and so on.

Hence
$$\cos z = a_0' + a_1\left(z - \frac{\pi}{2}\right) + a_2\left(z - \frac{\pi}{2}\right)^2 + ...$$

$$\cos z = 0 + 0 + 0 + 0 + ... + \left(z - \frac{\pi}{2}\right)a_1 + \left(z - \frac{\pi}{2}\right)^3 a_3 + ...$$

$$\cos z = \sum\left(z - \frac{\pi}{2}\right)^n a_n \text{ where } n \text{ is odd}$$

$$\cos z = \sum -\frac{\sin\frac{n\pi}{2}}{n!}\cdot\left(z - \frac{\pi}{2}\right)^n, \quad \text{where } n \text{ is odd integer.} \quad \textbf{Ans.}$$

Example 13: For the function $= \dfrac{2z^3 + 1}{z^2 + z}$, find

(*i*) A Taylor's series valid in the neighborhood of the point $z = i$.

(*ii*) A Laurent's series valid within the annulus of which center is origin.

Sol. (*i*)
$$f(z) = \frac{2z^3 + 1}{z^2 + z} = 2z - 2 + \frac{2z + 1}{z(1 + z)}$$
$$= 2(z - 1) + \frac{2z + 1}{z(1 + z)}$$

$$\Rightarrow \qquad f(z) = 2(z - 1) + \frac{1}{(1 + z)} + \frac{1}{z} \qquad \text{(By partial fraction)}$$

$$\therefore \qquad f^n(z) = (-1)^n n!\left[\frac{1}{(1 + z)^{n+1}} + \frac{1}{z^{(n+1)}}\right]$$

$$\therefore \qquad f^n(i) = (-1)^n n!\left[\frac{1}{(1 + i)^{n+1}} + \frac{1}{i^{(n+1)}}\right]$$

Now by Taylor's theorem

$$f(z) = \sum_{n=0}^{\infty} a_n (z-i)^n, \text{where } a_n = \frac{f^n(i)}{n!}$$

Now $$a_0 = f(i) = 2(i-1) + \frac{1}{(i+1)} + \frac{1}{i} = \frac{i}{2} - \frac{3}{2}$$

$$a_1 = f'(i) = 2 + (-1) + \left[\frac{1}{(i+1)^2} + \frac{1}{i^2}\right] = 3 + \frac{i}{2}$$

Hence $$f(z) = a_0 + a_1(z-i) + \sum_{n=2}^{\infty} a_n (z-i)^n$$

$$f(z) = \frac{i}{2} - \frac{3}{2} + \left(3 + \frac{i}{2}\right)(z-i) + \sum_{n=2}^{\infty} (-1)^n \left[\frac{1}{(1+i)^{n+1}} + \frac{1}{i^{n+1}}\right](z-i)^n. \textbf{ Ans.}$$

(*ii*) Now the center of circle is origin so the region of convergent can be taken as $0 < |z| < 1$.

We have $$f(z) = 2z - 2 + \frac{1}{1+z} + \frac{1}{z}$$

$$f(z) = 2z - 2 + (1+z)^{-1} + \frac{1}{z}$$

$$= -2 + 2z + \left[1 - z + z^2 - z^3 .....\right] + \frac{1}{z}$$

$$f(z) = -2 + 2z + \sum_{n=0}^{\infty} (-1)^n z^n + \frac{1}{z}. \quad \textbf{Ans.}$$

Example 14: Show that $\sin\left\{c\left(z + \frac{1}{z}\right)\right\}$ can be expanded in a series of the type

$$a_0 + a_1 z + a_2 z^2 + ... + \frac{b_1}{z} + \frac{b_2}{z^2} + ...$$

In which the coefficient of z^n and z^{-n}, are $\frac{1}{2\pi}\int_0^{2\pi} \sin(2c\cos\theta)\cos n\theta d\theta.$

Sol. The function $f(z) = \sin c\left(z + \frac{1}{z}\right)$ will be analytic at everywhere except at $z = 0$ *i.e.*, it will be analytic in the region $r \le |z| \le R$ where $r < R$. So we can expand by Laurent's theorem.

$$\sin c\left(z + \frac{1}{z}\right) = \sum_{n=0}^{\infty} a_n z^n + \sum_{n=1}^{\infty} b_n z^{-n}$$

where $$a_n = \frac{1}{2\pi i}\int_C \frac{f(z)}{(z-a)^{n+1}} dz \quad \text{and} \quad b_n = \frac{1}{2\pi i}\int_C \frac{f(z)}{(z-a)^{-n+1}} dz$$

$$\Rightarrow \qquad a_n = \frac{1}{2\pi i}\int_C \frac{\sin c\left(z+\frac{1}{z}\right)}{z^{n+1}}dz \text{ and } b_n = \frac{1}{2\pi i}\int_C \frac{\sin c\left(z+\frac{1}{z}\right)}{z^{-n+1}}dz \qquad \Big| \text{ As } a = 0$$

where C is any circle in the region with center at $z = 0$.

Now
$$a_n = \frac{1}{2\pi i}\int_C \frac{\sin c\left(z+\frac{1}{z}\right)}{z^{n+1}}dz$$

Let
$$z = e^{i\theta} \text{ as } |z| = 1$$

$$dz = ie^{i\theta}d\theta$$

$$\Rightarrow \qquad a_n = \frac{1}{2\pi i}\int_0^{2\pi} \frac{\sin c(e^{i\theta}+e^{-i\theta})}{e^{in\theta}\cdot e^{i\theta}} e^{i\theta} i\cdot d\theta$$

$$= \frac{1}{2\pi}\int_0^{2\pi} \sin(2c\cos\theta)\, e^{-in\theta}\cdot d\theta$$

$$= \frac{1}{2\pi}\int_0^{2\pi} \sin(2c\cos\theta)\cdot\cos n\theta\, d\theta - \frac{i}{2\pi}\int_0^{2\pi}\sin(2c\sin\theta)\cdot\cos n\theta\, d\theta$$

$$\Rightarrow \qquad a_n = \frac{1}{2\pi}\int_0^{2\pi} \sin(2c\cos\theta)\cdot\cos n\theta\, d\theta$$

if $n \to -n$

then
$$b_n = a_{-n} = \frac{1}{2\pi}\int_0^{2\pi} \sin(2c\cos\theta)\cdot\cos(-n\theta)\, d\theta$$

$$\Rightarrow \qquad b_n = a_{-n} = \frac{1}{2\pi}\int_0^{2\pi} \sin(2c\cos\theta)\cdot\cos n\theta\, d\theta$$

$\Big| \int_0^{2\pi}\sin(2c\cos\theta)\cdot\sin n\theta\, d\theta \to 0$ as $\int_0^{2\pi} F(\theta)d\theta \to 0$ by $F(2\pi-\theta) = -F(\theta)$ being satisfied

Hence
$$\sin c\left(z+\frac{1}{z}\right) = \sum_{n=0}^{\infty} a_n z^n + \sum_{n=1}^{\infty} b_n z^{-n}$$

where
$$a_n = b_n = \frac{1}{2\pi}\int_0^{2\pi} \sin(2c\cos\theta)\cdot\cos n\theta\, d\theta$$

Example 15: Prove that $\cosh\left(z+\frac{1}{z}\right) = a_0 + \sum_{n=1}^{\infty} a_n\left(z^n + \frac{1}{z^n}\right)$

where $a_n = \frac{1}{2\pi}\int_0^{2\pi} \cos n\theta \cosh(2\cos\theta)d\theta$

Sol. The function $f(z) = \cosh\left(z+\frac{1}{z}\right)$ is analytic at everywhere except at $z = 0$, *i.e.* in the region $r \le |z| \le R$. Hence the given function can be expanded in Laurent's series in the region $r \le |z| \le R$.

So $$f(z) = \cosh\left(z + \frac{1}{z}\right) = \sum_{n=0}^{\infty} a_n z^n + \sum_{n=1}^{\infty} b_n z^{-n} \qquad \ldots(1)$$

where $$a_n = \frac{1}{2\pi i}\int_C \frac{\cosh\left(z + \frac{1}{z}\right)}{z^{n+1}}\,dz$$

and $$b_n = \frac{1}{2\pi i}\int_C \frac{\cosh\left(z + \frac{1}{z}\right)}{z^{-n+1}}\,dz$$

where C is a circle with centre at $z = 0$. Let $|z| = 1$.

Then $z = e^{i\theta}$

So $$a_n = \frac{1}{2\pi i}\int_0^{2\pi} \frac{\cosh(e^{i\theta} + e^{-i\theta})}{e^{in\theta}\cdot e^{i\theta}} e^{i\theta} i \cdot d\theta$$

$$a_n = \frac{1}{2\pi}\int_0^{2\pi} \frac{\cosh(2\cos\theta)}{e^{in\theta}}\,d\theta$$

or $$a_n = \frac{1}{2\pi}\int_0^{2\pi} \cosh(2\cos\theta).\cos n\theta d\theta$$ Second integral vanishes by definite integral property

$$b_n = a_{-n} = \frac{1}{2\pi}\int_0^{2\pi} \cosh(2\cos\theta)\cdot\cos n\theta d\theta \quad \text{as } \cos(-\theta) = \cos\theta$$

$\Rightarrow$ $$b_n = a_n$$

Hence from equation (1)

$$f(z) = \cosh\left(z + \frac{1}{z}\right) = a_0 + \sum_{n=1}^{\infty} a_n z^n + \sum_{n=1}^{\infty} a_n \frac{1}{z^n}$$

$\Rightarrow$ $$\cosh\left(z + \frac{1}{z}\right) = a_0 + \sum_{n=1}^{\infty} a_n\left(z^n + \frac{1}{z^n}\right)$$ **Proved.**

Example 16: Show that if $c > 0$, then $e^{z+\left(\frac{c^3}{2z^2}\right)} = \sum_{n=-\infty}^{\infty} a_n z^n$

where $a_n = \dfrac{e^{\frac{-c}{2}}}{2\pi c^n}\displaystyle\int_0^{2\pi} e^{c(\cos\theta + \cos^2\theta)}.\cos[c\sin\theta(1 - \cos\theta) - n\theta]\,d\theta$

Sol. The function $f(z) = e^{z+(c^3/2z^2)}$ will be analytic at every where except the point $z = 0$ in the region $r < |z| < R$ so expanded the given function in Laurent's series.

$$e^{z+(c^3/2z^2)} = \sum_{n=0}^{\infty} a_n z^z + \sum_{n=1}^{\infty} b_n z^{-z} = \sum_{n=-\infty}^{\infty} a_n z^z \qquad \ldots(1)$$

where $$a_n = \frac{1}{2\pi i}\int_C \frac{e^{z+(c^3/2z^2)}}{z^{n+1}}dz$$

where C is a circle with center at $z = 0$.

Let $|z| = c$ so $z = ce^{i\theta}$

$$a_n = \frac{1}{2\pi i}\int_0^{2\pi} \frac{e^{\left(ce^{i\theta}+\frac{c}{2}e^{-2i\theta}\right)}}{ce^{i\theta}.c^n e^{in\theta}} ice^{i\theta}\cdot d\theta$$

$$a_n = \frac{1}{2\pi c^n}\int_0^{2\pi} e^{\left(ce^{i\theta}+\frac{c}{2}e^{-2i\theta}-in\theta\right)}d\theta$$

$$a_n = \frac{1}{2\pi c^n}\int_0^{2\pi} e^{\{c\cos\theta+ic\sin\theta+\frac{c}{2}\cos 2\theta-\frac{ic}{2}\sin 2\theta-in\theta\}}d\theta$$

$$= \frac{1}{2\pi c^n}\int_0^{2\pi} e^{c\left(\cos\theta+\frac{1}{2}c\cos 2\theta\right)} e^{\left\{i\left(c\sin\theta-\frac{c}{2}\sin 2\theta-n\theta\right)\right\}}d\theta$$

$$a_n = \frac{1}{2\pi c^n}\int_0^{2\pi} e^{c\left(\cos\theta+\frac{1}{2}\cos 2\theta\right)}.\left\{\cos\left(c\sin\theta-\frac{c}{2}\sin 2\theta-n\theta\right)+i\sin\left(c\sin\theta-\frac{c}{2}\sin 2\theta-n\theta\right)\right\}d\theta$$

$$a_n = \frac{1}{2\pi c^n}\int_0^{2\pi} e^{c\left(\cos\theta+\frac{1}{2}\cos 2\theta\right)}\left\{\cos\left(c\sin\theta-\frac{c}{2}\sin 2\theta-n\theta\right)\right\}d\theta$$

The 2nd term will be zero as $F(2\pi-\theta) = -F(\theta)$ being satisfied.

$$\therefore \quad a_n = \frac{1}{2\pi c^n}\int_0^{2\pi} e^{c\left(\cos\theta+\frac{1}{2}\cos 2\theta\right)}\cos\{c\sin\theta(1-\cos\theta)-n\theta\}d\theta$$

$$= \frac{1}{2\pi c^n}e^{-c/2}\int_0^{2\pi} e^{\frac{c}{2}}.e^{\frac{c}{2}\cos 2\theta} e^{c\cos\theta}\times\cos\{c\sin\theta(1-\cos\theta)-n\theta\}d\theta$$

$$= \frac{1}{2\pi c^n}e^{-c/2}\int_0^{2\pi} e^{\frac{c}{2}(1+\cos 2\theta)} e^{c\cos\theta}\times\cos\{c\sin\theta(1-\cos\theta)-n\theta\}d\theta$$

$$= \frac{e^{-c/2}}{2\pi c^n}\int_0^{2\pi} e^{c(\cos^2\theta+\cos\theta)}\times\cos\{c\sin\theta(1-\cos\theta)-n\theta\}d\theta$$

Hence $$e^{z+\frac{c^3}{2z^2}} = \sum_{n=-\infty}^{\infty} a_n z^n.$$

where $a_n = \dfrac{c^{-\frac{c}{2}}}{2\pi c^n}\displaystyle\int_0^{2\pi} e^{c(\cos\theta+\cos^2\theta)}\cdot\cos\{c\sin\theta(1-\cos\theta)-n\theta\}d\theta.$ **Proved.**

Example 17: If C is a closed contour around the origin, prove that

$$\left(\frac{a^n}{\lfloor n}\right)^2 = \frac{1}{2\pi i}\int_C \frac{a^n e^{az}}{\lfloor n\, z^{n+1}}\,dz$$

Hence deduce that

$$\sum_{n=0}^{\infty}\left(\frac{a^n}{n!}\right)^2 = \frac{1}{2\pi}\int_o^{2\pi} e^{2a\cos\theta}\,d\theta$$

Sol. Let $f(z) = e^{az}$

Then $\qquad f'(z) = ae^{az}, f''(z) = a^2 e^{az}, \ldots\ldots, f^n(z) = a^n e^{az}$

At $\qquad z = 0$

$$f^n(0) = a^n e^0 = a^n \quad \Rightarrow \quad a^n = f^n(0)$$

But $\qquad f^n(a) = \dfrac{n!}{2\pi i}\displaystyle\int_C \dfrac{f(z)}{(z-a)^{n+1}}\,dz$

At $\qquad a = 0,\ f^n(0) = \dfrac{n!}{2\pi i}\displaystyle\int_C \dfrac{f(z)}{z^{n+1}}\,dz$

or $\qquad a^n = \dfrac{n!}{2\pi i}\displaystyle\int_C \dfrac{f(z)}{z^{n+1}}\,dz$

or $\qquad \left(\dfrac{a^n}{n!}\right) = \dfrac{1}{2\pi i}\displaystyle\int_C \dfrac{f(z)}{z^{n+1}}\,dz$

or $\qquad \left(\dfrac{a^n}{n!}\right)^2 = \dfrac{1}{2\pi i}\displaystyle\int_C \dfrac{a^n e^{az}\,dz}{z^{n+1}.n!}.$ **Proved.**

Now $\qquad \displaystyle\sum_{n=0}^{\infty}\left(\frac{a^n}{n!}\right)^2 = \frac{1}{2\pi i}\int_C \frac{e^{az}}{z}\left[\sum_{n=0}^{\infty}\left(\frac{a}{z}\right)^n.\frac{1}{n!}\right]dz$

$$= \frac{1}{2\pi i}\int_C \frac{e^{az}}{z}\left[1 + \frac{a}{z} + \frac{1}{2!}\left(\frac{a}{z}\right)^2 + \ldots\right]dz$$

$$= \frac{1}{2\pi i}\int_C \frac{e^{az}}{z} e^{\frac{a}{z}}\,dz = \frac{1}{2\pi i}\int_C \frac{e^{a\left(z+\frac{1}{z}\right)}}{z}\,dz \qquad \left[e^x = 1 + \frac{x}{1!} + \frac{x^2}{2!} + \ldots\right]$$

Let $|z| = 1$ so $z = e^{i\theta}$

$$\sum_{n=0}^{\infty}\left(\frac{a^n}{n!}\right)^2 = \frac{1}{2\pi}\int_0^{2\pi} e^{2a\cos\theta}\,d\theta.$$ **Proved.**

Example 18: If the function $f(z)$ is analytic and one valued in $|z-a|<R$. Prove that for $0<r<R$.

$$f'(a)=\frac{1}{\pi r}\int_0^{2\pi} P(\theta)e^{-i\theta}d\theta, \text{ where } P(\theta) \text{ is the real part of } f(a+re^{i\theta})$$

Sol. Since the function $f(z)$ is regular in $|z-a|<R$, therefore it must be regular in $|z-a|<r$. The function $f(z)$ can be expanded by Taylor's theorem.

So
$$f(z)=\sum_{m=0}^{\infty} a_m (z-a)^m$$

$$f(z)=\sum_{m=0}^{\infty} a_m r^m e^{im\theta} \quad ...(1) \qquad \Big| \text{as } z-a=re^{i\theta}$$

We know that

$$f'(a)=\frac{1}{2\pi i}\int_C \frac{f(z)dz}{(z-a)^2} \quad ...(2)$$

But for the real part we will add the term $\dfrac{1}{2\pi i}\displaystyle\int \frac{\overline{f(z)}}{(z-a)^2}dz$ in above formula so we find the value of this term from equation (1)

$$\overline{f(z)}=\sum_{m=0}^{\infty} \bar{a}_m r^m e^{-im\theta}$$

Now
$$\frac{1}{2\pi i}\int_C \frac{\overline{f(z)}}{(z-a)^{n+1}}dz=\frac{1}{2\pi i}\int_0^{2\pi}\left(\sum_{m=0}^{\infty}\bar{a}_m r^m e^{-im\theta}\right)\frac{re^{i\theta}id\theta}{r^{n+1}e^{i(n+1)\theta}}$$

$$=\frac{1\times i}{2\pi i}\int_0^{2\pi}\sum_{m=0}^{\infty}\bar{a}_m r^{m-n}e^{-i(m+n)\theta}d\theta$$

$$=\frac{1}{2\pi}\sum_{m=0}^{\infty}\bar{a}_m r^{m-n}\int_0^{2\pi}e^{-i(m+n)\theta}d\theta$$

$$=\frac{1}{2\pi}\sum_{m=0}^{\infty}\bar{a}_m r^{m-n}\left[\frac{e^{-i(m+n)\theta}}{-(m+n)i}\right]_0^{2\pi}$$

$$=-\frac{1}{2\pi i}\sum_{m=0}^{\infty}\frac{\bar{a}_m r^{m-n}}{(m+n)}\left[\cos(m+n)\theta-i\sin(m+n)\theta\right]_0^{2\pi}$$

$$=-\frac{1}{2\pi i}\sum_{m=0}^{\infty}\frac{\bar{a}_m r^{m-n}}{(m+n)}\times 0=0$$

$$\Rightarrow \quad \frac{1}{2\pi i}\int_C\frac{\overline{f(z)}}{(z-a)^{n+1}}dz=0$$

If $n=1$

Then $$\frac{1}{2\pi i}\int_C \frac{\overline{f(z)}}{(z-a)^2}\,dz = 0$$

Adding this integral in (2), we get

$$f'(a) = \frac{1}{2\pi i}\int_C \frac{f(z)+\overline{f(z)}}{(z-a)^2}\,dz$$

$$= \frac{1}{2\pi i}\int_0^{2\pi} \frac{f(a+re^{i\theta})+\overline{f(a+re^{i\theta})}}{r^2e^{2i\theta}}\,rie^{i\theta}\,d\theta, \text{ as } z-a = re^{i\theta}$$

$$f'(a) = \frac{1}{2\pi}\int_0^{2\pi} \frac{2R.P.\text{ of } f(a+re^{i\theta})}{r^2e^{2i\theta}}\,re^{i\theta}\,d\theta \quad (\text{as } z+\bar{z} = 2R.P. \text{ of } z)$$

$\Rightarrow$ $$f'(a) = \frac{1}{\pi r}\int_0^{2\pi} P(\theta).e^{-i\theta}\,d\theta. \textbf{ Proved.}$$

Example 19: Find the Taylor's series expansion of a function of the complex variable. $f(z) = \dfrac{1}{(z-1)(z-3)}$ about the point $z = 4$. Find its region of convergence.

Sol. It is necessary that the function $f(z)$ will be analytic with in the circle at every point, if $z = 1$, $z = 3$ lie outside the circle so the radius of the circle must be unity. Hence the function $f(z)$ will be analytic with in the circle $|z-4| = 1$. The region of convergence is $|z-4| < 1$.

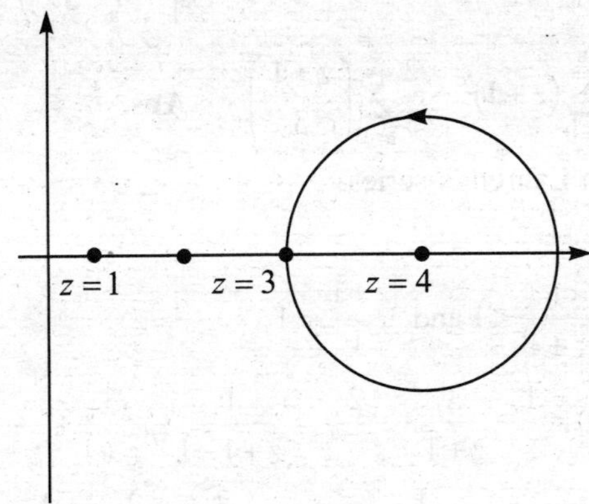

Now $$f(z) = \frac{1}{(z-1)(z-3)} = \frac{1}{2}\left[\frac{1}{(z-3)} - \frac{1}{(z-1)}\right]$$

$$= \frac{1}{2}\left[\frac{1}{(z-4+1)} - \frac{1}{(z-4+3)}\right]$$

$$= \frac{1}{2}\left(1+\overline{z-4}\right)^{-1} - \frac{1}{3}\left(1+\frac{z-4}{3}\right)^{-1}$$

$$f(z)=\frac{1}{2}\left\{1-(z-4)+(z-4)^2-...+...\right\}-\frac{1}{3}\left\{1-\left(\frac{z-4}{3}\right)+\left(\frac{z-4}{3}\right)^2-.......\right\}$$

$$\therefore \qquad f(z)=\frac{1}{6}-\frac{7}{18}(z-4)+\frac{25}{54}(z-4)^2-\frac{79}{162}(z-4)^3+... \quad \textbf{Ans.}$$

Example 20: Find Laurent's expansion for $f(z)=\dfrac{7z-2}{z^3-z^2+2z}$ in the regions given by

(*i*) $0<|z+1|<1$ (*ii*) $1<|z+1|<3$ (*iii*) $|z+1|>3$

Sol. $f(z)=\dfrac{7z-2}{z^3-z^2+2z}=\dfrac{7z-2}{z(z+1)(z-2)}$

or $$f(z)=\frac{1}{z}-\frac{3}{z+1}+\frac{2}{z-2}.$$

(*i*) When $0<|z+1|<1$, then

$$f(z)=\frac{1}{(z+1-1)}-\frac{3}{z+1}+\frac{2}{z+1-3}$$

$$=-\frac{3}{z+1}-\{1-(z+1)\}^{-1}-\frac{2}{3}\left\{1-\frac{z+1}{3}\right\}^{-1}$$

$$=-\frac{3}{z+1}-\left\{1+(z+1)+(z+1)^2+....\right\}-\frac{2}{3}\left\{1+\left(\frac{z+1}{3}\right)+\left(\frac{z+1}{3}\right)^2+....\right\}$$

$$f(z)=-\frac{3}{z+1}-\sum_{n=0}^{\infty}(z+1)^n-\frac{2}{3}\sum_{n=0}^{\infty}\left(\frac{z+1}{3}\right)^n. \quad \textbf{Ans.}$$

It is an expansion of $f(z)$ in Laurent's series.

(*ii*) When $1<|z+1|<3$ *i.e.* $\dfrac{1}{|z+1|}<1$ and $\dfrac{|z+1|}{3}<1$

So $$f(z)=\frac{1}{z}-\frac{3}{z+1}+\frac{2}{z-2}=\frac{1}{z+1-1}-\frac{3}{z+1}+\frac{2}{z+1-3}$$

$$f(z)=\frac{1}{(z+1)\left\{1-\dfrac{1}{z+1}\right\}}-\frac{3}{(z+1)}-\frac{2}{3}\frac{1}{\left(1-\dfrac{z+1}{3}\right)}$$

$$=\frac{1}{(z+1)}\left\{1-\frac{1}{z+1}\right\}^{-1}-\frac{3}{z+1}-\frac{2}{3}\left(1-\frac{z+1}{3}\right)^{-1}$$

$$=\frac{1}{(z+1)}\left\{1+\frac{1}{(z+1)}+\frac{1}{(z+1)^2}+......\right\}-\frac{3}{(z+1)}-\frac{2}{3}\sum_{n=0}^{\infty}\left(\frac{z+1}{3}\right)^n$$

$$= \frac{1}{(z+1)}\sum_{n=0}^{\infty}\left(\frac{1}{z+1}\right)^n - \frac{3}{(z+1)} - \frac{2}{3}\sum_{n=0}^{\infty}\left(\frac{z+1}{3}\right)^n. \quad \textbf{Ans.}$$

It is also an expansion in Laurent's series.

(*iii*) When $|z+1| > 3$ *i.e.* $\frac{3}{|z+1|} < 1$

$$f(z) = \frac{1}{(z+1)}\left\{1 - \frac{1}{(z+1)}\right\}^{-1} - \frac{3}{(z+1)} + \frac{2}{(z+1)}\left(1 - \frac{3}{z+1}\right)^{-1}$$

$$= \frac{1}{(z+1)}\sum_{n=0}^{\infty}\left(\frac{1}{z+1}\right)^n - \frac{3}{(z+1)} + \frac{2}{(z+1)}\sum_{n=0}^{\infty}\left(\frac{3}{z+1}\right)^n. \quad \textbf{Ans.}$$

It is an expansion in Laurent's series.

PROBLEM SET 3.1

1. Find the radius of convergence of the following power series:

(*i*) $\sum \frac{z^n}{n^n}$ (*ii*) $\sum (3+4i)^n z^n$ $\left[\textbf{Ans.}\ (i)\ \infty\ (ii)\ \frac{1}{5}\right]$

2. Find the radius of convergence of the series.

$$\frac{z}{2} + \frac{1.3}{2.5}z^2 + \frac{1.3.5}{2.5.8}z^3 + \ldots\ldots\ldots\ldots$$ $\left[\textbf{Ans.}\ \frac{3}{2}\right]$

3. Find the radius of convergence of the power series.

$$\sum_{n=1}^{\infty}\frac{(-1)^{n-1} z^{2n-1}}{(2n-1)!}$$ [**Ans.** ∞]

4. Find the radius of convergence of the series $\sum_{n=0}^{\infty}\frac{(n+1)z^n}{(n+2)(n+3)}$. [**Ans.** 1]

5. Obtain the expression for $\frac{(z-2)(z+2)}{(z+1)(z+4)}$ which are valid when

(*i*) $|z| < 1$ (*ii*) $1 < |z| < 4$ (*iii*) $|z| > 4$.

$\left[\textbf{Ans.}\ (i)\ -1 + \sum_{0}^{\infty}(-1)^{n+1}\left[1+4^{-n}\right]z^n \quad (ii)\ \sum_{n=1}^{\infty}(-1)^n\left[\frac{1}{z^n} - \left(\frac{z}{4}\right)^n\right] \quad (iii)\ 1 + \sum_{n=1}^{\infty}(-1)^n\frac{(1+4^n)}{z^n}\right]$

6. Expand $f(z)=\dfrac{z+3}{z\left(z^2-z-2\right)}$ in power of z; where

(*i*) $|z|<1$ (*ii*) $1<|z|<2$ (*iii*) $|z|>2$

$$\left[\textbf{Ans. } (i) -\frac{3}{2z}+\frac{2}{3}\sum_{n=1}^{\infty}(-1)^n z^n-\frac{5}{12}\sum_{n=0}^{\infty}\left(\frac{z}{2}\right)^n \quad (ii) -\frac{3}{2z}+\frac{2}{3z}\sum_{n=0}^{\infty}\frac{(-1)^n}{z^n}-\frac{5}{12}\sum_{n=0}^{\infty}\left(\frac{z}{2}\right)^n\right.$$

$$\left.(iii) -\frac{3}{2z}+\frac{2}{3z}\sum_{n=0}^{\infty}(-1)^n\frac{1}{z^n}-\frac{5}{6z}\sum_{n=0}^{\infty}\left(\frac{z}{2}\right)^n\right]$$

7. Obtain Taylor's series expansion of $f(z)=\dfrac{1}{z^2+4}$ about the point $z=-i$. Find the region of convergence.

$$\left[\textbf{Ans. } \frac{1}{4}\left[\frac{1}{3}\sum_{n=0}^{\infty}\left(\frac{z+i}{3i}\right)^n+\sum_{n=0}^{\infty}(-1)^n\left(\frac{z+i}{i}\right)^n\right] \text{ region of convergence is } |z+i|<1\right]$$

8. Find the Laurent's series expansion of the function

$$f(z)=\frac{z^2-6z-1}{(z-1)(z-3)(z+2)} \text{ in the region } 3<|z+2|<5.$$

$$\left[\textbf{Ans. } \frac{2}{(z+2)}+\frac{3}{(z+2)^2}+\frac{3}{(z+2)^3}\ldots+\frac{1}{5}\left[1+\frac{(z+2)}{5}+\frac{(z+2)^2}{5^2}+\ldots\right]\right]$$

9. Find the terms in the Laurent's expansion of $\dfrac{1}{z(e^z-1)}$ for the region $0<|z|<2\pi$.

$$\left[\textbf{Ans. } z^{-2}-\frac{1}{2}z^{-1}+\frac{1}{12}-\frac{z^2}{720}+\ldots\right]$$

10. Expand the following function in a Laurent's series about the point $z = 0$;

$$f(z)=\frac{1-\cos z}{z^3}. \qquad \left[\textbf{Ans. } \sum_{n=2}^{\infty}\frac{(-1)^n z^{2n-5}}{(2n-2)!}\right]$$

11. Show that when $|z+1|<1,\ z^{-2}=1+\sum_{n=1}^{\infty}(n+1)(z+1)^n$.

12. Find the expansion in powers of z of $\dfrac{1}{(2z-1)(z+3)}$ in the region $0<|z|<4$.

13. Represent a function $f(z)=z/(z-1)(z-2)$ by a series of positive and negative powers of $(z-1)$ which convergence to $f(z)$ when $0<|z-1|<2$.

$$\left[\textbf{Ans. } -\frac{1}{2(z-1)}-\frac{3}{4}\sum_{n=0}^{\infty}\left(\frac{z-1}{2}\right)^n\right]$$

14. Expand $f(z)=\dfrac{1}{(2z+1)^3}$ about (*a*) $z = 0$, (b) $z = 2$.

$$\left[\textbf{Ans. } \sum_{n=0}^{\infty}\frac{(-1)^n\lfloor(3+n-1)}{\lfloor n \quad \lfloor 2}(2z)^n, |2z|<1; \frac{1}{125}\sum_{n=0}^{\infty}(-1)^n\cdot\frac{(n+2)(n+1)}{2}\left(\frac{2}{5}\right)^n(z-2)^n\right]$$

15. Expand $f(z)=\dfrac{e^z}{z(z+1)}$ about $z_0 = 2$. $\left[\textbf{Ans. } \frac{1}{6}-\frac{5}{9}(z-2)+\frac{19}{27}(z-2)^2+\cdots, |z-2|<1\right]$

16. Expand $\dfrac{1}{(z+1)(z+3)}$ in Laurent series valid for

(*a*) $1<|z|<3$ (*b*) $|z|>3$ (*c*) $0<|z+1|<2$ (*d*) $|z|<1$.

$$\left[\textbf{Ans. } (a) -\frac{1}{2z^4}+\frac{1}{2z^3}-\frac{1}{2z^2}+\frac{1}{2z}-\frac{1}{6}+\cdots, (b)\frac{1}{z^2}-\frac{4}{z^3}+\frac{13}{z^4}-\frac{40}{z^5}+\cdots\right.$$

$$\left.(c)\ \frac{1}{2(z+1)}-\frac{1}{4}+\frac{(z+1)}{8}-\frac{1}{16}(z+1)^2+\cdots \quad (d)\ \frac{1}{3}-\frac{4}{9}z+\frac{13}{27}z^2-\frac{40}{81}z^3+\cdots\right]$$

17. Expand ze^{2z} in a Taylor's series about $z = -1$ and determine the region of convergence.

$$\left[\textbf{Ans. } \frac{1}{2^2}\left\{1+\left(1-\frac{2}{\lfloor 1}\right)(z+1)+\left(\frac{2}{\lfloor 1}-\frac{2^2}{\lfloor 2}\right)(z+1)^2+\cdots\right\}\right]$$

18. Find the expansions in powers of z for $\dfrac{1}{(z^2+1)(z^2+2)}$ when (*a*) $|z|<1$ (*b*) $1<|z|<\sqrt{2}$

(*c*) $|z|>\sqrt{2}$.

$$\left[\textbf{Ans. } (a)\sum_{n=0}^{\infty}(-1)^n\left(1-\frac{1}{2^{n+1}}\right)z^{2n}, \quad (b)\sum_{n=1}^{\infty}\frac{(-1)^n}{z^{2n+2}}+\sum_{n=0}^{\infty}\frac{(-1)^{n+1}z^{2n}}{2^{n+1}}\right.$$

$$\left.(c)\sum_{n=0}^{\infty}\frac{(-1)^n(1-2^n)}{z^{2n+2}}\right]$$

19. Expand $\dfrac{1}{z(z^2-3z+2)}$ as a power series valid in $1<|z|<2$. $\left[\textbf{Ans. } -\sum_{n=0}^{\infty}\left(\frac{1}{z^{n+2}}+\frac{z^{n-1}}{2^{n+1}}\right)\right]$

3.4 SINGULARITY

Let a function $f(z)$ is analytic at all points of a domain except some finite number of points. These points at which the function is not analytic are called singular points or singularities. For example, if $f(z)=\dfrac{1}{(z-2)}$, Then $z = 2$ is a singularity of $f(z)$.

3.4.1 Different Types of Singularities

(*i*) Isolated singularity: If $z = a$ is a singularity of $f(z)$ and if there is no other singularity within a small circle surrounding the point $z = a$, then $z = a$ is said to be an isolated singularity of the function $f(z)$; Example (*i*) The function $f(z) = 1/z$ is analytic everywhere except at $z = 0$, therefore $z = 0$ is an isolated singularity. (*ii*) The function $f(z) = \dfrac{z+2}{(z-1)(z-2)(z-3)}$ has three isolated singularities at $z = 1, 2$ and 3.

(*ii*) Non-isolated singularity: If there are other singularities surrounding $z = a$, then $z = a$ is called non-isolated singularity. Example, the function $f(z) = \dfrac{2}{\sin\dfrac{\pi}{z}}$ has an infinite number of isolated singularities. These singularities are at $z = \pm\dfrac{1}{n}$ where $n = 1, 2, 3.....$ the origin $z = 0$ *i.e.*, $n = \infty$ is also a singularity which called non-isolated singularity.

(*iii*) Removable singularity: In the Laurent's expansion $f(z) = \sum_{n=0}^{\infty} a_n (z-a)^n + \sum_{n=1}^{\infty} b_n (z-a)^{-n}$ the second part $\sum_{n=1}^{\infty} b_n (z-a)^{-n}$ is called the principal part. If the principal part of $f(z)$ contains no terms *i.e.*, if $b_n = 0$ for all n then the singularity $z = a$ is called removable singularity of $f(z)$. In this case $f(z) = \sum_{n=0}^{\infty} a_n (z-a)^n$.

An alternate definition: A singularity $z = a$ is called a removable singularity of $f(z)$ if $\lim_{z \to a} f(z)$ exists finitely.

Suppose $f(z) = \dfrac{\sin z}{z}$. Then $\lim_{z \to 0} \dfrac{\sin z}{z} = 1$

$\therefore$ $z = 0$ is a removable singularity.

$$\text{Again} \quad \frac{\sin z}{z} = \frac{1}{z}\left[z - \frac{z^3}{3} + \frac{z^5}{5} .....\right]$$

$$= 1 - \frac{z^2}{3} + \frac{z^4}{5} .....$$

Since no negative power of z occurs. Hence, $z = 0$ is a removable singularity by first definition.

(*iv*) Essential singularity (or isolated essential singularity): If $b_n \neq 0$ *i.e.*, the principal part contains an infinite number of terms then the singularity is called an isolated essential singularity.

Example: $z = 0$ is an isolated essential singularity of $e^{1/z}$, since the expansion of $e^{1/z} = 1 + \dfrac{1}{z} + \dfrac{1}{2!z^2} + ...$ is an infinite series of negative powers of z.

Pole: If the principal part contains a finite number of terms say m, then the singularity $z = a$ is called a pole of order m of $f(z)$. A pole of order one is called a simple pole.

Example: If $f(z) = \dfrac{1}{(z-5)^3 (z-4)^2}$ then $z = 5$ is a pole of order 3 and $z = 4$ is a pole of order 2.

Remark: It is noteworthy that every pole is a singularity but every singularity is not a pole.

Zero: A zero of an analytic function $f(z)$ is a value of z such that $f(z) = 0$.

3.5 LIMIT POINT OF ZEROS

The limit point of zeros of a function $f(z)$ is an isolated essential singularity.

Example: Consider the function $\sin \dfrac{1}{z-a}$

Its zeros are given by $\sin \dfrac{1}{z-a} = 0$ *i.e.* $\sin \dfrac{1}{z-a} = \sin n\pi$

$$\Rightarrow \qquad \frac{1}{z-a} = n\pi \quad \text{or} \quad z = a + \frac{1}{n\pi} \qquad (n = \pm 1, 1 \pm 2 ....)$$

Thus we see that there is a sequence of zeros of $f(z)$ which has $z = a$ its limit point. Hence $z = a$ is an essential singularity of the function $\sin \dfrac{1}{z-a}$.

3.6 LIMIT POINT OF POLES

The limit point of the poles of a function $f(z)$ is a non-isolated essential singularity.

Example: Consider the function $\tan 1/z$

So $$\tan \frac{1}{z} = \frac{\sin 1/z}{\cos 1/z}$$

For the poles $\cos \dfrac{1}{z} = 0 \qquad \Rightarrow \qquad \cos \dfrac{1}{z} = \cos\left(2n\pi + \dfrac{\pi}{2}\right)$

or $$z = \frac{1}{\left(2n + \frac{1}{2}\right)\pi} \{n = 0, \pm 1, \pm 2....\}$$

Obviously $z = 0$ is the limit point of these poles. Hence $z = 0$ is non-isolated essential singularity.

Example 1: Find the kind of singularities of the following:

(*a*) $\dfrac{\cot \pi z}{(z-a)^2}$ at $z = a$ and $z = \infty$ (*b*) $\tan \dfrac{1}{z}$ at $z = 0$

(*c*) $\operatorname{cosec}\dfrac{1}{z}$ at $z = 0$ (*d*) $\sin\left[\dfrac{1}{(1-z)}\right]$ at $z = 1$.

Sol. (*a*) $\dfrac{\cot \pi z}{(z-a)^2} = \dfrac{\cos \pi z}{\sin \pi z (z-a)^2}$

For the poles $\sin \pi z\,(z - a)^2 = 0$

$$\Rightarrow \quad \sin \pi z = 0, (z-a)^2 = 0$$

Now $\sin \pi z = 0$ gives $\pi z = n\pi$ or $z = n$, where $n = 0, \pm 1, \pm 2, \ldots \infty$

$\Rightarrow$ $z = \infty$ is a limit point of these poles. Hence $z = \infty$ is non-isolated essential singularity.

and $\quad (z-a)^2 = 0 \quad \Rightarrow \quad z = a$ (twice)

i.e., z = "a" is a pole of order two or $z = a$ is a singular point of the given function.

(*b*) $f(z) = \tan\dfrac{1}{z} = \dfrac{\sin\dfrac{1}{z}}{\cos\dfrac{1}{z}}$

For the poles $\quad \cos\dfrac{1}{z} = 0 = \cos\left(2n\pi \pm \dfrac{\pi}{2}\right)$

$$\Rightarrow \quad \frac{1}{z} = 2n\pi \pm \frac{\pi}{2}$$

$$\Rightarrow \quad z = \frac{1}{\left(2n \pm \frac{1}{2}\right)\pi}, \quad \text{where } n = 0, \pm 1 \ldots\ldots\infty$$

$\Rightarrow z = 0$ is the limit point of the poles. Hence, $z = 0$ is non-isolated essential singularity.

(*c*)

$$f(z) = \operatorname{cosec}\frac{1}{z} = \frac{1}{\sin 1/z}$$

For the poles $\quad \sin\dfrac{1}{z} = 0 = \sin n\pi$

$$\Rightarrow \quad z = \frac{1}{n\pi}, \quad \text{where } n = 0 \pm 1, \pm 2, \ldots\ldots\ldots\infty$$

$\Rightarrow$ $\quad z = 0$ is the limit point of the poles.

Hence, $z = 0$ is non-isolated essential singularity.

(*d*) $\quad f(z) = \sin\left(\dfrac{1}{1-z}\right)$

Zeros of $f(z)$ are given by putting $f(z) = 0$

$\therefore \quad \sin\left(\frac{1}{1-z}\right)=0 \quad \Rightarrow \quad \sin\left(\frac{1}{1-z}\right)=\sin n\pi$

$\Rightarrow \quad \frac{1}{1-z}=n\pi \quad \Rightarrow z=1-\frac{1}{n\pi}+1,(n=0,\pm 1,\ldots)$

$\Rightarrow$ $z=1$ is a limit point of these zeroes.

Hence $z = 1$ is a isolated essential singularity.

Example 2: What kind of singularity have the following:

(*a*) $\frac{1-e^z}{1+e^z}$ at $z=\infty$, (*b*) $\frac{z^2+4}{e^z}$ at $z=\infty$ (*c*) $\frac{z-2}{z^2}\sin\left(\frac{1}{1-z}\right)$

Sol. (*a*) Let $f(z)=\frac{1-e^z}{1+e^z}$

For the poles $1 + e^z = 0$

$$\Rightarrow \quad e^z=-1=e^{\pi i}$$

$$\Rightarrow \quad e^z=e^{(2n+1)\pi i}$$

$$\Rightarrow \quad z=(2n+1)\pi i, \text{ where } n=0\pm 1.....$$

General term $e^{\pi i}=e^{(2n+1)\pi i}$

$$\Rightarrow \quad z=(\infty+1)\pi i=\infty$$

Therefore, $z=\infty$ is the limit point of the poles, Hence $z=\infty$ is non-isolated singularity. **Ans.**

(*b*) Let $f(z)=\frac{(z^2+4)}{e^z}=(z^2+4)e^{-z}$

For the zeros $(z^2+4)e^{-z}=0$

But $e^{-z}\neq 0$ for all z.

Hence, $z^2+4=0 \quad \Rightarrow \quad z=\pm 2i$

Hence, $z = 2i, -2i$ are simple zeros.

(*c*) $f(z)=\frac{z-2}{z^2}\sin\left(\frac{1}{1-z}\right)$

The poles are obtained by putting $z^2 = 0$

$\Rightarrow \quad z=0,$ (is a pole of order two). **Ans.**

Now zeroes of $f(z)$, $(z-2)\sin\left(\frac{1}{1-z}\right)=0$

If $z-2=0 \quad \Rightarrow \quad z = 2$ is a simple zero. **Ans.**

and $\quad \sin\left(\frac{1}{1-z}\right)=0=\sin n\pi$

$\Rightarrow \quad \frac{1}{z-1}=n\pi \quad \Rightarrow \quad z=\frac{1}{n\pi}+1,(n=0,\pm 1,\ldots)$

$$z = \frac{1}{\infty} + 1 = 1$$

$\Rightarrow$ $z = 1$ is the limit point of the zeroes.

Hence, $z = 1$ is an isolated essential singularity.

Example 3: Show that the function e^{-1/z^2} has no singularities.

Sol. Let $f(z) = e^{-1/z^2} = \dfrac{1}{e^{1/z^2}}$

For the poles $e^{1/z^2} = 0$

which is not possible for any value of z real or complex.

Zeros of $f(z)$ are obtained as

$$e^{-1/z^2} = 0 = e^{-\infty} \quad \Rightarrow \quad \frac{1}{z^2} = \infty$$

$\Rightarrow$ $z = 0$ (twice).

Hence, $z = 0$ is a zero of order two *i.e.* there is no limit point of zeros.

Hence, there is no singularity.

Example 4: What kind of singularity have the following functions?

(*a*) $\dfrac{1}{1-e^z}$ at $z = 2\pi i$ (*b*) $\dfrac{1}{\sin z - \cos z}$ at $z = \pi/4$ (*c*) $\sec\dfrac{1}{z}$ at $z = 0$.

Sol. (*a*) Suppose $f(z) = \dfrac{1}{1-e^z}$

For the poles $1 - e^z = 0$

$\Rightarrow$ $e^z = 1 = e^{2n\pi i}$

$\Rightarrow$ $z = 2n\pi i$ where, $(n = 0, \pm 1, \pm 2, ...)$

$\Rightarrow$ $z = 2\pi i$ is a simple pole. **Ans.**

(*b*) For poles $\sin z - \cos z = 0$

or $$\tan z = 1 = \tan\left(n\pi + \frac{\pi}{4}\right)$$

$\Rightarrow$ $$z = n\pi + \frac{\pi}{4}, (n = 0, \pm 1, ...)$$

For $n = 0$, $z = \dfrac{\pi}{4}$ $\Rightarrow$ $z = \dfrac{\pi}{4}$ is a simple pole.

(*c*) $$f(z) = \sec\frac{1}{z} = \frac{1}{\cos 1/z}$$

For the poles $\cos 1/z = 0 = \cos\left(2n\pi + \dfrac{\pi}{2}\right)$

or $$\frac{1}{z} = 2n\pi + \frac{\pi}{2}$$

or $$z = \frac{1}{\left(2n + \frac{1}{2}\right)\pi}, (n = 0, \pm 1, \pm 2, ...)$$

$\Rightarrow \quad z = \frac{1}{\infty} = 0$

$\Rightarrow \quad z = 0$ is the limit point of the poles.

Hence, $z = 0$ is a non-isolated essential singularity.

PROBLEM SET 3.2

1. What kind of singularity have the following:

   (*i*) $\sin z - \cos z$ at $z = \infty$. [**Ans.** isolated essential singularity]

   (*ii*) $z \operatorname{cosec} z$ at $z = \infty$. [**Ans.** Non-isolated essential singularity]

2. What kind of singularity have the following functions:

   (*i*) $\tan 1/z$ at $z = 0$. [**Ans.** Non-isolated essential singularity]

   (*ii*) $\operatorname{cosec} 1/z$ at $z = 0$. [**Ans.** Non-isolated essential singularity]

   (*iii*) $e^{z/1-z}$ at $z = 1$. [**Ans.** Essential singularity]

3. Find the singularities of the function $\frac{e^{c/z-a}}{e^{z-a}-1}$, indicating the character of each singularity.

   [**Ans.** $z = a$ is an isolated essential singularity]

4. Classify the singularities of a function of a complex variable show that the only singularities of $\frac{\cot \pi z}{(z-a)^2}$ are poles. [**Ans.** $z = a$ is a pole of order 2]

5. Determine the nature of the pole at the origin of the function $\frac{e^z}{z \sin mz}$.

   [**Ans.** $z = 0$ is a simple pole, $z = \infty$ is non-isolated essential singularity]

3.7 RESIDUES

In complex analysis, the residue is a complex number which describe the behaviour of line integrals of analytic function around a singularity.

Residue at a pole: Let $f(z)$ is a single valued function has a pole of order m at $z = a$ then by the definition of pole the principal part of Laurent's expansion contains only m terms so that—

$$f(z) = \sum_{n=0}^{\infty} a_n (z-a)^n + \sum_{n=1}^{\infty} b_n (z-a)^{-n} \qquad ...(1)$$

where $a_n = \frac{1}{2\pi i}\int_C \frac{f(z)dz}{(z-a)^{n+1}}$, $b_n = \frac{1}{2\pi i}\int_{C_1} \frac{f(z)dz}{(z-a)^{-n+1}}$

C_1 being a circle $|z - a| = r$

For $n = 1$, $b_1 = \frac{1}{2\pi i}\int_{C_1} f(z)dz$ ...(2)

Hence the coefficient b_1 is called the residue of $f(z)$ at the pole $z = a$ and is denoted by Res$(z = a)$

For the pole $z = a$ (simple pole) the equation (1) can becomes

$$f(z) = \sum_{n=0}^{\infty} a_n (z-a)^n + \frac{b_1}{(z-a)}$$

$$\Rightarrow \quad \lim_{z\to a}(z-a)f(z) = b_1$$

Using equation (2), we get

$$\boxed{\text{Res}(z=a) = \lim_{z\to a}(z-a)f(z) = b_1 = \frac{1}{2\pi i}\int_{C_1} f(z)dz}$$

3.7.1 Residue at Infinity

If $f(z)$ has an isolated singularity at $z = \infty$ or is analytic there, then the residue at $z = \infty$ is defined as

$$\text{Res}(z=\infty) = \frac{1}{2\pi i}\int_C f(z)dz$$

The integral is taken in positive direction (anti-clockwise).

3.8 CAUCHY'S RESIDUE THEOREM

If $f(z)$ is analytic within and on a closed contour C, except at a finite number of poles. $z_1, z_2, \ldots, z_n$ within C then.

$$\int_C f(z)dz = 2\pi i(R_1 + R_2 + \ldots + R_n) = 2\pi i(\text{sum of residues})$$

Proof: Let $C_1, C_2, \ldots, C_n$, be small circles with centre at $z_1, z_2, \ldots, z_n$, respectively and radii so small that they lie within closed curve C and do not overlap. $f(z)$ is analytic within the annulus bounded by these circles and the curve C.

Hence by Cauchy's theorem for multi-connected region, we have

$$\int_C f(z)dz = \int_{C_1} f(z)dz + \int_{C_2} f(z)dz + \ldots + \int_{C_n} f(z)dz \quad \ldots(1)$$

But $b_1 = \text{Res}(z = z_1) = R_1 = \frac{1}{2\pi i}\int_{C_1} f(z)dz$

or $\int_{C_1} f(z)dz = 2\pi i R_1$

Similarly $\int_{C_2} f(z)dz = 2\pi i R_2$

$$\vdots$$

$$\int_{C_n} f(z)dz = 2\pi i R_n$$

Using these values in equation (1), we get

$$\boxed{\int_C f(z)dz = 2\pi i\left(R_1 + R_2 + ... + R_n\right)}$$

3.9 METHOD OF FINDING THE RESIDUES

(i) Residue of $f(z)$ at a simple pole $z = a$

$$\text{Res}(z = a) = \lim_{z \to a}(z - a)f(z)$$

(ii) If $f(z)$ is of the form

$$f(z) = \frac{\phi z}{\psi(z)}, \text{ where } \psi(a) = 0 \text{ but } \phi(a) \neq 0$$

$$\text{Res}(z = a) = \lim_{z \to a}(z - a)f(z) = \lim_{z \to a}(z - a)\frac{\phi(z)}{\psi(z)}$$

$$= \lim_{z \to a}\frac{(z - a)\left[\phi(a) + (z - a)\phi'(a) + ...\right]}{\psi(a) + (z - a)\psi'(a) + \ldots} \quad \text{(By Taylor's Theorem)}$$

$$= \lim_{z \to a}\frac{\left[\phi(a) + (z - a)\phi'(a) + ...\right]}{\frac{\psi(a)}{(z - a)} + \psi'(a) + \frac{(z - a)}{2!}\psi''(a) + ...} \quad \Big| \text{ As } \psi(a) = 0$$

$$\Rightarrow \qquad \text{Res}(z = a) = \frac{\phi(a)}{\psi'(a)} \quad \text{or} \quad \boxed{\text{Res}(z = a) = \frac{\phi(z)}{\psi'(z)} \text{ as } z \to a}$$

(iii) Residue at a pole of order m

Let $f(z) = \dfrac{\phi(z)}{(z - a)^m}$, where $\phi(z)$ is analytic

We know that $\text{Res }(z = a) = \dfrac{1}{2\pi i}\displaystyle\int_C f(z)dz$

So $$\text{Res }(z = a) = \frac{1}{2\pi i}\int_C \frac{\phi(z)}{(z - a)^m}dz \qquad ...(1)$$

By Cauchy's integral formula, we have

$$\phi^{m-1}(a) = \frac{(m - 1)!}{2\pi i}\int_C \frac{\phi(z)}{(z - a)^m}dz \qquad ...(2)$$

$$\left(f^n(a) = \frac{n!}{2\pi i} \int_C \frac{f(z)dz}{(z-a)^{n+1}} \right)$$

Using equation (2) in equation (1), we get

$$\text{Res}(z = a) = \frac{1}{(m-1)!} \phi^{m-1}(a)$$

or
$$\text{Res}(z = a) = \frac{1}{(m-1)!} \left\{ \frac{d^{m-1}}{dz^{m-1}} \phi(z) \right\} \quad \text{as } z \to a$$

or
$$\boxed{\text{Res}(z = a) = \frac{1}{(m-1)!} \frac{d^{m-1}}{dz^{m-1}} \{(z-a)^m f(z)\} \quad \text{as } z \to a}$$

(*iv*) Residue at a pole $z = a$ of any order

We have seen that the residue of $f(z)$ at $z = a$ is the coefficient of $\frac{1}{z-a}$ in Laurent's expansion of $f(z)$ and therefore coefficient of $\frac{1}{t}$ in the expansion of $f(a + t)$ as a power series.

Example 1: Find the residue of $\frac{z^2}{z^2 + a^2}$ at $z = ia$

Sol. We have $f(z) = \frac{z^2}{z^2 + a^2} = \frac{z^2}{(z+ai)(z-ai)}$

for the poles $(z+ai)(z-ai) = 0$

$$\Rightarrow \qquad z = ai, z = -ai$$

$$\text{Res}(z = ai) = \lim_{z \to ia} (z - ia) \cdot f(z)$$

$$= \lim_{z \to ia} (z - ia) \cdot \frac{z^2}{(z+ai)(z-ai)}$$

$$= \lim_{z \to ia} \frac{z^2}{z + ai} = \frac{-a^2}{2ai} = \frac{1}{2} ia. \qquad \textbf{Ans.}$$

Example 2: Find the residue of $f(z) = \frac{z^3}{(z-1)^4 (z-2)(z-3)}$ at $z = 1, 2, 3$ and evaluate $\int_C f(z)dz$, where C is the circle $|z| = \frac{5}{2}$.

Sol. For the poles $(z-1)^4 (z-2)(z-3) = 0$

$z = 1$ is a pole of order 4, $z = 2$ and $z = 3$ are simple pole.

(*i*) Residue at
$$(z = 1) = \frac{1}{3!} \left[\frac{d^3}{dz^3} \left\{ \frac{(z-1)^4 \cdot z^3}{(z-1)^4 (z-2)(z-3)} \right\} \right]_{z=1}$$

$$= \frac{1}{3!}\left[\frac{d^3}{dz^3}\left\{\frac{z^3}{(z-2)(z-3)}\right\}\right]_{z=1}$$

$$= \frac{1}{6}\left[\frac{d^3}{dz^3}\left\{z+5-\frac{8}{z-2}+\frac{27}{z-3}\right\}\right]_{z=1}$$

$$= \frac{1}{6}\left[\frac{d^2}{dz^2}\left\{1-\frac{27}{(z-3)^2}+\frac{8}{(z-2)^2}\right\}\right]_{z=1}$$

$$= \frac{1}{6}\left[\frac{d}{dz}\left\{\frac{54}{(z-3)^3}-\frac{16}{(z-2)^3}\right\}\right]_{z=1}$$

$$= \frac{1}{6}\left[\frac{-162}{(z-3)^4}+\frac{48}{(z-2)^4}\right]_{z=1} = -\frac{27}{16}+8 = \frac{101}{16}$$

$\Rightarrow$ $$R_1 = \frac{101}{16}. \quad \textbf{Ans.}$$

(*ii*) $\text{Res}\,(z=2) = \lim_{z\to 2}(z-2)\dfrac{z^3}{(z-1)^4(z-2)(z-3)}$

$$= \lim_{z\to 2}\frac{z^3}{(z-1)^4(z-3)} = -8$$

$\Rightarrow$ $$R_2 = -8. \quad \textbf{Ans.}$$

(*iii*) $\text{Res}\,(z=3) = \lim_{z\to 3}(z-3)\dfrac{z^3}{(z-1)^4(z-2)(z-3)} = \dfrac{27}{16}.$ **Ans.**

The given curve is $|z| = 5/2$.
Hence only the poles $z = 1$ and $z = 2$ lie inside the circle.
Hence by Cauchy's Residue theorem

$$\int_C f(z)dz = 2\pi i(R_1 + R_2)$$

$$= 2\pi i\left(\frac{101}{16}-8\right)$$

$$= 2\pi i\left(-\frac{27}{16}\right) = -\frac{27\pi i}{8}. \quad \textbf{Ans.}$$

Example 3: Find the residue at $z = 0$ of the following functions:

(*a*) $\dfrac{1+e^z}{\sin z + z\cos z}$ (*b*) $z\,\cos 1/z$

Sol. (*a*)
$$\text{Res}(z=0) = \lim_{z\to 0} z\frac{1+e^z}{\sin z + z\cos z}$$

$$= \lim_{z\to 0}\frac{1+e^z}{\frac{\sin z}{z}+\cos z} = \frac{1+1}{1+1} = 1, \qquad \left|\text{as } \lim_{z\to 0}\frac{\sin z}{z} = 1\right.$$

(*b*)
$$z\cos\frac{1}{z} = z\left\{1 - \frac{1}{z^2 2!} + \frac{1}{z^4 4!} - .....\right\}$$

$$= z - \frac{1}{z2!} + \frac{1}{z^3 4!} - ..$$

Coefficient of $\frac{1}{z} = -\frac{1}{2}$

$$\text{Res}(\text{at } z=0) = -\frac{1}{2}. \quad \textbf{Ans.}$$

Example 4: Evaluate by the method of calculus of residues.

$\int_C \frac{dz}{(z-1)(z+1)}$, where C is a circle $|z| = 3$.

Sol.
$$\text{Let } f(z) = \frac{1}{(z-1)(z+1)}$$

For the poles $(z - 1)(z + 1) = 0$

$\Rightarrow$ $z = 1, -1$

but C is a circle of a radius 3 with center at $z = 0$, therefore, both the poles $z = 1$ and $z = -1$ lie within the circle.

$$R_1 = \text{Res}\,(z=1) = \lim_{z\to 1}\frac{(z-1)}{(z-1)(z+1)} = \frac{1}{2}$$

$$R_2 = \text{Res}\,(z=-1) = \lim_{z\to -1}\frac{(z+1)}{(z-1)(z+1)} = -\frac{1}{2}$$

By Cauchy's residue theorem

$$\int_C f(z)dz = 2\pi i\{R_1 + R_2\}$$

$$\Rightarrow \quad \int_C \frac{dz}{(z-1)(z+1)} = 2\pi i\left(\frac{1}{2} - \frac{1}{2}\right) = 0. \quad \textbf{Ans.}$$

Example 5: If $f(z) = \frac{(z-1)}{(z+1)^2(z-2)}$ then evaluate $\int_C f(z)dz$ where C is the circle $|z-i| = 2$.

Sol. For the poles $(z + 1)^2 (z - 2) = 0$

$\Rightarrow$ $z = -1$(order 2) and $z = 2$ (simple pole)

Hence, the pole $z = -1$ lie with in the circle whenever the pole $z = 2$ lies outside the circle.

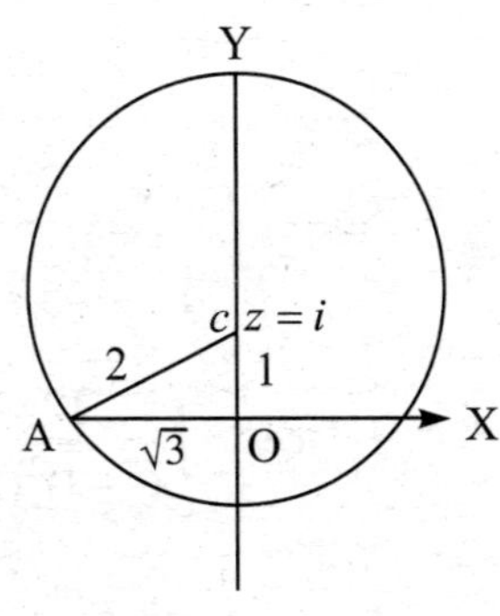

$$\therefore \qquad \text{Res}(z=-1) = \left[\frac{d}{dz}\left\{(z+1)^2 \frac{(z-1)}{(z+1)^2(z-2)}\right\}\right]_{z=-1}$$

$$= \left\{\frac{d}{dz}\frac{(z-1)}{(z-2)}\right\}_{z=-1}$$

$$= \left[\frac{-(z-1)}{(z-2)^2} + \frac{1}{(z-2)}\right]_{z=-1} = \frac{2}{9} - \frac{1}{3} = -\frac{1}{9}$$

$$\Rightarrow \qquad R_1 = -\frac{1}{9} \text{ and } R_2 = 0$$

By Cauchy's residue theorem

$$\int_C f(z)dz = 2\pi i\left\{-\frac{1}{9} + 0\right\} = -\frac{2\pi i}{9}. \quad \textbf{Ans.}$$

Example 6: Find the value of the integral by the method of residue $\int_C \frac{12z-7}{(z-1)^2(2z+3)}dz$, where C is the circle.

(*i*) $|z+i| = \sqrt{3}$ (*ii*) $|z| = 2$

Sol. (*i*) For the poles $(z-1)^2(2z+3) = 0 \Rightarrow z = -3/2, z = 1$ (of order 2) the pole $z = 1$ lies only within the circle. So

$$\text{Res}(z=1) = \left[\frac{d}{dz}\left\{(z-1)^2 \frac{(12z-7)}{(z-1)^2(2z+3)}\right\}\right]_{z=1}$$

$$= \left[\frac{d}{dz}\left\{\frac{(12z-7)}{(2z+3)}\right\}\right]_{z=1}$$

$$= \left[\frac{-2(12z-7)}{(2z+3)^2} + \frac{12}{2z+3}\right]_{z=1} = \left[\frac{-(2\times 5)}{25} + \frac{12}{5}\right]$$

$$\text{Res}(z=1) = \frac{10}{5} = 2.$$

Hence $$\int_C \frac{12z-7}{(z-1)^2(2z+3)}dz = 2\pi i \times 2 = 4\pi i. \quad \textbf{Ans.}$$

(*ii*) In this region both the pole $z = -3/2, 1$ lie within the circle.

$$\therefore \qquad \text{Res}(z=-3/2) = \lim_{z \to -\frac{3}{2}} \frac{(z+3/2)(12z-7)}{(z-1)^2 \cdot 2(z+3/2)}$$

So $$= \lim_{z \to \frac{-3}{2}} \frac{(12z-7)}{(z-1)^2 \cdot 2} = \frac{12 \times \left(-\frac{3}{2}\right) - 7}{2\left(\frac{-3}{2}-1\right)^2} = \frac{-18-7}{\frac{1}{2} \times 25} = -\frac{25 \times 2}{25} = -2.$$

Hence $$\int_C \frac{12z-7}{(z-1)^2(2z+3)} dz = 2\pi i(2-2) = 0. \quad \textbf{Ans.}$$

Example 7: Find the value of $\int_C \frac{1}{z^2 \sin z} dz$ where C is triangle with vertices (0, 1) (2, –2) (7,1)

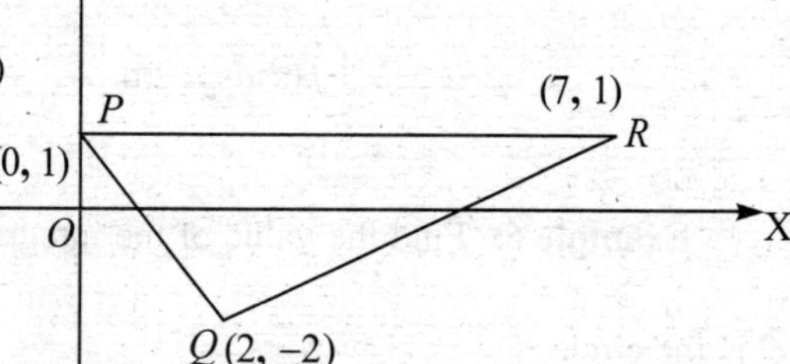

Sol. Let $f(z) = \int_C \frac{1}{z^2 \sin z} dz$

For the poles $z^2 \sin z = 0$

$\Rightarrow$ $z = 0$ (is a pole of order 2)

and $\sin z = \sin n\pi \Rightarrow z = n\pi, (n = 0, \pm 1, \pm 2....)$ here only the pole $z = \pi$, lies within the region, residue at $z = \pi$

$$\text{Res } (z = \pi) = \lim_{z \to \pi} \frac{z - \pi}{z^2 \sin z} = \frac{0}{0}$$

By L-Hospital rule, Res $(z = \pi) = \lim_{z \to \pi} \frac{1}{2z \sin z + z^2 \cos z}$

$$= \frac{1}{2\pi \sin \pi + \pi^2 \times \cos \pi} = -\frac{1}{\pi^2}$$

By Cauchy's residue theorem

$$\int_C \frac{1}{z^2 \sin z} dz = 2\pi i \left(-\frac{1}{\pi^2}\right) = -\frac{2i}{\pi}. \quad \textbf{Ans.}$$

Example 8: Evaluate $\int_C \frac{e^z}{\cos \pi z} dz$, where C is the unit circle $|z| = 1$.

Sol. Let $f(z) = \frac{e^z}{\cos \pi z}$ for poles $\cos \pi z = 0 = \cos(2n+1)\frac{\pi}{2}$

$$\Rightarrow \quad z = (2n+1)\frac{1}{2}, \{n = 0, \pm 1, \pm 2\}$$

$$\Rightarrow \quad z = \pm\frac{1}{2}, \pm\frac{3}{2}, \pm\frac{5}{2}, \ldots\ldots\ldots\ldots$$

Here only the poles $z = \pm\frac{1}{2}$ lie within the circle $|z| = 1$.

So, $R_1 = \text{Res}\left(z=\frac{1}{2}\right) = \lim_{z\to\frac{1}{2}} \frac{\left(z-\frac{1}{2}\right)e^z}{\cos\pi z} = \frac{0}{0}$

Applying L-Hospital rule

$$R_1 = \text{Res}\left(z=\frac{1}{2}\right) = \lim_{z\to\frac{1}{2}} \frac{\left(z-\frac{1}{2}\right)e^z + e^z}{-\pi\sin\pi z} = \frac{-e^{1/2}}{\pi}$$

Similarly $\qquad R_2 = \text{Res}\left(z=-\frac{1}{2}\right) = \frac{e^{-1/2}}{\pi}$

By Cauchy's residue theorem we have

$$\int_C \frac{e^z}{\cos\pi z}dz = 2\pi i\left\{\frac{-e^{1/2}+e^{-1/2}}{\pi}\right\} = -2i(e^{1/2}-e^{-1/2})$$

$\Rightarrow \qquad \int_C \frac{e^z}{\cos\pi z}dz = -4i\sinh\frac{1}{2}.$ **Ans.** $\qquad \sinh hx = \frac{e^x - e^{-x}}{2}$

Example 9: Evaluate $\int_C f(z)dz$, where C is the circle $|z-1| = 2$ and the function $f(z) = \frac{1}{z^2\sinh z}$.

Sol. For the poles $z^2 \sinh z = 0$

$\Rightarrow \qquad \sinh z = 0 \quad \Rightarrow \quad z = n\pi$

And $\qquad z^2 = 0$

$\Rightarrow \qquad z = 0$ (is a pole of order 2)

Here the poles $z = n\pi$ lie outside the circle.

Y

$|z-1|=2$

C

O'

O

z=1

X

$\therefore$ Res $(z = 0)$ = the coefficient of $\frac{1}{z}$ in the expansion of $f(z)$.

Expansion of $\qquad f(z) = \frac{1}{z^2\cdot\frac{1}{2}\left\{2z+\frac{2z^3}{3!}+\frac{2z^5}{5!}+....\right\}}$

$$= \frac{1}{z^3}\frac{1}{\left\{1+\frac{z^2}{3!}+\frac{z^4}{5!}+....\right\}}$$

$\qquad \sinh\theta = \frac{1}{2}\left[2\theta + \frac{2\theta^3}{3!}+...\right]$

Expansion of $\qquad f(z) = \frac{1}{z^3}\left\{1+\frac{z^2}{3!}+\frac{z^4}{5!}+....\right\}^{-1}$

$$= \frac{1}{z^3} - \frac{1}{3!z} - \frac{z}{5!} \ldots\ldots$$

$$\Rightarrow \qquad \text{Res}(z = 0) = -\frac{1}{3!} = -\frac{1}{6}$$

Hence $$\int_C f(z)dz = 2\pi i \times -\frac{1}{6} = -\frac{\pi i}{3}. \quad \textbf{Ans.}$$

Example 10: Find the poles and residue at each pole of the function

$$f(z) = \frac{z^2}{(z+1)^2(z-2)}$$

Futher, state Residue Theorem and hence evaluate:

$$\int_C f(z)dz, \quad \text{where} \quad C: |z| = \frac{5}{2}$$

Sol. We have $$f(z) = \frac{z^2}{(z+1)^2(z-2)}$$

For the poles : $(z+1)^2(z-2) = 0$

$\Rightarrow$ $z = -1$ (order 2) and $z = 2$ (simple pole)

$\therefore$ Residue of $f(z)$ at $z = -1$ is

$$R_1 = \frac{1}{|(2-1)}\left[\frac{d}{dz}\left\{(z+1)^2 \cdot \frac{z^2}{(z+1)^2(z-2)}\right\}\right]_{z=-1}$$

$$= \left[\frac{d}{dz}\left(\frac{z^2}{z-2}\right)\right]_{z=-1} = \left[\frac{(z-2)2z - z^2}{(z-2)^2}\right]_{z=-1} = \frac{5}{9}$$

And Residue of $f(z)$ at $z = 2$ is

$$R_2 = \lim_{z\to 2}(z-2) \cdot \frac{z^2}{(z+1)^2(z-2)} = \lim_{z\to 2}\frac{z^2}{(z+1)^2} = \frac{4}{9}$$

Since both the poles $z = -1$ and $z = 2$ lie inside the circle $|z| = \frac{5}{2}$ hence, by Cauchy's residue theorem

$$\oint_C f(z)dz = 2\pi i(R_1 + R_2 + \ldots + R_n)$$

$$= 2\pi i(\text{sum of residues})$$

$$= 2\pi i\left(\frac{5}{9} + \frac{4}{9}\right) = 2\pi i. \quad \textbf{Ans.}$$

Example 11: Use residue theorem to evaluate $\int_C \frac{3z^2+z-1}{(z^2-1)(z-3)} dz$ around the circle $|z| = 2.$

Sol. Let $$f(z) = \frac{3z^2+z-1}{(z^2-1)(z-3)}$$

Here $f(z)$ has simple poles 1, – 1, 3 and only 1, – 1 lie inside $|z| = 2$

$$\therefore \quad \text{Res } (z=1) = \left[\frac{\phi(z)}{\psi'(z)}\right]_{z=1} = \left[\frac{3z^2+z-1}{3z^2-6z-1}\right]_{z=1} = -\frac{3}{4}$$

Similarly $$\text{Res } (z=-1) = \frac{3-1-1}{3+6-1} = \frac{1}{8}$$

By residue theorem

$$\int_C \frac{3z^2+z-1}{(z^2-1)(z-3)} dz = 2\pi i\left(-\frac{3}{4}+\frac{1}{8}\right) = 2\pi i\left(-\frac{5}{8}\right) = -\frac{5\pi i}{4}. \quad \textbf{Ans.}$$

Example 12: Evaluate $\int_C \frac{\sin z}{z\cos z} dz$, where C is a circle $|z| = 2.$

Sol. For poles $z\cos z = 0 \quad \Rightarrow \quad z = 0$ and $\cos z = 0$

$$\therefore \quad \cos z = \cos\frac{n\pi}{2} \Rightarrow z = \frac{n\pi}{2} (n = \pm 1, \pm 2, \ldots)$$

Here only the poles at $z = 0$, $z = +\frac{\pi}{2}$ and $z = -\frac{\pi}{2}$ lie within the circle

So, $$\text{Res } (z=0) = \lim_{z\to 0} z.\frac{\sin z}{z\cos z} = \frac{\sin 0}{\cos 0} = 0$$

$$\Rightarrow \quad R_1 = 0$$

$$R_2 = \text{Res}\left(z = \frac{\pi}{2}\right) = \lim_{z\to\frac{\pi}{2}}\left(z-\frac{\pi}{2}\right).\frac{\sin z}{z\cos z} = \frac{0}{0}$$

By L. Hospital Rule

$$R_2 = \lim_{z\to\frac{\pi}{2}} \frac{\left(z-\frac{\pi}{2}\right).\cos z + \sin z}{\cos z - z\sin z} = \frac{1}{-\frac{\pi}{2}} \quad i.e., \; R_2 = -\frac{2}{\pi}$$

Similarly $$R_3 = \text{Res}\left(z = -\frac{\pi}{2}\right) = \frac{2}{\pi}$$

Hence $\int_C \frac{\sin z}{z\cos z}\,dz = 2\pi i\{R_1 + R_2 + R_3\}$

$$= 2\pi i\left\{0 - \frac{2}{\pi} + \frac{2}{\pi}\right\} = 0. \quad \textbf{Ans.}$$

Example 13: Evaluate $\int_C z^4 e^{1/z}\,dz$, where C is a circle $|z| = 1$.

Sol. Let $f(z) = z^4 e^{1/z} = \frac{z^4}{e^{-1/z}}$

For the poles $e^{-1/z} = 0$

It is only possible when $z = 0$. $\quad\left|\; e^{-1/z} = \frac{1}{e^{1/z}} = \frac{1}{e^{1/0}} = \frac{1}{e^{\infty}} = \frac{1}{\infty} = 0\right.$

Res(z = 0) = the coefficient of $\frac{1}{z}$ in expansion of $f(z)$

$$= z^4\{e^{1/z}\} = z^4\left\{1 + \frac{1}{z} + \frac{1}{2!z^2} + \frac{1}{3!z^3} + \frac{1}{4!z^4} + \frac{1}{5!z^5} \ldots\right\}$$

$$\text{Res}(z = 0) = \frac{1}{5!} = \frac{1}{120}$$

Hence $\int_C z^4 e^{1/z}\,dz = 2\pi i \times \frac{1}{120} = \frac{\pi i}{60}.$ **Ans.**

Example 14: Evaluate $\int_C \frac{dz}{z\sin z}$, where C is the unit circle $|z| = 1$.

Sol. Let $f(z) = \frac{1}{z\sin z}$,

For the poles $z\sin z = 0$

$\Rightarrow \quad z = 0$ and $\sin z = 0$

If $\quad \sin z = \sin n\pi$

$\Rightarrow \quad z = n\pi\,(n = 0, \pm 1, \pm 2, \pm \ldots..)$

$z = 0$, is only a single pole which lies within the circle.

So, Res (z = 0) = the coefficient of $\frac{1}{z}$ in the expansion of $f(z)$

$$= \frac{1}{z\left[z - \frac{z^3}{3!} + \frac{z^5}{5!} \ldots\ldots\right]} = \frac{1}{z^2\left[1 - \frac{z^2}{3!} + \frac{z^4}{5!} \ldots\ldots\right]}$$

$$= \frac{1}{z^2}\left[1 - \left(\frac{z^2}{6} - \frac{z^4}{120}\ldots\right)\right]^{-1}$$

$$= \frac{1}{z^2}\left[1 + \left(\frac{z^2}{6} - \frac{z^4}{120}\ldots\right) + \left(\frac{z^2}{6} - \frac{z^4}{120}\ldots\right)^2 + \ldots\right]$$

$$= \frac{1}{z^2}\left[1 + \frac{z^2}{6} - \frac{z^4}{120} + \frac{z^4}{36} + \ldots\right] = \frac{1}{z^2} + \frac{1}{6} - \frac{z^2}{120} + \frac{z^2}{36} + \ldots.$$

$$= \frac{1}{z^2} + \frac{1}{6} - \frac{7}{360}z^2 + \ldots..$$

The coefficient of $\frac{1}{z} = 0$, so Res $(z = 0) = 0$

Hence $\int_C \frac{dz}{z\sin z} = 2\pi i \times 0 = 0.$ **Ans.**

Example 15: Evaluate $\int_C \frac{z^2 e^{zt}}{z^2+1}dz$ where C is a circle $|z| = 2$.

Sol. Let $f(z) = \frac{z^2 e^{zt}}{z^2+1}$, for the poles $z^2 + 1$

$\Rightarrow z^2 = -1 \quad \Rightarrow \quad z^2 = i^2$ or $z = \pm i$

Both the poles $z = i, -i$ lies within the circle $|z| = 2$

$$\text{Res } (z = i) = \lim_{z\to i}(z-i)\cdot\frac{z^2 e^{zt}}{(z-i)(z+i)} = \lim_{z\to i}\frac{z^2 e^{zt}}{(z+i)} = \frac{-e^{it}}{2i} \quad i.e.\ R_1 = -\frac{e^{it}}{2i}$$

Similarly Res $(z = -i)$ *i.e.* $R_2 = \frac{e^{-it}}{2i}$.

Hence
$$\int_C \frac{z^2 e^{zt}}{(z^2+1)}dz = 2\pi i\left\{-\frac{e^{it}}{2i} + \frac{e^{-it}}{2i}\right\} = -2\pi i\left\{\frac{e^{it} - e^{-it}}{2i}\right\}$$

$$\Rightarrow \quad \int_C \frac{z^2 e^{zt}}{(z^2+1)}dz = -2\pi i \sin t. \quad \textbf{Ans.}$$

Example 16: Find the value of the integral $\int_C \frac{\cot hz\, dz}{(z-i)}$, where C is a circle $|z| = 2$.

Sol. Let
$$f(z) = \frac{\cot hz}{(z-i)} = \frac{(e^z + e^{-z})}{(e^z - e^{-z})(z-i)}$$
As $\cot h\theta = \frac{e^\theta + e^{-\theta}}{e^\theta - e^{-\theta}}$

For the poles $(e^z - e^{-z})(z-i) = 0$

$$\Rightarrow \qquad (e^z - e^{-z}) = 0 \text{ and } (z-i) = 0$$

$$e^z = e^{-z} \text{ and } z = i$$

$$\Rightarrow \qquad e^{2z} = 1 = \cos 0 + i \sin 0 = e^{i0} \Rightarrow z = 0$$

The poles are $z = 0, i$

Both the poles $z = 0$ and $z = i$ lie within the circle $|z| = 2$,

So, $\text{Res}(z=0)$, we apply the method $\dfrac{\phi(z)}{\psi(z)}$

$$\therefore \qquad \frac{\phi(z)}{\psi(z)} = \frac{\left(\dfrac{e^z + e^{-z}}{z-i}\right)}{e^z - e^{-z}} \text{ and } \frac{\phi(z)}{\psi'(z)} = \frac{\left(\dfrac{e^z + e^{-z}}{z-i}\right)}{e^z + e^{-z}} = \frac{1}{z-i}$$

$$R_1 = \left[\frac{\phi(z)}{\psi'(z)}\right]_{z=0} = \left[\frac{1}{z-i}\right]_{z=0} = \frac{1}{-i} = \frac{i}{-i^2} = i$$

and $\text{Res}\,(z = i)$

$$R_2 = \lim_{z \to i}(z-i)\frac{e^z + e^{-z}}{(z-i)(e^z - e^{-z})} = \frac{e^i + e^{-i}}{e^i - e^{-i}} = \cot hi$$

Hence $$\int_C \frac{\cot hz\, dz}{(z-i)} = 2\pi i(i + \cot hi). \quad \textbf{Ans.}$$

PROBLEM SET 3.3

1. Determine the residue at each pole—

(*a*) $\dfrac{z^2}{(z-1)(z-2)^2}$ (*b*) $\dfrac{1}{z^4+1}$, (*c*) $\dfrac{1-e^{2z}}{z^4}$ (*d*) $\dfrac{z}{\sin z}$ and (*e*) $\cot z$.

[**Ans.** (*a*) $R_1 = 1$, $R_2 = 0$, (*b*) $-\dfrac{1}{4}e^{m\pi i/4}$ (*c*) $-\dfrac{4}{3}$ (*d*) $(-1)^n n\pi$, (*e*) 1]

2. Determine the poles and the residue at each pole of the function.

$$f(z) = \frac{z^2}{(z-1)^2(z+2)}.$$

$\left[\textbf{Ans. } R_1 = \dfrac{4}{9} \text{ and } R_2 = \dfrac{5}{9}\right]$

Find the value of

3. $\dfrac{z^2}{z^2+a^2}$ at $z = ai$. $\left[\textbf{Ans. } \dfrac{1}{2}ia\right]$

4. $\tan z$ at all poles. [**Ans.** –1]

5. $\dfrac{ze^z}{(z-a)^3}$ at its pole. $\left[\textbf{Ans.}\ \ \dfrac{1}{2}(a+2)e^a\right]$ **6.** $z^2e^{1/z}$ at the point $z=0$. $\left[\textbf{Ans.}\ \ \dfrac{1}{6}\right]$

7. $\dfrac{1+e^z}{\sin z+z\cos z}$ at $z=0$. **[Ans.** 1] **8.** $\dfrac{1}{z(e^z-1)}$ at its poles. $\left[\textbf{Ans.}\ \ -\dfrac{1}{2}\right]$

9. Evaluate the following integrals using Cauchy's Residue theorem.

(*i*) $\displaystyle\int_C \frac{\cos\pi z^2+\sin\pi z^2}{(z+1)(z+3)}dz;$ $\quad C=|z|=3.$ $\left[\textbf{Ans.}\ \ -\dfrac{\pi i}{4}\right]$

(*ii*) $\displaystyle\int_C \frac{3z^2+z+1}{(z^2-1)(z+3)}dz;$ $\quad C=|z|=2.$ $\left[\textbf{Ans.}\ \ -\dfrac{\pi i}{4}\right]$

10. (*i*) $\displaystyle\int_C \frac{e^{2z}+z^2}{(z-1)^5}dz;$ $\quad C=|z|=2.$ $\left[\textbf{Ans.}\ \ \dfrac{4\pi e^2 i}{3}\right]$

(*ii*) $\displaystyle\int_C \frac{4z^2-4z+1}{(z-2)(z^2+4)}dz;$ $\quad C=\ |z|=1.$ **[Ans.** 0]

11. $\displaystyle\int_C \frac{\sin\pi z^2+\cos\pi z^2}{(z-1)^2(z-2)}dz;$ where C is a circle $|z|=3$ $[\textbf{Ans.}\ \ 4\pi i(\pi+1)]$

12. $\displaystyle\int_C \frac{z\cos z}{\left(z-\frac{\pi}{2}\right)^3}dz;$ where C is a circle $|z-1|=1.$ $[\textbf{Ans.}\ \ 2\pi i]$

13. $\displaystyle\int_C \frac{(1-2z)}{z(z-1)(z-2)}dz;$ where $|z|=1.5.$ $[\textbf{Ans.}\ 3\pi i]$

14. $\displaystyle\int_C \frac{4z^2-4z+1}{(z-2)(4+z^2)}dz;$ $C=\ |z|=1.$ **[Ans.** 0]

15. $\displaystyle\int_C \frac{z\sec z}{(1-z)^2}dz;$ where C is a circle $|z|=2.$ $[\textbf{Ans.}\ 2\pi i\sec 1(1+\tan 1)]$

16. $\displaystyle\int_C \frac{\sin z}{z^6}dz;$ where C is a circle $|z|=2.$ $\left[\textbf{Ans.}\ \ \dfrac{\pi i}{60}\right]$

17. Prove that $\displaystyle\int_C ze^{1/2}dz=\pi i,$ where C is the circle $|z|=5.$

18. Prove that $\displaystyle\int_C \frac{e^z}{\cosh z}dz=8\pi i,$ where C is the circle $|z|=5.$

19. Evaluate $\int_C \frac{e^{zt}}{z^2(z^2+2z+2)}dz$, where C is the circle $|z|=3$.

$$\left[\textbf{Ans. } 2\pi i\left\{\frac{t-1}{2}+\frac{e^{-t}}{4}(\sin t+\cos t)\right\}\right]$$

20. Evaluate $\int_C \frac{3z^3+2}{(z-1)(z^2+4)}dz$, where C is (*a*) $|z-2|=2$, (*b*) $|z|=4$. [**Ans.** πi]

21. Evaluate $\int_C \frac{e^{zt}}{z(z^2+1)}dz$ $(t>0)$, where C is the square with vertices at $1+i, -1+i, -1-i, 1-i$.

[**Ans.** $2\pi i(1-\cos t)$]

3.10 EVALUATION OF REAL INTEGRALS BY CONTOUR INTEGRATION

Contour integration: A large number of real definite integrals, whose evaluation by usual methods becomes some times very difficult and can be easily evaluated by using Cauchy's theorem of residues. For this, we take closed curve C, find the poles of the function $f(z)$ and calculate residue at those poles which lie within C. Then using Cauchy's theorem of residues, we have

$$\int_C f(z)dz = 2\pi i\,(\text{Sum of residues}).$$

The curve is usually called contour and the method of integrating along a contour is called contour integration.

3.10.1 Evaluation the Integrals of the Type $\int_0^{2\pi} f(\cos\theta, \sin\theta)\,d\theta$ by Contour Integration where $f(\cos\theta, \sin\theta)$ is a Rational Function

If we take $z = e^{i\theta}$, then

$$\int_0^{2\pi} f(\cos\theta,\sin\theta)d\theta = \int_0^{2\pi} f\left(\frac{e^{i\theta}+e^{-i\theta}}{2}, \frac{e^{i\theta}-e^{-i\theta}}{2i}\right)d\theta$$

$$= \int_C f\left\{\frac{1}{2}\left(z+\frac{1}{z}\right), \frac{1}{2i}\left(z-\frac{1}{z}\right)\right\}\frac{dz}{zi}$$

$$dz = ie^{i\theta}d\theta \Rightarrow dz = iz\,d\theta \quad \text{or } d\theta = \frac{dz}{zi}$$

$$= \int_C f(z)dz$$

where $f(z)$ is rational function of z and C is the unit circle $|z|=1$.

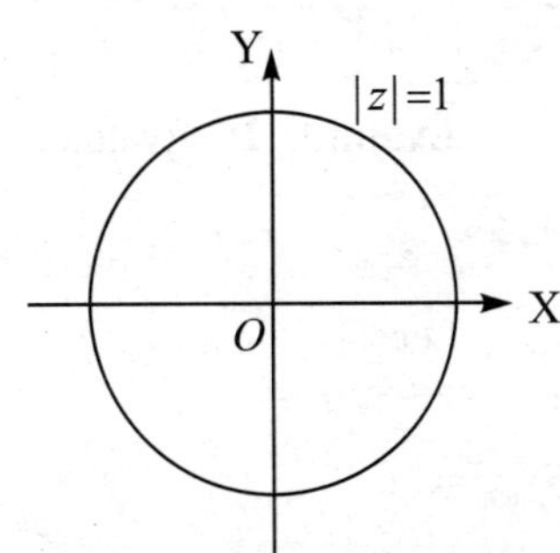

Example 1: Prove that $\int_0^{2\pi} \frac{\cos 2\theta}{5+4\cos\theta} d\theta = \frac{\pi}{6}.$

Sol. Let $I = \int_0^{2\pi} \frac{\cos 2\theta}{5+4\cos\theta} d\theta = R.P. \int_0^{2\pi} \frac{e^{2i\theta} d\theta}{5+2(e^{i\theta}+e^{-i\theta})}$

Putting $z = e^{i\theta}$, $dz = e^{i\theta} i d\theta$ and $d\theta = \frac{dz}{zi}$

$$I = R.P. \frac{1}{i} \int_C \frac{z^2 dz}{5+\left(2z+\frac{2}{z}\right)} \frac{1}{z} dz, \quad \text{where } |z| = 1.$$

or $$I = R.P. \frac{1}{i} \int_C \frac{z^2 dz}{2z^2+5z+2}$$

$$I = R.P. \frac{1}{i} \int_C f(z) dz \qquad ...(1)$$

where, $$f(z) = \frac{z^2}{2z^2+5z+2} = \frac{z^2}{(z+2)(2z+1)}$$

For the poles $(z+2)(2z+1) = 0$

$\Rightarrow$ $$z = -\frac{1}{2}, -2$$

But only the pole $z = -\frac{1}{2}$, lies within the circle $|z| = 1$

So $$\text{Res}\left(z = -\frac{1}{2}\right) = \lim_{z \to -\frac{1}{2}} \left(z+\frac{1}{2}\right) \frac{z^2}{(z+2)2\left(z+\frac{1}{2}\right)}$$

$$= \lim_{z \to -\frac{1}{2}} \frac{z^2}{2(z+2)} = \frac{\frac{1}{4}}{2\left(-\frac{1}{2}+2\right)} = \frac{1\times 2}{8\times 3} = \frac{1}{12}$$

Hence, $\int_C f(z) dz = 2\pi i \times \frac{1}{12} = \frac{\pi i}{6}$

From equation (1), $I = R.P. \frac{1}{i}\left(\frac{\pi i}{6}\right) = \frac{\pi}{6}$. **Proved.**

Example 2: Evaluate $\int_0^{\pi} \frac{a\,d\phi}{a^2+\sin^2\phi}$

Let $$I = \int_0^{\pi} \frac{a\,d\phi}{a^2+\sin^2\phi} = \int_0^{\pi} \frac{a\,d\phi}{a^2+\frac{1-\cos 2\phi}{2}}$$

$$= 2a\int_0^{\pi} \frac{d\phi}{(2a^2+1)-\cos 2\phi}$$

$$= \frac{2a}{2}\int_0^{2\pi} \frac{d\theta}{(2a^2+1)-\cos\theta}, \text{ Putting } 2\phi = \theta$$

$$= 2a\int_0^{2\pi} \frac{d\theta}{(4a^2+2)-(e^{i\theta}+e^{-i\theta})}$$

$$= 2a\int_C \frac{1}{(4a^2+2)-\left(z+\frac{1}{z}\right)}\frac{dz}{iz}, \text{ writing } z = e^{i\theta} \text{ or } d\theta = \frac{dz}{iz}$$

where $C \equiv |z| = 1$

$$I = \frac{2a}{i}\int_C \frac{dz}{(4a^2z+2z-z^2-1)} = 2ai\int_C \frac{dz}{z^2-2z(1+2a^2)+1}$$

$$\Rightarrow \qquad I = 2ai\int_C f(z)dz \qquad \ldots(1)$$

where $f(z) = \frac{1}{z^2-2z(1+2a^2)+1} = \frac{1}{(z-\alpha)(z-\beta)}$

where $\alpha = (1+2a^2)+2a\sqrt{1+a^2}$, $\beta = (1+2a^2)-2a\sqrt{1+a^2}$

It is obvious that $|\alpha|>1$ and as $|\alpha\beta| = 1$ $\therefore$ $|\beta|<1$

$\therefore$ only β lies within the circle $|z| = 1$.

$$\therefore \qquad \text{Res}(z=\beta) = \lim_{z\to\beta}(z-\beta)\frac{1}{(z-\alpha)(z-\beta)} = \frac{1}{\beta-\alpha}$$

$$\therefore \qquad \text{Res}(z=\beta) = \frac{1}{(1+2a^2)-2a\sqrt{1+a^2}-(1+2a^2)-2a\sqrt{1+a^2}}$$

$$= \frac{1}{-4a\sqrt{1+a^2}}$$

From equation (1), $$I = 2ai\int_C f(z)dz = 2ai\left[2\pi i \times \left(\frac{1}{-4a\sqrt{1+a^2}}\right)\right]$$

$$= \frac{\pi}{\sqrt{1+a^2}}. \textbf{ Ans.}$$

Example 3: Use method of contour integration to prove that

$$\int_0^{2\pi} \frac{d\theta}{1+a^2-2a\cos\theta} = \frac{2\pi}{1-a^2}; 0 < a < 1.$$

Sol. Let $$I = \int_0^{2\pi} \frac{d\theta}{1+a^2-2a\cos\theta}$$

$$= \int_0^{2\pi} \frac{d\theta}{1+a^2-2a\left(\dfrac{e^{i\theta}+e^{-i\theta}}{2}\right)} = \int_0^{2\pi} \frac{d\theta}{1+a^2-a(e^{i\theta}+e^{-i\theta})}$$

Putting $z = e^{i\theta}$, $dz = e^{i\theta} i d\theta$, $d\theta = \dfrac{dz}{zi}$

$$I = \frac{1}{i}\int_C \frac{dz}{(1+a^2)z - az^2 - a} = -\frac{1}{ia}\int_C \frac{dz}{(z-a)\left(z-\dfrac{1}{a}\right)}$$

$\Rightarrow$ $$I = \int_C f(z)\, dz$$

where $$f(z) = \frac{-1}{ia(z-a)\left(z-\dfrac{1}{a}\right)} \text{ and } |z| = 1$$

For the poles $(z-a)\left(z-\dfrac{1}{a}\right) = 0$

$\Rightarrow$ $$z = a,\ z = \frac{1}{a}$$

But $0 < a < 1$, so $z = a$ only lies inside C.

$\therefore \text{Res}(z = a) = \lim\limits_{z\to a}(z-a)f(z)$

$$= \lim_{z\to a}(z-a)\left\{-\frac{1}{ia(z-a)\left(z-\dfrac{1}{a}\right)}\right\}$$

$$= -\operatorname*{Lim}_{z\to a}\frac{1}{ia\left(z-\frac{1}{a}\right)} = \frac{1}{i(1-a^2)}$$

Hence by Cauchy's residue theorem, we have

$$I = 2\pi i \times \frac{1}{i(1-a^2)} = \frac{2\pi}{1-a^2}. \quad \textbf{Proved.}$$

Example 4: Prove that $\int_0^{2\pi} \frac{\sin^2\theta\, d\theta}{a+b\cos\theta} = \frac{2\pi}{b^2}\left[a-\sqrt{a^2-b^2}\right]$, where $a > b > 0$.

Sol. Let $I = \int_0^{2\pi} \frac{\sin^2\theta}{a+b\cos\theta}d\theta = \int_0^{2\pi}\frac{1-\cos 2\theta}{2a+2b\cos\theta}.d\theta$

$$I = R.P.\int_0^{2\pi}\frac{1-e^{2i\theta}}{2a+2b\left(\frac{e^{i\theta}+e^{-i\theta}}{2}\right)}d\theta$$

Putting $z = e^{i\theta}$ so that $dz = ie^{i\theta}d\theta$, we get

$$I = R.P.\int_C \frac{\left(1-z^2\right)}{2a+b\left(z+\frac{1}{z}\right)}\frac{dz}{iz}, \text{ where } C \equiv |z| = 1.$$

$$= R.P.\frac{1}{i}\int_C\frac{\left(1-z^2\right)}{2az+bz^2+b}dz = R.P.\frac{1}{ib}\int_C\frac{1-z^2}{z^2+\frac{2a}{b}z+1}dz$$

$$I = R.P.\frac{1}{ib}\int_C f(z)dz \quad \ldots(1)$$

where $f(z) = \frac{1-z^2}{z^2+\frac{2a}{b}z+1}$.

For the poles $z^2+\frac{2a}{b}z+1 = 0$

$$z = \frac{-\frac{2a}{b}\pm\sqrt{\frac{4a^2}{b^2}-4}}{2} = \frac{-a\pm(a^2-b^2)^{1/2}}{b}$$

$$\Rightarrow \quad \alpha = \frac{-a+(a^2-b^2)^{1/2}}{b} \text{ and } \beta = \frac{-a-(a^2-b^2)^{1/2}}{b}$$

Since $|\alpha\beta| = 1,\ |\beta| > 1$, so $|\alpha| < 1$.

Therefore, only the pole $z = \alpha$ lies within the circle $|z| = 1$.

$$\therefore \quad \text{Res}(z=\alpha) = \lim_{z\to\alpha}(z-\alpha)f(z) = \lim_{z\to\alpha}(z-\alpha).\left(\frac{1-z^2}{(z-\alpha)(z-\beta)}\right)$$

$$= \lim_{z\to\alpha}\frac{(1-z^2)}{(z-\beta)} = \frac{1-\alpha^2}{\alpha-\beta} = \frac{\alpha\left(\frac{1}{\alpha}-\alpha\right)}{(\alpha-\beta)}$$

$$= \frac{\alpha(\beta-\alpha)}{(\alpha-\beta)} = \frac{-\alpha(\beta-\alpha)}{(\beta-\alpha)} = -\alpha \qquad [\alpha\beta = 1]$$

$$\text{Res}(z=\alpha) = \frac{a-(a^2-b^2)^{1/2}}{b}$$

$$\therefore \quad \int_C f(z)dz = \frac{2\pi i}{b}\{a-(a^2-b^2)^{1/2}\}$$

From equation (1),
$$I = R.P.\frac{1}{ib}\times\frac{2\pi i}{b}\{a-(a^2-b^2)^{1/2}\}$$

$$= \frac{2\pi}{b^2}\{a-\sqrt{(a^2-b^2)}\}.$$ **Proved.**

Example 5: Apply calculus of residues to prove that–

$$\int_0^{2\pi} e^{\cos\theta}\cos(\sin\theta - n\theta)d\theta = \frac{2\pi}{n!}, \text{ where } n \text{ is positive integer.}$$

Sol. Let $I = \int_0^{2\pi} e^{\cos\theta}\cos(\sin\theta-n\theta)d\theta = R.P.\int_0^{2\pi} e^{\cos\theta}e^{i(\sin\theta-n\theta)}d\theta$

$$= R.P.\int_0^{2\pi} e^{\cos\theta+i\sin\theta-in\theta}d\theta = R.P.\int_0^{2\pi} e^{(e^{i\theta}-in\theta)}d\theta$$

$$= R.P.\int_0^{2\pi} e^{e^{i\theta}}e^{-in\theta}d\theta$$

Putting $z = e^{i\theta}, dz = ie^{i\theta}d\theta$.

So $d\theta = \dfrac{dz}{iz}$

$$\therefore \quad I = R.P.\int_C e^z z^{-n}\frac{dz}{zi} = R.P.\frac{1}{i}\int_C \frac{e^z}{z^{n+1}}dz$$

$$\Rightarrow \quad I = R.P.\frac{1}{i}\int_C f(z)dz \qquad ...(1)$$

where $C \equiv |z| = 1$

and
$$f(z) = \frac{e^z}{z^{n+1}}$$

For the poles $z^{n+1} = 0$

So $z = 0$ is a pole of order $(n+1)$

$$\therefore \quad \text{Res}(z=0) = \frac{1}{n!}\left[\frac{d^n}{dz^n}\left\{\frac{z^{(n+1)}e^z}{z^{n+1}}\right\}\right]_{z=0}$$

$$= \frac{1}{n!}\left[\frac{d^n}{dz^n}\{e^z\}\right]_{z=0} = \frac{1}{n!}\left[e^z\right]_{z=0} = \frac{1}{n!}$$

$$\text{Res}(z=0) = \frac{1}{n!}$$

By Cauchy's residues theorem, we get

$$\int_C f(z)dz = 2\pi i \times \frac{1}{n!}$$

From equation (1), $I = R.P.\frac{1}{i}2\pi i \times \frac{1}{n!} = \frac{2\pi}{n!}.$ **Proved.**

Example 6: Show that $\int_0^{\pi} \tan(\theta + ia)d\theta = i\pi$, where $R(a) > 0$

Sol. Let $I = \int_0^{\pi} \tan(\theta + ia)d\theta = \int_0^{\pi} \frac{\sin(\theta + ia)}{\cos(\theta + ia)}d\theta$

$$I = \int_0^{\pi} \frac{2\sin(\theta + ia)\cos(\theta - ia)}{2\cos(\theta + ia)\cos(\theta - ia)}d\theta$$

$$= \int_0^{\pi} \frac{(\sin 2\theta + \sin 2ia)}{(\cos 2\theta + \cos 2ia)}d\theta$$

| $2\sin A\cos B = \sin(A+B) + \sin(A-B)$ and $2\cos A\cos B = \cos(A+B) + \cos(A-B)$

$$= \frac{1}{2}\int_0^{2\pi} \frac{(\sin x + i\sin h2a)}{(\cos x + \cos h2a)}dx, \text{ where } 2\theta = x$$

Now putting $z = e^{ix}$

$$\Rightarrow \quad dz = ie^{ix} \cdot dx, \frac{dz}{iz} = dx, \text{ we get}$$

$$I = \frac{1}{2}\int_C \left[\frac{\frac{z - z^{-1}}{2i} + i\sinh 2a}{\frac{z + z^{-1}}{2} + \cosh 2a}\right]\frac{dz}{iz}, \text{ where } C \equiv |z| = 1$$

$$I = -\frac{1}{2}\int_C f(z)dz \quad ...(1), \quad \text{where } f(z) = \frac{z^2 - 1 - 2z\sinh 2a}{z(z^2 + 2z\cosh 2a + 1)}$$

For the poles $z(z^2 + 2z \cosh 2a + 1) = 0$

$$\Rightarrow \quad z = 0, z^2 + 2z\cosh 2a + 1 = 0$$

$$\Rightarrow \quad z = 0, z = -\cosh 2a \pm \sqrt{\cosh^2 2a - 1}$$

$$\Rightarrow \quad z = 0, z = -\cosh 2a \pm \sinh 2a \Rightarrow z = 0, \alpha, \beta$$

where $\alpha = -\cosh 2a + \sinh 2a$

$\beta = -\cosh 2a - \sinh 2a$

Evidently $|\alpha\beta| = 1, |\alpha| < 1$, so that $|\beta| > 1$.

Therefore $z = 0$, $z = \alpha$ only lie within the circle $|z| = 1$.

$$\text{Res } (z = 0) = \lim_{z\to 0} zf(z) = \lim_{z\to 0} \frac{z^2 - 2z\sinh 2a - 1}{z^2 + 2z\cosh 2a + 1} = \frac{0 - 0 - 1}{0 + 0 + 1} = -1$$

$$\Rightarrow \quad R_1 = -1$$

and
$$\text{Res } (z = \alpha) = \lim_{z\to\alpha}(z - \alpha)f(z) = \lim_{z\to\alpha} \frac{(z-\alpha)(z^2 - 1 - 2z\sinh 2a)}{z(z-\alpha)(z-\beta)}$$

$$\Rightarrow \quad \text{Res } (z = \alpha) = \frac{\alpha^2 - 2\alpha\sinh 2a - 1}{\alpha(\alpha - \beta)} = \frac{\alpha - 2\sinh 2a - \beta}{(\alpha - \beta)} \qquad \Big| \text{ As } \beta = \frac{1}{\alpha}$$

$$= \frac{2\sinh 2a - 2\sinh 2a}{2\sinh 2a} = 0$$

$$\Rightarrow \quad R_2 = 0$$

Hence
$$\int_C f(z)dz = 2\pi i \times (-1 + 0) = -2\pi i$$

From equation (1) $I = -\frac{1}{2} \times -2\pi i = \pi i.$ **Proved.**

Equation 7: Using the complex variables evaluate the real integral

$$\int_0^{2\pi} \frac{d\theta}{1 - 2p\sin\theta + p^2}, \text{ where } p^2 < 1.$$

Sol. Let $I = \int_0^{2\pi} \frac{d\theta}{1 - 2p\sin\theta + p^2} = \int_0^{2\pi} \frac{d\theta}{1 - 2p\left(\frac{e^{i\theta} - e^{-i\theta}}{2i}\right) + p^2}$

$$\Rightarrow \quad I = \int_0^{2\pi} \frac{d\theta}{1 + ip(e^{i\theta} - e^{-i\theta}) + p^2} \qquad ...(1)$$

Now putting $z = e^{i\theta}$ $\Rightarrow dz = ie^{i\theta}d\theta$

So that $d\theta = \dfrac{dz}{iz}$

$$I = \int_C \frac{dz}{zi - z^2p + p + p^2zi} = \int_C \frac{dz}{-z^2p + ip^2z + zi + p}$$

$$I = \int_C \frac{dz}{(iz+p)(izp+1)} \Rightarrow I = \int_C f(z)dz \qquad ...(2)$$

where $C \equiv |z| = 1$ and $f(z) = \dfrac{1}{(iz+p)(izp+1)}$

For the poles $(iz+p)(izp+1) = 0$

or $$z = -\frac{p}{i} = pi \text{ and } z = -\frac{1}{pi} = \frac{i}{p}$$

Since $|ip| < 1$ and $\left|\dfrac{i}{p}\right| > 1$ as $p^2 < 1$

So $z = pi$ is the only pole which inside the unit circle C.

$$\therefore \quad \text{Res}(z = pi) = \lim_{z\to pi}(z - pi)\frac{1}{(iz+p)(izp+1)} = \lim_{z\to pi}\frac{1}{i(izp+1)}$$

$$\text{Res}(z = pi) = \frac{1}{i}\frac{1}{(-p^2+1)}$$

Hence from equation (2) $\displaystyle\int_C f(z)dz = 2\pi i\left\{\frac{1}{i} \times \frac{1}{(1-p^2)}\right\}$

$$= \frac{2\pi}{(1-p^2)}. \quad \textbf{Ans.}$$

Example 8: Using complex variable prove that

$$\int_0^{\pi} \frac{\cos 2\theta d\theta}{1 - 2p\cos\theta + p^2} = \frac{\pi p^2}{1-p^2} (0 < p < 1)$$

Sol. Let $\displaystyle I = \int_0^{\pi} \frac{\cos 2\theta}{1 - 2p\cos\theta + p^2}d\theta = \frac{1}{2}\int_0^{2\pi} \frac{\cos 2\theta}{1 - 2p\left(\dfrac{e^{i\theta}+e^{-i\theta}}{2}\right) + p^2}d\theta$

$$= \frac{1}{2}\int_0^{2\pi} \frac{\cos 2\theta}{1 - p(e^{i\theta} + e^{-i\theta}) + p^2}d\theta$$

$$\Rightarrow \quad I = \frac{1}{2} \text{ real part of } \int_0^{2\pi} \frac{e^{2i\theta}}{1 - p(e^{i\theta} + e^{-i\theta}) + p^2}d\theta$$

Putting $z = e^{i\theta}, dz = e^{i\theta} i d\theta \quad \Rightarrow \quad d\theta = \frac{dz}{zi}$

$$I = \frac{1}{2} \text{ real part of } \int_C \frac{z^2}{(1-pz)\left(1-\frac{p}{z}\right)} \frac{dz}{zi}$$

$$= \frac{1}{2} \text{ real part of } \int_C \frac{-iz^2}{(1-pz)(z-p)} dz$$

$$= \frac{1}{2} \text{ real part of } \int_C f(z) dz \qquad \text{...(1)}$$

where $f(z) = \dfrac{-iz^2}{(1-pz)(z-p)}$.

For the poles $(1-pz)(z-p) = 0$

$\Rightarrow$ $$z = \frac{1}{p}, z = p$$

Only the pole $z = p$ lies within the circle $|z| = 1$ as $p < 1$

so $$\text{Res } (z = p) = \lim_{z \to p} (z-p) \frac{-iz^2}{(1-pz)(z-p)} = -\frac{ip^2}{1-p^2}$$

Hence $\displaystyle\int_C f(z)\, dz = 2\pi i \left\{ -\frac{ip^2}{1-p^2} \right\} = \frac{2\pi p^2}{1-p^2}$.

From equation (1), we get $I = \dfrac{1}{2}$ real part of $\displaystyle\int_C f(z) dz$

$$= \frac{1}{2} \times \frac{2\pi p^2}{(1-p^2)} = \frac{\pi p^2}{1-p^2}. \qquad \textbf{Proved.}$$

Example 9: Prove that $\displaystyle\int_0^{2\pi} \frac{\cos^2 3\theta}{1-2p\cos 2\theta + p^2} d\theta = \pi \frac{(1-p+p^2)}{(1-p)}, 0 < p < 1$

Sol. Let $I = \displaystyle\int_0^{2\pi} \frac{\cos^2 3\theta}{1-2p\cos 2\theta + p^2} d\theta = \frac{1}{2} \int_0^{2\pi} \frac{(1+\cos 6\theta)}{1-2p\cos 2\theta + p^2} d\theta$, as $\cos^2 3\theta = \dfrac{1+\cos 6\theta}{2}$

$$I = \frac{1}{2} \text{ real part of } \int_0^{2\pi} \frac{1+e^{6i\theta}}{1-p(e^{2i\theta}+e^{-2i\theta})+p^2} d\theta$$

Putting $z = e^{i\theta}, dz = e^{i\theta} i d\theta \quad \Rightarrow \quad d\theta = \frac{dz}{zi}$

$$I = \frac{1}{2} \text{ real part of } \int_C \frac{1+z^6}{1-p\left(z^2+\frac{1}{z^2}\right)+p^2} \frac{dz}{zi}$$

$$= \frac{1}{2} \text{ real part of } \frac{1}{i}\int_C \frac{z(1+z^6)dz}{(1-pz^2)(z^2-p)}$$

$$= \frac{1}{2} \text{ real part of } \frac{1}{i}\int_C f(z)dz, \text{ where } C \equiv |z| = 1$$

and $$f(z) = \frac{z(1+z^6)}{(1-pz^2)(z^2-p)}$$

For the poles $(1-pz^2)(z^2-p) = 0$

$\Rightarrow$ $$z = \pm\frac{1}{\sqrt{p}}, \text{ and } z = \pm\sqrt{p}$$

The only pole which lie within C are $z = \pm\sqrt{p}$, as $p < 1$.

$\therefore$ $$\text{Res}(z=\sqrt{p}) \lim_{z\to\sqrt{p}} (z-\sqrt{p})\frac{z(1+z^6)}{(1-pz^2)(z^2-p)} = \frac{1}{2}\cdot\frac{1+p^3}{1-p^2} = R_1$$

and $$\text{Res}(z=-\sqrt{p}) \lim_{z\to-\sqrt{p}} (z+\sqrt{p})\frac{z(1+z^6)}{(1-pz^2)(z^2-p)} = \frac{1}{2}\cdot\frac{1+p^3}{1-p^2} = R_2$$

$\Rightarrow$ $$R_1 + R_2 = \frac{1}{2}\cdot\frac{1+p^3}{1-p^2} + \frac{1}{2}\cdot\frac{1+p^3}{1-p^2} = \frac{1+p^3}{1-p^2}$$

Hence by Cauchy's residue theorem

$$\int_C f(z)dz = 2\pi i\left\{\frac{1+p^3}{1-p^2}\right\}$$

$\therefore$ $$I = \frac{1}{2} \text{ real part of } \frac{1}{i}\int_C f(z)dz$$

$$= \frac{1}{2} \text{ real part of } \frac{1}{i} \times 2\pi i\left\{\frac{1+p^3}{1-p^2}\right\}$$

$$= \frac{1}{2} \text{ real part of } 2\pi.\frac{1+p^3}{1-p^2}$$ | $a^3+b^3 = (a+b)(a^2-ab+b^2)$

$$= \pi.\frac{1+p^3}{1-p^2} = \pi.\frac{1-p+p^2}{1-p}.$$ **Proved.**

Example 10: Using complex variable technique evaluate the real integral $\int_0^{2\pi} \frac{\sin^2 \theta}{5+4\cos 3\theta} d\theta$.

Sol.
$$I = \int_0^{2\pi} \frac{\sin^2 \theta}{5+4\cos 3\theta} d\theta = \int_0^{2\pi} \frac{\left(\frac{1-\cos 2\theta}{2}\right) d\theta}{5+4\cos 3\theta}$$

$$I = \frac{1}{2} \text{ real part of } \int_0^{2\pi} \frac{1-e^{2i\theta}}{5+2(e^{3i\theta}+e^{-3i\theta})} d\theta$$

Putting $z = e^{i\theta}, dz = e^{i\theta} i d\theta \quad \Rightarrow \quad d\theta = \frac{dz}{zi}$

$$I = \frac{1}{2} \text{ real part of } \frac{1}{i} \int_C \frac{1-z^2}{5+2\left(z^3+\frac{1}{z^3}\right)} \frac{dz}{z}$$

$$= \frac{1}{2} \text{ real part of } \frac{1}{i} \int_C \frac{(1-z^2)z^2}{2z^6+5z^3+2} dz$$

$$\Rightarrow \qquad I = \frac{1}{2} \text{ real part of } \frac{1}{i} \int_C f(z) dz. \qquad \ldots(1)$$

where
$$f(z) = \frac{z^2(1-z^2)}{2z^6+5z^3+2} \text{ and } C \equiv |z| = 1$$

For the poles $2z^6 + 5z^3 + 2 = 0$

$$2z^6 + 4z^3 + z^3 + 2 = 0$$

$$2z^3(z^3+2) + 1(z^3+2) = 0$$

$$= (z^3+2)(2z^3+1) = 0$$

$$\Rightarrow \qquad z^3 + 2 = 0 \text{ and } 2z^3 + 1 = 0$$

$$\Rightarrow \qquad z^3 = -2 \text{ and } z^3 = -\frac{1}{2}$$

$$z = 2^{1/3} e^{(2n+1)\frac{\pi i}{3}} (n = 0, 1, 2) \text{ and } z = \left(\frac{1}{2}\right)^{\frac{1}{3}} e^{(2n+1)\pi i/3} (n = 0,1,2)$$

Therefore, only the pole $z = -\left(\frac{1}{2}\right)^{\frac{1}{3}}$ lies within the circle $|z| = 1$.

$$\text{So, Res}\left(z=-\left(\frac{1}{2}\right)^{\frac{1}{3}}\right)=\left\{\frac{\phi(z)}{\psi'(z)}\right\}_{z\to a} \quad \text{as } \psi(a)=0$$

$$\text{Res}\left(z=-\left(\frac{1}{2}\right)^{\frac{1}{3}}\right)=\left\{\frac{(1-z^2)z^2}{12z^5+15z^2}\right\}_{z=-\left(\frac{1}{2}\right)^{\frac{1}{3}}}$$

$$\text{Res}\left(z=-\left(\frac{1}{2}\right)^{\frac{1}{3}}\right)=\left\{1-\frac{1}{2^{2/3}}\right\}\left(\frac{1}{2}\right)^{2/3}\Bigg/\left\{-12\times\left(\frac{1}{2}\right)^{5/3}+15\left(\frac{1}{2}\right)^{2/3}\right\}$$

$$=\frac{\dfrac{\{2^{2/3}-1\}}{2^{2/3}}\times\dfrac{1}{2^{2/3}}}{3\left(\dfrac{1}{2}\right)^{2/3}\left\{-\dfrac{4}{2}+5\right\}}$$

$$=\frac{\dfrac{1}{2^{2/3}}\times\{2^{2/3}-1\}}{9}$$

$$\text{Res}\left\{z=-\left(\frac{1}{2}\right)^{1/3}\right\}=\frac{\{2^{2/3}-1\}}{9.2^{2/3}}$$

Hence by Cauchy's residue theorem, we have

$$\int_C f(z)dz=\frac{2\pi i}{9.2^{2/3}}\{2^{2/3}-1\}=\frac{2^{1/3}}{9}\pi i\{2^{2/3}-1\}$$

$$\Rightarrow \qquad \int_C f(z)dz=\frac{2^{1/3}\pi i}{9}\{2^{2/3}-1\}$$

From equation (1)

$$I=\frac{1}{2}\text{ real part of }\frac{1}{i}\int_C f(z)dz$$

$$=\frac{1}{2}\text{ real part of }\frac{1}{i}\times\frac{2^{1/3}\pi i}{9}\{2^{2/3}-1\}$$

$$=\frac{1}{2}\times\frac{2^{1/3}\pi}{9}\{2^{2/3}-1\}$$

$$=\frac{\pi}{9\times 2^{2/3}}\{2^{2/3}-1\}. \quad \textbf{Ans.}$$

PROBLEM SET 3.4

Evaluate the following:

1. $\int_0^{2\pi} \frac{d\theta}{2+\cos\theta}$ $\left[\textbf{Ans. } 2\pi/\sqrt{3}\right]$ **2.** $\int_0^{\pi} \frac{d\theta}{a+b\cos\theta}$ where a > |b| $\left[\textbf{Ans. } 2\pi/\sqrt{a^2-b^2}\right]$

3. $\int_0^{2\pi} \frac{d\theta}{\sqrt{2}-\cos\theta}$ [**Ans.** π] **4.** $\int_0^{2\pi} \frac{\sin^2\theta\, d\theta}{a+b\cos 2\theta}$ $\left[\textbf{Ans. } \frac{2\pi}{b^2}\{a-\sqrt{a^2-b^2}\}, a>b>0\right]$

5. $\int_0^{2\pi} \frac{d\theta}{(a+b\cos\theta)^2}$ $\left[\textbf{Ans. } \left(\frac{2\pi a}{(a^2-b^2)^{3/2}}\right)\right]$ **6.** $\int_0^{2\pi} \cos^{2n}\theta d\theta, n\in I$ $\left[\textbf{Ans. } \pi 2n!/(2)^{2n-1}(n!)^2\right]$

7. Prove that $\int_0^{2\pi} e^{-\cos\theta}\cos(n\theta+\sin\theta)d\theta = \frac{2\pi(-1)^n}{n!}$

8. Evaluate $\int_0^{2\pi} \frac{d\theta}{\frac{5}{4}+\sin\theta}$ $\left[\textbf{Ans. } \frac{8\pi}{3}\right]$ **9.** $\int_{-\pi}^{\pi} \frac{a\cos\theta}{a+\cos\theta}d\theta$ $\left[\textbf{Ans. } 2\pi a\left(1-\frac{a}{\sqrt{a^2-1}}\right)\right]$

10. Evaluate (i) $\int_0^{2\pi} \frac{\sin n\theta}{1+2a\cos\theta+a^2}d\theta$

(ii) $\int_0^{2\pi} \frac{\cos n\theta}{1+2a\cos\theta+a^2}d\theta$ where $a^2<1$ and n is positive integer.

$\left[\textbf{Ans. } (i)\ 0,\ (ii)\ \frac{2\pi(-1)^n a^n}{(1-a^2)}\right]$

3.11 EVALUATION OF $\int_{-\infty}^{\infty} f(x)dx$

Consider the integral $\int_C f(z)dz$ taken round a closed curve C consisting of the upper half C_R of the circle $|z| = R$, and part of real axis from $-R$ to R. If there are no poles of $f(z)$ on the real axis, the circle $|z| = R$ which is arbitrary can be taken such that there is no singularity on its circumference C_R in the upper half of the plane, but possibly some poles inside the contour C specified above.

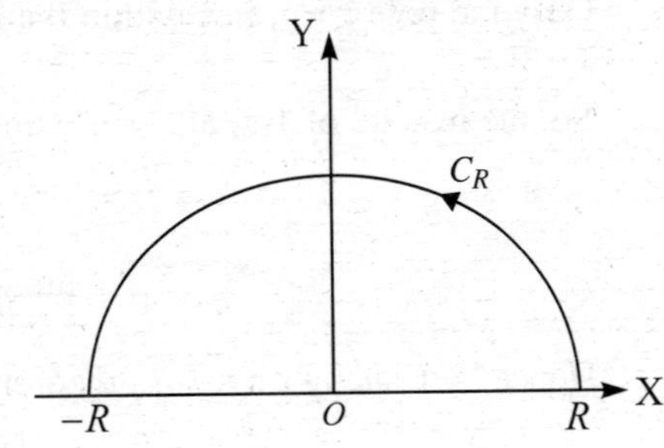

By Cauchy's theorem of residue.

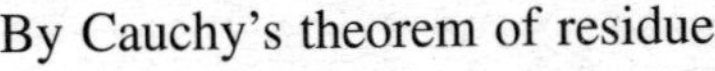

$$\int_C f(z)dz = 2\pi i\{\text{sum of the residues of } f(z) \text{ at the poles within } C\}$$

i.e., $$\int_{-R}^{R} f(x)dx + \int_{C_R} f(z)dx = 2\pi i\{\text{sum of the residues}\}$$

$$\therefore \lim_{R\to\infty}\int_{-R}^{R} f(x)dx + \lim_{R\to\infty}\int_{C_R} f(z)dz = 2\pi i\{\text{sum of the residues}\}$$

or $$\int_{-\infty}^{\infty} f(x)dx + \lim_{R\to\infty}\int_{C_R} f(z)dz = 2\pi i\{\text{sum of the residues}\}$$

If $R \to \infty$, then the value of the integral over the semi circle C_R approach zero

i.e., $\int_{C_R} f(z)dz \to 0$ as $R \to \infty$, then we have

$$\boxed{\lim_{R\to\infty}\int_{-\text{R}}^{\text{R}} f(x)dx = 2\pi i(\text{sum of residues at poles})}$$

Example 1: Apply the calculus of residue to find the value of an integral $\int_0^{\infty}\frac{\cos ax}{(x^2+1)}dx.$

Sol. $$\int_0^{\infty}\frac{\cos ax}{(x^2+1)}dx = \text{real part of } \int_0^{\infty}\frac{e^{iax}}{(x^2+1)}dx \qquad ...(1)$$

Now consider the integral $\int_C f(z)dz$, where $f(z) = \frac{e^{iaz}}{(z^2+1)}dz$

Taken round the contour C consisting of the upper half of a large circle and real axis from $-R$ to R.

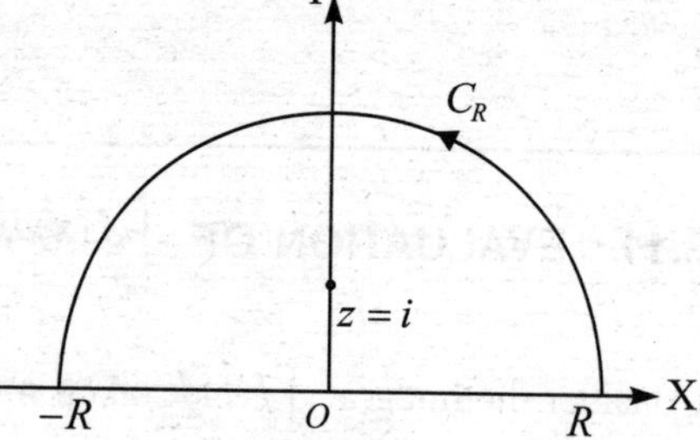

For the poles $(z^2 + 1) = 0 \quad \Rightarrow z^2 = i^2$

$\Rightarrow \quad z = \pm i$

Only the pole $z = i$, lies within the region.

So, the residue of $f(z)$ at $z = i$, $= \lim_{z\to i}(z-i)\frac{e^{iaz}}{(z+i)(z-i)}$

$$= \lim_{z\to i}\frac{e^{iaz}}{(z+i)} = \frac{e^{-a}}{2i}$$

Hence by Cauchy's residues theorem, we have

$$\int_C f(z)dz = 2\pi i\{\text{Sum of the residues of } f(z) \text{ within contour}\}$$

i.e. $$\int_{-R}^{R} f(x)dx + \int_{C_R} f(z)dz = 2\pi i \times \frac{e^{-a}}{2i} = \pi e^{-a}$$

Taking lim R $\to \infty$, we get

$$\lim_{R\to\infty}\int_{-R}^{R} f(x)dx + \lim_{R\to\infty}\int_{C_R} f(z)dz = \pi e^{-a}$$

$$\Rightarrow \qquad \int_{-\infty}^{\infty} \frac{e^{iax}}{x^2+1}dx = \pi e^{-a} \qquad \Bigg|\ \text{As } \lim_{R\to\infty}\int_{C_R} f(z)dz = 0 \qquad \ldots(2)$$

Now real part of $\displaystyle\int_{-\infty}^{\infty} \frac{e^{iax}}{x^2+1}dx = \pi e^{-a}$

or $$\int_{-\infty}^{\infty} \frac{\cos ax}{(x^2+1)}dx = \pi e^{-a}$$

Hence

$$\int_{0}^{\infty} \frac{\cos ax}{(x^2+1)}dx = \frac{\pi e^{-a}}{2}. \quad \textbf{Ans.}$$

Example 2: Apply the method of calculus of residues to prove that $\displaystyle\int_0^{\infty} \frac{\log(1+x^2)}{(1+x^2)}dx = \pi \log 2.$

Sol. Consider the integral $\displaystyle\int_C \frac{\log(z+i)}{(1+z^2)}dz$, where $f(z) = \dfrac{\log(z+i)}{(1+z^2)}$ and C is closed contour consisting of real axis and the upper half C_R of a large semi circle $|z| = R$.

Now for the poles

$$z^2+1=0 \qquad \Rightarrow z = \pm i$$

Only the pole $z = i$ lies within the contour.

$\therefore$ The residue at $(z = i) = \displaystyle\lim_{z\to i}(z-i)\frac{\log(z+i)}{(z-i)(z+i)} = \frac{\log 2i}{2i}$

$$\text{Res }(z=i) = \frac{1}{2i}\log_e\left\{2e^{i\pi/2}\right\}$$

$$= \frac{1}{2i}\left\{\log_e 2 + \log_e e^{i\pi/2}\right\}$$

$$= \frac{1}{2i}\left\{\log_e 2 + \frac{i\pi}{2}\right\}$$

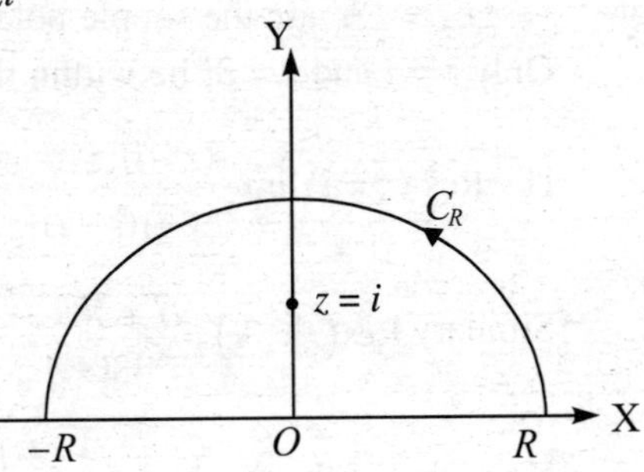

Hence by Cauchy's theorem, we have

$$\int_C f(z)dz = 2\pi i \times \frac{1}{2i}\left\{\log_e 2 + \frac{i\pi}{2}\right\}$$

$$\Rightarrow \quad \int_{-R}^{R} f(x)dx + \int_{C_R} f(z)dz = \pi\left\{\log_e 2 + \frac{i\pi}{2}\right\}$$

Taking lim R $\to \infty$, we get

$$\lim_{R\to\infty}\int_{-R}^{R} f(x)dx + \lim_{R\to\infty}\int_{C_R} f(z)dz = \pi\left\{\log_e 2 + \frac{i\pi}{2}\right\}$$

i.e.,
$$\int_{-\infty}^{\infty}\frac{\log(x+i)}{(1+x^2)}dx = \pi\left\{\log_e 2 + \frac{i\pi}{2}\right\} \quad ...(1)$$

$$\Rightarrow \quad \int_{-\infty}^{\infty}\frac{\frac{1}{2}\log(x^2+1)+i\tan^{-1}\frac{1}{x}}{(x^2+1)}dx = \pi\left(\log_e 2 + \frac{i\pi}{2}\right), \left|\text{As } \log(\alpha+i\beta) = \frac{1}{2}\log(\alpha^2+\beta^2)+i\tan^{-1}\frac{\beta}{\alpha}\right.$$

Equating only real part, we get

$$\frac{1}{2}\int_{-\infty}^{\infty}\frac{\log(x^2+1)}{(x^2+1)}dx = \pi\log_e 2$$

or
$$\int_{0}^{\infty}\frac{\log(1+x^2)}{(1+x^2)}dx = \pi\log_e 2. \quad \textbf{Proved.}$$

Example 3: Prove that $\int_{-\infty}^{\infty}\frac{x^2-x+2}{x^4+10x^2+9}dx = 5\pi/12.$

Sol. Consider the integral $\int_C f(z)dz$, where $f(z) = \frac{z^2-z+2}{z^4+10z^2+9}$ and C is a closed contour which consisting of the upper half of a large circle $|z| = R$ and the real axis from $-R$ and R.

For the poles $z^4+10z^2+9=0$

i.e., $(z^2+1)(z^2+9)=0$

or $z=\pm i, z=\pm 3i$ are the simple poles

Only $z = i$ and $z = 3i$ lie within the contour.

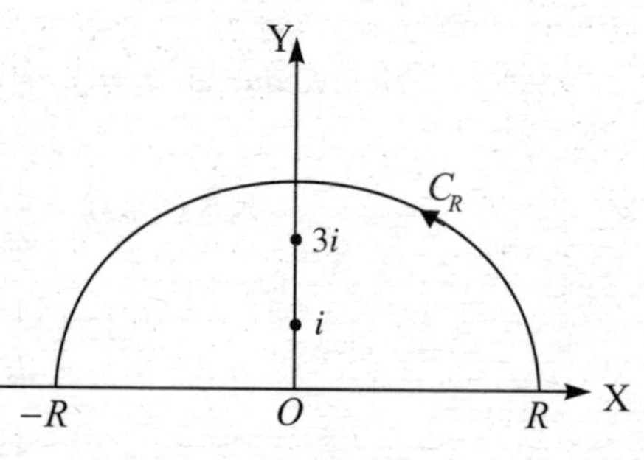

$\therefore$ The Res $(z = i)$ $\lim_{z\to i}\frac{(z-i)(z^2-z+2)}{(z-i)(z+i)(z^2+9)} = \frac{1-i}{16i}$

Similary $\text{Res}(z=3i) = \frac{7+3i}{48i}$

Now sum of the residues $= \frac{1-i}{16i} + \frac{7+3i}{48i} = \frac{5}{24i}$

Hence by Cauchy's residue theorem, we have

$$\int_C f(z) = 2\pi i \times (\text{sum of residues}) = 2\pi i \times \frac{5}{24i} = \frac{5\pi}{12}$$

i.e., $$\int_{-R}^{R} f(x)dx + \int_{C_R} f(z)dz = \frac{5\pi}{12}$$

Taking lim R → ∞, we get

$$\lim_{R\to\infty}\int_{-R}^{R} f(x)dx = \frac{5\pi}{12} \qquad \Big|\ \text{As } \lim_{R\to\infty}\int_{C_r} f(z)dz = 0$$

$$\Rightarrow \qquad \int_{-\infty}^{\infty} \frac{x^2 - x + 2}{x^4 + 10x^2 + 9}dx = \frac{5\pi}{12}\cdot \qquad \textbf{Proved.}$$

Example 4: Using complex variable techniques, evaluate the integral $\int_0^{\infty}\frac{dx}{x^2+16}$.

Sol. Consider the integral $\int_C \frac{dz}{z^2+16}$, where $f(z) = \frac{1}{z^2+16}$ and C is a closed contour which consisting of the upper half of a large circle $|z| = R$ and the real axis from $-R$ to R.

For the poles $z^4 + 16 = 0$ *i.e.* $z^4 = -16$.

$$z^4 = 16[\cos(2n+1)\pi + i\sin(2n+1)\pi]$$

$$z = 2[\cos(2n+1)\pi + i\sin(2n+1)\pi]^{1/4}$$

$$z = 2\left[\cos(2n+1)\frac{\pi}{4} + i\sin(2n+1)\frac{\pi}{4}\right]$$

$$z = 2e^{i(2n+1)\frac{\pi}{4}} \quad \text{where } (n = 0, 1, 2, 3)$$

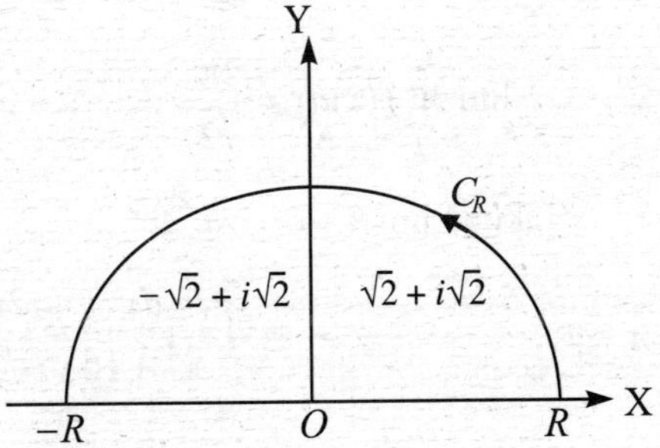

Y

C_R

$-\sqrt{2}+i\sqrt{2}$ $\sqrt{2}+i\sqrt{2}$

$-R$ O R X

$$\text{for } n = 0,\ z_1 = 2\left(\frac{1}{\sqrt{2}} + \frac{i}{\sqrt{2}}\right) = \sqrt{2} + i\sqrt{2}$$

$$\text{for } n = 1,\ z_2 = 2\left(-\frac{1}{\sqrt{2}} + i\frac{1}{\sqrt{2}}\right) = -\sqrt{2} + i\sqrt{2}$$

$$\text{for } n = 2,\ z_3 = 2\left(-\frac{1}{\sqrt{2}} - i\frac{i}{\sqrt{2}}\right) = -\sqrt{2} - i\sqrt{2}$$

$$\text{for } n = 3,\ z_4 = 2\left(\frac{1}{\sqrt{2}} - i\frac{i}{\sqrt{2}}\right) = \sqrt{2} - i\sqrt{2}$$

Only the poles z_1 and z_2 lie within the contour

$$\therefore \qquad \text{Res}(z = e^{i\pi/4}) = \left[\frac{1}{\frac{d}{dz}(z^2+16)}\right]_{z=e^{i\pi/4}} = \left[\frac{1}{2z}\right]_{z=e^{i\pi/4}} = \frac{1}{2\cdot e^{i\pi/4}}$$

$$= \frac{1}{2}\left[\cos\frac{\pi}{4} - i\sin\frac{\pi}{4}\right]$$

$$= \frac{1}{2}\left[\frac{1}{\sqrt{2}} - i\frac{1}{\sqrt{2}}\right]$$

Similarly $\text{Res}(z = e^{3i\pi/4}) = \frac{1}{2}\left[-\frac{1}{\sqrt{2}} - i\frac{1}{\sqrt{2}}\right]$

Sum of the residues $= \frac{1}{2}\left[\frac{1}{\sqrt{2}} - i\frac{1}{\sqrt{2}} - \frac{1}{\sqrt{2}} - i\frac{1}{\sqrt{2}}\right] = -\frac{i}{\sqrt{2}}$

Hence by Cauchy's residue theorem, we have

$$\int_C f(z)dz = 2\pi i \times \left(\frac{-i}{\sqrt{2}}\right) = \frac{2\pi}{\sqrt{2}}$$

i.e., $\int_{-R}^{R} f(x)dx + \int_{C_R} f(z)dz = \frac{2\pi}{\sqrt{2}}$

$$\lim_{R\to\infty}\int_{-R}^{R} f(x)dx = \frac{2\pi}{\sqrt{2}}$$

Taking $\lim R \to \infty$, we get

or $$\int_{-\infty}^{\infty}\frac{dx}{x^4+16} = \frac{2\pi}{\sqrt{2}} \quad \text{or} \quad 2\int_0^{\infty}\frac{dx}{x^4+16} = \frac{2\pi}{\sqrt{2}}$$

or $$\int_0^{\infty}\frac{dx}{x^4+16} = \frac{\pi}{\sqrt{2}}.$$ **Ans.**

Example 5: Using the calculus of residue to prove that

$$\int_0^{\infty}\frac{x\sin ax}{x^2+k^2}dx = \frac{1}{2}\pi e^{-ka} \text{ where } k > 0$$

Sol. Consider the integral $\int_C f(z)\,dz$, where $f(z) = \frac{z\sin az}{z^2+k^2} = I.P. \text{ of } \frac{ze^{iaz}}{z^2+k^2}$ and C is a closed contour which consisting of the real axis form $-R$ to R and the upper half of the large circle $|z| = R$.

For the poles $z^2 + k^2 = 0 \quad \Rightarrow \quad z = \pm ki$

Only the pole $z = ki$ lies within the closed contour.

$$\therefore \text{Res}(z = ki) = \lim_{z\to ki}(z-ki)\frac{ze^{iaz}}{(z-ki)(z+ki)} = \frac{e^{-ka}}{2}$$

Hence by Cauchy's residues theorem

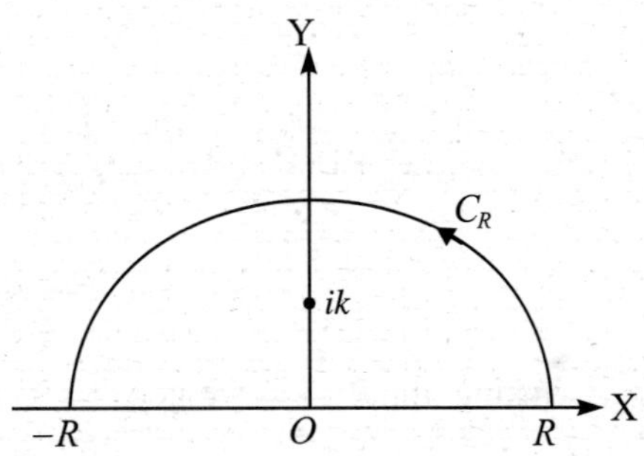

$$\int_C f(z)dz = 2\pi i \times \frac{e^{-ka}}{2} = \pi i e^{-ka}$$

i.e., $$\int_{-R}^{R} f(x)dx + \int_{C_R} f(z)dz = \pi i e^{-ka}$$

Taking $\lim R \to \infty$, we get

$$\lim_{R\to\infty} \int_{-R}^{R} f(x)dx + \lim_{R\to\infty} \int_{C_R} f(z) + z = \pi i e^{-ka}$$

or $$I.P. \text{ of } \int_{-\infty}^{\infty} \frac{xe^{iax}}{x^2 + k^2} = \pi i e^{-ka} \qquad ...(1)$$

$$\Rightarrow \qquad \int_{-\infty}^{\infty} \frac{x \sin ax}{x^2 + k^2} dx = \pi i e^{-ka}$$

$$\Rightarrow \qquad \int_{0}^{\infty} \frac{x \sin ax}{x^2 + k^2} dx = \frac{\pi}{2} e^{-ka}. \quad \textbf{Proved.}$$

Example 6: Using residue theorem, evaluate

$$\int_{-\infty}^{\infty} \frac{x^2 dx}{(x^2 + 1)(x^2 + 4)}$$

Sol. Let $$f(z) = \frac{z^2}{(z^2 + 1)(z^2 + 4)}$$

For the poles $= (z^2 + 1)(z^2 + 4) = 0 \Rightarrow z = i, i, 2i, -2i$

Choose the contour C as shown in the figure

The poles i and $2i$ lie in the upper half of the circle $|z| = R$.

$$\therefore \qquad \text{Res } (z = i) = \left[\frac{\phi(z)}{\psi'(z)}\right]_{z=i} = \left[\frac{z^2}{\frac{d}{dz}\{(z^2 + 1)(z^2 + 4)\}}\right]_{z=i}$$

$$= \left[\frac{z^2}{4z^3 + 10z}\right]_{z=i} = \frac{-1}{-4i + 10i} = \frac{-1}{6i} = \frac{i}{6}$$

Similarly $$\text{Res } (z = 2i) = -\frac{i}{3}$$

By Cauchy's residue theorem

$$\int_{-R}^{R} \frac{x^2}{(x^2+1)(x^2+4)} dx + \int_{C_R} \frac{x^2}{(z^2+1)(z^2+4)} dx = 2\pi i \text{ (sum of residues)} = 2\pi i \left(\frac{i}{6} - \frac{i}{3}\right)$$

$$= 2\pi i \left(\frac{-i}{6}\right) = \frac{\pi}{3}$$

Taking $\lim R \to \infty$, we get

$$\int_{-\infty}^{\infty} \frac{x^2}{(x^2+1)(x^2+4)} dx + 0 = \frac{\pi}{3}$$

or $$\int_{-\infty}^{\infty} \frac{x^2}{(x^2+1)(x^2+4)} dx = \frac{\pi}{3}\cdot \quad \textbf{Ans.}$$

Example 7: By Method of contour integration, prove that $\int_{-\infty}^{\infty} \frac{dx}{(1+x^2)^2} = \frac{\pi}{4}$.

Sol. Consider the integral $\int_C \frac{dz}{(1+z^2)^2}$, where $f(z) = \frac{1}{(1+z^2)^2}$ and C is a closed contour consisting of upper half of a large circle $|z| = R$ and the real axis from $-R$ to R.

For the poles $(z^2+1)^2 = 0$

$\Rightarrow$ $(z^2+1)^2 = 0$

$\Rightarrow$ $z = \pm i$ (order 2)

Only the pole $z = i$ (of order 2) lies within the closed contour

$$\therefore \quad \text{Res } (z = i) = \left[\frac{d}{dz}\left\{\frac{(z-i)^2}{(z^2+1)^2}\right\}\right]_{z\to i}$$

$$= \left[\frac{d}{dz}\left\{\frac{(z-i)^2}{(z+i)^2(z-i)^2}\right\}\right]_{z\to i}$$

$$= \left[\frac{d}{dz}\left\{\frac{1}{(z+i)^2}\right\}\right]_{z\to i}$$

$$= \left[-\frac{1\times 2}{(z+i)^3}\right]_{z\to i} = -\frac{2}{8i^3} = \frac{1}{4i}$$

Hence by Cauchy's residues theorem

$$\int_C f(z)dz = 2\pi i \times \frac{1}{4i} = \frac{\pi}{2}$$

or $$\int_{-R}^{R} f(x)dx + \int_{C_R} f(z) = \frac{\pi}{2}$$

Taking $\lim R \to \infty$, we get

$$\lim_{R\to\infty}\int_{-R}^{R} f(x)dx + \lim_{R\to\infty}\int_{C_R} f(x)dz = \frac{\pi}{2}$$

or $$\int_{-\infty}^{\infty} f(x)dx = \frac{\pi}{2}$$ As $\lim_{R\to\infty}\int_{C_R} f(z)dz = 0$

or $$\int_{-\infty}^{\infty} \frac{dx}{(1+x^2)^2} = \frac{\pi}{2} \text{ or } \int_0^{\infty} \frac{dx}{(1+x^2)^2} = \frac{\pi}{4}.$$ **Proved.**

Example 8: Using contour integration evaluate

$$\int_0^{\infty} \frac{dx}{1+x^4}$$

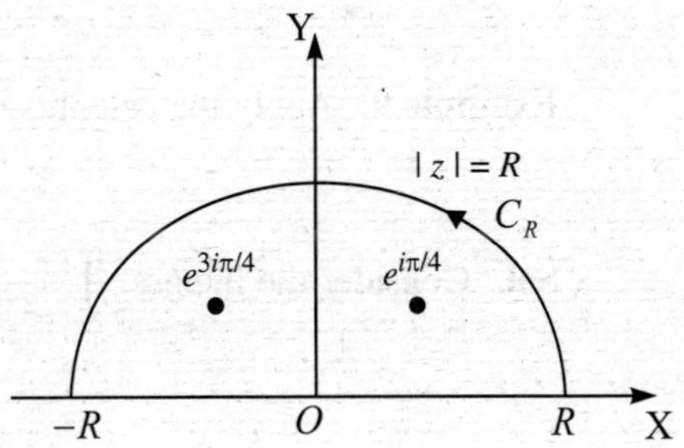

Sol. Let $$f(z) = \frac{1}{1+z^4}$$

For the poles $z^4 + 1 = 0 \Rightarrow z^4 = -1$

$$\Rightarrow z = \{\cos(2n+1)\pi + i\sin(2n+1)\pi\}^{1/4}$$

or $z = \cos(2n+1)\frac{\pi}{4} + i\sin(2n+1)\frac{\pi}{4}$, where n = 0, 1, 2, 3.

For n = 0, $$z = \cos\frac{\pi}{4} + i\sin\frac{\pi}{4} \pm \frac{1}{\sqrt{2}} + i\cdot\frac{1}{\sqrt{2}} = e^{i\pi/4}$$

For n = 1, $$z = \cos\frac{3\pi}{4} + i\sin\frac{3\pi}{4} = -\frac{1}{\sqrt{2}} + i\cdot\frac{1}{\sqrt{2}} = e^{3i\pi/4}$$

For n = 2, $$z = \cos\frac{5\pi}{4} + i\sin\frac{5\pi}{4} = -\frac{1}{\sqrt{2}} - i\frac{1}{\sqrt{2}} = e^{5i\pi/4}$$

For n = 3, $$z = \cos\frac{7\pi}{4} + i\sin\frac{7\pi}{4} = \frac{1}{\sqrt{2}} - i\frac{1}{\sqrt{2}} = e^{7i\pi/4}$$

Here only the poles corresponding to n = 0 and n = 1, i.e., $z = e^{i\pi/4}$ and $z = e^{3i\pi/4}$ lie in the upper helf of z-plane.

$\therefore$ $$\text{Res}(z = e^{i\pi/4}) = \lim_{z \to e^{i\pi/4}} \frac{z - e^{i\pi/4}}{z^4 + 1} = \lim_{z \to e^{i\pi/4}} \cdot \frac{1}{4z^3} \quad \text{(By L' Hospital's rule)}$$

$$= \frac{1}{4} e^{-3i\pi/4}$$

Similarly, $\text{Res}\ (z = e^{3i\pi/4}) = \frac{1}{4} e^{-9i\pi/4}$

By Cauchy's residue theorem

$$\int_{-R}^{R} \frac{dx}{1+x^4} + \int_{C_R} \frac{dz}{1+z^4} = \frac{2\pi i}{4}(e^{-3i\pi/4} + e^{-9i\pi/4}) = \frac{\pi i}{2}(-i\sqrt{2}) = \frac{\pi\sqrt{2}}{2}$$

Taking limit $R \to \infty$, we get

$$\int_{-\infty}^{\infty} \frac{dx}{1+x^4} + \underset{R \to \infty}{Lt} \int_{C_R} \frac{dz}{1+z^4} = \frac{\pi\sqrt{2}}{2}$$

or $$\int_{-\infty}^{\infty} \frac{dx}{1+x^4} = \frac{\pi\sqrt{2}}{2} \quad \text{or} \quad \int_{0}^{\infty} \frac{dx}{1+x^4} = \frac{\pi\sqrt{2}}{4}. \quad \textbf{Ans.}$$

Example 9: Apply the calculus of residues to evaluate $\int_{-\infty}^{\infty} \frac{\cos x dx}{(x^2+a^2)(x^2+b^2)}$, $a > b > 0$

Sol. Consider the integral $\int_C \frac{\cos z dz}{(z^2+a^2)(z^2+b^2)} dz$ = Real part of $\int_C \frac{e^{iz} dz}{(z^2+a^2)(z^2+b^2)}$

Let $f(z) = \frac{e^{iz}}{(z^2+a^2)(z^2+b^2)}$ and C is a closed contour of the upper half of a large circle $|z| = R$ and the real axis from $-R$; to R for the poles $(z^2+a^2)(z^2+b^2) = 0$

$\Rightarrow$ $$z = \pm ai \text{ and } z = \pm bi$$

Only the poles $z = ai$ and $z = bi$ lie within the contour.

$$\text{Res}\ (z = ai) = \lim_{z \to ai} (z - ai) \frac{e^{iz}}{(z-ai)(z+ai)(z^2+b^2)} = \frac{-e^{-a}}{2ia(a^2-b^2)}$$

Similarly, $\text{Res}\ (z = bi) = \frac{e^{-b}}{2ib(a^2-b^2)}$

Sum of the residues $\frac{1}{2i(a^2-b^2)}\left(\frac{e^{-b}}{b} - \frac{e^{-a}}{a}\right)$

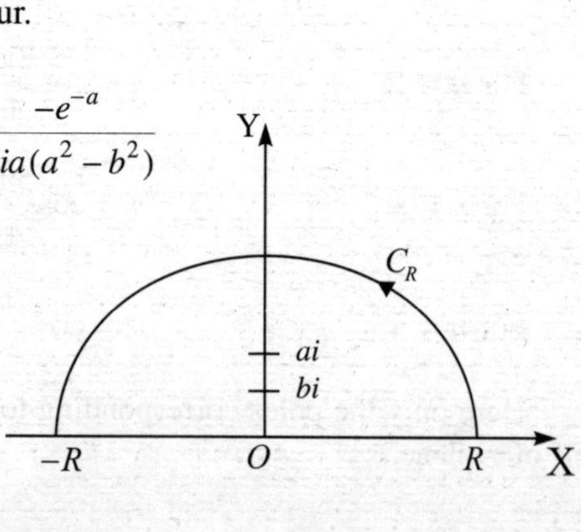

Y
C_R
ai
bi
$-R$ O R X

Hence by Cauchy's residue theorem

$$\int_C f(z)dz = 2\pi i \times \frac{1}{2i(a^2-b^2)}\left(\frac{e^{-b}}{b}-\frac{e^{-a}}{a}\right) = \frac{\pi}{(a^2-b^2)}\left(\frac{e^{-b}}{b}-\frac{e^{-a}}{a}\right)$$

i.e.,
$$\int_{-R}^{R} f(x)dx + \int_{C_R} f(z)dz = \frac{\pi}{(a^2-b^2)}\left(\frac{e^{-b}}{b}-\frac{e^{-a}}{a}\right)$$

Taking $\lim R \to \infty$, we get

$$\lim_{R\to\infty}\int_{-R}^{R} f(x)dx + \lim_{R\to\infty}\int_{C_R}^{R} f(z)dz = \frac{\pi}{a^2-b^2}\left(\frac{e^{-b}}{b}-\frac{e^{-a}}{a}\right)$$

or
$$\int_{-\infty}^{\infty} f(x)dx = \frac{\pi}{a^2-b^2}\left(\frac{e^{-b}}{b}-\frac{e^{-a}}{a}\right)$$ As $\lim_{R\to\infty}\int_{C_R} f(z)dz = 0$

$$\int_{-\infty}^{\infty}\frac{e^{ix}dx}{(x^2+a^2)(x^2+b^2)} = \frac{\pi}{(a^2-b^2)}\left(\frac{e^{-b}}{b}-\frac{e^{-a}}{a}\right)$$

or real part of $\int_{-\infty}^{\infty}\frac{e^{ix}}{(x^2+a^2)(x^2+b^2)}dx = \frac{\pi}{(a^2-b^2)}\left(\frac{e^{-b}}{b}-\frac{e^{-a}}{a}\right)$

$$\Rightarrow \quad \int_{-\infty}^{\infty}\frac{\cos x}{(x^2+a^2)(x^2+b^2)}dx = \frac{\pi}{(a^2-b^2)}\left(\frac{e^{-b}}{b}-\frac{e^{-a}}{a}\right).$$ **Proved.**

Example 10: Use the method of contour integration to prove that

$$\int_{-\infty}^{\infty}\frac{\cos x^2+\sin x^2-1}{x^2}dx = 0.$$

Sol. Consider the integral $\int_C \frac{e^{iz^2}-1}{z^2}dz$, where and C is a closed contour consisting of the upper half of a large circle $|z| = R$ and real axis from $-R$ to R.

Since $z = 0$, is a simple pole of order 2.

So
$$\text{Res}(z=0) = \left\{\frac{d}{dz}z^2\left(\frac{e^{iz^2}-1}{z^2}\right)\right\}_{z\to 0}$$

$$= \left\{\frac{d}{dz}(e^{iz^2}-1)\right\}_{z\to 0} = \left\{e^{iz^2}\times 2zi\right\}_{z\to 0} = 0$$

Hence by Cauchy's residue theorem

$$\int_C f(z)dz = 2\pi i \times 0 = 0$$

i.e., $$\int_{-R}^{R} f(x)dx + \int_{C_R} f(z)dz = 0$$

Taking $\lim R \to \infty$, we get

$$\lim_{R\to\infty} \int_{-R}^{R} f(x)dx + \lim_{R\to\infty} \int_{C_R} f(z)dz = 0$$

or $$\int_{-\infty}^{\infty} f(x)dx = 0 \qquad \left| \text{As } \lim_{R\to\infty} \int_{C_R} f(z)dz = 0 \right.$$

or $$\int_{-\infty}^{\infty} \frac{e^{ix^2} - 1}{x^2} dx = 0$$

$$\Rightarrow \int_{-\infty}^{\infty} \frac{(\cos x^2 + i \sin x^2) - 1}{x^2} dx = 0$$

Equating real and imaginary part, we have

$$\int_{-\infty}^{\infty} \frac{\cos x^2 - 1}{x^2} dx = 0 \text{ and } \int_{-\infty}^{\infty} \frac{\sin x^2}{x^2} dx = 0$$

Adding these we have $\int_{-\infty}^{\infty} \frac{\sin x^2 + \cos x^2 - 1}{x^2} dx = 0.$ **Proved.**

Example 11: Show by the method of contour integration that

$$\int_0^{\infty} \frac{\cos mx}{(a^2 + x^2)^2} dx = \frac{\pi}{4a^3}(1 + ma)e^{-ma} (a > 0), (m > 0)$$

Sol. Consider the integral $\int_C \frac{\cos mz}{(z^2 + a^2)^2} dz$ = real part of $\int_C \frac{e^{imz}}{(z^2 + a^2)^2} dz$ where $f(z) = \frac{e^{imz}}{(z^2 + a^2)^2}$ and C is a closed contour consisting of the upper half of a large circle $|z| = R$ and real axis from $-R$ to R.

For the poles $(z^2 + a^2)^2 = 0$

$\Rightarrow$ $z = ai$ and $z = -ai$, are two poles each of order 2.

Only the pole $z = ai$ lies within the contour.

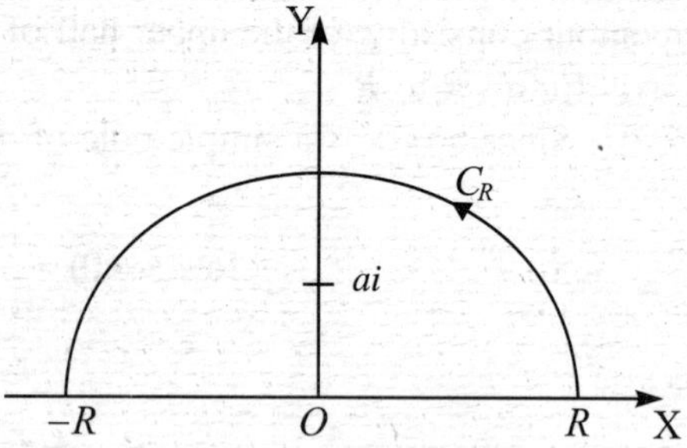

Y
C_R
ai
$-R$ O R X

So $$\text{Res}(z = ai) = \left\{ \frac{d}{dz} (z - ai)^2 \left(\frac{e^{imz}}{(z + ai)^2 (z - ai)^2} \right) \right\}_{z\to ai}$$

$$= \left\{ \frac{d}{dz} \left(\frac{e^{imz}}{(z + ai)^2} \right) \right\}_{z\to ai} = \left[\frac{-2e^{imz}}{(z + ai)^3} + \frac{ime^{imz}}{(z + ai)^2} \right]_{z\to ai}$$

$$= \frac{2e^{-am}}{8ia^3} - \frac{ime^{-am}}{4a^2} = \frac{e^{-am}}{4ia^3} + \frac{me^{-am}}{4ia^2} = \frac{e^{-am}}{4ia^3}\{1 + am\}$$

Hence by Cauchy's residue theorem, we have

$$\int_C f(z)dz = 2\pi i \times \frac{e^{-am}}{4ia^3}(1 + am) = \frac{\pi}{2a^3}(1 + am)e^{-ma}$$

i.e.,
$$\int_{-R}^{R} f(x)dx + \int_{C_R} f(z)dz = \frac{\pi}{2a^3}(1 + am)e^{-ma}$$

Taking $\lim R \to \infty$, we get

$$\lim_{R\to\infty} \int_{-R}^{R} f(x)dx + \lim_{R\to\infty} \int_{C_R} f(z)dz = \frac{\pi}{2a^3}(1 + am)e^{-ma}$$

or
$$\int_{-\infty}^{\infty} f(x)dx = \frac{\pi}{2a^3}(1 + am)e^{-ma}$$

or
$$\int_{-\infty}^{\infty} \frac{e^{imx}}{(a^2 + x^2)^2} dx + 0 = \frac{\pi}{2a^3}(1 + ma)e^{-ma}$$

or
$$\int_{-\infty}^{\infty} \frac{\cos mx + i \sin mx}{(a^2 + x^2)^2} dx = \frac{\pi}{2a^3}(1 + ma)e^{-ma}$$

Equating only real part, we get

$$\int_{-\infty}^{\infty} \frac{\cos mx}{(a^2 + x^2)^2} dx = \frac{\pi}{2a^3}(1 + ma)e^{-ma}$$

$$\int_{0}^{\infty} \frac{\cos mx}{(a^2 + x^2)^2} dx = \frac{\pi}{4a^3}(1 + ma)e^{-ma}.$$ **Proved.**

Example 12: By method of contour integration find the value of the integral

$$\int_{-\infty}^{\infty} \frac{\sin mx}{(x^2 + a^2)} dx$$

Sol. Consider the integral $\int_C \frac{\sin mz}{(z^2 + a^2)} dz = i.p.$ of $\int_C \frac{e^{imz}}{(z^2 + a^2)} dz$, where $f(z) = \frac{e^{imz}}{(z^2 + a^2)}$ and C is a closed contour consisting of the upper half C_R of a large circle $|z| = R$ and real axis from $-R$ to R.

For the poles $(z^2 + a^2) = 0$

$$z = \pm ai$$

Only the pole $z = ai$ lies within the contour.

So, $$\text{Res}(z = ai) = \lim_{z\to ai}(z-ai)\frac{e^{imz}}{(z-ai)(z+ai)}$$

$$= \lim_{z\to ai}\frac{e^{imz}}{(z+ai)} = \frac{e^{-ma}}{2ai}$$

Hence by Cauchy's residues theorem, we have

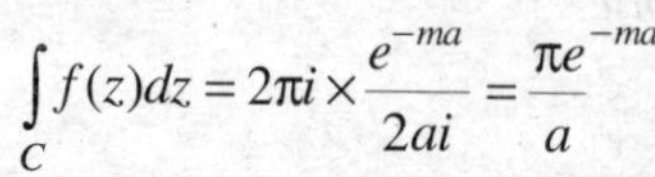

$$\int_C f(z)dz = 2\pi i \times \frac{e^{-ma}}{2ai} = \frac{\pi e^{-ma}}{a}$$

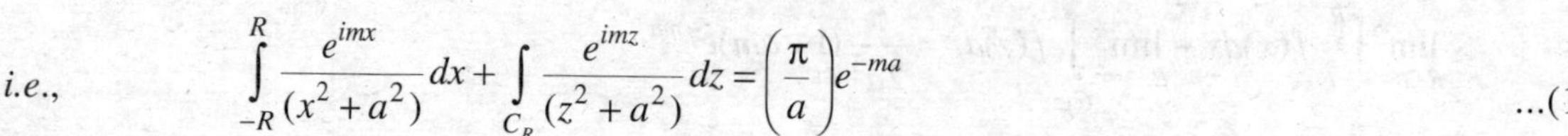

i.e., $$\int_{-R}^{R}\frac{e^{imx}}{(x^2+a^2)}dx + \int_{C_R}\frac{e^{imz}}{(z^2+a^2)}dz = \left(\frac{\pi}{a}\right)e^{-ma} \qquad ...(1)$$

making as $R\to\infty$ $$\lim_{R\to\infty}\int_{C_R}\frac{e^{imz}}{(z^2+a^2)}dz = 0$$

From (1) $$\int_{-\infty}^{\infty}\frac{e^{imx}}{(x^2+a^2)}dx = \left(\frac{\pi}{a}\right)e^{-ma}$$

or $$\text{I.P of} \int_{-\infty}^{\infty}\frac{e^{imx}}{(x^2+a^2)}dx = 0 \Rightarrow \int_{-\infty}^{\infty}\frac{\sin mx}{(x^2+a^2)}dx = 0. \quad \textbf{Ans.}$$

PROBLEM SET 3.5

Evaluate the following by contour integration:

1. $\displaystyle\int_{-\infty}^{\infty}\frac{dx}{x^4+1}$ $\left[\textbf{Ans. } \dfrac{\pi}{\sqrt{2}}\right]$

2. $\displaystyle\int_{-\infty}^{\infty}\frac{x\sin\pi x}{x^2+2x+5}dx$ $\left[\textbf{Ans. } -\dfrac{2+i}{4}\sin 2i\pi\right]$

3. $\displaystyle\int_{0}^{\infty}\frac{x^2}{x^6+1}dx$ $\left[\textbf{Ans. } \dfrac{\pi}{6}\right]$

4. $\displaystyle\int_{-\infty}^{\infty}\frac{\cos x}{(x^2+a^2)(x^2+b^2)}dx, a>b>0$ $\left[\textbf{Ans. } \dfrac{\pi}{(a^2-b^2)}\left(\dfrac{e^{-b}}{b}-\dfrac{e^{-a}}{a}\right)\right]$

5. $\displaystyle\int_{0}^{\infty}\frac{x^3\sin x}{(x^2+a^2)}dx$ $\left[\textbf{Ans. } -\dfrac{\pi}{4}(a-2)e^{-a}\right]$

6. $\displaystyle\int_{0}^{\infty}\frac{\cos x}{(a^2+x^2)^2}dx$ $\left[\textbf{Ans. } \dfrac{\pi}{4a^3}(1+a)e^{-a}\right]$

7. $\displaystyle\int_{-\infty}^{\infty}\frac{x^2dx}{(x^2+a^2)^3}$ $\left[\textbf{Ans. } \dfrac{\pi}{8a^3}\right]$

8. $\displaystyle\int_{-\infty}^{\infty}\frac{\sin x}{x^2+4x+5}dx$ $\left[\textbf{Ans. } -\dfrac{\pi}{e}\sin 2\right]$

9. $\int_{-\infty}^{\infty} \frac{x^2 - x + 2}{x^4 + 10x^2 + 9} dx$ $\left[\textbf{Ans. } \frac{5\pi}{12}\right]$ **10.** $\int_{-\infty}^{\infty} \frac{dx}{(x^2+1)^2}$ $\left[\textbf{Ans. } \frac{\pi}{2}\right]$

11. $\int_{-\infty}^{\infty} \frac{\cos 3x}{(x^2+1)(x^2+4)} dx$ $\left[\textbf{Ans. } \frac{\pi}{3}\left(\frac{e^{-3}}{1} - \frac{e^{-6}}{2}\right)\right]$ **12.** $\int_{0}^{\infty} \frac{\cos mx}{(x^2+a^2)} dx \;\; m > 0$ $\left[\textbf{Ans. } \frac{\pi}{2a} e^{-ma}\right]$

13. Prove that $\int_0^{\infty} \frac{dx}{(x^2+1)^3} = \frac{\pi}{16}$ **14.** Prove that $\int_0^{\infty} \frac{(2x^2-1)dx}{x^4+5x^2+4} = \frac{\pi}{4}$

15. Evaluate $\int_0^{\infty} \frac{dx}{(x^2+9)(x^2+4)^2}$ $\left[\textbf{Ans. } \frac{\pi}{200}, R_1 = -\frac{13i}{200}, R_2 = -\frac{3}{50i}\right]$

16. Evaluate $\int_{-\infty}^{\infty} \frac{x \sin \pi x}{x^2 + 2x + 5} dx$ $\left[\textbf{Ans. } -\pi e^{-2\pi}\right]$

17. Evaluate $\int_0^{\infty} \frac{x \sin x}{x^2 + a^2} dx, a > 0$ $\left[\textbf{Ans. } \frac{\pi}{2} e^{-a}\right]$

18. Evaluate $\int_{-\infty}^{\infty} \frac{x \cos x}{x^2 - 2x + 10} dx$ $\left[\textbf{Ans. } \pi \frac{(\cos 1 - 3 \sin 1)}{(3e^3)}\right]$

3.12 INDENTED SEMI-CIRCULAR CONTOUR OR POLES LIE ON THE REAL AXIS

When the poles of $f(z)$ lie on the real axis and also within the semi-circular region, then those which lie on the real axis can be avoided by drawing small semi-circles C_{r_1}, C_{r_2} etc. about those poles as contur and small radii r_1 and r_2 in the upper half of the plane. This method is said to be indenting the semi-circular contour when the semi-circle of radius R has been indented, then $f(z)$ is regular along this modified contour C and the integral $\int_C f(z)dz$ can be evaluated by Cauchy's residue theorem.

There are some useful result which help in the evaluation of integrals.

Result 1: If $f(z)$ has a simple pole at $z = a$ on the real axis, then

$$\lim_{r \to 0} \int_{C_r} f(z)dz = \pi u(\text{Res } z = a)$$

Result 2: If $f(z)$ has general simple poles on the real axis, then

$$\int_{-\infty}^{\infty} f(x)dx = 2\pi i \text{ (Sum of residues at poles in uper half)} + \pi i \text{ (Sum of residues at poles on real axis)}$$

Example: Show that if $a \geq b \geq 0$, then $\int_0^\infty \frac{\cos 2ax - \cos 2bx}{x^2} dx = \pi(b-a)$

Sol. Consider the integral $\int_C f(z)dz$, where $f(z) = \frac{e^{2iaz} - e^{2ibz}}{z^2}$ taken round the contour C consisting of the upper half of the large circle $|z| = R$ and real axis from $-R$ to R indented at $z = 0$ by a small circle C_r of radius r.

$z = 0$ (order 2) lies on the real axis

$$\therefore \qquad \text{Res } (z=0) = \left[\frac{d}{dz}\left\{z^2 \cdot \frac{e^{2iaz} - e^{2ibz}}{z^2}\right\}\right]_{z=0}$$

$$= \left[2ia\, e^{2iaz} - 2ib\, e^{2ibz}\right]_{z=0}$$

$$= 2i(a-b)$$

Thus $\int_{-\infty}^{\infty} f(x)dx = 2\pi i(0) + \pi i \cdot 2i(a-b)$

$$\Rightarrow \quad \int_{-\infty}^{\infty} \frac{e^{2iax} - e^{2ibx}}{x^2} dx = -2\pi(a-b)$$

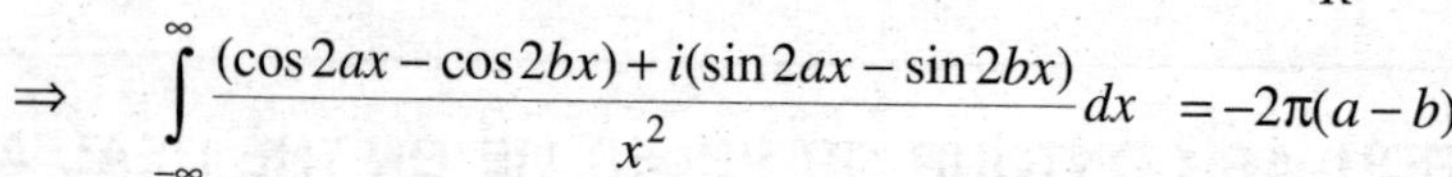
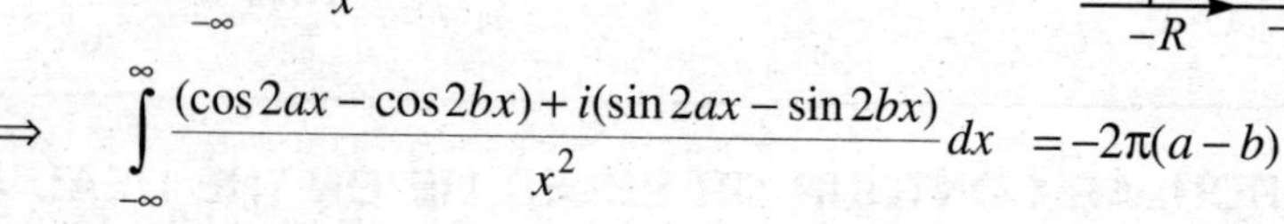

$$\Rightarrow \quad \int_{-\infty}^{\infty} \frac{(\cos 2ax - \cos 2bx) + i(\sin 2ax - \sin 2bx)}{x^2} dx = -2\pi(a-b)$$

Equating real part, we get

$$\int_{-\infty}^{\infty} \frac{\cos 2ax - \cos 2bx}{x^2} dx = -2\pi(a-b)$$

or $$\int_0^{\infty} \frac{\cos 2ax - \cos 2bx}{x^2} dx = \pi(b-a). \qquad \textbf{Proved.}$$

3.13 BRANCH POINT

A point is called a branch point of function $f(z)$ if some of the branches interchange as the independent variable z describe a close path about it.

Example 1: Consider a multiple valued function

$$w = z^{1/2}, z = re^{i\theta}$$

Sol. $$w = r^{1/2} e^{i\theta/2}$$

Make a complete circuit in anticlockwise direction around the origin starting from the point P. At P, Let $\theta = \theta_1$.

After a complete circuit, when we have reached at P_1.

$$\Rightarrow \qquad \theta = \theta_1 + 2\pi$$

$$w = r^{1/2}\exp.\left[\frac{(\theta_1 + 2\pi)i}{2}\right]$$

$$= r^{1/2}\exp.\left[\frac{\theta_1 i}{2} + \pi i\right]$$

$$= -r^{1/2}\exp.\left(\frac{i\theta_1}{2}\right)$$

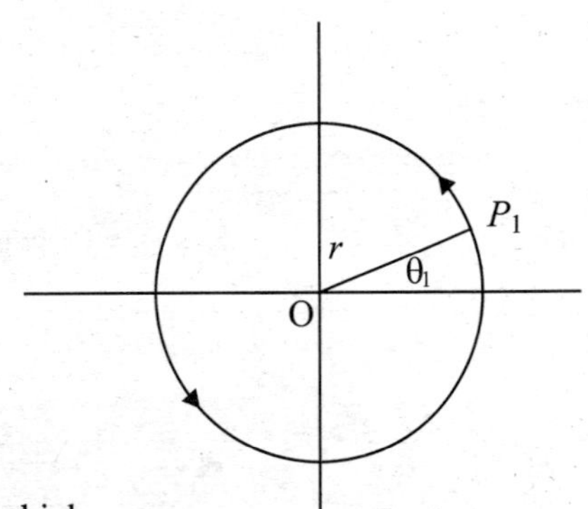

This shows that the same value of w is not obtained with which we have started. After second complete circuit when we have reached at P_1.

$$\theta = \theta_1 + 4\pi$$

$$w = r^{1/2}e^{i(\theta_1 + 4\pi/2)} = r^{1/2}e^{i\theta_1/2}e^{2\pi i} = r^{1/2}e^{i\theta_1/2}$$

This shows that the same value is obtained with which we have started.

Thus, we are on one branch of the function $w = z^{1/2}$, if $0 \le \theta \le 2\pi$ and on the second branch of the function if

$$2\pi \le \theta \le 4\pi$$

Each branch of the function is single valued. In order to keep the function single valued an artificial barrier OQ (where Q is supposed to be at infinity) is made. This barrier is known as branch line or branch cut or cross cut. The point O is called the branch point.

Example 2: By integration round a suitable contour, prove that $\int_0^\infty \frac{(\log x)^2}{(1+x^2)}dx = \frac{\pi^3}{8}$.

Sol. Consider the integral $\int_C f(z)dz$, where $f(z) = \frac{(\log z)^2}{(1+z^2)}$, taken round the contour C consisting of the upper half of a large circle $|z| = R$ of real axis from $-R$ to R indented at $z = 0$ by a small circle C_r of radius r.

$z = 0$ is the branch point of $f(z)$, from new lines for the poles.

$1 + z^2 = 0 \quad i.e.\ z = \pm i$

Thus the only singularity within the contour is a simple pole at $z = i$ the residue at which is

$$\lim_{z \to i}\left\{(z-i)\frac{(\log z)^2}{(1+z^2)}\right\}$$

$$= \lim_{z \to i}\left\{\frac{(\log z)^2}{z+i}\right\}$$

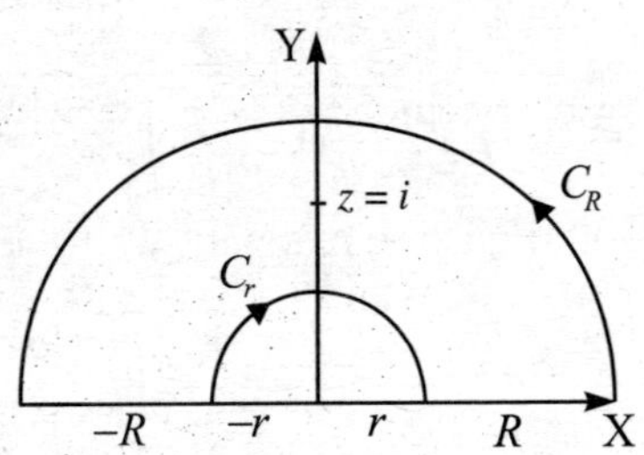

$$= \frac{(\log i)^2}{2i} = \frac{\left(\log e^{i\pi/2}\right)^2}{2i} = \frac{\left(i\frac{\pi}{2}\right)^2}{2i} = -\frac{\pi^2}{8i}$$

and $$\text{Res } (z=0) = \lim_{z\to 0} \frac{z(\log z)^2}{1+z^2} = \lim_{z\to 0} \frac{z\cdot\{\log(1+z-1)\}^2}{1+z^2}$$

$$= \lim_{z\to 0} \frac{z\left[(z-1) - \frac{(z-1)^2}{2} + \frac{(z-1)^3}{3} \ldots\ldots\right]}{1+z^2}$$

$$\left| \text{As } \log(1+x) = x - \frac{x^2}{2} + \frac{x^3}{3} \ldots\ldots \right.$$

or $$\text{Res } (z=0) = 0$$

Hence $$\int_{-\infty}^{\infty} f(x)dx = 2\pi i\left(-\frac{\pi^2}{8i}\right) + \pi i(0)$$

or $$\int_{-\infty}^{\infty} \frac{(\log x)^2}{(1+x^2)} dx = -\frac{\pi^3}{4}$$

$$\int_{-\infty}^{0} \frac{(\log x)^2}{(1+x^2)} dx + \int_{0}^{\infty} \frac{(\log x)^2}{(1+x^2)} dx = -\frac{\pi^3}{4}$$

Replace x by $-x$ in the first integral, we get

$$\int_{0}^{\infty} \frac{\{\log(-x)\}^2}{(1+x^2)} dx + \int_{0}^{\infty} \frac{(\log x)^2}{(1+x^2)} dx = -\frac{\pi^3}{4}$$

$$\int_{0}^{\infty} \frac{\{\log(xe^{i\pi})\}^2}{(1+x^2)} dx + \int_{0}^{\infty} \frac{(\log x)^2}{(1+x^2)} dx = -\frac{\pi^3}{4}$$

or $$\int_0^{\infty} \frac{(\log x + i\pi)^2}{1+x^2} dx + \int_0^{\infty} \frac{(\log x)^2}{1+x^2} dx = -\frac{\pi^3}{4}$$

or $$\int_0^{\infty} \frac{2(\log x)^2 - \pi^2 + 2\pi i \log x}{1+x^2} dx = -\frac{\pi^3}{4}$$

Equating real part, we have

$$2\int_0^{\infty} \frac{(\log x)^2}{1+x^2} dx - \pi^2 \int_0^{\infty} \frac{dx}{1+x^2} dx = -\frac{\pi^3}{4}$$

i.e., $$2\int_0^{\infty} \frac{(\log x)^2}{1+x^2} dx - \pi^2 \frac{\pi}{2} = -\frac{\pi^3}{4}$$

or $$2\int_0^\infty \frac{(\log x)^2}{1+x^2}\,dx = \frac{\pi^3}{2} - \frac{\pi^3}{4} = \frac{\pi^3}{4}$$

$\therefore$ $$\int_0^\infty \frac{(\log x)^2}{1+x^2}\,dx = \frac{\pi^3}{8}.$$ **Proved.**

Example 3: Prove by the contour integration

$$\int_0^\infty \frac{x^{a-1}}{1-x}\,dx = \pi \cot a\pi \qquad 0 < a < 1$$

and $$\int_0^\infty \frac{x^{a-1}}{1+x} = \frac{\pi}{\sin a\pi}$$

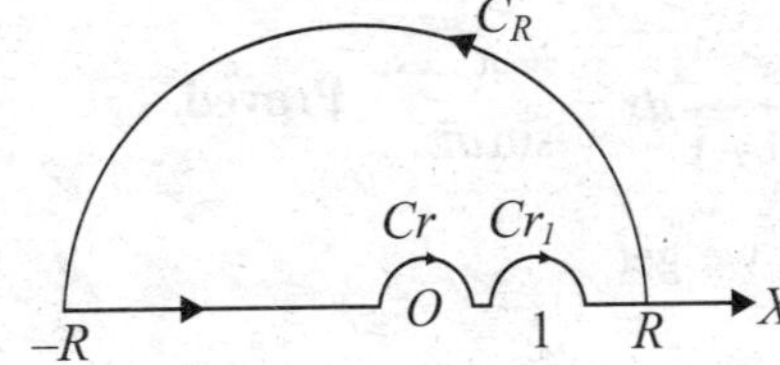

Sol. Consider the integral $\int_C f(z)dz$,

where $$f(z) = \frac{z^{a-1}}{1-z}$$

Taken round the closed contour C consisting of real axis from $-R$ to R, and upper half of a large circle $|z| = R$ indented at $z = 0$, $z = 1$, the radii of indentations being r and r_1 respectively.

$f(z)$ has two poles $z = 0$, $z = 1$ lie on real axis

$\therefore$ $$\text{Res } (z = 0) = \lim_{z\to 0} z \cdot \frac{z^a}{z(1-z)} = 0$$

and $$\text{Res } (z = 1) = \lim_{z\to 1} (z-1) \cdot \frac{z^a}{-z(z-1)} = -1$$

$\therefore$ $$\int_{-\infty}^{\infty} \frac{x^{a-1}}{1-x}\,dx = 2\pi i(0) + \pi i(-1)$$

$\Rightarrow$ $$\int_{-\infty}^{\infty} \frac{x^{a-1}}{1-x}\,dx = -\pi i$$

or $$\int_{-\infty}^0 \frac{x^{a-1}}{1-x}\,dx + \int_0^\infty \frac{x^{a-1}}{1-x}\,dx = -\pi i$$

Putting $-x$ for x in the first integral, we have

$$\int_0^\infty \frac{(-1)^{a-1}\, x^{a-1}}{1+x}\,dx + \int_0^\infty \frac{x^{a-1}}{1-x}\,dx = -\pi i \qquad \left|\ \text{As } \int_\infty^0 f(x)dx = -\int_0^\infty f(x)dx\right.$$

or $$-\int_0^\infty \frac{(e^{i\pi})^a\, x^{a-1}}{1+x}\,dx + \int_0^\infty \frac{x^{a-1}}{1-x}\,dx = -\pi i \qquad \left|\ \text{As } e^{i\pi} = -1\right.$$

$$-\int_0^\infty (\cos\pi + i\sin\pi)^a \cdot \frac{x^{a-1}}{1+x}\,dx + \int_0^\infty \frac{x^{a-1}}{1-x}\,dx = -\pi i$$

$$\Rightarrow \qquad -\int_0^\infty \left(\cos a\pi \cdot \frac{x^{a-1}}{1+x} - \frac{x^{a-1}}{1-x}\right)dx - \int_0^\infty \sin a\pi \cdot \frac{x^{a-1}}{1+x}\,dx = -\pi i$$

Equating imaginary and real parts, we have

$$\cos a\pi \int_0^\infty \frac{x^{a-1}}{1+x}\,dx - \int_0^\infty \frac{x^{a-1}}{1-x}\,dx = 0 \qquad ...(i)$$

and $$\int_0^\infty \frac{x^{a-1}}{1+x}\,dx = \frac{\pi}{\sin a\pi}. \quad \textbf{Proved.} \qquad ...(ii)$$

Using (*ii*) in (*i*), we get

$$\cos a\pi \cdot \frac{\pi}{\sin a\pi} - \int_0^\infty \frac{x^{a-1}}{1-x}\,dx = 0$$

$$\Rightarrow \qquad \int_0^\infty \frac{x^{a-1}}{1-x}\,dx = \pi\cot a\pi. \quad \textbf{Proved.}$$

Example 4: Apply the calculus of residues to prove that $\int_0^\infty \frac{\sin\pi x}{x(1-x^2)}\,dx = \pi.$

Sol. Consider the integral

$$\int_C \frac{\sin\pi z}{z(1-z^2)}\,dz = I.P. \text{ of } \int_C \frac{e^{i\pi z}}{z(1-z^2)}\,dz$$

where $f(z) = \frac{e^{i\pi z}}{z(1-z^2)}$

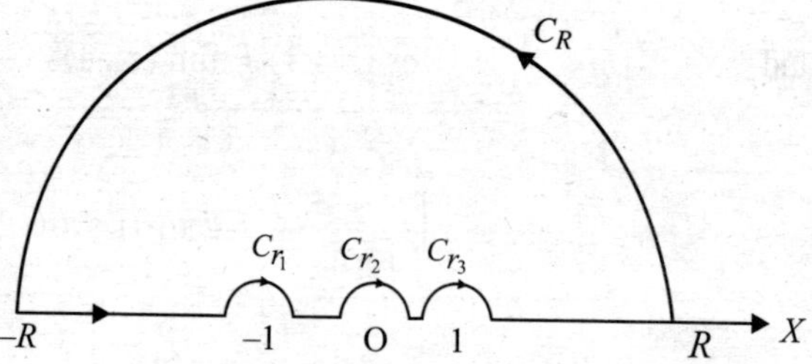

Taken round a contour C_R consisting of the upper half of the large circle C_R, $|z| = R$ and the real axis from $-R$ to R indented at $z = -1$, and $z = 1$. The small circles are C_{r_1}, C_{r_2} and C_{r_3}, their centers are at the points $z = -1$, $z = 0$ and $z = 1$ and small radii r_1, r_2, r_3 respectively.

For the poles

$$z(1-z^2) = 0$$

$$\Rightarrow \qquad z = 0, z = \pm 1$$

All the three poles lie on the real axis. And no pole lies in the upper half.

$$\text{Res}(z=0) = \lim_{z\to 0} z \cdot \frac{e^{i\pi z}}{z(1-z^2)} = 1$$

$$\text{Res}(z=1) = \lim_{z\to 1} (z-1) \cdot \frac{e^{i\pi z}}{z(1-z)(1+z)} = -\frac{e^{i\pi}}{2} = -\frac{1}{2}(\cos\pi + i\sin\pi) = \frac{1}{2}$$

and $$\text{Res}\ (z=-1)=\lim_{z\to 0}(z+1)\cdot\frac{e^{i\pi z}}{z(1-z)(1+z)}=-\frac{1}{2}e^{-i\pi}=\frac{1}{2}$$

Thus $$\int_{-\infty}^{\infty} f(x)dx = 2\pi i(0)+\pi i\left(1+\frac{1}{2}+\frac{1}{2}\right)$$

$$= 2\pi i$$

or $$\int_{-\infty}^{\infty}\frac{e^{i\pi x}}{x\left(1-x^2\right)}dx = 2\pi i$$

Equating imaginary parts, we have

$$\int_{-\infty}^{\infty}\frac{\sin \pi x}{x\left(1-x^2\right)}dx = 2\pi \quad \text{or} \quad \int_{0}^{\infty}\frac{\sin \pi x}{x\left(1-x^2\right)}dx = \pi. \qquad \textbf{Proved.}$$

Example 5(a): Use the method of contour integration to prove that

$$\int_0^{\infty}\frac{x^b}{1+x^2}dx=\frac{\pi}{2}\sec\frac{\pi b}{2},(-1<b<1)$$

Sol. Consider the integral $\int_C f(z)dz$ where $f(z)=\dfrac{z^b}{z^2+1}$.

Poles of $f(z)$ are given by $1 + z^2 = 0$ at $z=\pm i$.

The only pole $z = i$ lies in the upper half.

∴ Residue at $z = i$ is $\lim_{z\to i}(z-i)f(z)$

$$=\lim_{z\to i}(z-i)\left(\frac{z^b}{(z-i)(z+i)}\right)=\frac{(i)^b}{2i}$$

$$=\frac{1}{2i}\left(e^{i\pi/2}\right)^b=\frac{1}{2i}e^{i\pi b/2}$$

Hence by Cauchy's residue theorem, we have

$$\int_{-R}^{R} f(x)dx+\int_{C_R} f(z)dz = 2\pi i\times\frac{1}{2i}e^{i\pi b/2}$$

Taking $\lim R\to\infty$, we get

$$\lim_{R\to\infty}\int_{-R}^{R} f(x)dx+\lim_{R\to\infty}\int_{-R}^{R} f(z)dz = \pi e^{i\pi b/2}$$

or $$\int_{-\infty}^{\infty} f(x)dx = \pi e^{i\pi b/2}$$

or $$\int_{-\infty}^{0} f(x)dx+\int_{0}^{\infty} f(x)dx = \pi e^{i\pi b/2}$$

or $$\int_{-\infty}^{0} \frac{x^b}{1+x^2}dx + \int_0^{\infty} \frac{x^b}{1+x^2}dx = \pi e^{i\pi b/2}$$

Putting $-x$ for x in the first integral, we have

or $$\int_0^{\infty} \frac{(-x)^b}{1+x^2}dx + \int_0^{\infty} \frac{x^b}{1+x^2}dx = \pi e^{i\pi b/2}$$

or $$\int_0^{\infty} \frac{\left(xe^{\pi i}\right)^b}{1+x^2}dx + \int_0^{\infty} \frac{x^b}{1+x^2}dx = \pi e^{i\pi b/2} \quad \text{since } e^{\pi i} = -1$$

or $$\int_0^{\infty} \frac{x^b}{1+x^2}\left(1+e^{i\pi b}\right)dx = \pi e^{i\pi b/2}$$

Equating real parts, we have

$$\int_0^{\infty} \frac{x^b}{1+x^2} \times (1+\cos b\pi)\,dx = \pi \cos\frac{b\pi}{2}$$

or $$\int_0^{\infty} \frac{x^b}{1+x^2}dx = \frac{\pi\cos\frac{1}{2}b\pi}{1+\cos b\pi} = \frac{\pi}{2}\sec\frac{b\pi}{2}.$$ **Proved.**

Example 5(b): Use the method of contour integration to prove that

$$\int_0^{\infty} \frac{x^{a-1}}{1+x^2}dx = \frac{1}{2}\pi\text{cosec}\left(\frac{\pi a}{2}\right), 0 < a < 2.$$

Sol. Let $f(z) = \dfrac{z^{a-1}}{1+z^2}$ where, $0 < a < 2$.

Then $z = 0$ is a singularity of $f(z)$, for $0 < a < 1$ and poles are given by $1 + z^2 = 0$, *i.e.* $z = \pm i$ are two simple poles.

Choose the closed contour C, consisting of the real axis from $-R$ to R and the upper half of a large circle $|z| = R$ indented at $z = 0$ let r be the radius of indentation.

The pole $z = i$ only lies within C.

Residue at the simple pole $z = i$ is

$$\lim_{z\to i}(z-i)f(z)$$

$$= \lim_{z\to i}(z-i)\frac{z^{a-1}}{z^2+1} = \lim_{z\to i}(z-i)\frac{z^{a-1}}{(z-i)(z+i)}$$

$$\text{Res }(z = i) = \frac{(i)^{a-1}}{2i} = -\frac{1}{2}(i)^a = -\frac{1}{2}\left(e^{i\pi/2}\right)^a = -\frac{1}{2}e^{ia\pi/2}$$

And $$\text{Res }(z = 0) = \lim_{z\to 0} zf(z) = \lim_{z\to 0} z\frac{z^{a-1}}{1+z^2} = \lim_{z\to 0}\frac{z^a}{1+z^2} = 0$$

$\therefore$ $$\int_{-\infty}^{\infty} f(x)dx = 2\pi i\left(-\frac{1}{2}e^{ia\pi/2}\right) + \pi i(0)$$

$\Rightarrow$ $$\int_{-\infty}^{\infty} \frac{x^{a-1}}{1+x^2}dx = -\pi i e^{ia\pi/2}$$

or $$\int_{-\infty}^{0} f(x)dx + \int_{0}^{\infty} f(x)dx = -\pi i e^{i\pi a/2}$$

or $$\int_{-\infty}^{0} \frac{x^{a-1}}{1+x^2}dx + \int_{0}^{\infty} \frac{x^{a-1}}{1+x^2}dx = -\pi i e^{i\pi a/2} \quad \text{...(1)}$$

But $$\int_{-\infty}^{0} \frac{x^{a-1}}{1+x^2}dx = \int_{0}^{\infty} \frac{(-x)^{a-1}}{1+x^2}dx \text{ writing } -x \text{ for } x$$

$$= -\int_{0}^{\infty} \frac{(-1)^a x^{a-1}}{1+x^2}dx = -\int_{0}^{\infty} \frac{\left(e^{i\pi}\right)^a x^{a-1}}{1+x^2}dx \text{ since } e^{i\pi} = -1$$

$\Rightarrow$ $$\int_{-\infty}^{0} \frac{x^{a-1}}{1+x^2}dx = -e^{i\pi a}\int_{0}^{\infty} \frac{x^{a-1}}{1+x^2}dx$$

From relation (1), we get

$$-e^{i\pi a}\int_{0}^{\infty} \frac{x^{a-1}}{1+x^2}dx + \int_{0}^{\infty} \frac{x^{a-1}}{1+x^2}dx = -\pi i e^{i\pi a/2}$$

or $$\left(1-e^{i\pi a}\right)\int_{0}^{\infty} \frac{x^{a-1}}{1+x^2}dx = -\pi i e^{i\pi a/2}$$

or $$\frac{\left(e^{i\pi a/2} - e^{-i\pi a/2}\right)}{2i}\int_{0}^{\infty} \frac{x^{a-1}}{1+x^2}dx = \frac{1}{2}\pi$$

or $$\sin\left(\frac{\pi a}{2}\right)\int_{0}^{\infty} \frac{x^{a-1}}{1+x^2}dx = \frac{1}{2}\pi$$

or $$\int_{0}^{\infty} \frac{x^{a-1}}{1+x^2}dx = \frac{1}{2}\pi \operatorname{cosec}\left(\frac{\pi a}{2}\right). \quad \textbf{Proved.}$$

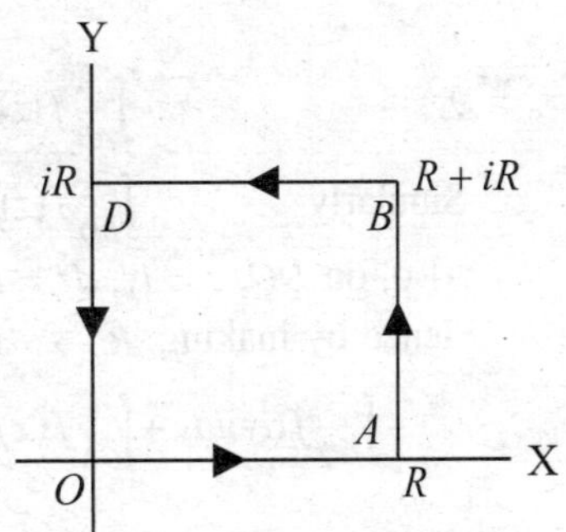

Example 6: Integrating $\dfrac{e^{iz}}{z+a}$ along the boundary of the square defined by

$$x = 0,\ x = R,\ y = 0,\ y = R,$$

Prove that $$\int_{0}^{\infty} \frac{\cos x}{x+a}dx = \int_{0}^{\pi} \frac{xe^{-ax}}{1+x^2}dx \quad a > 0$$

$$\int_{0}^{\infty} \frac{\sin x}{x+a}dx = \int_{0}^{\pi} \frac{e^{-ax}}{1+x^2}dx.$$

Sol. Consider the integral $\int_C f(z)dz$

where $f(z)=\dfrac{e^{iz}}{z+a}$ taken round the enclosed contour C, consisting of the perimeter of the given square.

Poles of $f(z)$ are given by $z + a = 0$ *i.e.* $z = -a$ is the simple pole which too is outside the given contour.

Hence by Cauchy's theorem, we have

$$\int_C f(z)dz=0$$

i.e.,

$$\left\{\int_{AB}+\int_{BD}+\int_{DO}+\int_{OA}\right\}f(z)dz=0 \qquad ...(1)$$

Now, on AB, $z = R + iy$, $dz = idy$ and y goes from 0 to R.

So

$$\left|\int_{AB} f(z)dz\right| \le \int_{AB}|f(z)||dz| = \int_{AB}\frac{|e^{iz}|}{|z+a|}|dz|$$

$$\le \int_{AB}\frac{|e^{iz}|}{|z|-a}|dz|$$

$$= \int_0^R \frac{|e^{i(R+iy)}|}{|R+iy|-a}|idy| \le \int_0^R \frac{e^{-y}}{|R|-|iy|-a}\cdot dy$$

$$= \int_0^R \frac{e^{-y}}{R-y-a}dy$$

$$= -\int_{R-a}^{-a}\frac{e^{t+a-R}}{t}dt \qquad \text{[Put } R-y-a=t]$$

$$= \frac{e^a}{e^R}\int_{-a}^{R-a}\frac{e^t}{t}dt \qquad \text{which} \to 0 \text{ as } R\to\infty$$

$$\left(\text{Since } \int_{-a}^{R-a}\frac{e^t}{t}dt \text{ is finite}\right)$$

$\therefore$

$$\int_{AB} f(z)dz=0 \quad \text{when } R\to\infty$$

Similarly

$$\int_{BD} f(z)dz=0 \quad \text{when } R\to\infty$$

Also, on DO, $z = iy$, $dz = idy$; and on OA, $z = x$, $dz = dx$.

Hence by making $R\to\infty$ relation (1) becomes

$$\int_{DO} f(iy)idy+\int_{OA} f(x)dx=0$$

i.e., $$\int_{\infty}^{0} \frac{e^{i(iy)}}{iy+a} i dy + \int_0^{\infty} \frac{e^{ix}}{x+a} dx = 0$$

or $$\int_0^{\infty} \frac{e^{ix}}{x+a} dx = i\int_0^{\infty} \frac{e^{-y}}{a+iy} dy$$

$$= i\int_0^{\infty} \frac{e^{-ax} a dx}{a(1+ix)} \qquad \text{[Putting } y = ax\text{]}$$

$$= i\int_0^{\infty} \frac{(1-ix)e^{-ax}}{1+x^2} dx$$

$$\Rightarrow \quad \int_0^{\infty} \frac{e^{-ix}}{x+a} dx = \int_0^{\infty} \frac{(i+x)e^{-ax}}{(1+x^2)} dx$$

Equating real and imaginary parts, we have

$$\int_0^{\infty} \frac{\cos x}{x+a} dx = \int_0^{\infty} \frac{xe^{-ax}}{1+x^2} dx$$

and $$\int_0^{\infty} \frac{\sin x}{x+a} dx = \int_0^{\infty} \frac{e^{-ax}}{1+x^2} dx. \quad \textbf{Proved.}$$

Example 7: Apply the calculus of residues to prove that

$$\int_0^{\infty} \frac{\cosh ax}{\cosh \pi x} dx = \frac{1}{2} \sec(a/2) \text{ where } -\pi < a < \pi$$

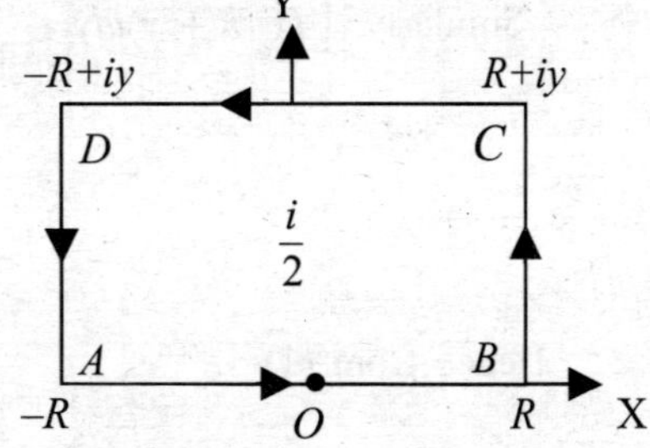

taken round the rectangular contour consisting of the real axis, the lines $x = \pm R$ and the line $y = 1$.

Sol. The poles of $f(z)$ are given by

$e^{2\pi z} + 1 = 0$, *i.e.* $e^{2\pi z} = -1 = e^{\pi(2n\pi+\pi)}$

i.e. $2z = i\ (2n + 1)$ where n is any integer.

The only pole which lies within the contour is a simple pole at $z = \frac{i}{2}$

The residue of $f(z)$ at $\left(z = \frac{i}{2}\right)$

$$\left[\frac{e^{az}}{(d/dz)\cosh \pi z}\right]_{z=i/2} = \frac{e^{ai/2}}{\pi i}$$

Hence by Cauchy's residue theorem, we have

$$\int_C f(z)dz = 0, or \left\{\int_{AB} + \int_{BC} + \int_{CD} + \int_{DA}\right\} f(z)dz = 2\pi i \times \frac{e^{ia/2}}{\pi i}$$

On AB, $z = x$, $dz = dx$.

On BC, $z = R + iy$, $dz = idy$

On CD $z = x + i$, $dz = dx$.

On DA, $z = -R + iy$, $dz = idy$

$$\therefore \int_{-R}^{R} f(x)dx + \int_0^1 f(R+iy)idy + \int_R^{-R} f(x+i)dx + \int_1^0 f(-R+iy)idy = 2\pi i \times \frac{e^{ai/2}}{\pi i} \quad ...(1)$$

Now $$\left|\int_0^1 f(R+iy)idy\right| \le \int_0^1 |f(R+iy)||i||dy| = \int_0^1 \left|\frac{2e^{(a+\pi)(R+iy)}}{e^{2\pi(R+iy)+1}}\right||dy|$$

$$= 2\int_0^1 \frac{\left|e^{(a+\pi)(R-iy)}\right|}{\left|e^{2\pi(R+iy)}+1\right|}dy \le 2\int_0^1 \frac{e^{(a+\pi)R}}{\left|e^{2\pi(R+iy)}\right|-1}dy$$

$$= 2\int_0^1 \frac{e^{(a+\pi)R}}{e^{2\pi R}-1}dy = \frac{2e^{(a+\pi)R}}{e^{2\pi R}-1}$$

$$= \frac{2}{e^{(\pi-a)R} - e^{-(\pi+a)R}} \to 0 \text{ as R} \to \infty \text{ since } -\pi < a < \pi$$

Similarly $$\left|\int_1^0 f(-R+iy)idy\right| \le \frac{2e^{-(a+\pi)R}}{1-e^{-2\pi R}}$$

$$= \frac{2}{e^{(a+\pi)R} - e^{(a-\pi)R}} \text{ which} \to 0 \text{ as } R \to \infty$$

Hence from (1)

$$\int_{-\infty}^{\infty} f(x+i)dx + \int_{\infty}^{-\infty} f(x+i)dx = 2\pi i \times \frac{e^{ai/2}}{\pi i}$$

or $$\int_{-\infty}^{\infty} \frac{e^{ax}}{\cosh \pi x}dx + \int_{\infty}^{-\infty} \frac{e^{a(\pi+i)}}{\cosh \pi(x+i)}dx = 2e^{ai/2}$$

i.e., $$\int_{-\infty}^{\infty} \frac{e^{ax}}{\cosh \pi x}dx - \int_{-\infty}^{\infty} \frac{e^{ai}e^{ax}}{-\cosh \pi x}dx = 2e^{ai/2}$$

$$\left[\text{Since } \cosh \pi(x+i) = \cos i\pi(x+i) = \cos(\pi xi - \pi) = -\cos \pi xi = -\cosh \pi x\right]$$

i.e.,
$$\left(1+e^{ai}\right)\int_{-\infty}^{\infty}\frac{e^{ax}}{\cosh \pi x}dx = 2e^{ai/2}$$

or
$$\int_{-\infty}^{\infty}\frac{e^{ax}dx}{\cosh \pi x}=\frac{2e^{ai/2}}{1+e^{ia}}$$

i.e.,
$$\int_{-\infty}^{\infty}\frac{e^{ax}}{\cosh \pi x}dx=\frac{1}{\frac{1}{2}\left(e^{ai/2}+e^{-ai/2}\right)}=\sec\left(\frac{a}{2}\right)$$

Putting $-a$ for a in the result, we have

$$\int_{-\infty}^{\infty}\frac{e^{-ax}}{\cosh \pi x}dx=\sec\left(\frac{a}{2}\right)$$

Adding the last two results, we have

$$\int_{-\infty}^{\infty}\frac{e^{ax}+e^{-ax}}{\cosh \pi x}dx=2\sec\left(\frac{a}{2}\right)$$

or
$$\int_{-\infty}^{\infty}\frac{\cosh ax}{\cosh \pi x}dx=\sec\frac{a}{2} \text{ or } \int_{0}^{\infty}\frac{\cosh ax}{\cosh \pi x}dx=\frac{1}{2}\sec\left(\frac{a}{2}\right).$$ **Proved.**

Example 8: Apply the calculus of residue to prove that

$$\int_{-\infty}^{\infty}\frac{e^{ax}}{e^{x}+1}dx=\pi \operatorname{cosec} a\pi, \qquad (0<a<1)$$

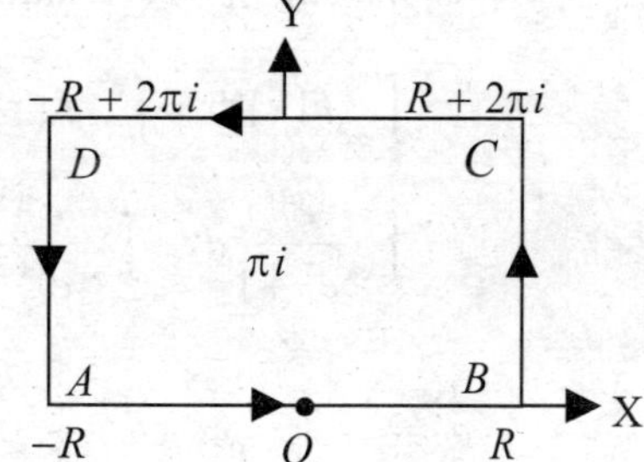

Sol. Consider $\int_C f(z)dz$ where $f(z)=\dfrac{e^{az}}{1+e^{z}}$

which the contour C is the perimeter of a rectangle whose sides are the lines, real axis $x=\pm R$ and $y=2\pi$

The poles of $f(z)$ are given by $1+e^{z}=0$

i.e., $e^{z}=-1=e^{i(2n+1)\pi}$

i.e., $z=i(2n+1)\pi$ where n is zero or any integer positive or negative. Out of these poles only the simple pole $z=\pi i$ lies within the contour.

At $z=\pi i$ the residue

$$=\left[\frac{e^{az}}{\left(\frac{d}{dz}\right)\left(1+e^{z}\right)}\right]_{z=\pi i}=-e^{a\pi i}$$

Hence the Cauchy's residue theorem

$$\int_{-R}^{R} f(x)dx + \int_{0}^{\pi} f(R+iy)idy + \int_{R}^{-R} f(x+2\pi i)dx + \int_{2\pi}^{0} f(-R+iy)idy = 2\pi i\left(-e^{a\pi i}\right) \quad ...(1)$$

$$\text{But } \left|\int_0^{2\pi} f(R+iy)idy\right| \le \int_0^{2\pi} |f(R+iy)||idy| = \int_0^{2\pi} \left|\frac{e^{a(R+iy)}}{1+e^{R+iy}}\right| |idy|$$

$$\le \int_0^{2\pi} \frac{\left|e^{a(R+iy)}\right|}{\left|e^{R+iy}\right|-1} dy \quad \le \int_0^{2\pi} \frac{e^{aR}}{e^R-1} dy$$

$$= \frac{e^{aR}}{e^R-1} 2\pi = \frac{2\pi}{e^{(1-a)R} - e^{-aR}} \to 0 \text{ as } R \to \infty \text{ since } 0 < a < 1.$$

$$\text{Similarly } \left|\int_{2\pi}^{0} f(-R+iy)idy\right| \le \frac{e^{-aR}}{1-e^{-R}} 2\pi$$

$$= \frac{2\pi}{e^{aR} - e^{-(1-a)R}} \text{ which } \to 0 \text{ as } R \to \infty \text{ since } 0 < a < 1.$$

Hence making $R \to \infty$ relation (1) residues

$$\int_{-\infty}^{\infty} f(x)dx = \int_{\infty}^{-\infty} f(x+2\pi i)dx = -2\pi i.e^{a\pi i}$$

i.e.,

$$\int_{-\infty}^{\infty} \frac{e^{ax}}{e^x+1} dx - \int_{-\infty}^{\infty} \frac{e^{a(x+2\pi i)}}{e^{x+2\pi i}+1} dx = -2\pi i.e^{a\pi i}$$

or

$$\int_{-\infty}^{\infty} \frac{e^{ax}}{e^x+1} dx - \int_{-\infty}^{\infty} \frac{e^{2a\pi i} e^{ax}}{e^x+1} dx = -2\pi i e^{a\pi i}$$

or

$$\left(1-e^{2a\pi i}\right) \int_{-\infty}^{\infty} \frac{e^{ax}}{e^x+1} dx = -2\pi i e^{a\pi i}$$

or

$$\int_{-\infty}^{\infty} \frac{e^{ax}dx}{e^x+1} = -\frac{2\pi i e^{a\pi i}}{1-e^{2a\pi i}}$$

$$= \frac{2i\pi}{e^{a\pi i} - e^{-a\pi i}}$$

$$= \frac{\pi}{\sin a\pi}$$

$$= \pi \operatorname{cosec} a\pi.$$ **Proved.**

PROBLEM SET 3.6

Using the method of contour integration evaluate:

1. $\int_0^{\infty} \frac{x^2}{(x^2+1)^2} dx$ $\left[\textbf{Ans. } \frac{\pi}{2}\right]$

2. $\int_0^{\infty} \frac{\log x}{1+x^2} dx$ $\left[\textbf{Ans. } \frac{\pi^3}{8}\right]$

3. $\int_0^{\infty} \frac{\sin mx}{x} dx$ $\left[\textbf{Ans. } \frac{\pi}{2}\right]$

4. $\int_0^{\infty} \frac{x^{a-1}}{1-x} dx; 0 < a < 1$ $[\textbf{Ans. } \pi \cot a\pi]$

5. $\int_0^{\infty} \frac{\cos x}{a^2 - x^2} dx$ $\left[\textbf{Ans. } \frac{\pi}{2a} \sin a\right]$

6. $\int_0^{\infty} \frac{\sin mx}{x(x^2+a^2)} dx$ $\left[\textbf{Ans. } \frac{\pi}{2a^2}\left(1-e^{-ma}\right)\right]$

7. $\int_0^{\infty} \frac{x^4}{x^6-1} dx$ $\left[\textbf{Ans. } \frac{\pi}{6}\sqrt{3}\right]$

8. $\int_0^{\infty} \frac{\log(1+x^2)}{x^{1+\alpha}} dx;\ 0 < \alpha < 1$ $[\textbf{Ans. } 0]$

9. By integrating e^{-z^2} round the rectangle whose vertices are 0, R, $R + ia$, ia, show that

$$(i)\ \int_0^{\infty} e^{-x^2} \sin 2ax\, dx = e^{-a^2} \int_0^{a} e^{y^2} dy \quad \text{and} \quad (ii)\ \int_0^{\infty} e^{-x^2} \cos 2ax\, dx = \frac{e^{-a^2}}{2} \sqrt{\pi}.$$

10. Evaluate $\int_{-\infty}^{\infty} \frac{e^{ax}}{e^x+1} dx$ $\left[\textbf{Ans. } \frac{\pi}{\sin a\pi}\right]$

11. Prove that $\int_{-\infty}^{\infty} \frac{\sin x}{x^2-2x+5} dx = \frac{\pi \sin 1}{2e^2}$.

12. Prove that $\int_{-\infty}^{\infty} \frac{\sin x}{x^2+4x+5} dx = -\frac{\pi \sin 2}{e}$.

3.14 CONFORMAL REPRESENTATION

3.14.1 Transformation or Mapping

The equations

$$\left.\begin{aligned} u &= u(x, y) \\ v &= v(x, y) \end{aligned}\right\} \qquad \text{...(1)}$$

define a transformation or mapping by means of which a correspondence between uv-plane and xy-plane can be established. If corresponding to each point of xy-plane, there is a unique point of uv-plane and conversely, then such transformation is called one-one transformation. The corresponding points in the two planes are called images of each other. Hence, by means of equation (1) a region or curve of xy-plane is said to transform or mapped on or represented by the corresponding curve of the uv-plane.

3.14.2 Conformal Transformation

Suppose the transformation $u = u(x, y)$, $v = v(x, y)$ maps the two curves c_1 c_2 intersecting at the point $P(z_0)$ of z plane on the two curves c_1' and c_2' intersecting at $P'(w_0)$ of w-plane.

If the angle between c_1 and c_2 at z_0 is equal to the angle

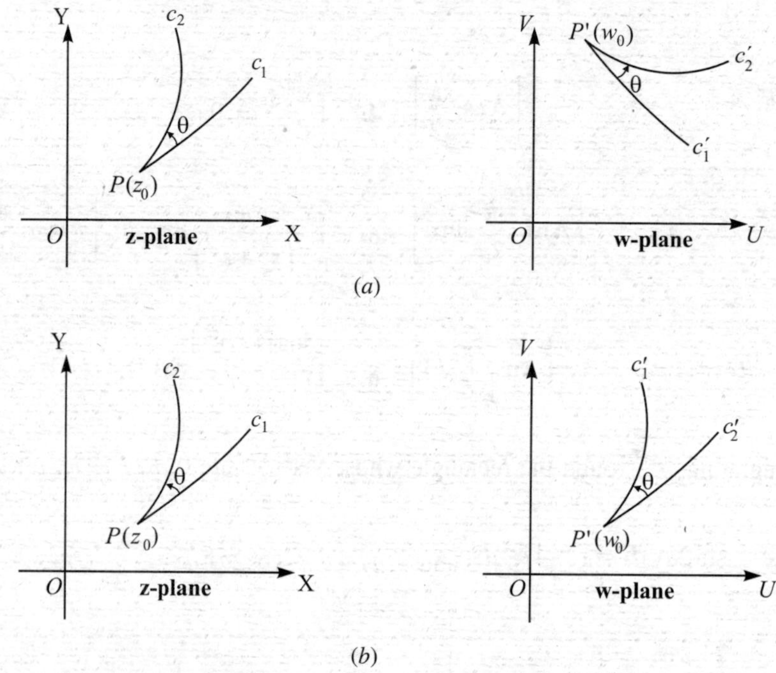

(*a*)

(*b*)

between c_1' and c_2' at w_0, then the transformation is called **isogonal**. If the sense of rotation as well as the magnitude of the angle is preserved, the transformation is called **conformal**.

In Fig. (*a*) the transformation is conformal and in Fig. (*b*) the transformation is isogonal.

$w = \bar{z}$ is isogonal transformation.

3.15 SOME GENERAL TRANSFORMATIONS

1. Translation: The map $w = z + \beta$ corresponds to a translation. For, by this transformation, the figure in w-plane is the same as figure in z-plane with a different origin.

Example 1: Let a rectangular domain R be bounded by $x = 0$, $y = 0$, $x = 2$, $y = 1$. Determine the region R' of w-plane into which R is mapped under the transformation. $w = z + (1 - 2i)$

Sol. $w = z + (1 - 2i)$ gives $u + iv = x + iy + (1 - 2i)$

$\Rightarrow$ $u = x + 1,\ v = y - 2$

By the map $u = x + 1$, the lines $x = 0$, $x = 2$ are mapped respectively on the lines $u = 1$, $u = 3$.

By the map $v = y - 2$, the lines $y = 0$, $y = 1$ are mapped on $v = -2$, $v = -1$ respectively.

The required image is rectangle R' bounded by $u = 1$, $u = 3$, $v = -2$, $v = -1$ in w-plane as shown in the figure.

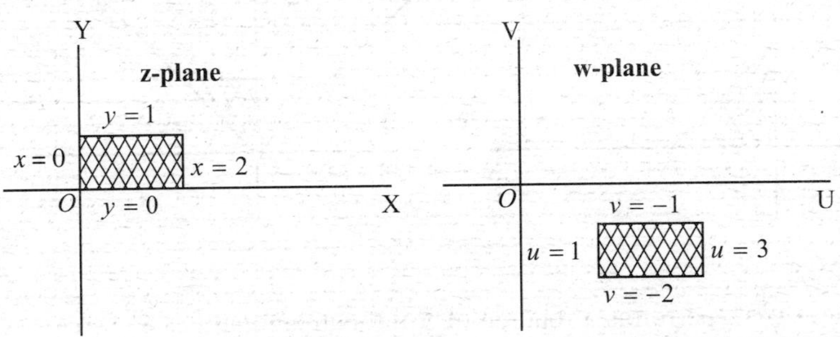

Similarly we can show that each point of R is mapped into one and only one point of R' and conversely. Hence the transformation performs a translation of rectangle. The given map is of the type $w = z + \beta$, where $\beta = 1 - 2i$. Here the rectangle R is translated in the direction of vector $\beta = 1 - 2i$.

2. Rotation: By the transformation $w = ze^{i\theta_0}$, figures in z-plane are rotated through an angle θ_0, if $\theta_0 > 0$, the rotation is anti-clockwise and if $\theta_0 < 0$ the rotation is clockwise.

Example 2: Consider the transformation $w = ze^{i\pi/4}$ and determine the region R' in w-plane corresponding to the triangular region R bounded by the lines $x = 0$, $y = 0$ and $x + y=1$ in z-plane.

Sol. $w = ze^{i\pi/4}$ gives $u + iv = (x + iy)\dfrac{(1+i)}{\sqrt{2}}$

$$\Rightarrow \qquad u = \frac{1}{\sqrt{2}}(x - y), v = \frac{1}{\sqrt{2}}(x + y) \qquad \ldots(1)$$

Putting $x = 0$ in the (1)

$$u = \frac{-1}{\sqrt{2}} y, v = \frac{1}{\sqrt{2}} y \quad \text{or} \quad v = -u$$

Putting $y = 0$ in the (1)

$$u = \frac{1}{\sqrt{2}} x, v = \frac{1}{\sqrt{2}} x \quad \text{or} \quad v = u$$

Putting $x + y = 1$ in the set (1)

$$v = \frac{1}{\sqrt{2}}$$

Hence, the triangular region R in z-plane is mapped on a triangular region R' of w-plane bounded by the lines.

$$v = u, v = -u, v = \frac{1}{\sqrt{2}}$$

The two regions are shown in the following figure.

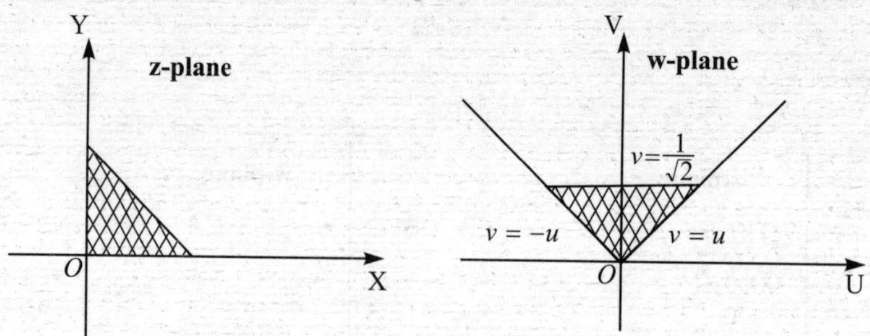

The mapping $w = ze^{i\pi/4}$ performs a rotation of R through an angle $\pi/4$.

3. Magnification (Stretching): Consider the map $w = az$, where a is real. The two figures in z-plane and w-plane are similar and similarly situated about their respective origins but the figure in w-plane is a times the figure in z-plane, such map is called magnification.

Example 3: Consider the transformation $w = 2z$, and determine the region R' of w-plane into which the triangular region R enclosed by the lines $x = 0$, $y = 0$, $x + y = 1$ in the z-plane is mapped under the map.

Sol. $w = 2z$ gives $u + iv = 2x + i2y$.

$$\Rightarrow \qquad u = 2x, v = 2y$$

$$x = 0,\ u = 2x \quad \Rightarrow \quad u = 0$$

$$y = 0,\ v = 2y \quad \Rightarrow \quad v = 0$$

$$x + y = 1,\ u = 2x,\ v = 2y$$

$$\Rightarrow \quad u + v = 2(x + y) = 2.\ 1 = 2$$

$$\Rightarrow \qquad u + v = 2$$

Hence the region R' is triangular bounded by the lines $u = 0$, $v = 0$, $u + v = 2$. The two regions are shown in the figure.

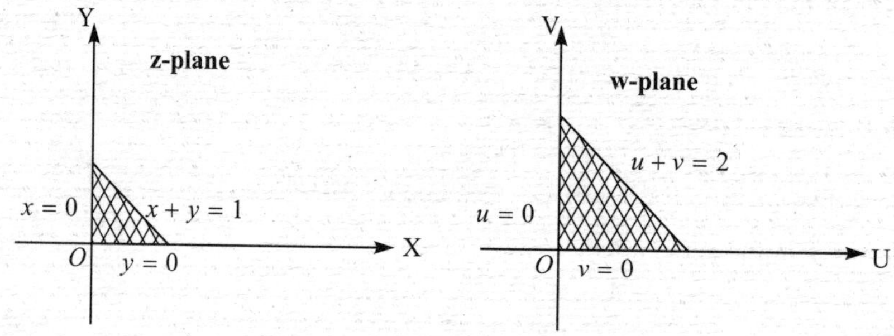

This transformation $w = 2z$ performs a magnification of R into R'.

4. Inversion: By means of the transformation $w = \frac{1}{z}$, figures in z-plane are mapped upon the resiprocal figures in w-plane.

Example 4: Consider the map $w = 1/z$ and determine the region R' in w-plane of the infinite strip R bounded by $\frac{1}{4} < y < \frac{1}{2}$

Sol. $w = \frac{1}{z}$ gives $u + iv = \frac{x - iy}{x^2 + y^2}$

$$\Rightarrow \quad u = \frac{x}{x^2 + y^2}, v = \frac{-y}{x^2 + y^2} \quad \Rightarrow \frac{u}{v} = -\frac{x}{y} \quad \Rightarrow x = -\frac{uy}{v}$$

$$v = \frac{-y}{x^2 + y^2} \quad \Rightarrow v = -\frac{v^2}{y(u^2 + v^2)}$$

$$\Rightarrow \quad y = \frac{-v}{u^2 + v^2} \quad \text{...(1)}$$

$$y < \frac{1}{2} \quad \Rightarrow \quad \frac{-v}{u^2 + v^2} < \frac{1}{2} \quad \Rightarrow -2v < u^2 + v^2 \quad \Rightarrow \quad u^2 + (v+1)^2 > 1 \quad \text{...(2)}$$

$$y > \frac{1}{4} \quad \Rightarrow \quad \frac{-v}{u^2 + v^2} > \frac{1}{4} \quad \Rightarrow -4v > u^2 + v^2 \quad \Rightarrow \quad u^2 + (v+2)^2 < 2^2 \quad \text{...(3)}$$

Finally, $\frac{1}{4} < y < \frac{1}{2} \Rightarrow u^2 + (v+2)^2 < 2^2$ and $u^2 + (v+1)^2 > 1$.

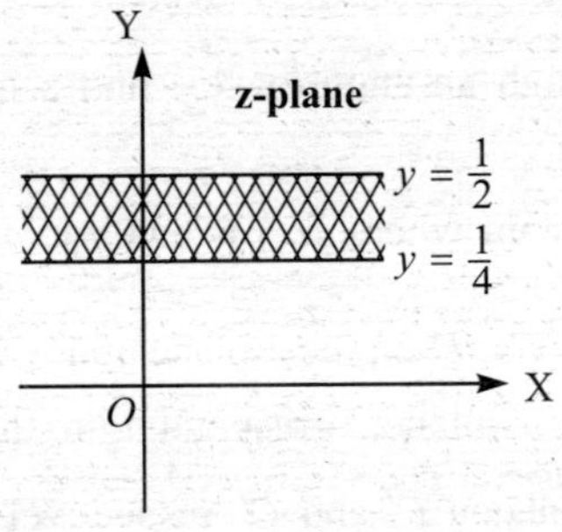

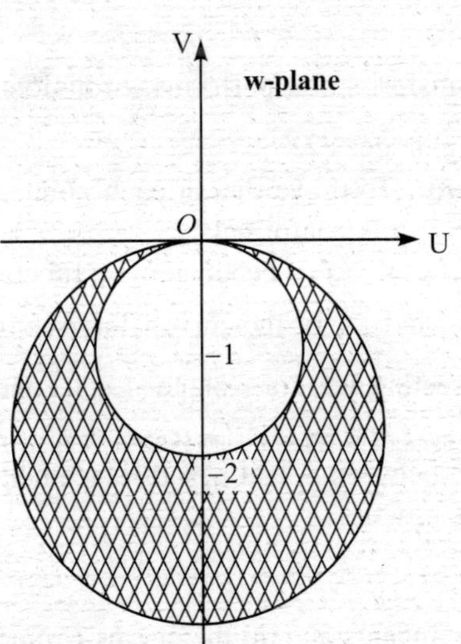

This shows that region R' in w-plane is bounded by the two circles.

$$u^2 + (v+2)^2 = 2^2 \text{ and } u^2 + (v+1)^2 = 1.$$

s.t. the region is exterior to the circle $u^2 + (v+1)^2 = 1$ and interior to the circle $u^2 + (v+2)^2 = 4$. The two domains are shown in the figures.

Example 5: (Rotation and magnification): $w = az$ where a is complex number.

Consider the map $w = ze^{i\pi/4}\sqrt{2}$ and determine the region R' of w-plane into which the rectangular region R bounded by $x = 0$, $y = 0$, $x = 2$, $y = 3$ in z-plane is mapped under this map.

Sol. We have

$$u + iv = (x + iy)\left(\frac{1+i}{\sqrt{2}}\right)\sqrt{2}$$

$$\Rightarrow \qquad u = x - y,\ v = x + y. \qquad \ldots(1)$$

Putting $x = 0$ in (1), $u = -y$, $v = y$ so that $v = -u$

Putting $y = 0$ in (1), $u = x$, $v = x$ so that $v = u$

Putting $y = 3$ in (1), $u = x - 3$, $v = x + 3$ so that $u + 3 = v - 3$ or $v = u + 6$

Putting $x = 2$ in (1), $u = 2 - y$, $v = 2 + y$ or $2 - u = v - 2$ or $v = 4 - u$

Thus the region R' is rectangle bounded by the lines $v = u$, $v = -u$, $v = u + 6$, $v = 4 - u$. The two regions are shown in the following figure.

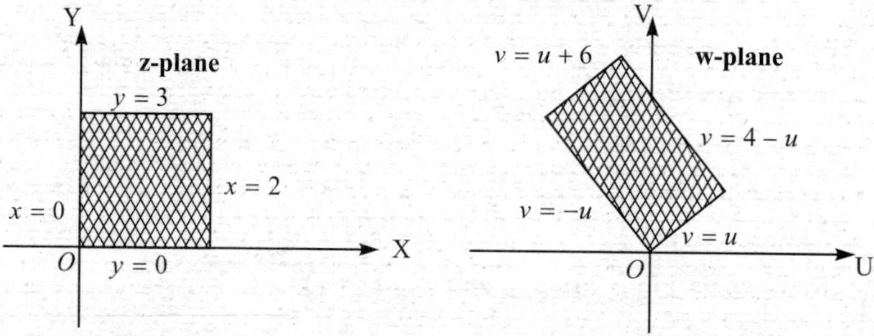

This transformation performs a rotation of R through an angle of 45° and a magnification of length of magnitude $\sqrt{2}$.

Theorem: To prove that at each points z of a domain where $f(z)$ is analytic and $f'(z) \neq 0$, the mapping $w = f(z)$ is conformal.

Proof: Let $w = f(z)$ be an analytic function in a domain D on the z-plane and z_0 be an interior of D. Also let C_1' and C_2' be the curves (intersecting at w_0) in w-plane corresponding to the two curves C_1 and C_2 (intersecting at z_0) of z-plane. Let w_1 and w_2 be points on C_1' and C_2' respectively corresponding to the points z_1, z_2 on C_1 and C_2 respectively s.t.

distance between z_1 and z_0 = distance between z_2 and $z_0 = r$, say

So we can write

$$z_1 - z_0 = re^{i\theta_1}, z_2 - z_0 = re^{i\theta_2}$$

Let the tangents at z_0 to the curves C_1 and C_2 make angles α_1 and α_2 with the real axis so that

$$\theta_1 \to \alpha_1, \theta_2 \to \alpha_2 \ \text{ as } r \to 0$$

Also let the tangents at w_0 to the curves C_1' and C_2' make angles β_1 and β_2 with the real axis and let

$$w_1 - w_0 = \rho_1 e^{i\phi_1}, w_2 - w_0 = \rho_2 e^{i\phi_2}$$

where $\qquad \phi_1 \to \beta_1$ as $\rho_1 \to 0$ and $\phi_2 \to \beta_2$ as $\rho_2 \to 0$

$$f'(z_0) = \lim_{z \to z_0} \frac{f(z) - f(z_0)}{z - z_0}$$

$$= \lim_{z_1 \to z_0} \frac{w_1 - w_2}{z_1 - z_0} = \lim_{z_1 \to z_0} \frac{\rho_1 e^{i\phi} 1}{r e^{i\theta} 1}$$

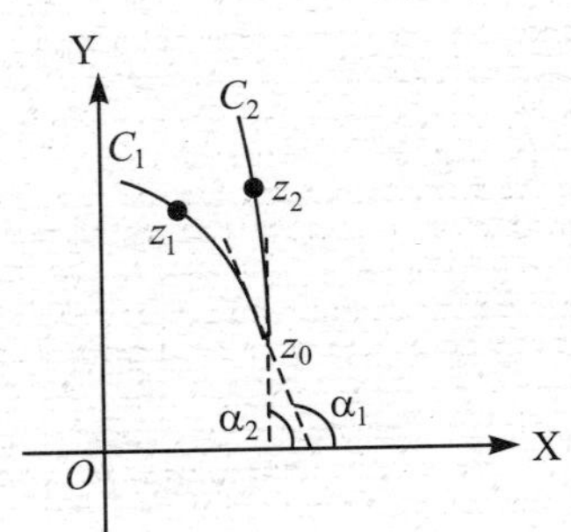

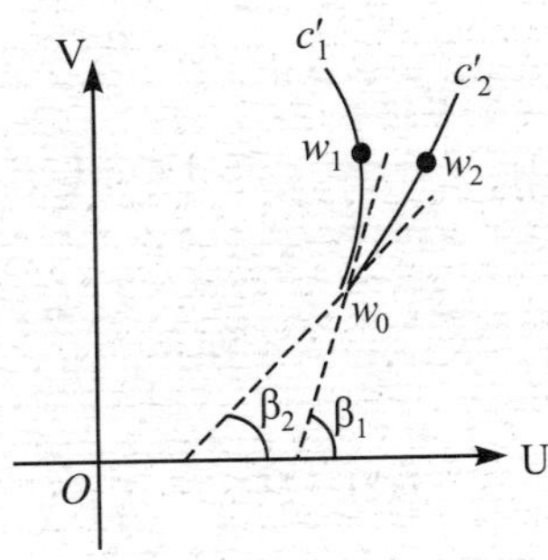

or $$f'(z_0) = \lim_{z_1 \to z_0} \frac{\rho_1}{r} e^{i(\phi_1 - \theta_1)}$$

By assumption $f'(z_0) \neq 0$. Write $f'(z_0) = \mathrm{R}e^{i\lambda}$

Then $$\mathrm{R}e^{i\lambda} = \lim_{z_1 \to z_0} \frac{\rho_1}{r} e^{i(\phi_1 - \theta_1)}$$

Equating modulus and argument,

$$R = \lim_{z \to z_0} \left(\frac{\rho_1}{r} \right), \lambda = \lim(\phi_1 - \theta_1) = \beta_1 - \alpha_1$$

or $$\beta_1 = \lambda + \alpha_1$$

Similarly we can show that $\beta_2 = \lambda + \alpha_2$.

From which $\beta_1 - \beta_2 = \alpha_1 - \alpha_2$.

This proves that the angle between C_1' C_2' at w_0 is equal in magnitude and as well as in sign to the angle between the curves C_1 and C_2. Therefore, the transformation is conformal.

Theorem: To study the conformal property when $f'(z) = 0$.

If $f(z)$ is a regular function of z.

Proof: We know that the transformation $w = f(z)$ is conformal at $z = z_0$. If $f(z)$ is analytic at $z = z_0$ and $f'(z_o) \neq 0$. Let us examine the case when $f'(z_0) = 0$. Suppose further that $f(z_0)$ has a zero of order n–1 at z_0 so that

$$f'(z) = (z - z_0)^{n-1} \phi(z) \text{ where } \phi(z) \text{ is analytic and } \phi(z_0) \neq 0.$$

$$\Rightarrow \qquad f'(z_0) = f''(z_0) = \ldots = f^n(z_0) = 0, f^{(n+1)}(z_0) \neq 0 \qquad \ldots(1)$$

Expanding $f(z)$ by Taylor's theorem in the neighborhood of (z_0), we have

$$f(z) = \sum_{m=0}^{\infty} a_m (z - z_0)^m \qquad ...(2)$$

where
$$a_m = \frac{f^m(z_0)}{m!} \qquad ...(3)$$

Applying (1) to (3),

$$a_m = 0 \text{ for } m = 1, 2....n$$

Then by (2)
$$f(z) = a_0 (z - z_0)^o + \sum_{m=n+1}^{\infty} a_m (z - z_0)^m$$

But
$$a_0 = \frac{f(z_0)}{0!} = f(z_0)$$

$\therefore$
$$f(z) - f(z_0) = a_{n+1}(z - z_0)^{n+1} + ... \qquad (\text{where } a_{n+1} \neq 0)$$

$\Rightarrow$
$$w_1 - w_0 = a_{n+1}(z_1 - z_0)^{n+1}$$

Taking $w_1 - w_0 = \rho_1 e^{i\phi_1}$, $z_1 - z_0 = r_1 e^{i\theta_1}$ $a_{n+1} = ae^{i\lambda}$, we get

$$\rho_1 e^{i\phi_1} = a r_1^{n+1} e^{i\{(n+1)\theta_1 + \lambda\}}$$

$\Rightarrow$
$$\lim \phi_1 = \lim\{(n+1)\theta_1 + \lambda\} = (n+1)\alpha_1 + \lambda$$

Similarly,
$$\lim \phi_2 = (n+1)\alpha_2 + \lambda$$

$\Rightarrow$
$$\lim(\phi_2 - \phi_1) = (n+1)(\alpha_2 - \alpha_1)$$

Thus curves C'_1 and C'_2 have definite tangents at w_0 but the angle between them is $(n+1)$ times the angle between the curves C_1 and C_2 passing through z_0. Consequently conformal property does not hold at z_0.

Example 6: Find the image of $|z - 3i| = 3$ under the mapping $w = \frac{1}{z}$. [*U.T.U. 2008*]

Sol. The mapping $w = \frac{1}{z}$ or $z = \frac{1}{w}$

or
$$x + iy = \frac{1}{(u + iv)} = \frac{(u - iv)}{(u + iv)(u - iv)}$$

$$(x + iy) = \frac{u - iv}{u^2 + v^2}$$

Equating real and imaginary part

$$x = \frac{u}{u^2 + v^2}$$

$$y = -\frac{v}{u^2 + v^2}$$

Now the circle is $|z-3i|=3$

$$|x+iy-3i|=3$$

or
$$x^2+(y-3)^2=9$$

Putting the value of x and y in above equation.

$$\frac{u^2}{(u^2+v^2)^2}+\left(-\frac{v}{u^2+v^2}-3\right)^2=9$$

$\Rightarrow$
$$\frac{u^2}{(u^2+v^2)^2}+\frac{(-v-3u^2-3v^2)^2}{(u^2+v^2)^2}=9$$

$\Rightarrow$
$$u^2+(-v-3u^2-3v^2)^2=9(u^2+v^2)^2$$

$\Rightarrow$
$$u^2+v^2+9u^4+9v^4+6u^2v+6v^3+18u^2v^2$$
$$=9u^4+18u^2v^2+9v^4$$

$\Rightarrow$
$$u^2+v^2+6u^2v+6v^3=0$$

$\Rightarrow$
$$u^2+v^2+6v(u^2+v^2)=0$$

$\Rightarrow$
$$(u^2+v^2)(6v+1)=0$$

$\Rightarrow$ $6v+1=0$, is the image of the circle $|z-3i|=3$. **Ans.**

Example 7: Show that under the transformation $w=\frac{1}{z}$, the image of the hyperbola $x^2-y^2=1$ is the lemniscates $R^2=\cos 2\phi$.

Sol. The equation of the circle is $x^2-y^2=1$ ...(1)

Putting $x=r\cos\theta,\ y=r\sin\theta$

Then we get $r^2(\cos^2\theta-\sin^2\theta)=1$

$$r^2\cos 2\theta=1 \quad ...(2)$$

Now
$$w=\frac{1}{z} \text{ or } z=\frac{1}{w}$$

or
$$re^{i\theta}=\frac{1}{Re^{i\phi}}$$

or
$$re^{i\theta}=\frac{1}{R}e^{-i\phi}$$

$\Rightarrow$
$$r=\frac{1}{R} \text{ and } \theta=-\phi$$

Using these values in equation (2), we get $\frac{1}{R^2}\cos 2\phi = 1$

$$\cos 2\phi = R^2. \textbf{ Proved.}$$

3.16 THE LINEAR TRANSFORMATION

A transformation of the form $w = az + b$ is called a linear transformation, where a and b are complex constants.

3.17 BI-LINEAR TRANSFORMATION OR MOBIUS TRANSFORMATION

A transformation of the form

$$w = \frac{az+b}{cz+d} \qquad ...(1)$$

is called a bi-linear transformation or linear fractional transformation, where a, b, c, d are complex constants. Such type of transformation was first studied by Mobius and hence it is some times called Mobius transformation. (1) is expressible as

$$cwz + wd - az - b = 0$$

since it is linear in both w and z. That is why it is called bi-linear the relation (1) can be written as $w = \frac{a(z+b/a)}{c(z+d/c)}$ from this we notice that if $\frac{b}{a} = \frac{d}{c}$, *i.e.* $ad - bc = 0$, then for every value of z there are the same values of w. And if $ad - bc \neq 0$, then there correspond different values of w to different values of z.

The expression $ad - bc$ is called the determinant of the transform.

3.18 CROSS RATIO

If there are four points z_1, z_2, z_3 and z_4 taken in order, then the ratio $\frac{(z_1 - z_2)(z_3 - z_4)}{(z_2 - z_3)(z_4 - z_1)}$ is called the cross ratio of z_1, z_2, z_3 and z_4.

Theorem: If under the bi-linear transformation w_1, w_2, w_3 and w_4 are the images of z_1, z_2, z_3 and z_4 respectively, to show that cross ratio of w_1, w_2, w_3 and w_4 is equal to the cross ratio of z_1, z_2, z_3 and z_4.

Proof: Let the bi-linear transformation be

$$w = \frac{az+b}{cz+d}$$

$$w_1 = \frac{az_1+b}{cz_1+d}, w_2 = \frac{az_2+b}{cz_2+d}, w_3 = \frac{az_3+b}{cz_3+d} \text{ and } w_4 = \frac{az_4+b}{cz_4+d}$$

$$\therefore \qquad w_1 - w_2 = \frac{az_1+b}{cz_1+d} - \frac{az_2+b}{cz_2+d} = \frac{(ad-bc)}{(cz_1+d)} \cdot \frac{(z_1 - z_2)}{(cz_2+d)}$$

Similarly $$w_2 - w_3 = \frac{(ad-bc)}{(cz_2+d)} \cdot \frac{(z_2 - z_3)}{(cz_3+d)}$$

$$w_3 - w_4 = \frac{(ad-bc)}{(cz_3+d)} \cdot \frac{(z_3 - z_4)}{(cz_4+d)}$$

$$w_4 - w_1 = \frac{(ad-bc)}{(cz_4+d)} \cdot \frac{(z_4 - z_1)}{(cz_1+d)}$$

Hence, we get

$$\frac{(w_1 - w_2)(w_3 - w_4)}{(w_2 - w_3)(w_4 - w_1)} = \frac{(z_1 - z_2)(z_3 - z_4)}{(z_2 - z_3)(z_4 - z_1)}$$

Example 8: Show that the transformation $w = \dfrac{5-4z}{4z-2}$ transforms the circle $|z| = 1$, into a circle of radius unity in w-plane and find the center of the circle.

Sol. We have $w = \dfrac{5-4z}{4z-2}$ from which

$$4zw - 2w = 5 - 4z \qquad \text{or} \qquad z = \frac{2w+5}{4w+4}$$

$$|z| = z\bar{z} = 1 \text{ corresponds to } \left[\frac{2w+5}{4(w+1)}\right]\left[\frac{2\bar{w}+5}{4(\bar{w}+1)}\right] = 1$$

or $$4w\bar{w} + 25 + 10(w + \bar{w}) = 16(w\bar{w} + 1 + w + \bar{w}) \qquad ...(1)$$

But $$w = u + iv,\ \bar{w} = u - iv,\ w\bar{w} = u^2 + v^2, w + \bar{w} = 2u$$

Hence (1)

$$\Rightarrow \qquad u^2 + v^2 + u - (3/4) = 0 \qquad ...(2)$$

Comparing this with $u^2 + v^2 + 2gu + 2fv + c = 0$, we get center $= (-g, -f) = \left(-\frac{1}{2}, 0\right)$ and radius $= \left(g^2 + f^2 - c\right)^{1/2}$

$$\therefore \text{ radius} = \left[\frac{1}{4} + 0 + \frac{3}{4}\right]^{1/2} = 1$$

Hence (2) represents circle with its center at $\left(-\frac{1}{2}, 0\right)$ and of radius one in w-plane.

Example 9: Show that the relation

$$w = \frac{iz+2}{4z+i}$$

transforms the real axis in the z-plane into a circle in the w-plane. Find the center and the radius of the circle and the point in the z-plane which is mapped at the center of the circle.

Sol. The given transformation is $w = \dfrac{iz+2}{4z+i}$ ...(1)

Solving for z. we have $$z = \frac{2-iw}{4w-i} \qquad ...(2)$$

Now $$z = \frac{2-iw}{4w-i} = \frac{2-iw}{4w-i} \cdot \frac{4\bar{w}+i}{4\bar{w}+i}$$

To make the denominator real multiply above and below by $(4\bar{w}+i)$

Thus $x+iy = \dfrac{8\bar{w}+w-4iw\bar{w}+2i}{16w\bar{w}-4(w-\bar{w})/i+1}$ the denominator is now real.

The real axis in the z-plane is $y = 0$. So corresponding to $y = 0$, we must have imaginary part on the right side of the above relation also equal to zero, that is

$$-8v+v-4w\bar{w}+2=0$$

i.e., $$-8v+v-4(u^2+v^2)+2=0$$

i.e., $$4(u^2+v^2)+7v-2=0$$

which is a circle in w-plane.

i.e. $$u^2+v^2+\frac{7}{4}v-\frac{1}{2}=0$$

Hence, the real axis in the z-plane transforms into a circle in the w-plane.

The center of the circle is $\left(0,-\dfrac{7}{8}\right)$ and the radius $=\sqrt{\dfrac{49}{64}+\dfrac{1}{2}}=\dfrac{9}{8}$

If we denote the center by w_0, then $w_0 = u_0 + iv_0 = -\dfrac{7}{8}i$.

If z_0 corresponds to w_0 then from (2), we have

$$z_0 = \frac{2-iw_0}{4w_0-i} = \frac{2-\frac{7}{8}}{-\frac{7}{2}i-i} = \frac{1}{4}i.$$

Example 10: Find the bi-linear transformation that maps the points $z_1 = -i,\ z_2 = 0,\ z_3 = i$ into the points $w_1 = -1,\ w_2 = i,\ w_3 = 1$ respectively. Into what curve the y-axis transformed to this transformation?

Sol. We know that

$$\frac{(w-w_1)(w_2-w_3)}{(w-w_3)(w_2-w_1)} = \frac{(z-z_1)(z_2-z_3)}{(z-z_3)(z_2-z_1)} \qquad ...(1)$$

Putting all the given values in equation (1).

$$\frac{(w+1)(i-1)}{(w-1)(i+1)} = \frac{(z+i)(0-i)}{(z-i)(0+i)}$$

$$\frac{wi-w+i-1}{wi+w-i-1} = \frac{-z-i}{z-i}$$

$$\Rightarrow \quad \frac{wi-w+i-1+wi+w-i-1}{wi-w+i-1-wi-w+i+1}=\frac{-z-i+z-i}{-z-i-z+i} \qquad \text{(Applying C and D rule)}$$

$$\Rightarrow \quad \frac{wi-1}{-w+i}=\frac{i}{z} \qquad \Rightarrow \quad w(iz+i)=z-1$$

$$\Rightarrow \quad w=\frac{z-1}{i(z+1)}=\frac{i\left(\frac{z}{i}+i\right)}{i(z+1)}$$

$$\Rightarrow \quad w=\frac{(-iz+i)}{(z+1)}$$

$$w=\frac{-i(x+iy)+i}{x+iy+1} \qquad \Rightarrow \quad u+iv=\frac{(-ix+y+i)(x+1-iy)}{(x+1+iy)(x+1-iy)}$$

$$=\frac{-ix^2-ix-iy^2+ix+i+xy-xy+y+y}{(x+1)^2+y^2}=\frac{-i(x^2+y^2-1)+2y}{x^2+y^2+2x+1}$$

Equating real and imaginary parts, we have

$$u=\frac{2y}{x^2+y^2+2x+1} \text{ and } v=\frac{-x^2-y^2+1}{x^2+y^2+2x+1}$$

Equation of y-axis is $x = 0$

Hence $$u=\frac{2y}{y^2+1} \qquad ...(2)$$

$$v=\frac{-y^2+1}{y^2+1} \qquad ...(3)$$

$$\Rightarrow \quad y=\sqrt{\frac{1-v}{1+v}}$$

On putting the value of y in equation (2), we get

$$u=\frac{2\sqrt{\frac{1-v}{1+v}}}{\frac{1-v}{1+v}+1}=2\frac{\sqrt{\frac{1-v}{1+v}}(1+v)}{1-v+1+v}=\sqrt{1-v^2}$$

$$u^2=1-v^2 \text{ or } u^2+v^2=1$$

Hence y-axis in the z-plane is transformed to the circle $u^2+v^2=1$ with center at origin and radius equal to 1 in w-plane.

Example 11: Find the bi-linear transformation which maps the points $z = 1, i, -1$ into the points $w = i, 0, -i$.

Hence find the image of $|z| < 1$.

Sol. We know that

$$\frac{(w-w_1)(w_2-w_3)}{(w-w_3)(w_2-w_1)}=\frac{(z-z_1)(z_2-z_3)}{(z-z_3)(z_2-z_1)}$$

$$\Rightarrow \quad \frac{(w-i)(0+i)}{(w+i)(0-i)}=\frac{(z-1)(i+1)}{(z+1)(i-1)}$$

$$\Rightarrow \quad \frac{w-i}{w+i}=i\left(\frac{z-1}{z+1}\right)$$

$$\Rightarrow \quad \frac{w}{-i}=\frac{iz-i+z+1}{iz-i-z-1}=\frac{(i+1)z-(i-1)}{(i-1)z-(i+1)}=-i\left(\frac{z-i}{z+i}\right) \text{(Applying C and D rule)}$$

$$\Rightarrow \quad w=\frac{i-z}{i+z}=\frac{i(iz+1)}{i(-iz+1)}$$

$$\Rightarrow \quad w=\frac{(iz+1)}{(-iz+1)}$$

$$u+iv=\frac{i(x+iy)+1}{-i(x+iy)+1}=\frac{(ix-y+1)(ix+y+1)}{(-ix+y+1)(ix+y+1)}=\frac{-x^2-y^2+1+2ix}{x^2+(y+1)^2}$$

Equating real part, we get

$$u=\frac{-x^2-y^2+1}{x^2+(y+1)^2} \qquad ...(1)$$

But $\qquad |z|<1 \Rightarrow x^2+y^2<1 \qquad$ or $\qquad 1-x^2-y^2>0$

or from equation (1), $u > 0$ as denominator is positive. **Ans.**

Example 12: Find the fixed points and the normal form of the following bi-linear transformations.

(*a*) $w=\dfrac{3z-4}{z-1}$ and (*b*) $w=\dfrac{z-1}{z+1}$

Discuss the nature of these transformations.

Sol. (*a*) The fixed points are obtained by

$$z=\frac{3z-4}{z-1} \quad \text{or} \quad z^2-4z+4=0 \quad \text{or} \quad (z-2)^2=0 \quad \Rightarrow z=2$$

$z = 2$ is the only fixed point. This transformation is parabolic.

Normal Form

$$w=\frac{3z-4}{z-1} \quad \text{or} \quad wz-w=3z-4$$

$$\Rightarrow \quad \frac{1}{w-2}=\frac{1}{z-2}+1. \qquad \textbf{Ans.}$$

(*b*) The fixed points are obtained by

$$z = \frac{z-1}{z+1} \quad \text{or} \quad z^2 + 1 = 0 \quad \Rightarrow \quad z = \pm i$$

Hence $\pm i$ are the two fixed points.

Normal Form

$$w = \frac{z-1}{z+1}$$

$$\Rightarrow \qquad w - i = \frac{z-1}{z+1} - i \text{ and } w + i = \frac{z-1}{z+1} + i$$

$$\frac{w-i}{w+i} = \frac{z-1-iz-i}{z-1+iz+i} = \frac{(1-i)(z-i)}{(1+i)(z+i)} = \frac{(-i^2-i)(z-i)}{(1+i)(z+i)} \qquad \Big| \text{ As } i^2 = -1$$

$$\frac{w-i}{w+i} = -i\left(\frac{z-i}{z+i}\right) = a\left(\frac{z-i}{z+i}\right), \text{ where } a = -i$$

The transformation is elliptic. **Ans.**

Example 13: Prove that in general the bi-linear transformation

$$w = \frac{az+b}{cz+d}$$

there are two values of z (invariant points) for which $w = z$, but there is only one if $(a-d)^2 + 4bc = 0$. Show that, if there are distinct invariant points p and q, the transformation may be put in the form

$$\frac{w-p}{w-q} = k\frac{z-p}{z-q}$$

and that if there is only one invariant point p, the transformation may be put in the form $\frac{1}{w-p} = \frac{1}{z-p} + k$.

Sol. The bi-linear transformation is $w = \frac{az+b}{cz+d}$

If $w = z$, then it gives $\qquad z = \frac{az+b}{cz+d}$

i.e., $\qquad cz^2 + (d-a)z - b = 0$

Let p, q be the roots of this equation, then solving the above equation, we have

$$p = \frac{a-d+\sqrt{\{(a-d)^2 + 4bc\}}}{2c}$$

and

$$q = \frac{(a-d)-\sqrt{\{(a-d)^2 + 4bc\}}}{2c}$$

Thus p and q as expressed above are the two invariant points of the transformation. They reduce to one if,

$$(a-d)^2+4bc=0\,.$$

and then
$$p=q=\frac{a-d}{2c} \qquad \text{...(1)}$$

when p and q are distinct points.

We have $cp^2+(d-a)p-b=0$ *i.e.* $b-pd=cp^2-ap$

and $cq^2+(d-a)q-b=0$ *i.e.* $b-qd=cq^2-aq$ ...(2)

Now
$$\frac{w-p}{w-q}=\frac{\dfrac{az+b}{cz+d}-p}{\dfrac{az+b}{cz+d}-q}=\frac{(a-cp)z+b-pd}{(a-cq)z+b-qd}$$

$$=\frac{(a-cp)z+cp^2-ap}{(a-cq)z+cq^2-aq} \quad \text{from (2)}$$

$$=\frac{(a-cp)(z-p)}{(a-cq)(z-q)}$$

$$=k\frac{z-p}{z-q} \quad \text{where } k=\frac{a-cp}{a-cq}$$

i.e.,
$$\frac{w-p}{w-q}=k\frac{z-p}{z-q}$$

and when there is only one invariant point say p. Then as shown above in (1) and (2), we have

$$p=\frac{a-d}{2c} \quad \text{and } b-pd=cp^2-ap$$

Now
$$\frac{1}{w-p}=\frac{1}{\dfrac{az+b}{cz+d}-p}=\frac{cz+d}{(a-cp)z+b-pd}$$

$$=\frac{cz+d}{(a-cp)z+cp^2-ap}$$

$$=\frac{cz+d}{(z-cp)(z-b)}$$

$$=\frac{(a-cp)+c(z-p)}{(a-cp)(z-p)}$$

$$= \frac{1}{(z-p)} + \frac{c}{(a-cp)}$$

$$= \frac{1}{(z-p)} + k, \quad \text{where } k = \frac{c}{(a-cp)}$$

$$\Rightarrow \quad \frac{1}{w-p} = \frac{1}{z-p} + k. \quad \textbf{Proved.}$$

Example 14: Find the bilinear transformation which maps $z = 1, i, -1$ onto the points $w = i, 0, -i$ respectively. Hence find (*i*) the image of $|z| < 1$ (*ii*) the invariant points of the transformation.

Sol. We know that

$$\frac{(w-w_1)(w_2-w_3)}{(w-w_3)(w_2-w_1)} = \frac{(z-z_1)(z_2-z_3)}{(z-z_3)(z_2-z_1)}$$

$$\therefore \quad \frac{(w-i)(i)}{(w+i)(-i)} = \frac{(z-1)(i+1)}{(z+1)(i-1)}$$

$$\Rightarrow \quad \frac{w-i}{w+i} = i\left(\frac{z-1}{z+1}\right)$$

$$\Rightarrow \quad \frac{2w}{-2i} = \frac{iz-i+z+1}{iz-i-z-1} = \frac{(i+1)z-(i-1)}{(i-1)z-(i+1)} = -i\left(\frac{z-i}{z+i}\right) \quad \Big| \text{ Applying } C \text{ and } D \text{ rule}$$

$$\Rightarrow \quad w = \frac{i-z}{i+z} \qquad \text{...(1)}$$

which is the required bi-linear transformation.

From (1), we get

$$z = i\left(\frac{1-w}{1+w}\right)$$

(*i*) $\therefore$ $|z| < 1$ is mapped into the region

$$\left| i\left(\frac{1-w}{1+w}\right)\right| < 1$$

$$\Rightarrow \quad \frac{|i||1-w|}{|1+w|} < 1$$

$$\Rightarrow \quad |1-w| < |1+w|$$

$$\Rightarrow \quad |1-u-iv| < |1+u+iv|$$

$$\Rightarrow \quad (1-u)^2 + v^2 < (1+u)^2 + v^2$$

$$\Rightarrow \quad 1+u^2+v^2-2u < 1+u^2+v^2+2u$$

$$\Rightarrow \quad u > 0. \qquad \textbf{Ans.}$$

(*ii*) The invariant points obtained as follows:

From (1) $$z = \frac{i-z}{i+z} \Rightarrow zi + z^2 = i - z$$

$$\Rightarrow z^2 + z(i+1) - i = 0$$

$$\Rightarrow \quad z = \frac{-(i+1) \pm \sqrt{i^2 + 2i + 1 + 4i}}{2}$$

$$z = \frac{-(i+1) \pm \sqrt{6i}}{2}$$

$$\Rightarrow \quad z_1 = \frac{-(i+1) + \sqrt{6i}}{2}, z_2 = \frac{-(i+1) - \sqrt{6i}}{2}.$$ **Ans.**

Example 15: Find transformation which maps outside $|z| = 1$ on the half plane $R(w) \geq 0$ so that the points $z = 1, -i, -1$ corresponds $w = i, 0, -i$ respectively.

Sol. Let $z_1 = 1,\ z_2 = -i,\ z_3 = -1,\ w_1 = i,\ w_2 = 0,\ w_3 = -i.$

Since cross ratio is preserved under a bi-linear transformation.

So $$\frac{(w-w_1)(w_2-w_3)}{(w_1-w_2)(w_3-w)} = \frac{(z-z_1)(z_2-z_3)}{(z_1-z_2)(z_3-z)}$$

Putting the values,

$$\frac{(w-i)(0+i)}{(i-0)(-i-w)} = \frac{(z-1)(-i+1)}{(1+i)(-1-z)}$$

or $$\frac{w-i}{w+i} = -i\left(\frac{z-1}{z+1}\right) = \frac{-iz+i}{z+1}$$

or $$\frac{(w-i)+(w+i)}{(w-i)-(w+i)} = \frac{-iz+i+z+1}{-iz+i-(z+1)}$$

or $$\frac{2w}{-2i} = \frac{z(1-i)+(1+i)}{-z(1+i)-(1-i)}$$

or $$w = i\left(\frac{1-i}{1+i}\right)\left[\frac{z+\left(\frac{1+i}{1-i}\right)}{z+\left(\frac{1-i}{1+i}\right)}\right] = i(-i)\left(\frac{z+i}{z-i}\right)$$

or $$w = \frac{z+i}{z-i} \qquad \text{...(1)}$$

This is the required transformation.

By (1), $$w(z-i) = z+i \quad \text{or} \quad z = i\left(\frac{w+1}{w-1}\right) \qquad \text{...(2)}$$

$|z| \geq 1$ is transformed into $\left|\dfrac{w+1}{w-1}\right|.|i| \geq 1$

or $$|w+1|^2 \geq |w-1|^2 \text{ or } (u+1)^2+v^2 \geq (u-1)^2+v^2$$

or $$R(w)=u \geq 0$$

Thus exterior of the circle $|z|=1$ is transformed into half plane $R(w) \geq 0$. **Proved.**

Deduction: Find the images of concentric circles $|z|=r,\ r>1$ relative to the above transformation.

$$|i|.\left|\frac{w+1}{w-1}\right|=|z|=r \text{ or } |1+w|^2=|1-w|^2 r^2$$

or $$(1+u)^2+v^2=r^2\left[(1-u)^2+v^2\right]$$

or $$u^2+v^2+2\left(\frac{1+r^2}{1-r^2}\right)u+1=0$$

Thus the image of $|z|=r$ is a circle in w-plane of radius.

$$=\left(g^2+f^2-c\right)^{1/2}=\left[\left(\frac{1+r^2}{1-r^2}\right)^2+0-1\right]^{1/2}=\frac{2r}{1-r^2}$$

and center at $$(-g,-f)=\left(-\frac{1+r^2}{1-r^2},0\right).$$

Example 16: Find the bi-linear transformation which maps the points $z_1=0,\ z_2=-1,\ z_3=i,$ onto $w_1=i,\ w_2=0,\ w_3=\infty$. Also find the image of the unit circle $|z|=1$.

Sol. The cross ratio preserved under a bi-linear transformation.

So $$\frac{(w-w_1)(w_2-w_3)}{(w_1-w_2)(w_3-w)}=\frac{(z-z_1)(z_2-z_3)}{(z_1-z_2)(z_3-z)}$$

$$\Rightarrow \quad \frac{(w-w_1)\left(\dfrac{w_2}{w_3}-1\right)}{(w_1-w_2)\left(1-\dfrac{w}{w_3}\right)}=\frac{(z-z_1)(z_2-z_3)}{(z_1-z_2)(z_3-z)}$$

$$\Rightarrow \quad \frac{(w-i)(0-1)}{(i-0)(1-0)}=\frac{(z-0)(-1-i)}{(0+1)(i-z)}$$

$$\Rightarrow \quad -\frac{w-i}{i}=-z\left(\frac{1+i}{i-z}\right)$$

$$wi-wz+1+iz=+iz-z$$

$$\Rightarrow \quad w(-i+z)=z+1$$

$$\Rightarrow \qquad w = \frac{z+1}{z-i} \qquad ...(1)$$

This is the required bi-linear transformation

From (1) $$z = \frac{iw+1}{w-1}$$

But $|z| = 1$

$$\Rightarrow \quad \left|\frac{iw+1}{w-1}\right| = 1 \quad \Rightarrow \quad |iw+1| = |w-1|$$

$$\Rightarrow \quad |i(u+iv)+1| = |u+iv-1|$$

$$\Rightarrow \quad |1-v+iu| = |iv+u-1|$$

$$\Rightarrow \quad (1-v)^2 + u^2 = (u-1)^2 + v^2$$

$$\Rightarrow \quad u - v = 0$$

or $$u = v \quad \text{or} \quad v = u$$

Therefore the image of the unit circle $|z| = 1$ in z-plane is a straight line making an angle $\pi/4$ with u-axis and passing through origin in w-plane.

Example 17: Show that the transformation $w = \frac{2z+3}{z-4}$ transforms the circle $x^2 + y^2 - 4x = 0$ into the straight line $4u + 3 = 0$ and explain the curve obtained is not a circle.

Sol. The given transformation is $w = \frac{2z+3}{z-4}$ ...(1)

Solving for z, we have
$$\left.\begin{aligned} z &= \frac{4w+3}{w-2} \\ \Rightarrow \quad \bar{z} &= \frac{4\bar{w}+3}{\bar{w}-2} \end{aligned}\right\} \qquad ...(2)$$

The given equation is

$$x^2 + y^2 - 4x = 0$$

$$\Rightarrow \quad (x+iy)(x-iy) - 4x = 0 \qquad \left(\begin{aligned} z+\bar{z} &= x+iy \\ +x-iy &= 2x \end{aligned}\right)$$

$$z\bar{z} - 4x = 0$$

or $$z\bar{z} - 2(z+\bar{z}) = 0 \qquad ...(3)$$

Putting the value of equation (2) in the equation (3), then we get

$$\frac{4w+3}{w-2} \cdot \frac{4\bar{w}+3}{\bar{w}-2} - 2\left(\frac{4w+3}{w-2} + \frac{4\bar{w}+3}{\bar{w}-2}\right) = 0$$

or $$2(w+\bar{w}) + 3 = 0$$

or $$4u + 3 = 0 \qquad |\text{As } w + \bar{w} = 2u|$$

which is a straight line to the w-plane

Explanation: This is possible under bi-linear transformation. Since we regard a straight line as a particular case of circle.

Example 18: Find the image of $x^2+y^2-4y+2=0$ under the mapping $w=\dfrac{z-i}{iz-1}$.

Sol. $$w(iz-1)=z-i$$

$$\Rightarrow \quad z=\frac{w-i}{iw-1}$$

$$x+iy=\frac{u+i(v-1)}{iu-(v+1)} \quad ...(1)$$

$$x-iy=\frac{u-i(v-1)}{-iu-(v+1)} \quad ...(2)$$

$$\therefore \quad x^2+y^2=\frac{u^2+(v-1)^2}{u^2+(v+1)^2}$$

Subtracting (2) from (1)

$$2iy=\frac{-2iu^2-2i(v^2-1)}{u^2+(v+1)^2}$$

Hence, $x^2+y^2-4y+2=0$ transforms into

$$\frac{u^2+(v-1)^2}{u^2+(v+1)^2}+4\frac{u^2+(v^2-1)}{u^2+(v+1)^2}+2=0$$

$$\Rightarrow u^2+(v-1)^2+4\left[u^2+(v^2-1)\right]+2\left[u^2+(v-1)^2\right]=0$$

$$\Rightarrow \quad 7(u^2+v^2)+2v-1=0$$ which is the required image. **Ans.**

Example 19: Show that under the transformation $w=\dfrac{z-i}{z+i}$, the real axis in z-plane mapped into the circle $|w|=1$. What portion of the z-plane corresponds to the interior of the circle?

Sol. Since $w=\dfrac{z-i}{z+i}$

Therefore, real axis in z-plane (y = 0) *i.e.* $z = x$ transforms into

$$w=\frac{z-i}{z+i}=\frac{(x-i)^2}{x^2+1}=\frac{x^2-1-2ix}{x^2+1}$$

$$|w|=\sqrt{\left(\frac{x^2-1}{x^2+1}\right)^2+\left(\frac{2x}{x^2+1}\right)^2}=1$$

Hence the real axis in z-plane transforms into the circle $|w|=1$

Put $z = i$, we get $w = 0$. Hence the half z-plane above the real axis corresponds to the interior of the circle $|w| = 1$. **Proved.**

Example 20: If a is any real positive number show that the transformation $w = \dfrac{z-a}{z+a}$ transforms conformally the plane $x > 0$ to the unit circle $|w| < 1$.

Sol. Since
$$w = \frac{z-a}{z+a} \quad ...(1)$$

and
$$|w| < 1$$

So
$$\left|\frac{z-a}{z+a}\right| < 1$$

$$\Rightarrow \quad |z-a| < |z+a|$$

$$\Rightarrow \quad |x-a+iy| < |x+a+iy|$$

$$\Rightarrow \quad (x-a)^2 + y^2 < (x+a)^2 + y^2$$

$$\Rightarrow \quad -2ax < 2ax$$

$$\Rightarrow \quad 4ax > 0$$

$$x > 0 \quad (a \text{ positive})$$

Hence the transformation (1) transforms conformally the plane $x > 0$ to the unit circle $|w| < 1$.

Example 21: Show that $w = \dfrac{i-z}{i+z}$ maps the real axis of the z-plane into the circle $|w| = 1$ and the half plane $y > 0$ into the interior of the unit circle $|w| = 1$ in the w-plane.

Sol. Since
$$w = \frac{i-z}{i+z} \quad ...(1)$$

(*i*)
$$|w| = 1 \quad \Rightarrow \quad \left|\frac{i-z}{i+z}\right| = 1$$

$$\Rightarrow \quad |i-z| = |i+z|$$

$$\Rightarrow \quad |i-x-iy| = |i+x+iy|$$

$$\Rightarrow \quad x^2 + (1-y)^2 = x^2 + (1+y)^2$$

$$\Rightarrow \quad 1 + y^2 - 2y = 1 + y^2 + 2y$$

$$\Rightarrow \quad 4y = 0 \quad \Rightarrow \quad y = 0 \qquad \text{(real axis of } z\text{-plane)}$$

Hence the given transformation maps the real axis of z-plane into the unit circle $|w| = 1$.

(*ii*)
$$|w| < 1 \quad \Rightarrow \quad \left|\frac{i-z}{i+z}\right| < 1$$

$$\Rightarrow \qquad |i-z| < |i+z|$$

$$\Rightarrow \qquad |i-x-iy| < |i+x+iy|$$

$$\Rightarrow \qquad x^2+(1-y)^2 < x^2+(1+y)^2$$

$$\Rightarrow \qquad x^2+1+y^2-2y < x^2+1+y^2+2y$$

$$\Rightarrow \qquad 4y > 0 \qquad \Rightarrow y > 0$$

Hence the given transformation maps the half-plane $y > 0$ of the z-plane into the interior of the unit circle $|w| = 1$ in the w-plane. **Proved.**

Example 22: Show that the transformation $w = i\left(\frac{1-z}{1+z}\right)$ transforms the circle $|z| = 1$ onto the real axis of the w-plane and the interior of the circle into the upper half of the w-plane.

Sol. Since $w = i\left(\frac{1-z}{1+z}\right) \Rightarrow \frac{w}{i} = \frac{1-z}{1+z}$

$$\Rightarrow \qquad \frac{w+i}{w-i} = \frac{2}{-2z} \qquad \Rightarrow \qquad z = \frac{i-w}{i+w} \qquad \text{(By C and D rule)}$$

(*i*) But $$|z| = 1 \Rightarrow \left|\frac{i-w}{i+w}\right| = 1$$

$$\Rightarrow \qquad |i-w| = |i+w|$$

$$\Rightarrow \qquad |i-u-iv| = |i+u+iv|$$

$$\Rightarrow \qquad u^2+(1-v)^2 = u^2+(1+v)^2$$

$$\Rightarrow \qquad 1+v^2-2v = 1+v^2+2v$$

$$\Rightarrow \qquad 4v = 0 \qquad \Rightarrow v = 0 \qquad \text{(real axis of } w\text{-plane)}$$

Hence the given transformation transforms the circle $|z| = 1$ onto the real axis of the w-plane.

(*ii*) $$|z| < 1 \qquad \Rightarrow \qquad \left|\frac{i-w}{i+w}\right| < 1$$

$$\Rightarrow \qquad |i-w| < |i+w|$$

$$\Rightarrow \qquad |i-u-iv| < |i-u+iv|$$

$$\Rightarrow \qquad u^2+(1-v)^2 < u^2+(1+v)^2$$

$$\Rightarrow \qquad 1+v^2-2v < 1+v^2+2v$$

$$\Rightarrow \qquad 4v > 0 \qquad \Rightarrow v > 0$$

Hence, the given transformation transforms the circle $|z| = 1$ into the upper half of w-plane.

Proved.

Example 23: Find the condition that the transformation $w = \dfrac{az+b}{cz+d}$ transforms the unit circle in the w-plane into straight line in the z-plane.

Sol. The transformation is

$$w = \frac{az+b}{cz+d} = \frac{a}{c} \cdot \frac{\left(z + \dfrac{b}{a}\right)}{\left(z + \dfrac{d}{c}\right)}$$

Under this transformation, the unit circle $|w| = 1$ in w-plane transforms into

$$1 = \left|\frac{a}{c}\right| \left| \frac{z + \dfrac{b}{a}}{z + \dfrac{d}{c}} \right| \Rightarrow \left| \frac{z + \dfrac{b}{a}}{z + \dfrac{d}{c}} \right| = \left|\frac{c}{a}\right| \quad \text{in } z\text{-plane}$$

This represents a line if $\left|\dfrac{c}{a}\right| = 1$ or $|a| = |c|$

The corresponding line is $\left| \dfrac{z + \dfrac{b}{a}}{z + \dfrac{d}{c}} \right| = 1.$ **Ans.**

Example 24: Find the invariant points of the transformation $w = -\left(\dfrac{2z+4i}{iz+1}\right)$. Prove also that these two points together with any point z and its image w, form a set of four points having a constant cross ratio.

Sol. The invariant points of the transformation are given by

$$z = -\left(\frac{2z+4i}{iz+1}\right) \Rightarrow iz^2 + 3z + 4i = 0$$

$$\Rightarrow \qquad z^2 - 3iz + 4i = 0 \qquad \Rightarrow \qquad (z-4i)(z+i) = 0$$

$\therefore$ The invariant points are $4i$ and $-i$.

Taking $z_1 = z, z_2 = w = -\left(\dfrac{2z+4i}{iz+1}\right), z_3 = 4i$ and $z_4 = -i$, the cross-ratio of the four points z_1, z_2, z_3 and z_4 is given by

$$(z_1, z_2, z_3, z_4) = \frac{\left(z + \dfrac{2z+4i}{iz+1}\right)(4i+i)}{\left(4i + \dfrac{2z+4i}{iz+1}\right)(z+i)}$$

$$= \frac{5i\left(z^2+3z+4i\right)}{(z+i)(-2z+8i)} = \frac{5i\left(iz^2+3z+4i\right)}{2i\left(iz^2+3z+4i\right)}$$

$$= \frac{5}{2} (\text{constant independent of } z). \quad \textbf{Ans.}$$

Example 25: Prove that $w = \frac{z}{1-z}$ maps the upper half of the z-plane onto the upper half of the w-plane. What is the image of the circle $|z| = 1$ under this transformation?

Sol. Since
$$w = \frac{z}{1-z} \qquad \text{...(1)}$$

$$\Rightarrow \quad z = \frac{w}{1+w} \qquad \text{...(2)}$$

$$\Rightarrow \quad x + iy = \frac{u+iv}{(1+u)+iv}$$

$$= \frac{(u+iv)\{(1+u)-iv\}}{(1+u)^2 + v^2}$$

$$\Rightarrow \quad x + iy = \left\{\frac{u(1+u)+v^2}{(1+u)^2+v^2}\right\} + i\left\{\frac{v(1+u)-uv}{(1+u)^2+v^2}\right\}$$

$$= \left\{\frac{u(1+u)+v^2}{(1+u)^2+v^2}\right\} + \frac{iv}{(1+u)^2+v^2} \qquad \text{...(3)}$$

$$\Rightarrow \quad y = \frac{v}{(1+u)^2+v^2}$$

Now, upper half of z-plane is $y > 0$

$$\Rightarrow \quad \frac{v}{(1+u)^2+v^2} > 0$$

$$\Rightarrow \quad v > 0$$

which is upper half of w-plane. Hence the given transformation (1) maps the upper half of z-plane onto the upper half of w-plane. **Proved.**

Now,
$$|z| = 1 \quad \Rightarrow \left|\frac{w}{1+w}\right| = 1 \qquad \text{(Form 2)}$$

$$\Rightarrow \quad |w| = |1+w|$$

$$\Rightarrow \quad |u+iv| = |1+u+iv|$$

$$\Rightarrow \quad u^2 + v^2 = (1+u)^2 + v^2$$

$$\Rightarrow \quad u^2 + v^2 = 1 + u^2 + 2u + v^2$$

$$\Rightarrow \quad 2u + 1 = 0$$

which is a straight line.

Hence, the image of the circle $|z| = 1$ is the straight line $2u + 1 = 0$. **Ans.**

Example 26: Show that the map of the real axis of the z-plane on the w-plane by the transformation $w = \dfrac{1}{z+i}$ is a circle and find its center and radius.

Sol. Since $$w = \frac{1}{z+i} \quad \text{...(1)}$$

So $$z = \frac{1-iw}{w} \quad \text{...(2)}$$

$$z = \left\{\frac{1-i(u+iv)}{u+iv}\right\}.\frac{u-iv}{u-iv} = \left\{\frac{1+v-iu}{u^2+v^2}\right\}(u-iv)$$

$$\Rightarrow \quad x+iy = \left[\frac{u}{u^2+v^2}\right] + i\left[\frac{-u^2-v(1+v)}{u^2+v^2}\right]$$

Equating real and imaginary parts.

$$x = \frac{u}{u^2+v^2}, y = -\left(\frac{u^2+v^2+v}{u^2+v^2}\right)$$

Real axis of the z-plane is, $y = 0$

$$\Rightarrow \quad -\left(\frac{u^2+v^2+v}{u^2+v^2}\right) = 0 \quad \Rightarrow \quad u^2+v^2+v = 0. \quad \textbf{Proved.} \quad \text{...(3)}$$

Comparing equation (3) with $u^2+v^2+2gu+2fv+c=0$, we get

$$g = 0, f = \frac{1}{2}, c = 0$$

$$\text{center} = (-g - f) = \left(0, -\frac{1}{2}\right)$$

and $$\text{radius} = \sqrt{g^2+f^2-c} = \sqrt{0+\frac{1}{4}-0} = \frac{1}{2}$$

Hence, the equation (3) represents a circle with its center at $\left(0, \frac{-1}{2}\right)$ and radius $\frac{1}{2}$ in w-plane.

Ans.

Example 27: Show that both the transformations, $w = \dfrac{z+i}{z-i}$ and $w = \dfrac{i+z}{i-z}$ transform $|w| \le 1$ into the lower half plane $I(z) \le 0$.

Sol. If $w = \dfrac{z+i}{z-i}$

then $$w\bar{w} - 1 = \frac{z+i}{z-i} \cdot \frac{\bar{z}-i}{\bar{z}+i} - 1$$

$$= \frac{i+z}{i-z} \cdot \frac{-i+\bar{z}}{-i+\bar{z}} - 1 \text{ same if } w = \frac{i+z}{i-z}$$

$$= \frac{(z+i)(\bar{z}-i) - (z-i)(\bar{z}+i)}{(\bar{z}+i)(z-i)} = \frac{-2i(z-\bar{z})}{|z-i|^2}$$

$$= \frac{4I(z)}{|z-i|^2} \text{ the denominator being positive.}$$

or $$|w|^2 - 1 = \frac{4I(z)}{|z-i|^2} \text{ for both the transformations.}$$

Hence under both transformations $|w|^2 - 1 \leq 0$ according as $I(z) \leq 0$.

This means that the circle $|w| = 1$ transforms into the real axis, and interior of this circle transforms into the lower half z-plane. **Proved.**

Example 28: In the transformation $z = \dfrac{i-w}{i+w}$, show that the positive half of the w-plane given by $v \geq 0$ corresponds to the circle $|z| \leq 1$ in the z-plane.

Sol. The given transformation is $z = \dfrac{i-w}{i+w}$

i.e., $$z = \frac{i-(u+iv)}{i+(u+iv)} \quad \text{or} \quad z = \frac{-u+i(1-v)}{u+i(1+v)}$$

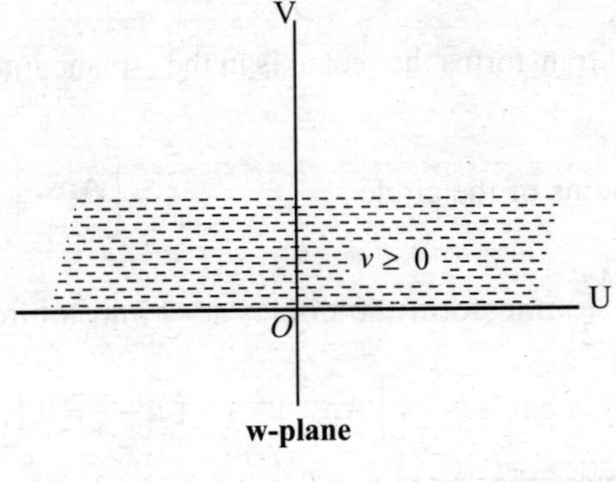

w-plane

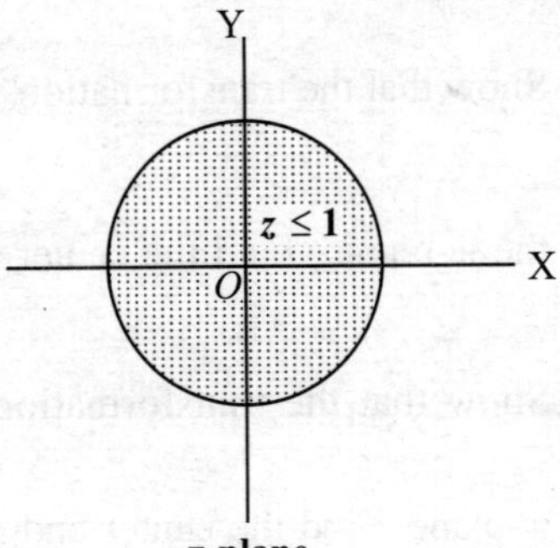

z-plane

$$\therefore \quad |z| = \frac{|-u+i(1-v)|}{|u+i(1+v)|} = \sqrt{\frac{u^2+(1-v)^2}{u^2+(1+v)^2}}$$

i.e., $$|z| = \sqrt{\frac{u^2 + (1-v)^2}{u^2 + (1+v)^2}} \qquad ...(1)$$

Clearly $|z| = 1$ when $v = 0$

Thus real axis in the w-plane corresponds to the unit circle $|z| = 1$ in the z-plane.

And $|z| < 1$ when $v > 0$ is (positive)

Hence the region upper half of the w-plane corresponds to the region interior to the circle $|z| = 1$ in the z-plane. **Proved.**

PROBLEM SET 3.7

1. Discuss the application of the transformation $w = \dfrac{iz+1}{z+i}$ to the areas in the z-plane which are respectively inside and outside the unit circle with its center at the origin.
2. Find the bi-linear transformation which maps $z_1 = 0$, $z_2 = 1$, $z_3 = \infty$, onto $w_1 = i$, $w_2 = -1$, $w_3 = -i$, respectively. $\left[\textbf{Ans. } w = \dfrac{z-i}{iz-1}\right]$
3. Find the image of $|z - 2i| = 2$ under the mapping $w = \dfrac{1}{z}$. $\left[\textbf{Ans. } v + \dfrac{1}{4} = 0 \text{ or } 4v + 1 = 0\right]$
4. Find the bi-linear transformation which maps 1, i, –1 to 2, i, –2 respectively. Find the fixed and critical points of the transformation. **[Ans.** i, $2i$]
5. Show that the transformation $w = \dfrac{i(1-z)}{(1+z)}$ maps the circle $|z| = 1$ into the real axis of the w-plane and the interior of the circle $|z| < 1$ into the upper half of the w-plane.
6. Show that the transformation $w = \dfrac{iz+2}{4z+i}$ transforms the real axis in the z-plane into circle in the w-plane. Find the center and the radius of the circle. $\left[\textbf{Ans. } \left(0, \dfrac{7}{8}\right), \dfrac{9}{8}\right]$
7. Show that the transformation $w = \dfrac{5-4z}{4z-2}$ transform the circle $|z| = 1$ into a circle in the w-plane. Find the center and radius of this circle. $\left[\textbf{Ans. } u^2 + v + u - \tfrac{3}{4} = 0, \left(-\tfrac{1}{2}, 0\right), 1\right]$
8. Find the bi-linear transformation which maps the points $z_1 = 0, z_2 = -i, z_3 = 2i$ into the points $w = 5i, \infty, -\tfrac{i}{3}$ respectively. $\left[\textbf{Ans. } w = \dfrac{3z-5i}{iz-1}\right]$

UNIT IV

Statistics and Probability

4.1 MOMENTS

In statistics the term moment is used to describe the properties of distribution empirical or theoretical. They represent a convenient and unifying method for summarizing many descriptive statistical measures such as measures of central tendency, measures of dispersions and measures of skewness and kurtosis.

Definition: The moments of a frequency are the mean values of the powers of the variable x. The rth moment μ_r for an individual series $x_1, x_2, x_3, \ldots\ldots, x_n$ about the mean $\bar{x}$ is given by

$$\mu_r = \frac{1}{n}\sum (x_i - \bar{x})^r \; ; \; r = 0, 1, 2, 3\ldots..$$

or
$$\mu_r = \frac{1}{n}\sum d^r$$

and the moments for a frequency distribution of a variable x about the mean $\bar{x}$ is defined as

$$\mu_r = \frac{1}{N}\sum f_i (x_i - \bar{x})^r \; ; \; r = 0, 1, 2, 3\ldots..$$

or
$$\mu_r = \frac{1}{N}\sum f_i d^r, \quad \text{where} \quad N = \sum f_i$$

(1) For $r = 0$,
$$\mu_0 = \frac{1}{N}\sum f_i (x_i - \bar{x})^0 = \frac{1}{N}\sum f_i = 1$$

$$\Rightarrow \quad \boxed{\mu_0 = 1}$$

(2) For $r = 1$,
$$\mu_1 = \frac{1}{N}\sum f_i (x_i - \bar{x}) = \frac{1}{N}\sum f_i x_i - \bar{x}$$

$$\mu_1 = \bar{x} - \bar{x} = 0$$

$$\Rightarrow \quad \boxed{\mu_1 = 0}$$

(3) For $r = 2$, $$\mu_2 = \frac{1}{N}\sum f_i (x_i - \bar{x})^2$$

$$\mu_2 = (\text{S.D.})^2 = \text{variance} = \sigma^2$$

$\Rightarrow$ $$\boxed{\text{variance} = \sigma^2}$$

4.1.1 Moments about a Number

The moment about any number ($x = a$) is defined as

$$\mu'_r = \frac{1}{N}\sum f_i (x_i - a)^r, \; r = 0, 1, 2, \ldots\ldots$$

or $$\mu'_r = \frac{1}{N}\sum f_i d_i^r c^r \text{ where } c = \text{Equal interval and } d_i^r = \frac{(x_i - a)^r}{c}$$

1. For $r = 0$, $$\mu'_0 = \frac{1}{N}\sum f_i (x_i - a)^0 = 1$$

$$\boxed{\mu'_0 = 1}$$

2. For $r = 1$, $$\mu'_1 = \frac{1}{N}\sum f_i (x_1 - a)$$

$$= \frac{1}{N}\sum f_i x_i - \frac{\sum f_i}{N} a$$

$$= \bar{x} - a$$

and so on $$\boxed{\mu'_1 = \bar{x} - a}$$

4.1.2 Relation between μ_r and μ'_r [*U.T.U. 2008*]

Since $$\mu_r = \frac{1}{N}\sum f_i (x_i - \bar{x})^r = \frac{1}{N}\sum f_i \{(x_i - a) - (\bar{x} - a)\}^r$$

$$\mu_r = \frac{1}{N}\sum f_i \{(x_i - a) - \mu'_1\}^r \qquad (\mu'_1 = \bar{x} - a)$$

$$= \frac{1}{N}\sum f_i \left\{(x_i - a)^r - {}^rc_1 (x_i - a)^{r-1}\mu'_1 + {}^rc_2 (x_i - a)^{r-2}\mu_1'^2 + \ldots\ldots\ldots + (-1)^r \mu_1'^r\right\}$$

$\Rightarrow$ $$\mu_r = \mu'_r - {}^rc_1 \mu'_{r-1}\mu'_1 + {}^rc_2 \mu'_{r-2}\mu_1'^2 + \ldots + (-1)^r \mu_1'^r$$

For $r = 2, 3, 4, \ldots\ldots$

$$\boxed{\mu_2 = \mu'_2 - \mu_1'^2}$$

$$\boxed{\mu_3 = \mu_3' - 3\mu_2'\mu_1' + 2\mu_1'^3}$$

$$\boxed{\mu_4 = \mu_4' - 4\mu_3'\mu_1' + 6\mu_2'\mu_1'^2 - 3\mu_1'^4}$$

4.1.3 Moments about the Origin

The moment about origin is defined as

$$v_r = \frac{1}{N}\sum f_i x_i^r$$

For $r = 0, 1, 2, 3, \ldots$, we get

$$v_0 = \frac{1}{N}\sum f_i x_1^0 = \frac{N}{N} = 1 \qquad \Big| \text{As } \sum f_i = N$$

$$v_1 = \frac{1}{N}\sum f_i x_i = \bar{X}$$

$$v_2 = \frac{1}{N}\sum f_i x_i^2$$

and so on.

4.1.4 Relation between μ_r, μ'_r and ν_r

We have $$v_r = \frac{1}{N}\sum f_i x_i^r = \frac{1}{N}\sum f_i (x_1 - a + a)^r$$

$$= \frac{1}{N}\sum f_i \left[(x_1 - a)^r + {}^rC_1 (x_1 - a)^{r-1} \cdot a + \ldots + a^r\right]$$

$$v_r = \mu_r' + {}^r c_1 \mu_{r-a}' a + \ldots + a^r \qquad \ldots(1)$$

The equation (1) is the relation between v_r and u'_r

For $r = 1 = 1, 2, \ldots$, we get

$$v_1 = \mu_1' + a\mu_0' \Rightarrow \boxed{v_1 = \mu_1' + a} \qquad \Big| \text{As } \mu_0' = 1$$

$$v_2 = \mu_2' + 2au_1' + a^2 u_0' \Rightarrow \boxed{v_2 = \mu_2' + 2a\mu_1' + a^2} \quad \text{and so on.}$$

If we take $a = \bar{x}$ in equation (1), we get

$$v_1 = \mu_1 + \mu_0 \bar{x} \Rightarrow \boxed{v_1 = \bar{x}} \quad (\text{As } \mu_1 = 0, \mu_0 = 1)$$

$$v_2 = \mu_2 + {}^2c_1 \mu_1 \bar{x} + {}^2c_2 \mu_0 \bar{x}^2 \Rightarrow \boxed{v_2 = \mu_2 + \bar{x}^2}$$

Similarly

$$v_3 = \mu_3 + 3\mu_2\bar{x} + \bar{x}^3$$

$$v_4 = \mu_4 + 4\mu_3\bar{x} + 6\mu_2\bar{x}^2 + \bar{x}^4$$

4.2 MOMENT GENERATING FUNCTIONS

The moment generating function of a random variable x is denoted by $M(t)$ and defined as follows:

$$M(t) = E(e^{tx})$$

where t is a real const.

The function $M(t)$ generates moments, that is why it is called the moment generating function.

(*a*) **Moment generating function for discrete random variable:** The moment generating function of the discrete probability distribution of a random variable x about any point a is defined as

$$M_a(t) = E(e^{t(x-a)}) = \sum_{i=1}^{n} e^{t(x-a)} \cdot p_i$$

(*b*) **For continuous probability distribution:** The moment generating function for continuous probability distribution is defined as:

$$M(t) = E(e^{tx}) = \int e^{tx} f(x)\,dx$$

Here $E(e^{tx})$ is the moment generating function of $f(x)$.

Therefore $\boxed{M_a(t) = \int e^{t(x-a)} f(x)dx. \text{ if } a = 0, \text{ then } M_o(t) = \int e^{xt} f(x)dx}$

Moment generating function about various points (For discrete random variable)

(*i*) $M_o(t) = \sum e^{tx} \cdot pi$, is called m.g.f. about origin

(*ii*) $M_a(t) = \sum e^{t(x-a)} \cdot pi$, is called m.g.f. about any point a

(*iii*) $M_{\bar{x}}(t) = \sum e^{t(x-\bar{x})} \cdot pi$, is called m.g.f. about mean.

Result 1: The rth moment of x about any point a is the coefficient of $\frac{t^r}{\lfloor r}$ in the expansion of

$$\sum_i e^{t(x-a)} \cdot pi$$

i.e., $\mu'_r =$ Coefficient of $\frac{t^r}{\lfloor r}$

We know that $M_a(t) = \sum_i e^{t(x-a)} \cdot pi$

$$M_a(t) = \sum_i \left[1 + \frac{t(x-a)}{\lfloor 1} + \frac{t^2(x-a)^2}{\lfloor 2} + \ldots + \frac{t^r(x-a)^2}{\lfloor r} + \ldots\right] pi$$

$$= \sum_i pi + t\sum pi(x-a) + \frac{t^2}{\lfloor 2}\sum pi(x-a)^2 + \ldots + \frac{t^r}{\lfloor r}\sum pi(x-a)^r$$

$$= 1 + \mu_1' t + \mu_2' \frac{t^2}{\lfloor 2} + \ldots + \mu_r' \cdot \frac{t^r}{\lfloor r} + \ldots$$

Coefficient of $\frac{t^r}{\lfloor r}$ is μ_r'

$$\Rightarrow \qquad \boxed{\mu_r' = \text{Coeff. of } \frac{r^r}{\lfloor r}}$$

Result 2: Similarly, central moment μ_r is also the coefficient of $\frac{t^r}{\lfloor r}$ in the expansion of $\sum_i e^{t(x-\bar{x})} \cdot pi$

Generation of moments by m.g.f.

There are two methods to find moments by m.g.f.

1. By expansion: Result 1 and Result 2 shows the moments.
2. By differentiation:

We know that

$$M_a(t) = 1 + \mu_1' \cdot t + \mu_2' \frac{t^2}{\lfloor 2} + \ldots + \mu_r' \cdot \frac{t^r}{\lfloor r}$$

Differentiating r times w.r. to 't' and putting $t = 0$, we get

$$\left[\frac{d^r}{dt^r} M_a(t)\right]_{t=0} = \mu_r'.$$

Example 1: Find the moment generating function for triangular distribution defined by

$$f(x) = \begin{Bmatrix} x, 0 \le x \le 1 \\ 2-x, 1 \le x \ge 2 \end{Bmatrix}.$$

Sol. Here $\qquad M_0 = \int_0^1 xe^{tx}\,dx + \int_1^2 (2-x)e^{tx}\,dx$

$$= \left[\left(\frac{xe^{tx}}{t}\right)_0^1 - \int_0^1 1 \cdot \frac{e^{tx}}{t}\,dx + \left\{(2-x)\frac{e^{tx}}{t}\right\}_1^2 - \int_1^2 (-1) \cdot \frac{e^{tx}}{t} \cdot dx\right]$$

$$= \frac{e^t}{t} - \frac{1}{t}\left(\frac{e^{tx}}{t}\right)_0^1 - \frac{e^t}{t} + \frac{1}{t}\left(\frac{e^{tx}}{t}\right)_1^2$$

$$= -\frac{1}{t^2}(e^t - 1) + \frac{1}{t^2}(e^{2t} - e^t) = \frac{1}{t^2}(e^{2t} - 2e^t + 1)$$

$$= \frac{\left(e^t - 1\right)^2}{t^2}. \quad \textbf{Ans.}$$

Example 2: Find the moment generating function of the exponential distribution.

$$f(x) = \frac{1}{c} e^{-\frac{x}{c}}, 0 \le x \le \infty, c > 0$$

Sol. The moment generating about origin is given by

$$M_o(t) = \int_0^\infty e^{tx} \cdot \frac{1}{c} e^{-\frac{x}{c}} dx = \frac{1}{c}\int_0^\infty e^{\left(t - \frac{1}{c}\right)x} dx = (1 - ct)^{-1} = 1 + ct + c^2t^2 + \ldots. \quad \textbf{Ans.}$$

Example 3: Prove that moment generating function about 'a' is equal to e^{-at} times the m.g.f. about the origin *i.e.*, $M_a(t) = e^{-at} M_o(t)$.

Sol. We know that

$$M_a(t) = \sum_i e^{t(x_i - a)} \cdot p_i = \sum_i e^{tx_i} \cdot e^{-ta} \cdot p_i$$

$$= e^{-at} \cdot \sum_i e^{tx} \cdot p_i$$

or $$M_a(t) = e^{-at} \cdot M_o(t) \qquad \Big| \text{ As } M_o(t) = \sum e^{tx} \cdot p_i. \quad \textbf{Ans.}$$

Example 4: Find the m.g.f. of the random variable x if

$$f(x) = \frac{1}{3}, -1 < x < 2$$

$$= 0, \quad \text{elsewhere.}$$

Sol. $$M(t) = \int_{-\infty}^{-1} f(x)e^{tx} \cdot dx + \int_{-1}^{2} f(x)e^{tx} \cdot dx + \int_{2}^{\infty} f(x)e^{tx} \cdot dx$$

$$= 0 + \int_{-1}^{2} \frac{1}{3} e^{tx} \cdot dx + 0$$

$$= \frac{1}{3}\left[\frac{e^{tx}}{t}\right]_{-1}^{2} = \frac{1}{3t}(e^{2t} - e^{-t})$$

Here $$M(t) = \frac{1}{3i}(e^{2t} - e^{-t}), t \ne 0$$

If $t = 0$, then

$$M_o(t) = \int_{-1}^{2} \frac{1}{3} e^o dx = \frac{1}{3}[x]_{-1}^{2} = \frac{1}{3}[2+1] = 1$$

Thus the m.g.f. is $\quad M(t) = \dfrac{e^{2t} - e^{-t}}{3t}, t \neq 0$

$$= 1 \qquad t = 0. \quad \textbf{Ans.}$$

Example 5: If $\mu_r = \lfloor r \cdot 3^r$, then find m.g.f. about mean.

Sol. We have $\mu_r = \lfloor r \cdot 3^r$

$\therefore$ Putting $\quad r = 1, 2, 3, \ldots$

$$\mu_1 = 3$$

$$\mu_2 = \lfloor 2 \cdot 3^2$$

$$\mu_3 = \lfloor 3 \cdot 3^3 \text{ and so an}$$

we know that

$$M_{\bar{x}}(t) = 1 + \mu_1 t + \mu_2 \frac{t^2}{\lfloor 2} + \mu_3 \frac{t^3}{\lfloor 3} + \ldots$$

or

$$M_{\bar{x}}(t) = 1 + 3t + (\lfloor 2 \cdot 3^2) \cdot \frac{t^2}{\lfloor 2} + (\lfloor 3 \cdot 3^3) \cdot \frac{t^3}{\lfloor 3} + \ldots$$

$$= 1 + 3t + (3t)^2 + \ldots$$

$\Rightarrow$

$$M_{\bar{x}}(t) = (1 - 3t)^{-1}. \quad \textbf{Ans.}$$

Example 6: Find the m.g.f. of a random variable whose probability function is given by $P(x) = p(1-p)^{x-1}; x = 1, 2, \ldots$.

Sol. We know that

$$M_x(t) = \sum_{x=1}^{\infty} e^{tx} \cdot p(1-p)^{x-1}$$

$$= \frac{p}{(1-p)} \sum_{x=1}^{\infty} e^{tx} \cdot (1-p)^x$$

$$= \frac{p}{1-p}[e^t(1-p) + e^{2t}(1-p)^2 + \ldots]$$

$$= \frac{p}{1-p}[\{1+e^t(1-p)+e^{2t}(1-p)^2+\ldots\}-1]$$

$$M_x(t) = \frac{p}{1-p}[\{1-(1-p)e^t\}^{-1}-1]$$

$$= \frac{p}{q}[(1-qe^t)^{-1}-1] \quad \Big| \text{ As } 1-p=q. \qquad \textbf{Ans.}$$

Example 7: Find the first four moments about mean for the following individual series 1, 5, 6, 8, 9, 13.

Sol. The mean of the given series is

$$\bar{x} = \frac{1+5+6+8+9+13}{6} = \frac{42}{6} = 7$$

Now, we construct the following table:

x	$x-\bar{x}$	$(x-\bar{x})^2$	$(x-\bar{x})^3$	$(x-\bar{x})^4$
1	– 6	36	– 216	1296
5	– 2	4	– 8	16
6	– 1	1	– 1	1
8	1	1	1	1
9	2	4	8	16
13	6	36	216	1296
	$\Sigma(x-\bar{x}) = 0$	$\Sigma(x-\bar{x})^2 = 82$	$\Sigma(x-\bar{x})^3 = 0$	$\Sigma(x-\bar{x})^4 = 2626$

Here $n = 6$

$\therefore$

$$\mu_1 = \frac{\Sigma(x-\bar{x})}{n} = \frac{0}{6} = 0$$

$$\mu_2 = \frac{\Sigma(x-\bar{x})^2}{n} = \frac{82}{6} = 13.67$$

$$\mu_3 = \frac{\Sigma(x-\bar{x})^3}{n} = \frac{0}{6} = 0$$

$$\mu_4 = \frac{\Sigma(x-\bar{x})^4}{n} = \frac{2626}{6} = 437.67.$$

Example 8: Calculate the variance and third central moment from the following data:

x_i	0	1	2	3	4	5	6	7	8
f_i	1	9	26	59	72	52	29	7	1

Sol. Here we find the moments about any point and then we find the central moments (moments about mean) otherwise the calculation will be so lengthy.

Let $a = 4$

x_i	f_i	$(x_i - 4)$	$f_i(x_i - 4)$	$f_i(x_i - 4)^2$	$f_i(x_i - 4)^3$
0	1	– 4	– 4	16	– 64
1	9	– 3	– 27	81	– 243
2	26	– 2	– 52	104	– 208
3	59	– 1	– 59	59	– 59
4	72	0	0	0	0
5	52	1	52	52	52
6	29	2	58	116	232
7	7	3	21	63	189
8	1	4	4	16	64
	$N = 256$		$\Sigma f_i(x_i - 4) = -7$	$\Sigma f_i(x_i - 4)^2 = 507$	$\Sigma f_i(x_i - 4)^3 = -37$

The moments about $a = 4$ obtained as follows:

$$\mu_1' = \frac{1}{N}\sum_{i=1}^{n} f_i(x_i - 4) = \frac{-7}{256} = -0.027$$

$$\mu_2' = \frac{1}{N}\sum_{i=1}^{n} f_i(x_i - 4)^2 = \frac{507}{256} = 1.9804$$

$$\mu_3' = \frac{1}{N}\sum_{i=1}^{n} f_i(x_i - 4)^3 = \frac{-37}{256} = -0.144$$

Now we find the central moments (moments about mean) $\mu_1 = 0$ (always)

$$\mu_2 = \mu_2' - \mu_1'^2 = 1.9804 - (-0.027)^2 = \mathbf{1.97967}$$

$\therefore$ $$\text{Variance} = \mu_2 = \mathbf{1.97967}$$

$$\mu_3 = \mu_3' - 3\mu_2'\mu_1' + 2\mu_1'^3 = -0.144 - 3(1.9804)(-0.027) + 2(-0.027)^3 = \mathbf{0.3043}$$

Sheppard's Correction for Moments

Correction which must be applied to the measured moments μ_r obtained from normally distributed data which have been stored in order to obtain correct moments μ_r^*. The corrected interpretations of the second, third and fourth moments are then

$$\mu_2^* = \mu_2 - \frac{h^2}{12} \quad \text{...(i)}$$

$$\mu_3^* = \mu_3 \quad \text{...(ii)}$$

$$\mu_4^* = \mu_4 - \frac{1}{2}\mu_2 + \frac{7}{240}h^2 \quad \text{...(iii)}$$

where h is the class interval.

Example 9: Find out the first four moments about $a = 45$ from the following data regarding ages of 250 persons.

Ages between	0 – 10	10 – 20	20 – 30	30 – 40	40 – 50	50 – 60	60 – 70	70 – 80	80 – 90
Frequency	15	20	25	24	12	31	61	12	50

Sol.

Ages between	*Frequency* f_i	x	$u_i = \frac{x-45}{10}$	f_iu_i	$f_iu_i^2$	$f_iu_i^3$	$f_iu_i^4$
0 – 10	15	5	– 4	– 60	240	– 960	3840
10 – 20	20	15	– 3	– 60	180	– 540	1620
20 – 30	25	25	– 2	– 50	100	– 200	400
30 – 40	24	35	– 1	– 24	24	– 24	24
40 – 50	12	45	0	0	0	0	0
50 – 60	31	55	1	31	31	31	31
60 – 70	61	65	2	122	244	488	976
70 – 80	12	75	3	36	108	324	972
80 – 90	50	85	4	200	800	3200	12800
	$N = 250$	–	–	$\Sigma f_iu_i = 195$	$\Sigma f_iu_i^2 = 1727$	$\Sigma f_iu_i^3 = 2319$	$\Sigma f_iu_i^4 = 20663$

Now $$\mu_1' = \frac{1}{N}\sum_{i=1}^{n} f_iu_i \times h = \frac{195}{250} \times 10 = \mathbf{7.8}$$

$$\mu'_2 = \frac{1}{N}\sum_{i=1}^{n} f_i u_i^2 \times h^2 = \frac{1727}{250} \times 100 = \mathbf{690.8}$$

$$\mu'_3 = \frac{1}{N}\sum_{i=1}^{n} f_i u_i^3 \times h^3 = \frac{2319}{250} \times 1000 = \mathbf{9276}$$

$$\mu'_4 = \frac{1}{N}\sum_{i=1}^{n} f_i u_i^4 \times h^4 = \frac{20663}{250} \times 10000 = \mathbf{826520}.$$

Example 10: The first four moments of a distribution about $x = 2$ are 1, 2.5, 5.5 and 16, find the four moments about

(*i*) The mean (*ii*) Zero.

Sol. We have $\mu'_1 = 1, \mu'_2 = 2.5, \mu'_3 = 5.5, \mu'_4 = 16$ where $\boldsymbol{a} = \mathbf{2}$.

(*i*) **Moments about mean**

$$\mu_1 = \mathbf{0}$$

$$\mu_2 = \mu'_2 - \mu_1'^2 = 2.5 - 1 = \mathbf{1.5}$$

$$\mu_3 = \mu'_3 - 3\mu'_2\mu'_1 + 2\mu_1'^3 = 5.5 - 3(2.5) + 2 = \mathbf{0}$$

$$\mu_4 = \mu'_4 - 4\mu'_3\mu'_1 + 6\mu'_2\mu_1'^2 - 3\mu_1'^4 = 16 - 4(5.5) + 6(2.5) - 3 = \mathbf{6}$$

(*ii*) **Moments about zero**

$$\nu_1 = \bar{x} = a + \mu'_1 = 2 + 1 = \mathbf{3}$$

$$\nu_2 = \mu_2 + \nu_1^2 = 1.5 + 9 = \mathbf{10.5}$$

$$\nu_3 = \mu_3 + 3\mu_2\nu_1 + \nu_1^3 = 0 + 3(1.5)(3) + 27 = \mathbf{40.5}$$

$$\nu_4 = \mu_4 + 4\mu_3\nu_1 + 6\mu_2\nu_1'^2 + \nu_1^4 = 6 + 4(0)(3) + 6(1.5)(9) + 81 = \mathbf{168}.$$

Example 11: Find the corrected values of the following moments:

$\mu_2 = 164, \mu_3 = 360, \mu_4 = 6200$ and the width of class $h = 10$.

Sol. We have $\mu_2 = 164, \mu_3 = 360, \mu_4 = 6200$ and $h = 10$.

It μ_2^*, μ_3^* and μ_4^* are the corrected values of μ_2, μ_3 and μ_4 respectively then

$$\mu_2^* = \mu_2 - \frac{h^2}{12} = 164 - \frac{100}{12} = \mathbf{156.67}$$

$$\mu_3^* = \mu_3 = \mathbf{360}$$

$$\mu_4^* = \mu_4 - \frac{1}{2}\mu_2 + \frac{7}{240}h^2 = 6200 - \frac{164}{2} + \frac{7}{240} \times 100 = \mathbf{6120.92}$$

Pearson's β and γ coefficients

Karl Pearson defined the following four coefficients, based upon four moments about the mean.

β Coefficients:
$$\begin{cases} \beta_1 = \dfrac{\mu_3^2}{\mu_2^3} = \dfrac{\mu_3^2}{\sigma^6} \\ \beta_2 = \dfrac{\mu_4}{\mu_2^2} = \dfrac{\mu_4}{\sigma^4} \end{cases}$$

γ Coefficients:
$$\begin{cases} r_1 = \sqrt{\beta_1} = \dfrac{\mu_3}{\sigma^3} \\ r_2 = \beta_2 - 3 = \dfrac{\mu_4}{\mu_2^2} - 3 \end{cases}$$

These are pure numbers and they provide information about the shape of the curve obtained from the frequency distribution.

4.3 SKEWNESS

Skewness is a measure of symmetry, or more precisely, the lack of symmetry. A distribution, or data set, is symmetric if it looks the same to the left and right of the centre point where the mean, median and mode are equal to each other. If a distribution fails to be symmetric then we say that it is **skewed distribution.** The skewness for a normal distribution is zero.

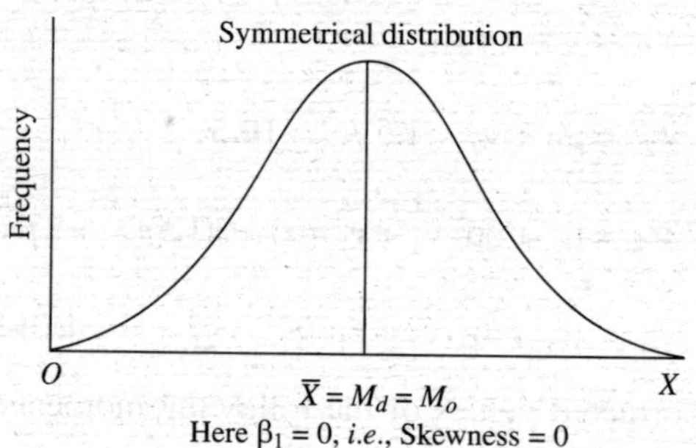

Here $\beta_1 = 0$, *i.e.*, Skewness = 0

If the curves of a variable x_i are distributed symmetrically about the mean which is taken as the origin then for every positive value of $x - \overline{x}$ there correspond a negative equal value. Hence when these values are clubbed (Unite with a common purpose) they continue their signs and cancels on addition

$$\therefore \qquad \mu_3 = \frac{1}{N}\sum_{i=1}^{n} f_i(x_i - \overline{x})^3 = 0$$

Hence
$$\boxed{\beta_1 = \frac{\mu_3^2}{\mu_2^3} = 0}$$

From the above discussion we see that β_1 can be taken as measure of skewness.
This is called **coefficient of skewness.**

4.3.1 Types of Skewness

1. **Positive skewness:** If there are extreme values towards the positive end of a distribution, the distribution is said to be **positively skewed.** In a positively skewed distribution, the mean is greater than the mode.

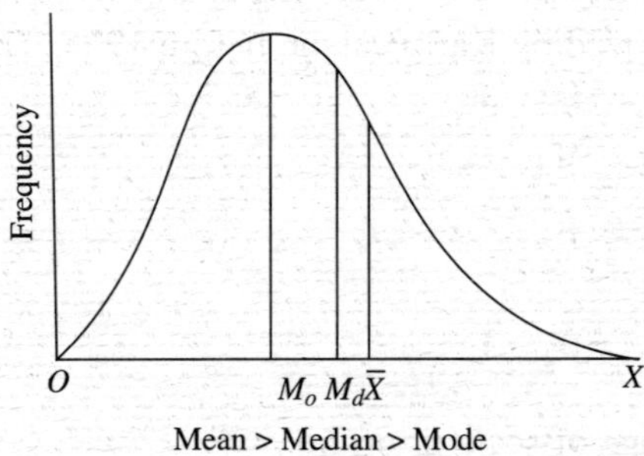

Mean > Median > Mode

2. **Negative skewness:** If there are extreme values towards the negative end of a distribution, the distribution is said to be **negatively skewed.** In a negatively skewed distribution, the mean is less than the mode.

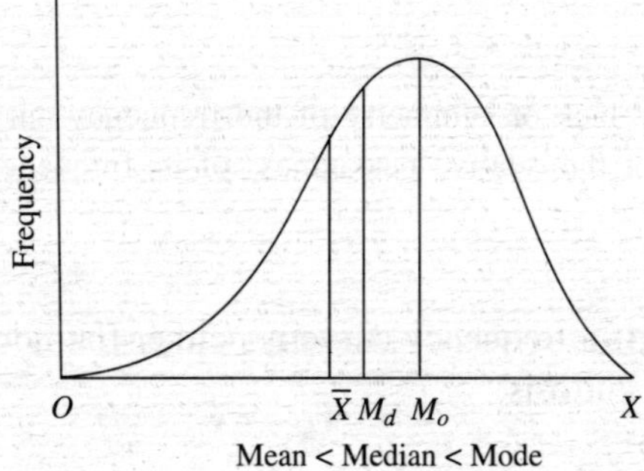

Mean < Median < Mode

4.3.2 Measurement of Skewness

1. If $\beta_1 = 0$, then there is no skewness.
2. If $\beta_1 > 0$, then there is positive skewness.
3. If $\beta_1 < 0$, then there is negative skewness.

Karl Pearson Coefficient of Skewness S_{KP}

The Karl Pearson's coefficient skewness is defined as

$$S_{KP} = \frac{\text{Mean} - \text{Mode}}{\text{s.d.}}$$

We know that

$$\text{Mean} - \text{Mode} = 3(\text{Mean} - \text{Median})$$

Hence Karl Pearson's coefficient of skewness is defined in terms of median as

$$S_{KP} = \frac{3(\text{Mean} - \text{Median})}{\text{s.d.}}.$$

4.4 KURTOSIS

Kurtosis (Greek word, meaning bulging) is a measure of the peakedness of its frequency curve of the probability distribution of a real-valued random variable. Thus Kurtosis enables us to have an idea about the flatness or peakedness of a frequency curve.

4.4.1 Measurement of Kurtosis

Measurement of kurtosis popularized by Karl Pearson before 1905. The usual measure is β_2 given by

$$\beta_2 = \frac{\mu_4}{\mu_2^2}$$

or

$$\gamma_2 = \beta_2 - 3$$

It is invariant under a change of scale or origin.

Alternative definition

For a normal distribution $\beta_2 = 3$; an alternative definition of kurtosis reduces the ratio by 3 to give a value of 0 for the normal.

Note: Skewness measures the lack of symmetry of the frequency curve of a distribution whenever kurtosis is a measure of the relative peakedness of its frequency curve.

4.4.2 Types of Kurtosis

1. Mesokurtic (zero kurtosis): If a frequency curve is neither flat nor peaked (normal curve), then it is said to be mesokurtic or zero kurtosis.

For Mesokurtic

$$\boxed{\beta_2 = 3 \text{ or } \gamma_2 = 0}$$

2. Platykurtic (negative kurtosis): If a frequency curve is flatter than the normal curve then it said to be platykurtic or negative kurtosis.

For Platykurtic

$$\boxed{\beta_2 < 3 \text{ or } \gamma_2 < 0}$$

3. Leptokurtic (positive kurtosis): If a frequency curve is more peaked than a normal curve then it said to be leptokurtic or positive kurtosis.

For Leptokurtic

$$\boxed{\beta_2 > 3 \text{ or } \gamma_2 > 0}$$

The graphical representation is given below:

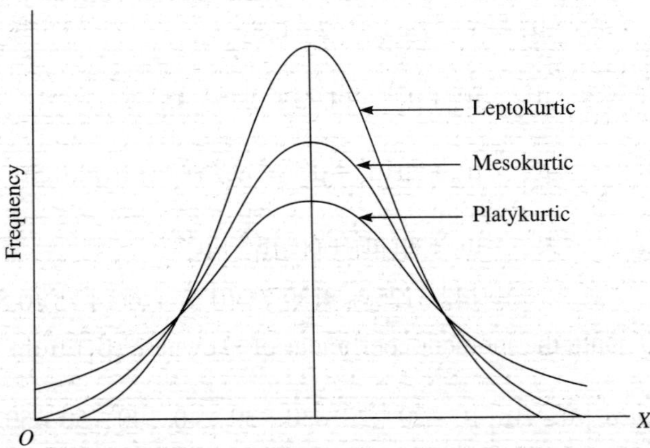

Example 12: The first four moments of a distribution about the value 4 of the variable are – 1.5, 17, – 30 and 108. Find the moments about mean, Karl Pearson coefficient β_1 and β_2. Find also the moments about (*i*) the origin (*ii*) about $x = 2$. [*U.T.U. 2008*]

Sol. We have $\mu_1' = -1.5, \mu_2' = 17, \mu_3' = -30, \mu_4' = 108$ where $\boldsymbol{a = 4}$

Moments about Mean

$$\mu_1 = 0$$

$$\mu_2 = \mu_2' - \mu_1'^2 = 17 - 2.25 = \mathbf{14.75}$$

$$\mu_3 = \mu_3' - 3\mu_2'\mu_1' + 2\mu_1'^3 = -30 - 3(17)(-1.5) + 2(-1.5)^3 = \mathbf{39.75}$$

$$\mu_4 = \mu_4' - 4\mu_3'\mu_1' + 6\mu_2'\mu_1'^2 - 3\mu_1'^4$$

$$= 108 - 4(-30)(-1.5) + 6(17)(-1.5)^2 - 3(1.5)^4 = \mathbf{142.3125}$$

$$\therefore \quad \beta_1 = \frac{\mu_3^2}{\mu_2^3} = \frac{(39.75)^2}{(14.75)^3} = \mathbf{0.4926}$$

$$\beta_2 = \frac{\mu_4}{\mu_2^2} = \frac{142.3125}{(14.75)^2} = \mathbf{0.6543.}$$

(i) Moments about origin

$$\nu_1 = \bar{x} = a + \mu_1' = 4 - 1.5 = \mathbf{2.5}$$

$$\nu_2 = \mu_2 + \nu_1^2 = 14.75 + (2.5)^2 = \mathbf{8.5}$$

$$\nu_3 = \mu_3 + 3\mu_2\nu_1 + \nu_1^3 = 39.75 + 3(14.75)(2.5) + (2.5)^3 = \mathbf{166}$$

$$\nu_4 = \mu_4 + 4\mu_3\nu_1 + 6\mu_2\nu_1^2 + \nu_1^4$$

$$= 142.3125 + 4(39.75)\ (2.5) + 6(14.75)\ (2.5)^2 + (2.5)^3 = \mathbf{1108.5625}$$

(ii) Moments about x = 2

Here $a = 2$

$$\therefore \quad \mu_1' = \bar{x} - a = 2.5 - 2 = \mathbf{0.5}$$

$$\mu_2' = \mu_2 + \mu_1'^2 = 14.75 + 0.25 = \mathbf{15}$$

$$\mu_3' = \mu_3 + 3\mu_2\mu_1' + \mu_1'^3 = 39.75 + 3(14.75)(0.5) + (0.5)^3 = \mathbf{62}$$

$$\mu_4' = \mu_4 + 4\mu_3\mu_1' + 6\mu_2\mu_1'^2 + \mu_1'^4$$

$$= 142.3125 + 4(39.75)(0.5) + 6(14.75)(0.5)^2 + (0.5)^4 = \mathbf{244.}$$

Example 13: Compute the moment coefficient of skewness (β_1) from the following data:

Marks obtained	0 – 10	10 – 20	20 – 30	30 – 40	40 – 50	50 – 60	60 – 70
Frequency	6	12	22	24	16	12	8

Sol. Here we construct the following table:

Let $a = 35$

Class intervals	*Frequency* (f_i)	*Mid values* (x_i)	$u_i = \dfrac{x_i - 35}{10}$	$f_i u_i$	$f_i u_i^2$	$f_i u_i^3$
0 – 10	6	5	– 3	– 18	54	– 162
10 – 20	12	15	– 2	– 24	48	– 96
20 – 30	22	25	– 1	– 22	22	– 22
30 – 40	24	35	0	0	0	0
40 – 50	16	45	1	16	16	16
50 – 60	12	55	2	24	48	96
60 – 70	8	65	3	24	72	216
Total	$N = 100$	–	–	$\Sigma f_i u_i = 0$	$\Sigma f_i u_i^2 = 260$	$\Sigma f_i u_i^3 = 48$

Now $$\mu_2' = \frac{1}{N}\sum_{i=1}^{n} f_i u_i^2 \times h^2 = \frac{260}{100} \times 100 = \mathbf{260}$$

$$\mu_3' = \frac{1}{N}\sum_{i=1}^{n} f_i u_i^3 \times h^3 = \frac{48}{100} \times 1000 = \mathbf{480}$$

Therefore $$\beta_1 = \frac{\mu_3'^2}{\mu_2'^3} = \frac{(480)^2}{(260)^3} = \mathbf{0.01}$$

$\Rightarrow \quad \beta_1 > 0.$

Hence the distribution is slightly positively skewed.

Example 14: Find the coefficient of skewness and kurtosis from the following data:

x_i	4.5	14.5	24.5	34.5	44.5	54.5	64.5	74.5	84.5	94.5
f_i	1	5	12	22	17	9	4	3	1	1

Sol. Let a = 44.5

x_i	*Frequency* (f_i)	$u_i = \dfrac{x-44.5}{10}$	$f_i u_i$	$f_i u_i^2$	$f_i u_i^3$	$f_i u_i^4$
4.5	1	– 4	– 4	16	– 64	256
14.5	5	– 3	– 15	45	– 135	405
24.5	12	– 2	– 24	48	– 96	192
34.5	22	– 1	– 22	22	– 22	22
44.5	17	0	0	0	0	0
54.5	9	1	9	9	9	9
64.5	4	2	8	16	32	64
74.5	3	3	9	27	81	243
84.5	1	4	4	16	64	256
94.5	1	5	5	25	125	625
Total	N = 75	–	$\Sigma f_i u_i$ = – 30	$\Sigma f_i u_i^2$ = 224	$\Sigma f_i u_i^3$ = – 6	$\Sigma f_i u_i^4$ = 2072

Now

$$\mu_1' = \frac{1}{N}\sum_{i=1}^{n} f_i u_i \times h = \frac{-30}{75}\times 10 = \mathbf{-4}$$

$$\mu_2' = \frac{1}{N}\sum_{i=1}^{n} f_i u_i^2 \times h^2 = \frac{224}{75}\times 100 = \mathbf{299}$$

$$\mu_3' = \frac{1}{N}\sum_{i=1}^{n} f_i u_i^3 \times h^3 = \frac{-6}{75}\times 1000 = \mathbf{-80}$$

$$\mu_4' = \frac{1}{N}\sum_{i=1}^{n} f_i u_i^4 \times h^4 = \frac{2072}{75}\times 10000 = \mathbf{276300}$$

Thus moments about mean

$$\mu_2 = \mu_2' - \mu_1'^2 = 299 - (-4)^2 = \mathbf{283}$$

$$\mu_3 = \mu'_{-3}\mu'_1 - 2\mu_1'^3 = -80 - 3(299)(-4) + 2(-4)^3 = \mathbf{3380}$$

$$\mu_4 = \mu'_4 - 4\mu'_3\mu'_1 + 6\mu'_2\mu_1'^2 - 3\mu_1'^4$$

$$= 276300 - 4(-80)(-4) + 6(299)(-4)^2 - 3(-4)^4 = \mathbf{302956}$$

$$\therefore \qquad \beta_1 = \frac{\mu_3^2}{\mu_2^3} = \frac{(3380)^2}{(283)^3} = \mathbf{0.504}$$

$$\beta_2 = \frac{\mu_4}{\mu_2^2} = \frac{302956}{(283)^3} = \mathbf{3.7827}$$

Hence $\beta_1 > 0$, the distribution is slightly positively skewed.

And $\beta_2 > 3$, the distribution is leptokuritc.

Example 15: The first four central moments of a distribution are 0, 2.5, 0.7 and 18.75. Examine the skewness and kurtosis of the distribution.

Sol. We have $\mu_1 = 0$, $\mu_2 = 2.5$, $\mu_3 = 0.7$ and $\mu_4 = 18.75$
To examine skewness, we compute

$$\beta_1 = \frac{\mu_3^2}{\mu_2^3} = \frac{(0.7)^2}{(2.5)^3} = \mathbf{0.031}$$

Since $\beta_1 > 0$ the distribution is positively skewed.
Next, kurtosis is given by the coefficient

$$\beta_2 = \frac{\mu_4}{\mu_2^2} = \frac{18.75}{(2.5)^2} = \mathbf{3.0}$$

Since $\beta_1 = 3$. Hence the curve is mesokurtic.

Example 16: Find the moment coefficient of kurtosis for the following data:

Class	40 – 45	45 – 50	50 – 55	55 – 60	60 – 65	65 – 70	70 – 75	75 – 80	80 – 85
Frequency	8	25	38	49	51	32	24	22	18

Sol. Here we are taking $a = 62.5$ and $h = 5$, get the following table:

Class	*Mid value* (x_i)	f_i	$u_i = \dfrac{x - 62.5}{5}$	f_iu_i	$f_iu_i^2$	$f_iu_i^3$	$f_iu_i^4$
40 – 45	42.5	8	– 4	– 32	128	– 512	2048
45 – 50	47.5	25	– 3	– 75	225	– 675	2025
50 – 55	52.5	38	– 2	– 76	152	– 304	608
55 – 60	57.5	49	– 1	– 49	49	– 49	49

60 – 65	32.5	51	0	0	0	0	0
65 – 70	67.5	32	1	32	32	32	32
70 – 75	72.5	24	2	48	96	192	384
75 – 80	77.5	22	3	66	198	594	1782
80 – 85	82.5	18	4	72	288	1152	4608
Total		$N = 267$	–	$\Sigma f_i u_i$ = – 14	$\Sigma f_i u_i^2$ = 1168	$\Sigma f_i u_i^3$ = 430	$\Sigma f_i u_i^4$ = 11536

Now
$$\mu_1' = \frac{1}{N}\sum_{i=1}^{n} f_i u_i \times h = \frac{-14}{267}\times 5 = -\mathbf{0.26}$$

$$\mu_2' = \frac{1}{N}\sum_{i=1}^{n} f_i u_i^2 \times h^2 = \frac{1168}{267}\times 5^2 = \mathbf{109.36}$$

$$\mu_3' = \frac{1}{N}\sum_{i=1}^{n} f_i u_i^3 \times h^3 = \frac{430}{267}\times 5^3 = \mathbf{201.31}$$

$$\mu_4' = \frac{1}{N}\sum_{i=1}^{n} f_i u_i^4 \times h^4 = \frac{11536}{267}\times 5^4 = \mathbf{27003.75}$$

Thus moments about mean

$$\mu_2 = \mu_2' - \mu_1'^2 = 109.36 - (-0.26)^2 = \mathbf{109.29}$$

$$\mu_4 = \mu_4' - 4\mu_3'\mu_1' + 6\mu_2'\mu_1'^2 - 3\mu_1'^4$$
$$= 27003.75 - 4 \times 201.31 \times (-0.26) + 6(109.36)(-0.26)^2$$
$$- 3(-0.26)^4 = \mathbf{27257.40}$$

$$\therefore \quad \beta_2 = \frac{\mu_4}{\mu_2^2} = \frac{26242.86}{(109.29)^2} = \mathbf{2.282}$$

Hence $\beta_2 < 3$, the distribution is platykurtic.

Example 17: For a frequency distribution $\left(\frac{f_i}{x_i}\right)$ prove that $\beta_2 \geq 1$.

Sol. We know that

$$\beta_2 = \frac{\mu_4}{\mu_2^2}$$

To prove $\beta_2 \geq 1$, we must prove that $\mu_4 \geq \mu_2^2$.

Now $$\mu_4 - \mu_2^2 = \frac{1}{N}\sum_{i=1}^{n} f_i(x_i - \bar{x})^4 - \left[\frac{1}{N}\sum_{i=1}^{n} f_i(x_i - \bar{x})^2\right]^2$$

Let $$z_i = (x_i - \bar{x})^2$$

Then $$\mu_4 - \mu_2^2 = \frac{1}{N}\sum_{i=1}^{n} f_i(z_i)^2 - \left[\frac{1}{N}\sum_{i=1}^{n} f_i z_i\right]^2$$

$\Rightarrow$ $$\mu_4 - \mu_2^2 = \sigma_z^2$$

Since $\sigma_z^2 \geq 0$ (always)

$\therefore$ $$\mu_4 - \mu_2^2 \geq 0 \Rightarrow \mu_4 \geq \mu_2^2.$$

Hence $\beta_2 \geq 1$ Proved.

Example 18: Find the first four moments of the following data. Also make Sheppard corrections.

Interval	10 – 20	20 – 30	30 – 40	40 – 50	50 – 60	60 – 70	70 – 80
Frequency	1	20	69	108	78	22	2

Sol.

Interval	f_i	x_i	$u_i = x_i - 45$	$u_i = \frac{x-45}{10}$	$f_i u_i$	$f_i u_i^2$	$f_i u_i^3$	$f_i u_i^4$
10 – 20	1	15	– 30	– 3	– 3	9	– 27	81
20 – 30	20	25	– 20	– 2	– 40	80	– 160	320
30 – 40	69	35	– 10	– 1	– 69	69	– 69	69
40 – 50	108	45	0	0	0	0	0	0
50 – 60	78	55	10	1	78	78	78	78
60 – 70	22	65	20	2	44	88	176	352
70 – 80	2	75	30	3	6	18	54	162
	$N = 300$	–	–	–	$\Sigma f_i u_i = 16$	$\Sigma f_i u_i^2 = 342$	$\Sigma f_i u_i^3 = 52$	$\Sigma f_i u_i^4 = 1062$

Now $$\mu_1' = \frac{1}{N}\sum f_i u_i \cdot 10 = \left(\frac{16}{300}\right)10 = \mathbf{0.53}$$

$$\mu_2' = \frac{1}{N}\sum f_i u_i^2 \cdot (10)^2 = \left(\frac{342}{300}\right)(10)^2 = \mathbf{114}$$

$$\mu_3' = \frac{1}{N}\sum f_i u_i^3 \cdot (10)^3 = \left(\frac{52}{300}\right)(10)^3 = \mathbf{173.33}$$

$$\mu_4' = \frac{1}{N}\sum f_i u_i^4 \cdot (10)^4 = \left(\frac{1062}{300}\right)(10)^4 = \mathbf{35400}$$

Moments about mean

$$\mu_1 = 0 \text{ (always)}$$

$$\mu_2 = \mu_2' - \mu_1'^2 = 114 - (0.53)^2 = \mathbf{113.72}$$

$$\mu_3 = \mu_3' - 3\mu_2'\mu_1' + 2\mu_1'^3 = 173.33 - 3(114)(0.53) + 2(0.53)^3 = \mathbf{-7.63}$$

$$\mu_4 = \mu_4' - 4\mu_3'\mu_1' + 6\mu_2'\mu_1'^2 - 3\mu_1'^4$$

$$= 35400 - 4(173.33)(0.53) + 6(114)(0.53)^2 - 3(0.53)^4 = \mathbf{35224.44}$$

Sheppard's correction, $h = 10$

$$\mu_1 \text{ (Corrected)} = \mu_1 = 0$$

$$\mu_2 \text{ (Corrected)} = \mu_2 - \frac{h^2}{12} 113.72 - \frac{(10)^2}{12} = 105.39$$

$$\mu_3 \text{ (Corrected)} = \mu_3 = -7.63$$

$$\mu_4 \text{ (Corrected)} = \mu_4 - \frac{1}{2}\mu_2 + \frac{7}{240}h^2$$

$$= 35224.44 - \frac{1}{2}(113.72) + \frac{7}{240}(10)^2 = 35170.4966. \quad \textbf{Ans.}$$

Example 19: Calculate the coefficient of the skewness from the following data:

Wages in rupees	0 –10	10 –20	20 –30	30 –40	40 –50	50 –60	60 –70
No. of labours	185	77	34	180	136	23	50

Sol. Let assumed mean = 35

Classes	f_i	*Mid value* x_i	$u_i = \frac{x_i - 35}{10}$	$f_i u_i$	$f_i u_i^2$	$f_i u_i^3$
0 – 10	185	5	– 3	– 555	1665	– 4995
10 – 20	77	15	– 2	– 154	308	– 616
20 – 30	34	25	– 1	– 34	34	– 34
30 – 40	180	35	0	0	0	0
40 – 50	136	45	1	136	136	136
50 – 60	23	55	2	46	92	184
60 – 70	50	65	3	150	450	1350
	$N = 685$			$\Sigma f_i u_i = -411$	$\Sigma f_i u_i^2 = 2685$	$\Sigma f_i u_i^3 = -3975$

Here $h = 10$

$$\mu_1' = \frac{1}{N}\sum f_i u_i \times h = \frac{-411}{685} \times 10 = -6$$

$$\mu_2' = \frac{1}{N}\sum f_i u_i^2 \times h^2 = \frac{2685}{685} \times 100 = 391.97$$

$$\mu_3' = \frac{1}{N}\sum f_i \mu_i^3 \times h^3 = \frac{-3975}{685} 1000 = -5802.9$$

$$\mu_1 = a + \mu_1' = 35 - 6 = 29$$

$$\mu_2 = \mu_2' - \mu_1'^2 = 391.97 - 36 = 355.97$$

$$\mu_3 = \mu_3' - 3\mu_2'\mu_1' + 2\mu_1'^3$$

$$= -5802.9 - 3 \times 391.97 \times (-6) + 2(-6)^3$$

$$= -5802.9 + 7055.46 - 432$$

$$= -6234.9 + 7055.46$$

$$= 820.56$$

$$\beta_1 = \frac{\mu_3^2}{\mu_2^3} = \frac{(820.56)^2}{(355.97)^3} = 0.0149$$

Since $\beta_1 > 0$.

Therefore the curve is positively skewed.

Example 20: In two frequency distributions the second moments about mean are 36 and 49 respectively, while third moments about mean are 43.2 and 85.75. Compare the skewness in the two frequency distributions.

Sol.

(*i*) In first frequency distribution, we have

$$\mu_2 = 36 \quad \text{and} \quad \mu_3 = 43.2$$

$$\therefore \qquad \beta_1 = \frac{\mu_3^2}{\mu_2^3} = \frac{(43.2)^2}{(36)^3} = 0.04$$

$$\Rightarrow \qquad \beta_1 > 0 \text{ (positive)}$$

Hence the distribution is positively skewed.

(*ii*) In second frequency distribution, we have

$$\mu_2 = 49, \mu_3 = 85.75$$

$$\therefore \qquad \beta_1 = \frac{\mu_3^2}{\mu_2^3} = \frac{(85.75)^2}{(49)^3} = 0.0625$$

$$\Rightarrow \qquad \beta_1 > 0 \text{ (positive)}$$

Hence again the distribution is positive skewed.

Example 21: For a distribution with the mean $\bar{x} = 10$, variance $\sigma^2 = 16$, $\gamma_1 = 1$ and $\beta_2 = 4$, find the first four moments about origin.

Sol. We have $\bar{x} = 10$; $\sigma^2 = \mu_2 = 16$, $\gamma_1 = 1$ and $\beta_2 = 4$

We know that

$$\gamma_1 = \sqrt{\beta_1} = \frac{\mu_3}{\mu_2^{3/2}} = 1 \qquad \left| \text{As } \beta_1 = \frac{\mu_3^2}{\mu_2^3} \right.$$

$$\Rightarrow \qquad \mu_3 = \mu_2^{3/2} = (16)^{3/2} = (4^2)^{3/2} = 4^2 = \mathbf{64}$$

and

$$\beta_2 = \frac{\mu_4}{\mu_2^2} = 4 \Rightarrow \mu_4 = 4(\mu_2^2) = 4 \times 16^2 = \mathbf{1024}$$

Next moments about origin are

$$\nu_1 = \bar{x} = 10$$

$$\nu_2 = \mu_2 + \nu_1^2 = 16 + 10^2 = 116 \qquad | \ \nu_1 = \bar{x}$$

$$\nu_3 = \mu_3 + 3\mu_2\nu_1 + (\nu_1)^3 = 64 + 3 \times 16 \times 10 + 10^3$$

$$= 64 + 480 + 1000 = 1544$$

$$\nu_4 = \mu_4 + 4\mu_3\nu_1 + 6\mu_2(\nu_1)^2 + (\nu_1)^4$$

$$= 1024 + 4 \times 64 \times 10 + 6 \times 16 \times 10^2 + 10^4$$

$$= 1024 + 2560 + 9600 + 10000 = \mathbf{23184}$$

Example 22: The first three moments about the origin are given by

$$\mu_1' = \frac{n+1}{2}, \mu_2' \frac{(n+1)(2n+1)}{6}, \mu_3' = \frac{n(n+1)^2}{4}$$

examine the skewness of the data.

Sol. We know that

$$\mu_3 = \mu_3' - 3\mu_2'\mu_1' + 2(\mu_1')^3$$

$$= \frac{n(n+1)^2}{4} - \frac{3(n+1)(2n+1)}{6} \cdot \frac{n+1}{2} + 2\left(\frac{n+1}{2}\right)^3$$

$$= \frac{(n+1)^2}{4}[n-(2n+1)+(n+1)]$$

$$= \frac{(n+1)^2}{4}[n-2n-1+n+1]$$

$$= \frac{(n+1)^2}{4}[0] = 0$$

$$\therefore \qquad \beta_1 = \frac{\mu_3^2}{\mu_2^3} = \frac{0}{\mu_2^3} = 0$$

Since $\beta_1 = 0$.

Thus there is no skewness in the data.

Example 23: The fourth moment mean of a frequency distribution is 48. What must be the value of its S.D. in order that the distribution be (*i*) Leptokurtic, (*ii*) Mesokurtic and (*iii*) Platykurtic?

Sol. We know that

$$\beta_2 = \frac{\mu_4}{\mu_2^2} = \frac{\mu_4}{\sigma^4} = \frac{48}{\sigma^4} \qquad \Big| \text{ As } \sigma^2 = \mu_2$$

(*i*) For Leptokurtic,

$$\beta_2 > 3$$

$$\Rightarrow \qquad \frac{48}{\sigma^4} > 3$$

$$\text{or} \qquad \sigma^4 < \frac{48}{3} \Rightarrow \sigma^4 < 16$$

$$\text{or} \qquad \sigma < 2$$

(*ii*) For Mesokurtic,

$$\beta_2 = 3$$

$$\Rightarrow \quad \frac{48}{\sigma^4} = 3 \Rightarrow \sigma = 2$$

(*iii*) For Platykurtic,

$$\beta_2 < 3$$

$$\Rightarrow \quad \sigma > 2.$$

PROBLEM SET 4.1

1. Compute the first four moments about mean from the following data:

Class intervals	0 – 10	10 – 20	20 – 30	30 – 40
Frequency (f)	1	3	4	2

[**Ans.** $\mu_1 = 0$, $\mu_2 = 81$, $\mu_3 = -144$ and $\mu_4 = 14817$]

2. Calculate the first four moments about mean for the following distribution. Also calculate β_1 and comment upon the nature of skewness.

Marks	0 – 20	20 – 40	40 – 60	60 – 80	80 – 100
Frequency	8	28	35	17	12

[**Ans.** $\mu_1 = 0$, $\mu_2 = 499.64$, $\mu_3 = 2579.57$, $\mu_4 = 589111.61$, $\beta_1 = 0.053$, skewness is positive]

3. Compute the first four central moments from the following data. Also find the two beta coefficients.

Value	5	10	15	20	25	30	35
Frequency	8	15	20	32	23	17	5

[**Ans.** 0, 59.99, – 50.18, 8356.64, 0.012 (negatively skewed), 2.32 platykurtic)]

4. The first four moments of a distribution are 1, 4, 10 and 46 respectively. Compute the moment coefficients of skewness and kurtosis and comment upon the nature of the distribution. [**Ans.** 0, 3. Thus the distribution is symmetrical and mesokurtic]

5. The first three moments of a distribution about the value 3 of a variable are 2, 10 and 30 respectively. Obtain $\bar{x}, \mu_2, \mu_3$ and hence β_1. Comment upon the nature of skewness.

[**Ans.** 5, 6, – 14, 0.907, the distribution is negatively skewed]

6. Calculate the measure of skewness and kurtosis on the basis of moments for the following distribution:

Marks more than	0	10	20	30	40
No. of students	10	8	6	3	1

[**Ans.** – 0.147]

7. Find the first four moments about the mean of the distribution also calculate β_1 and β_2.

x	2.0	2.5	3.0	3.5	4.0	4.5	5.0
f	5	38	65	92	70	40	10

[**Ans.** $\mu_1 = 0$, $\mu_2 = 18.13125$, $\mu_3 = 0.0791$, $\mu_4 = 8.033$, $\beta_1 = 0.001$, $\beta_2 = 2.44$]

8. Find first four central moments for the following frequency distribution:

Class	0 – 10	10 – 20	20 – 30	30 – 40
Frequency	1	3	4	2

[**Ans.** $\mu_1 = 0$, $\mu_2 = 81$, $\mu_3 = -144$, $\mu_4 = 14817$]

9. Find first three moments about 90 from the following frequency distribution:

x	82	88	90	91	92	95	97
f	7	11	15	8	4	3	2

[**Ans.** $\mu'_1 = -6.6$, $\mu'_2 = 13.98$, $\mu'_3 = -51.42$]

10. Find mean, variance and third central moments from the following data:

x	0	1	2	3	4	5	6	7	8
f	1	9	26	59	72	52	29	7	1

[**Ans.** $\bar{x} = 3.97$, variance $= \mu_2 = 1.979$, $\mu_3 = 0.0115$]

11. The first four moments of a distribution about the value 0 of a variable are – 0.20, 1.76, – 2.36 and 10.88. Find the moments about the mean and measure of skewness and kurtosis.
[**Ans.** $\mu_1 = 0$, $\mu_2 = 1.72$, $\mu_3 = -1.32$, $\mu_4 = 9.4192$, $\beta_1 = 0.34$, $\beta_2 = 3.18$]

12. The first four moments of a distribution are 0, 2.5, 0.7 and 18.75. Find the coefficient of skewness and kurtosis. [**Ans.** $\beta_1 = 0.03$, $\gamma_1 = 0.18$ (positive skew); $\beta_2 = 3$, $\gamma_2 = 0$ (Mesokurtic)]

13. Find the Karl Pearson's coefficient of skewness for the following data:

Wages in Rs.	10	11	12	13	14	15
Frequency	2	4	10	8	5	1

[**Ans.** – 0.417]

14. For a distribution the mean is 10, variance is 16, $\gamma_1 = 1$ and $\beta_2 = 4$. Find the first four moments about the origin. [**Ans.** 10; 116; 1544; 23184]

15. The first four moments of a distribution about $x = 5$ are 2, 20, 40 and 50. Show that the mean = 16, variance = 16, $\mu_3 = -64$, $\mu_4 = 162$, $\beta_1 = 1$ and $\beta_2 = 0.63$.

16. If $\mu_2 = 43.553$, $\mu_3 = -9.774$, $\mu_4 = 5508.567$ and $h = 3$; find the correct value of them.

17. The first four moments about $x = 5$ are 2, 20, 40 and 50. Find various characteristics.
[**Ans.** $\bar{x} = 7, \sigma = 4, \mu_1 = 0, \mu_2 = 16, \mu_3 = -64, \mu_4 = 162, \beta_1 = 1, \beta_2 = 0.63, \gamma_1 = 1$ and $\gamma_2 = -2.37$]

Moment Generating Function:

18. Find the moment generating of the random variable x whose probability density function is

$$f(x) = \begin{cases} \frac{x}{2}, & 0 \le x \le 2 \\ 0, & \text{otherwise} \end{cases}$$

$$\left[\textbf{Ans. } M(t) = \frac{e^{2t}}{t} - \frac{1}{2t^2} e^{2t}\right]$$

19. Find the moment generating function of $f(x) = ce^{-cx}$, $c > 0$, $0 \le x \le \infty$ and deduce that $\mu_2 = \frac{1}{c^2}$.
$$\left[\textbf{Ans. } \left(1 - \frac{t}{c}\right)^{-1}\right]$$

20. If $p(x) = \frac{1}{2^x}, x = 1, 2, 3, \ldots$ find the m.g.f. of x. Hence obtain the variance. $\left[\textbf{Ans. } \frac{e^t}{2} - e^t\right]$

21. If m.g.f. of a distribution about zero is $f(t)$, show that the m.g.f. about its mean $\bar{x}$ is $f(t) \cdot e^{-mt}$.

22. Find the moment generating function of the discrete Poisson distribution given by $p(x) = e^{-\lambda} \cdot \frac{\lambda^x}{\lfloor x}$. Also find the first and second moments about the mean.
[**Ans.** $\lambda(\lambda + 1)$, mean $= \lambda$, variance $= \lambda$]

23. Find the moment generating function of the random variable whose moments are $\mu'_r = \lfloor (r+1) \cdot 2^r$.
[**Ans.** $(1 - 2t)^{-2}$]

24. If $\mu'_r = \lfloor r \cdot 2^r$ then find m.g.f.
[**Ans.** $(1 - 2t)^{-1}$]

4.5 CORRELATION

Whenever two variables x and y are so related that increase in the one is accompanied by an increase or decrease in the other, then the variables are said to be correlated.

Example 1: The yield of crop varies with the amount of rainfall.
Example 2: Prices varies with demand.
Example 3: Cost of production varies with the cost of raw materials.

4.5.1 Various Types of Correlation

(a) Perfect correlation: If the two variables in such a manner that their ratio is always constant, then the correlation is said to be perfect. If corresponding to an increase (or decrease) in one variable, there is a proportional increase (or decrease) in the other variable, then the correlation is said to be *perfect positive*.

If corresponding to increase (or decrease) in one variable, there is a proportional decrease (or increase) in the other variable, then the correlation is said to *perfect negative.*

(b) Positive or Direct correlation: If an increase (or decrease) in one variable corresponds to an increase (or decrease) in the other, the correlation is said to be *positive.*

(c) Negative or Indirect correlation: If an increase (or decrease) in one variable corresponds to a decrease (or increase) in the other, the correlation is said to be *negative.*

(d) Independent variables: If the changes in one variable are not followed by changes in the other, we say that the variables are independent and they are not correlated.

4.5.2 Correlation Coefficient

The number showing the degree or extent to which x and y are related to each other, is called the correlation coefficient. It is denoted by $r(x, y)$ or r_{xy} or simply by r.

It has the following characteristics:

(*a*) $-1 \le r \le 1$

(*b*) If $r = -1$, there is a perfect negative correlation between x and y.

(*c*) If $r = 1$, there is a perfect positive correlation between x and y.

(*d*) If $r = 0$, then x and y are not correlated.

(*e*) If $0 < r \le 1$, there is positive correlation between x and y.

(*f*) If $-1 \le r < 0$, there is a negative correlation between x and y.

4.6 SCATTER OR DOT DIAGRAMS

Let (x_i, y_i), $1 \le i \le n$ be a bivariate distribution. If we plot the corresponding values of x and y, taking the values of x along x-axis and the values of y along y-axis, we obtain collection of dots, called dot diagrams or a scatter diagram.

(*a*) If the plotted points are in a straight line and show an upward trend, the correlation is to be perfect positive *e.g.*, radius and circumference of circles.

(*b*) If the plotted points are in a straight line and show a downward trend, the correlation is said to be perfect negative *e.g.*, production depends on the cost of raw material.

(*c*) If the plotted points are not on a straight line but are scattered around a straight line,variables are correlated. Closer the scatter of points around a straight line, higher is the degree of correlation.

(*d*) If the plotted points show no trend at all, then the variables are independent and are not correlated.

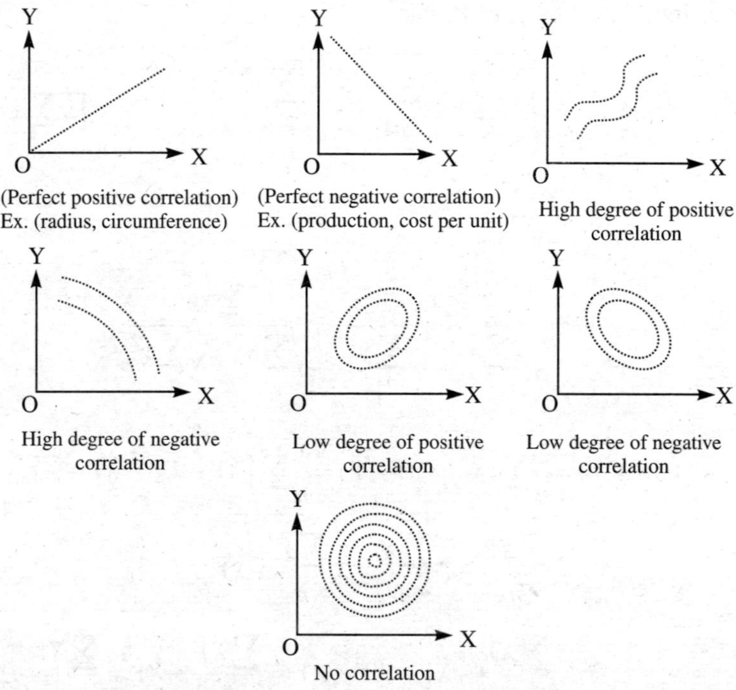

4.7 CORRELATION COEFFICIENT BY COVARIANCE OR KARL PEARSON'S METHOD

If $(X_1, Y_1), (X_2, Y_2), \ldots\ldots\ldots, (X_n, Y_n)$ be n pairs of observation on two variables X and Y, then the covariance of X and Y, written as cov(X, Y), is defined by

$$\text{cov}(X, Y) = \frac{1}{n}\sum(X - \bar{X})(Y - \bar{Y}).$$

Covariance indicates the joint variations between two variables.

The correlation coefficient or the coefficient of correlation or product moment correlation coefficient(r) between X and Y is defined as

$$r = \frac{\text{cov}(X, Y)}{\sigma_X \sigma_Y} = \frac{\sum\left[(X \quad \bar{X})(Y - \bar{Y})\right]}{n\sigma_X \sigma_Y}$$

where σ_X, σ_Y are standard deviations of X and Y respectively.

The formula for the correlation coefficient r may be written in different forms.

(a) If $x = X - \bar{X}$ = deviation of x from its mean and $y = Y - \bar{Y}$ = deviation of Y from its mean, then

$$r = \frac{\sum xy}{n\sigma_x \sigma_y} \qquad \ldots(1)$$

{$\therefore$ S.D. is independent of any change of origin}

This formula was introduced by Karl Pearson.

Now $$\sigma_x = \sqrt{\frac{\sum x^2}{n} - \left(\frac{\sum x}{n}\right)^2} = \sqrt{\frac{\sum x^2}{n}} \qquad \left\{\because \sum x = \sum X - \bar{X} = 0\right\}$$

Similarly, $$\sigma_y = \sqrt{\frac{\sum y^2}{n}},$$

$\therefore$ From (1), $$r = \frac{\sum xy}{n\sqrt{\frac{\sum x^2}{n}} \times \sqrt{\frac{\sum y^2}{n}}} = \frac{\sum xy}{\sqrt{\sum x^2} \times \sqrt{\sum y^2}}$$

(b) $$\text{cov}(X, Y) = \frac{1}{n}\sum (X - \bar{X})(Y - \bar{Y}) = \frac{1}{n}\sum\{XY - X\bar{Y} - \bar{X}Y + \bar{X}\bar{Y}\}$$

$$= \frac{\sum XY}{n} - \bar{Y}\frac{\sum X}{n} - \bar{X}\frac{\sum Y}{n} + \bar{X}\bar{Y}$$

$$= \frac{\sum XY}{n} - \bar{X}\bar{Y} - \bar{X}\bar{Y} + \bar{X}\bar{Y} = \frac{\sum XY}{n} - \bar{X}\bar{Y} = \frac{\sum XY}{n} - \left(\frac{\sum X}{n}\right)\left(\frac{\sum Y}{n}\right)$$

$$\therefore \quad r = \frac{\text{cov}(X, Y)}{\sigma_X \sigma_Y} = \frac{\frac{\sum XY}{n} - \left(\frac{\sum X}{n}\right)\left(\frac{\sum Y}{n}\right)}{\sqrt{\frac{\sum X^2}{n} - \left(\frac{\sum X}{n}\right)^2} \times \sqrt{\frac{\sum Y^2}{n} - \left(\frac{\sum Y}{n}\right)^2}}$$

Since the correlation coefficient is independent of the choice of origin the correct formula for correlation coefficient r, when the deviations are taken from assumed mean, is given

$$r = \frac{\frac{\sum xy}{n} - \left(\frac{\sum x}{n}\right)\left(\frac{\sum y}{n}\right)}{\sqrt{\frac{\sum x^2}{n} - \left(\frac{\sum x}{n}\right)^2} \times \sqrt{\frac{\sum y^2}{n} - \left(\frac{\sum y}{n}\right)^2}}$$

where $x = X - a$ and $y = Y - b$, a and b being assumed means of X and Y respectively.

This formula can also be written as

$$r = \frac{n\sum xy - \left(\sum x\right)\left(\sum y\right)}{\sqrt{n\sum x^2 - \left(\sum x\right)^2} \times \sqrt{n\sum y^2 - \left(\sum y\right)^2}} = \frac{\sum xy - \frac{\left(\sum x\right)\left(\sum y\right)}{n}}{\sqrt{\sum x^2 - \frac{\left(\sum x\right)^2}{n}} \times \sqrt{\sum y^2 - \frac{\left(\sum y\right)^2}{n}}}$$

(*c*) Karl Pearson's formula for correlation coefficient r with frequencies of both the variables X and Y is given by

$$\frac{\frac{\sum fdxdy}{N}-\left(\frac{\sum fdx}{N}\right)\left(\frac{\sum fdy}{N}\right)}{\sqrt{\frac{\sum fdx^2}{N}-\left(\frac{\sum fdx}{N}\right)^2}\times\sqrt{\frac{\sum fdy^2}{N}-\left(\frac{\sum fdy}{N}\right)^2}}=\frac{\sum fdxdy-\frac{\left(\sum fdx\right)\left(\sum fdy\right)}{N}}{\sqrt{\sum fdx^2-\frac{\left(\sum fdx\right)^2}{N}}\times\sqrt{\sum fdy^2-\frac{\left(\sum fdy\right)^2}{N}}}$$

where $dx=\frac{X-a}{c}$ and $dy=\frac{Y-b}{d}$ where a, b, c, and d are constants.

Theorem 1: The correlation coefficient is independent of the origin (i) and scale

Proof : Let $u_i=\frac{x_i-a}{h}$ and $v_i=\frac{y_i-b}{k}$ where $h, k>0$

$\therefore$ $x_i=a+hu_i$ and $y_i=b+kv_i$

Hence $\bar{x}=a+h\bar{u}$ and $\bar{y}=b+k\bar{v}$

$\therefore$ $x_i-\bar{x}=h(u_i-\bar{u})$ and $y_i-\bar{y}=k(v_i-\bar{v})$

Also $\sigma_x=h\sigma_u$ and $\sigma y=k\sigma_v$

$\therefore$ $r_{xy}=\frac{\Sigma(x_1-\bar{x})(y_i-\bar{y})}{n\sigma x\sigma y}=\frac{hk\Sigma(u_i-\bar{u})(v_i-\bar{v})}{n(h\sigma_u)(k\sigma_v)}=\frac{\Sigma(u_i-\bar{u})(v_i-\bar{v})}{n\sigma x\sigma v}=r_{uv}$. **Proved**

Theorem 2: The correlation coefficient r lies always between –1 and *i.e.* $-1\le r\le 1$

Proof : We know that

$$r_{xy}=\frac{\Sigma(x_1-\bar{x})(y_1-\bar{y})}{n\sigma x\sigma y}=\frac{\frac{1}{n}\Sigma(x_1-\bar{x})(y_1-\bar{y})}{\sqrt{\frac{\Sigma(x_1-\bar{x})^2}{n}}\cdot\sqrt{\frac{\Sigma(y_i-\bar{y})^2}{n}}}$$

Let $a_i=x_i-\bar{x}$ and $b_i=y_i-\bar{y}$

Then $r_{xy}^2=\frac{(\Sigma a_ib_i)^2}{(\Sigma a_i^2)(\Sigma b_i^2)}$

By Schwartz inequality we have $(\Sigma a_ib_i)^2\le(\Sigma a_i^2)(\Sigma b_i^2)$

$\therefore$ $r_{xy}^2\le 1$. Hence $|r_{xy}|\le 1$.

$\therefore$ $-1\le r\le 1$

Example 1: If x, y and z are uncorrelated variable each having same standard deviation obtain the coefficient of correlation between $x + y$ and $y + z$.

Sol. Given $\sigma_x = \sigma_y = \sigma_z = \sigma(\text{say})$

x and y are uncorrelated $\Rightarrow \Sigma(x-\bar{x})(y-\bar{y}) = 0$

y and z are uncorrelated $\Rightarrow \Sigma(y-\bar{y})(z-\bar{z}) = 0$

z and x are uncorrelated $\Rightarrow \Sigma(z-\bar{z})(z-\bar{x}) = 0$

Let $$u = x + y \quad \text{and} \quad v = y + z$$

$\therefore$ $$\bar{u} = \bar{x} + \bar{y} \quad \text{and} \quad \bar{v} = \bar{y} + \bar{z}$$

Now, $$\sigma_u^2 = \frac{1}{n}\Sigma(u-\bar{u})^2 = \frac{1}{n}\Sigma[(x-\bar{x})+(y-\bar{y})]^2$$

$$= \frac{1}{n}[\Sigma(x-\bar{x})^2 + \Sigma(y-\bar{x})^2 + 2\Sigma(x-\bar{x})(y-\bar{y})]$$

$$= \sigma_x^2 + \sigma_y^2 \quad (\text{since } \Sigma(x-\bar{x})(y-\bar{y}) = 0)$$

$$= 2\sigma^2$$

Similarly $\sigma_v^2 = 2\sigma^2$, Now, $\Sigma(u-\bar{u})(v-\bar{v}) = \Sigma[\{(x-\bar{x})+(y-\bar{y})\}\{(y-\bar{y})\}+(z-\bar{z})\}]$

$$= \Sigma(x-\bar{x})(y-\bar{y}) + \Sigma(y-\bar{y})^2 + \Sigma(x-\bar{x})(z-\bar{z}) + \Sigma(y-\bar{y})(z-\bar{z})$$

$$= 0 + n\sigma_y^2 + 0 + 0 = n\sigma_y^2 = n\sigma_y^2$$

$\therefore$ $$r_{uv} = \frac{\Sigma(u-\bar{u})(v-\bar{v})}{n\sigma u \sigma v} = \frac{n\sigma_y^2}{n} = \frac{n\sigma^2}{n 2\sigma^2} = \frac{1}{2}$$

Example 2: Find the coefficient of correlation from the following table:

X	1	2	3	4	5	6	7
Y	6	8	11	9	12	20	14

Sol.

X	Y	$x = X - \bar{X}$	$y = Y - \bar{Y}$	x^2	y^2	xy
1	6	–3	–4	9	16	12
2	8	–2	–2	4	4	4
3	11	–1	1	1	1	-1
4	9	0	–1	0	1	0
5	12	1	2	1	4	2
6	10	2	0	4	0	0
7	14	3	4	9	16	12
$\Sigma X = 28$	$\Sigma Y = 70$	0	0	$\Sigma x^2 = 28$	$\Sigma y^2 = 42$	$\Sigma xy = 29$

$$\bar{X} = \frac{\sum x}{n} = \frac{28}{7} = 4 \quad \text{and} \quad \bar{Y} = \frac{\sum y}{n} = \frac{70}{7} = 10$$

Karl Pearson's coefficient of correlation (r) is given by

$$r = \frac{\sum xy}{\sqrt{\sum x^2} \times \sqrt{\sum y^2}} = \frac{29}{\sqrt{28} \times \sqrt{42}} = 0.8457 \text{ (approx.)}$$

Example 3: Karl Pearson's coefficient of correlation between two variables X and Y is 0.28, their covariance is +7.6. If the variance of X is 9, find the S.D. of Y series.

Sol. Karl Pearson's coefficient of correlation r is given by

$$r = \frac{\text{cov}(X,Y)}{\sigma_X \sigma_Y} \qquad \ldots(1)$$

Here $r = 0.28$, cov(X, Y) = 7.6 and $\sigma_x^2 = 9$; $\therefore \sigma_x = 3$, using (1), $0.28 = \dfrac{7.6}{3\sigma_y}$

or $$0.84\,\sigma_y = 7.6 \quad \text{or} \quad \sigma_y = \frac{7.6}{0.84} = \frac{760}{84} = 9.048.$$

Hence, the S.D. of Y series is 9.048. **Ans.**

Example 4: Calculate the coefficient of correlation between X and Y series from the following data:

$$\sum_{i=1}^{15} (X_i - \bar{X})^2 = 136, \sum_{i=1}^{15} (Y_i - \bar{Y})^2 = 138 \text{ and } \sum_{i=1}^{15} (X_i - X)(Y_i - \bar{Y}) = 122.$$

Sol. We have

$$r = \frac{\sum xy}{\sqrt{\sum x^2} \times \sqrt{\sum y^2}}, \quad \text{where } x = X - \bar{X}, y = Y - \bar{Y}.$$

Here $$\sum xy = 122, \sum x^2 = 136 \quad \text{and} \quad \sum y^2 = 138.$$

$\therefore$ $$r = \frac{122}{\sqrt{136}\sqrt{138}} = \frac{122}{(11.6619)(11.7473)} = 0.89. \quad \textbf{Ans.}$$

Example 5: The following data regarding the heights (y) and weights (x) of 100 college students are given by:

$$\sum x = 15000, \ \sum x^2 = 2272500, \sum y = 6800, \sum y^2 = 463025 \text{ and } \sum xy = 1022250$$

Find the correlation coefficient between height and weight.

Sol. Here $n = 100$

Correlation coefficient r (x, y) is given by

$$r = \frac{n\sum xy - (\sum x)(\sum y)}{\sqrt{n\sum x^2 - (\sum x)^2} \times \sqrt{n\sum y^2 - (\sum y)^2}}$$

$$= \frac{(100 \times 1022250) - (15000 \times 6800)}{\sqrt{(100 \times 2272500) - (15000)^2}\sqrt{(100 \times 463025) - (6800)^2}}$$

$$= \frac{102225000 - 102000000}{\sqrt{227250000 - 225000000}\sqrt{46302500 - 46240000}}$$

$$= \frac{225000}{\sqrt{2250000}\sqrt{62500}} = \frac{225000}{1500 \times 250} = 0.6. \quad \textbf{Ans.}$$

Example 6: Calculate Pearson's coefficient of correlation between advertisement cost and sales the data given below:

Advertisement cost in '000' Rs.	39	65	62	90	82	75	25	98	36	78
Sales in lakh Rs.	47	53	58	86	62	68	60	91	51	84

Sol. Karl Pearson's coefficient of correlation is given by

$$r = \frac{\sum xy}{\sqrt{\sum x^2} \times \sqrt{\sum y^2}}, \quad \text{where } x = X - \bar{X},\ y = Y - \bar{Y}.$$

X	Y	$x = X - \bar{X}$	$y = Y - \bar{Y}$	x^2	y^2	xy
39	47	–26	–19	676	361	494
65	53	0	– 13	0	169	0
62	58	– 3	– 8	9	64	24
90	86	25	20	625	400	500
82	62	17	– 4	289	16	–68
75	68	10	2	100	4	20
25	60	–40	– 6	1600	36	240
98	91	33	25	1089	625	825
36	51	–29	–15	841	225	435
78	84	13	18	169	324	234
$\sum X = 650$	$\sum Y = 660$	0	0	$\sum x^2 = 5398$	$\sum y^2 = 2224$	$\sum xy = 2704$

$$\bar{X} = \frac{\sum x}{n} = \frac{650}{10} = 65;\ \bar{Y} = \frac{\sum y}{n} = \frac{660}{10} = 66. \quad \therefore r = \frac{2704}{\sqrt{5389}\sqrt{2224}}.$$

$= 0.78.$ **Ans.**

Example 7: Calculate Pearson's coefficient of correlation from the following taking 100 and 50 the assumed average of X and Y respectively:

X	104	111	104	114	118	117	105	108	106	100	104	105
Y	57	55	47	45	45	50	64	63	66	62	69	61

Sol.

X	Y	$x = X - a$	$y = Y - b$	x^2	y^2	xy
104	57	4	7	16	49	28
111	55	11	5	121	25	55
104	47	4	–3	16	9	–12
114	45	14	–5	196	25	–70
118	45	18	–5	324	25	–90
117	50=b	17	0	289	0	0
105	64	5	14	25	196	70
108	63	8	13	64	169	104
106	66	6	16	36	256	96
100=a	62	0	12	0	144	0
104	69	4	19	16	361	76
105	61	5	11	25	121	55
		$\Sigma x = 96$	$\Sigma y = 84$	$\Sigma x^2 = 1128$	$\Sigma y^2 = 1380$	$\Sigma xy = 312$

Pearson's coefficient of correlation (r) is given by

$$r = \frac{\sum xy - \frac{(\sum x)(\sum y)}{n}}{\sqrt{\sum x^2 - \frac{(\sum x)^2}{n}} \times \sqrt{\sum y^2 - \frac{(\sum y)^2}{n}}} = \frac{312 - \frac{96 \times 84}{12}}{\sqrt{1128 - \frac{96 \times 96}{12}} \times \sqrt{1380 - \frac{84 \times 84}{12}}}$$

$$= \frac{312 - 672}{\sqrt{1128 - 768} \times \sqrt{1380 - 588}} = \frac{-360}{\sqrt{360 \times 792}}.$$

$$= -\ 0.67 \text{ (approx.).}$$ **Ans.**

Example 8: A computer while calculating the correlation coefficient between two variables X and Y from 25 pairs observations obtained the following results:

$$n = 25,\ \sum X = 125, \sum Y = 100, \sum X^2 = 650, \sum Y^2 = 460 \quad \text{and} \quad \sum XY = 508.$$

It was, however, discovered at the time of checking that two pairs of observations were not correctly copied. They were taken as (6, 14) and (6, 8). Prove that the correct value of the correlation coefficient should be 2/3.

Sol. When the two incorrect pairs of observation are replaced by the correct pairs, the revised results for the whole series are:

$\sum X$ = 125 – (Sum of two incorrect values of X) + (Sum of the two correct values of X)

= 125 – (6 + 8) + (8 + 6) = 125.

Similarly,

$$\sum Y = 100 - (14+6) + (12+8) = 100$$

$$\sum X^2 = 650 - (36+64) + (64+36) = 650$$

$$\sum Y^2 = 460 - (14^2 + 6^2) + (12^2 + 8^2) = 460 - 232 + 208 = 436$$

and $$\sum XY = 508 - (14\times 6 + 8\times 6) + (8\times 12 + 6\times 8) = 508 - 132 + 144 = 520.$$

$\therefore$ Correct value of the correlation coefficient (r) is $r = \dfrac{\sum XY - \dfrac{(\sum X)(\sum Y)}{n}}{\sqrt{\sum X^2 - \dfrac{(\sum X)^2}{n}} \times \sqrt{\sum Y^2 - \dfrac{(Y)^2}{n}}}$

$$= \frac{520 - \dfrac{125\times 100}{25}}{\sqrt{650 - \dfrac{(125)^2}{25}} \times \sqrt{436 - \dfrac{(100)^2}{25}}}$$

$$= \frac{20}{\sqrt{25\times 36}} = \frac{20}{5\times 6} = \frac{2}{3}.$$ **Ans.**

Example 9: Establish the formula

$$r = \frac{\sigma_x^2 + \sigma_y^2 - \sigma_{x-y}^2}{2\sigma_x\sigma_y}$$

where r is the coefficient of correlation between x and y and $\sigma_x, \sigma_y, \sigma_{x-y}$ are concerned standard deviations. Hence evaluate r from the following data:

x:	21	23	30	54	57	58	72	78	81	90
y:	60	71	72	83	110	84	100	92	113	135

Sol. Suppose

$$z = x - y$$

$\therefore$ $$\bar{z} = \bar{x} - \bar{y} = (x - \bar{x}) - (y - \bar{y})$$

or $$(z - \bar{z})^2 = (x - \bar{x})^2 + (y - \bar{y})^2 - 2(x - \bar{x})(y - \bar{y})$$

or $$\Sigma(z - \bar{z})^2 = \Sigma(x - \bar{x})^2 + \Sigma(y - \bar{y})^2 - 2\Sigma(x - \bar{x})(y - \bar{y})$$

or $$\frac{\Sigma(z - \bar{z})^2}{n} = \frac{\Sigma(x - \bar{x})^2}{n} + \frac{\Sigma(y - \bar{y})^2}{n} - \frac{2\Sigma(x - \bar{x})(y - \bar{y})}{n}$$

$\Rightarrow$ $$\sigma_z^2 = \sigma_x^2 + \sigma_y^2 - 2r\sigma_x\sigma_y \qquad \left| \text{ As } r = \frac{\Sigma(x - \bar{x})(y - \bar{y})}{n\sigma_x\sigma_y} \right.$$

or $$\sigma_{x-y}^2 = \sigma_x^2 + \sigma_y^2 - 2r\sigma_x\sigma_y$$

or $$r = \frac{\sigma_x^2 + \sigma_y^2 - \sigma_{x-y}^2}{2\sigma x \sigma y} \qquad \text{...(1)} \quad \textbf{Proved.}$$

$$\bar{x} = \frac{\Sigma x}{n} = \frac{21 + 23 + 30 + 54 + 57 + 58 + 72 + 78 + 87 + 90}{10} = \frac{570}{10} = 57$$

$$\bar{y} = \frac{\Sigma y}{n} = \frac{60 + 71 + 72 + 83 + 110 + 84 + 100 + 92 + 113 + 135}{10}$$

$$= \frac{920}{10} = 92$$

x	y	$z = x - y$	$x - \bar{x}$	$y - \bar{y}$	$z - \bar{z}$	$(x - \bar{x})^2$	$(y - \bar{y})^2$	$(z - \bar{z})^2$
21	60	– 39	– 36	– 32	– 4	1296	1024	16
23	71	– 48	– 34	– 21	– 13	1156	441	169
30	72	– 42	– 27	– 20	– 7	729	400	49
54	83	– 29	– 3	–9	6	9	81	36
57	110	– 53	0	18	– 18	0	324	324
58	84	– 26	1	– 8	9	1	64	81
72	100	– 28	15	8	7	225	64	49
78	92	– 14	21	0	21	441	0	441
87	113	– 26	30	21	9	900	441	81
90	135	– 45	33	43	– 10	1089	1849	100
		Σz = – 350				$\Sigma(x-\bar{x})^2$ = 5846	$\Sigma(y-\bar{y})2$ = 4688	$\Sigma(z-\bar{z})^2$ = 1346

$$\bar{z} = \frac{\Sigma z}{n} = -\frac{350}{10} = -35$$

$$\sigma_x^2 = \frac{\Sigma(x - \bar{x})^2}{n} = \frac{1}{10} \times 5846 = 584.6$$

$$\sigma_y^2 = \frac{\Sigma(y - \bar{y})^2}{n} = \frac{1}{10} \times 4688 = 468.8$$

$$\sigma_{x-y}^2 = \sigma_z^2 = \frac{\sum\left(z - \bar{z}^2\right)}{n} = \frac{1}{10} \times 1346 = 134.6$$

Putting these values in relation (1), we get

$$r = \frac{584.6 + 468.8 - 134.6}{2\sqrt{584.6}\sqrt{468.8}} = \frac{918.8}{2(523.50786)}$$

or $r = 0.8775$ **Ans.**

Example 10: Find the coefficient of correlation between the grouped frequency distribution of two variable (Profits and Sales) given below in the form of a two-way frequency table:

Sales (in 0000 rupees) →

	80-90	90-100	100-110	110-120	120-130	Total
50-55	1	3	7	5	2	18
55-60	2	4	10	7	4	27
60-65	1	5	12	10	7	35
65-70	-	3	8	6	3	20
Total	4	15	37	28	16	100

Sol. **Sales** (in'0000 Rs.)→

Sales (in '0000' Rs.) →

Profit (in '000' Rs.) ↓

Class interval			80-90	90-100	100-110	110-120	120-130				
	Mid values (x)→		85	95	105	115	125				
Class Intervals	Mid values (y)↓	dx / dy	–2	–1	0	1	2	Total *f*	*fdy*	*fdy*²	*fdxdy*
50-55	52.5	–1	1 (2)	3 (3)	7 (0)	5 (–5)	2 (–4)	18	–18	18	–4
55-60	57.5	0	2 (0)	4 (0)	10 (0)	7 (0)	4 (0)	27	0	0	0
60-65	62.5	1	1 (–2)	5 (–5)	12 (0)	10 (10)	7 (14)	35	35	35	17
65-70	67.5	2	—	3 (–6)	8 (0)	6 (12)	3 (12)	20	40	80	18
		Total *f*	4	15	37	28	16	Σ*f*=100	Σ*fdy*=57	Σ*fdy*²=133	Σ*fdxdy*=31
		fdx	–8	–15	0	28	32	Σ*fdx*=37			
		*fdx*²	16	15	0	28	64	Σ*fdx*²=123			
		fdxdy	0	–8	0	17	22	Σ*fdxdy*=31			

Check

The correlation coefficient (r) is given by

$$r = \frac{\dfrac{\sum fdxdy}{N} - \left(\dfrac{\sum fdx}{N}\right)\left(\dfrac{\sum fdy}{N}\right)}{\sqrt{\dfrac{\sum fdx^2}{N} - \left(\dfrac{\sum fdx}{N}\right)^2} \times \sqrt{\dfrac{\sum fdy^2}{N} - \left(\dfrac{\sum fdy}{N}\right)^2}} = \frac{31 - \dfrac{37 \times 57}{100}}{\sqrt{123 - \dfrac{(37)^2}{100}} \times \sqrt{133 - \dfrac{(57)^2}{100}}}$$

$$= \frac{31 - 21.09}{\sqrt{123 - 13.69} \times \sqrt{133 - 32.49}} = \frac{9.91}{\sqrt{109.31 \times 100.51}}$$

$\therefore$ $$r = 0.09456 = 0.0946 \text{ (approx.)}$$

Notes:

1. $dx = \dfrac{x - 105}{10}$ and $dy = \dfrac{y - 57.5}{5}$.
2. Number written brackets in each cell is the product of that cell frequency and the corresponding values of *dx* and *dy*.
3. *fdxdy* represents algebraic sum of the number within brackets in any row or column.

4.8 RANK CORRELATION (OR CORRELATION BY RANK METHOD)

Simple correlation coefficient (or product-moment correlation coefficient) is based on the magnitudes of the variable. But in many situations it is not possible to find the magnitudes of the variable at all. For example, we cannot measure beauty or intelligence quantitatively. In this case, it is possible to rank the individuals in some order.Rank correlation is based on the rank or the order and not on the magnitude of the variable. It is more suitable if the individuals (or variables) can be arranged on order of merit of proficiency. If the rank to individuals assigned range from 1 to N, then the correlation coefficient between two series of ranks is called *Rank correlation coefficient.* Edward Spearman's formula for rank coefficient of correlation (r) is given by

$$r = 1 - \frac{6\sum D^2}{n(n^2-1)} \quad \text{or} \quad 1 - \frac{6\sum D^2}{n^3 - n}$$

where D is the difference between the rank of the series and N is the number of individuals in each series.

Example 11: Ten students got the following percentage of marks in Mathematics and Statistics:

Student (Roll No.)	1	2	3	4	5	6	7	8	9	10
Marks in mathematics	78	36	98	25	75	82	90	62	65	39
Marks in statistics	84	51	91	60	68	62	86	58	53	47

Sol. In Mathematics, students with Roll No. 3 gets the highest marks 98 and is ranked 1; Roll No. 7 securing 90 marks has rank 2 and so on. Similarly, we can find the ranks of students in Statistics.

Roll No.	*Mathematics*		*Statistics*		*Rank Diff.* $D = R_1 - R_2$	D^2
	Marks	*Rank* (R_1)	*Marks*	*Rank* (R_2)		
1	78	4	84	3	1	1
2	36	9	51	9	0	0
3	98	1	91	1	0	0
4	25	10	60	6	4	16
5	75	5	68	4	1	1
6	82	3	62	5	−2	4
7	90	2	86	2	0	0
8	62	7	58	7	0	0
9	65	6	53	8	−2	4
10	39	8	47	10	−2	4
Total						$\Sigma D^2 = 30$

Applying Edward Spearman's formula:

$$r = 1 - \frac{6\sum D^2}{n(n^2-1)} \qquad [\text{Here } n = 10, \Sigma D^2 = 30]$$

$$= 1 - \frac{6 \times 30}{10(10^2-1)} = 1 - \frac{18}{99} = 1 - \frac{2}{11} = \frac{9}{11} = 0.82..$$

4.9 EDWARD SPEARMAN'S FORMULA FOR TIED RANKS

If in a series two or more individuals (or items) have the same score, then we find the average of the ranks of these individuals and allot this average rank to each of them. [In example 12, as the score 98 occurs three times, there is a tie for the 7th place and in this case, we find the average rank $\frac{7+8+9}{3}=$ 8 of 7, 8, 9 and assign it to each 98].

In such cases, Edward Spearman's modified formula is

$$r=1-\frac{6\left[\sum D^2+\sum(t^3-t)/12\right]}{n(n^2-1)}$$

where t is the number of individuals (or items) involved in a tie in the first or second series.

Example 12: Find the rank correlation coefficient of the following data:

Series A	115	109	112	87	98	120	98	100	98	118
Series B	75	73	85	70	76	82	65	73	68	80

Sol. In a series A, the highest score is 120 and, therefore, its rank is 1; the rank of the second highest score 118 is 2. Proceeding in this manner, we find the rank of 98 is 7 occurs three times. As there is a tie for the 7th place, we find the average $\frac{7+8+9}{3}$ i.e., 8 and assign 8 as rank to each of the three 98. The score next to 98 is 87 and its rank is 10. Similarly, we can find the ranks of individuals (or items) in series B.

Section A		*Section B*		*Rank Diff.* $D=R_1-R_2$	D^2
Score	*Rank*(R_1)	*Score*	*Rank*(R_2)		
115	3	75	5	–2	4
109	5	73	6.5	–1.5	2.25
112	4	85	1	3	9
87	10	70	8	2	4
98	8	76	4	4	16
120	1	82	2	–1	1
98	8	65	10	–2	4
100	6	73	6.5	–0.5	0.25
98	8	68	9	–1	1
118	2	80	3	–1	1
Total					$42.50=\Sigma D^2$

By Edward Spearman's formula,

$$r=1-\frac{6\left[\sum D^2+\sum(t^3-t)/12\right]}{n(n^2-1)}$$

where t = Number of individuals involved in a tie.

Here $\Sigma D^2 = 42.50$, $n = 10$, $t = 3$ and 2

$$\sum\frac{t^3-t}{12}=\frac{3^3-3}{12}+\frac{2^3-2}{12}=\frac{24}{12}+\frac{6}{12}=2+0.5=2.5$$

$$\therefore \quad r = 1 - \frac{6\{42.5 + 2.5\}}{10.(10^2 - 1)} = 1 - \frac{6 \times 45}{10 \times 99} = 1 - \frac{3}{11} = \frac{8}{11} = 0.73.$$

Example 13: The coefficient of rank correlation between the marks in Statistics and Mathematics obtained by a certain group of students is $\frac{2}{3}$ and the sum of the squares of the difference in ranks is 55. Find the number of students in the group.

Sol. Edward Spearman's formula for rank correlation coefficient is

$$r = 1 - \frac{6\sum D^2}{n(n^2 - 1)}, \text{ here } r = 2/3, \Sigma D^2 = 55, n = ?$$

$$\therefore \quad \frac{2}{3} = 1 - \frac{6 \times 55}{n(n^2 - 1)} = 1 - \frac{330}{n(n^2 - 1)} \text{ or } \frac{330}{n(n^2 - 1)} = 1 - \frac{2}{3} = \frac{1}{3}$$

or $n(n^2 - 1) = 990$ or, $n(n - 1)(n + 1) = 990$ or, $(n - 1)n(n + 1) = 9.10.11$

Here $n = 10$ *i.e.*, the number of students in the group = 10.

Example 14: The coefficient of correlation of the marks obtained by 10 students in Mathematics and Statistics was found to be 0.5. It was then detected that the difference in ranks in the two subjects for one particular student was wrongly taken to be 3 in place of 7. What should be the correct rank correlation coefficient.

Sol. Edward Spearman's formula for rank correlation coefficient is

$$r = 1 - \frac{6\sum D^2}{n(n^2 - 1)}, \quad \text{here } r = 0.5, \ \Sigma D^2 = ?, n = 10$$

$$\therefore \quad 0.5 = 1 - \frac{6\sum D^2}{10(10^2 - 1)} = 1 - \frac{6\sum D^2}{10 \times 99} = 1 - \frac{\sum D^2}{165}$$

$$\text{or} \quad \frac{\sum D^2}{165} = 1 - 0.5 = 0.5 = \frac{1}{2}, \quad \text{or, } \sum D^2 = 82.5$$

Now, correct value of $\sum D^2$ = Incorrect $\sum D^2$ – (incorrect difference in ranks)^2 + (correct difference in ranks)^2

$$= 82.5 - (3)^2 + (7)^2 = 122.5$$

Hence the correct rank correlation coefficient $= 1 - \frac{6 \times 122.5}{10(10^2 - 1)} = 1 - \frac{122.5}{165} = \frac{42.5}{165} = 0.2576.$ **Ans.**

Example 15: Compute the rank correlation coefficient for the following data:

Persons	A	B	C	D	E	F	G	H	I	J
Rank in math	9	10	6	5	7	2	4	8	1	3
Rank in physics	1	2	3	4	5	6	7	8	9	10

Sol.

Persons	R_1	R_2	$D = R_1 - R_2$	D^2
A	9	1	8	64
B	10	2	8	64
C	6	3	3	9
D	5	4	1	1
E	7	5	2	4
F	2	6	– 4	16
G	4	7	– 3	9
H	8	8	0	0
I	1	9	– 8	64
J	3	10	– 7	49
				$\Sigma D^2 = 280$

$$r = 1 - \frac{6\sum D^2}{n(n^2-1)} = 1 - \frac{6 \times 280}{10(100-1)} = 1 - \frac{168}{99} = 1 - 1.697 = -0.697. \quad \textbf{Ans.}$$

Example 16: Ten competitors in a beauty contest were ranked by three judges in the following orders:

First judge	1	6	5	10	3	2	4	9	7	8
Second judge	3	5	8	4	7	10	2	1	6	9
Third judge	6	4	9	8	1	2	3	10	5	7

Use the method of rank correlation to determine which pair of judges has the nearest approach to common taste in beauty.

Sol. Let R_1, R_2, R_3 be the ranks given by the three judges.

Competitor	R_1	R_2	R_3	$D_{12} = R_1 - R_2$	$D_{13} = R_1 - R_3$	$D_{23} = R_2 - R_3$	D_{12}^2	D_{13}^2	D_{23}^2
A	1	3	6	–2	–5	–3	4	25	9
B	6	5	4	1	2	1	1	4	1
C	5	8	9	–3	–4	–1	9	16	1
D	10	4	8	6	2	–4	36	4	16
E	3	7	1	–4	2	6	16	4	36
F	2	10	2	–8	0	8	64	0	64
G	4	2	3	2	1	–1	4	1	1
H	9	1	10	8	–1	–9	64	1	81
I	7	6	5	1	2	1	1	4	1
J	8	9	7	–1	1	2	1	1	4
$n = 10$							$\sum D_{12}^2 = 200$	$\sum D_{13}^2 = 60$	$\sum D_{23}^2 = 214$

Rank correlation coefficient between first and second judges,

$$r_{k_{12}} = 1 - \left\{\frac{6\sum D^2}{n(n^2-1)}\right\} = 1 - \left\{\frac{6 \times 200}{10(100-1)}\right\} = -0.212$$

Rank correlation coefficient between first and third judges,

$$r_{k_{13}} = 1 - \left\{\frac{6\sum D^2}{n(n^2-1)}\right\} = 1 - \left\{\frac{6 \times 60}{10(100-1)}\right\} = 0.636$$

Rank correlation coefficient between second and third judges,

$$r_{k_{23}} = 1 - \left\{\frac{6\sum D^2}{n(n^2-1)}\right\} = 1 - \left\{\frac{6 \times 214}{10(100-1)}\right\} = -0.297$$

Correlation between first and second judges is negative *i.e.,* their opinions regarding beauty test are opposite to each other. Similarly opinions of second and third judges are opposite to each other, but the opinions of first and third are of similar type as their correlation is positive. It means that their liking and disliking are very much common.

4.10 REGRESSION

We know that in a functional relation between two variables, if we know the value of one variable, then the corresponding value of the other variable can be determined exactly.

But, in a statistical relationship between the two variables, when the value of one variable is known, we can simply estimate the corresponding value of another variable.

Regression analysis is the method used for estimating the unknown values of one variable, corresponding to the known values of another variables.

Dependent and independent variables: Suppose there is a relation between two variables. *The* variable whose value is known as independent variable, while the another one is called the dependent variable.

4.10.1 Line of Regression

Let $\{(x_i, y_i): 1 \le i \le n\}$ be a bivariate distribution. If we plot the corresponding values of x and y, taking of x-axis and the values of y along y-axis,we obtain a collection of dots, called the scatter-diagram.

If the scatter diagram indicates some relationship between x and y, then the dots of scatter-diagram will be connected round a line, called the *line of regression or the line of best fit.*

Regression line of y on x: *If we have to predict the values of y from given values of x, then the line of regression has an equation of the form* $y = a + bx$. *This is called the regression line of y on x.*

Regression line of x on y: *If we have to predict the values of x from given values of y, then the line of regression has an equation of the form,* $x = a + by$. *This is called the regression line of x on y.*

4.10.2 To Obtain the Equation of Line of Regression of y on x

Suppose that the line approximating the set of points $(x_1, y_1), (x_2, y_2), (x_3, y_3), \ldots\ldots\ldots, (x_n, y_n)$ has the equation

$$y = a + bx \qquad \ldots(1)$$

Let y_{λ_1} be the theoretical value for x_1 then $e_1 = y_1 - y_{\lambda_1}$

$$\Rightarrow \qquad e_1 = y_1 - (a + bx_1)$$

$$\Rightarrow \qquad e_1^2 = (y_1 - a - bx_1)^2; \text{ Similarly } e_2^2 = (y_2 - a - bx_2)^2 \text{ and so on.}$$

$$\therefore \qquad s = e_1^2 + e_2^2 + \ldots + e_n^2$$

$$s = \sum_{i=1}^{n} e_i^2$$

or

$$s = \sum_{i=1}^{n} (y_i - a - bx_i)^2$$

By the principle of least square, the value of s is minimum

$$\therefore \qquad \frac{\partial s}{\partial a} = 0 \text{ and } \frac{\partial s}{\partial b} = 0$$

$$\Rightarrow \qquad 2\sum_{i=1}^{n} (y_i - a - bx_i)(-1) = 0 \text{ and } 2\sum (y_i - a - bx_i)(-x_i) = 0$$

$$\Rightarrow \qquad \boxed{\Sigma y_i = na + b\Sigma x_i} \qquad \ldots(2)$$

and

$$\boxed{\sum x_i y_i = a\sum x_i b \sum x_i^2} \qquad \ldots(3)$$

Equations (2) and (3) are normal equations for this line solving (2) and (3) for a and b and putting these value in (1) we obtain the required equation of the line of regression of y on x.

4.10.3 To obtain the Equation of Line of Regression of x on y

Suppose that the line approximating the set of points $(x_1, y_1), (x_2, y_2), (x_3, y_3), \ldots\ldots\ldots, (x_n, y_n)$ has the equation

$$x = a + by \qquad \ldots(1)$$

Then,

$$\sum x_i = na + b\sum y_i \qquad \ldots(2)$$

and

$$\sum x_i y_i = a\sum y_i + b\sum y_i^2 \qquad \ldots(3)$$

These equations (2) and (3) are normal equations of this line.

Solving (2) and(3) for a and b and putting these values in (1), we get the equation of the line of regression of x on y.

Example 17: Find the line of regression of y on x for the following data:

x	10	9	8	7	6	4	3
y	8	12	7	10	8	9	6

Sol. We have $n = 7$. Now prepare the following table given below:

x_i	y_i	x_i^2	$x_i y_i$
10	8	100	80
9	12	81	108
8	7	64	56
7	10	49	70
6	8	36	48
4	9	16	36
3	6	9	18
$\Sigma x_i = 47$	$\Sigma y_i = 60$	$\Sigma x_i^2 = 355$	$\Sigma x_i y_i = 416$

Let the required line be, $y = a + bx$ ...(1)

The normal equations are

$$\sum y_i = na + b\sum x_i \qquad ...(2)$$

$$\sum x_i y_i = a\sum x_i + b\sum x_i^2 \qquad ...(3)$$

Putting the values from the table in (2) and (3), we get

$$60 = 7a + 47b$$

$$416 = 47a + 355b$$

Solving these equations, we get $a = 6.333$ and $b = 0.333$.

Putting these values in (1) the required equation is $y = 6.333 + 0.333x$.

Example 18: Find the line of regression of x on y for the following data:

x	6	2	10	4	8
y	9	11	5	8	7

Sol. We have $n = 7$. Now prepare the following table given below:

x_i	y_i	x_i^2	$x_i y_i$
6	9	36	54
2	11	4	22
10	5	100	50
4	8	16	32
8	7	64	56
$\Sigma x_i = 30$	$\Sigma y_i = 40$	$\Sigma x_i^2 = 220$	$\Sigma x_i y_i = 214$

Let the required line be, $x = a + by$ ...(1)

$\therefore$ The normal equations are

$$\sum x_i = na + b\sum y_i \quad ...(2)$$

$$\sum x_i y_i = a\sum y_i + b\sum y_i^2 \quad ...(3)$$

Putting the values from the table in (2) and (3), we get

$$30 = 5a + 40b$$

$$214 = 40a + 220b$$

Solving these equations, we get a = 3.92 and b = 0.26.

Putting these values in (1) the required equation is $x = 3.92 + 0.26y$.

Example 19: Find the regression line of y on x for the following data:

x	1	3	4	6	8	9	11	14
y	1	2	4	4	5	7	8	9

Estimate the value of y, where x =10.

Sol. We have n = 8. Now prepare the following table given below:

x_i	y_i	x_i^2	$x_i y_i$
1	1	1	1
3	2	9	6
4	4	16	16
6	4	36	24
8	5	64	40
9	7	81	63
11	8	121	88
14	9	196	126
$\Sigma x_i = 56$	$\Sigma y_i = 40$	$\Sigma x_i^2 = 524$	$\Sigma x_i y_i = 364$

Let the required line be, $y = a + bx$ ....(1)

Then, $y_i = a + bx_i$ and $x_i y_i = ax_i + bx_i^2$ for each i.

$\therefore$ The normal equations are

$$\sum y_i = na + b\sum x_i \quad ...(2)$$

$$\sum x_i y_i = a\sum x_i + b\sum x_i^2 \quad ...(3)$$

Putting the values from the table in (2) and (3), we get

$$40 = 8a + 56b$$

$$364 = 56a + 524b$$

Solving these equations, we get $a = \dfrac{6}{11}$ and $b = \dfrac{7}{11}$.

Putting these values in (1) the required equation is

$$y = \frac{6}{11} + \frac{7}{11}x \text{ or } 7x - 11y + 6 = 0. \quad \textbf{Ans.}$$

If $$x = 10, y = \frac{6}{11} + \frac{7}{11}(10) = \frac{76}{11} = 6\frac{10}{11}. \quad \textbf{Ans.}$$

Example 20: In a study between the amount of rainfall and the quantity of air pollution removed the following data were collected.

Daily rainfall in 0.01 *cm*	4.3	4.5	5.9	5.6	6.1	5.2	3.8	2.1
Pollution removed (mg/m^3*)*	12.6	12.1	11.6	11.8	11.4	11.8	13.2	14.1

Find the regression line of y on x.

Sol. We have $n = 8$. Now prepare the following table given below:

x_i (metre)	y_i	x_i^2	$x_i y_i$
4.3	12.6	18.49	54.18
4.5	12.1	20.25	54.45
5.9	11.6	34.81	68.44
5.6	11.8	31.36	66.08
6.1	11.4	37.21	69.54
5.2	11.8	27.04	61.36
3.8	13.2	14.44	50.16
2.1	14.1	4.41	29.61
$\Sigma x_i = 37.5$	$\Sigma y_i = 98.6$	$\Sigma x_i^2 = 188.01$	$\Sigma x_i y_i = 453.82$

Let the required line be, $y = a + bx$ ....(1)

Then, $y_i = a + bx_i$ and $x_i y_i = ax_i + bx_i^2$ for each i.

∴ The normal equations are

$$\sum y_i = na + b\sum x_i \quad ...(2)$$

$$\sum x_i y_i = a\sum x_i + b\sum x_i^2 \quad ...(3)$$

Putting the values from the table in (2) and (3), we get

$$98.6 = 8a + 37.5b$$

$$453.82 = 37.5a + 188.01b$$

Solving these equations, we get $a = 15.49$ and $b = -0.675$

Putting these values in (1) the required equation is $y = 15.49 - 0.675x$.

4.11 ANOTHER FORM OF EQUATIONS OF LINES OF REGRESSION

Theorem 1: Show that the equation of the line of regression of y on x is given by

$$y-\bar{y}=r.\frac{\sigma_y}{\sigma_x}(x-\bar{x})$$

where $\bar{x}$ and $\bar{y}$ are the means of x-axis series and y-axis series respectively r is the coefficient of correlation between x and y; σ_x and σ_y are the standard deviations of x-series and y-series respectively.

Proof: Suppose that the line approximating the set of points (x_1, y_1), (x_2, y_2), (x_3, y_3),...... (x_n, y_n), has the equation $y=a+bx.$ ...(1)

Then $y_i=a+bx_i$ and $x_iy_i=ax_i+bx_i^2$ for each $i=1, 2,\ldots, n$

$$\sum y_i=na+b\sum x_i \quad \text{...(2)}$$

$$\sum x_i y_i=a\sum x_i+b\sum x_i^2 \quad \text{...(3)}$$

From (2), we have $\dfrac{\sum y_i}{n}=a+b.\dfrac{\sum x_i}{n}$

or $\bar{y}=a+b\bar{x}$ ...(4)

Thus, it follows that $(\bar{x},\bar{y})$ lies on the line, $y=a+bx$.

Shifting the origin to $(\bar{x},\bar{y})$, (2) becomes

$$\sum(y_i-\bar{y})=na+b\sum(x_i-\bar{x}) \text{ or } a=0 \quad \left[\because \sum(x_i-\bar{x})=\sum(y_i-\bar{y})=0\right]$$

Shifting the origin to $(\bar{x},\bar{y})$ and taking $a = 0$

(1) becomes $(y-\bar{y})=b(x-\bar{x})$ ...(5)

(2) becomes $\sum(x_i-\bar{x})(y_i-\bar{y})=b\sum(x_i-\bar{x})^2$ ...(6)

From (6), we have

$$b=\frac{\sum(x_i-\bar{x})(y_i-\bar{y})}{\sum(x_i-\bar{x})^2}=\frac{\sum(dx_i.dy_i)}{\sum(dx_i)^2}$$

$$=\frac{\sum(dx_i.dy_i)}{n.(\sigma_x)^2}=r.\frac{\sigma_y}{\sigma_x} \quad \left[\therefore r=\frac{\sum(dx_i.dy_i)}{n(\sigma_x.\sigma_y)}\right]$$

Putting this value of b in (5).

The required equation of the line of regression of y on x is

$$y-\bar{y}=r.\frac{\sigma_y}{\sigma_x}(x-\bar{x})$$

***Coefficient of regression of y on x*:** The real number $b = r.\frac{\sigma_y}{\sigma_x}$ is called the coefficient of regression of y on x and it is denoted by b_{yx}. Thus $b_{yx} = r.\frac{\sigma_y}{\sigma_x}$.

Theorem 2: The equation of the line of regression of x on y is given by

$$(x - \bar{x}) = r.\frac{\sigma_x}{\sigma_y}(y - \bar{y}).$$

Proof: Proceed as in theorem 1.

Coefficient of regression of x on y: The real number, $r.\frac{\sigma_x}{\sigma_y}$ is called the coefficient of regression of x on y. and it is denoted by b_{xy}. Thus $b_{xy} = r \cdot \frac{\sigma x}{\sigma y}$

Theorem 3: Prove that

(a) The coefficient of regression of y on x is given by

$$b_{yx} = \frac{\sum x_i y_i - \frac{(\sum x_i)(\sum y_i)}{n}}{\left\{\sum x_i^2 - \frac{(\sum x_i)^2}{n}\right\}}$$

(b) The coefficient of regression of x on y is given by

$$b_{xy} = \frac{\sum x_i y_i - \frac{(\sum x_i)(\sum y_i)}{n}}{\left\{\sum y_i^2 - \frac{(\sum y_i)^2}{n}\right\}}$$

Proof: By definition, we have

$$b_{yx} = r.\frac{\sigma_y}{\sigma_x} = r.\frac{\sigma_y.\sigma_x}{(\sigma_x)^2}$$

$$= \frac{\text{cov}(x, y)}{(\sigma_x)^2} \qquad \left[\because r.\sigma_y.\sigma_x = \text{cov}(x, y)\right]$$

$$= \frac{\sum x_i y_i - \frac{(\sum x_i)(\sum y_i)}{n}}{\left\{\sum x_i^2 - \frac{(\sum x_i)^2}{n}\right\}} = \frac{n\sum xy - \sum x \sum y}{n\sum x^2 - (\sum x)^2}$$

(*b*) Similarly $$b_{xy} = \frac{n\sum xy - \sum x \sum y}{n\sum y^2 - \left(\sum y\right)^2}$$ [Drop the suffix]

Example 21: Find the regression coefficient b_{yx}, between x and y for the following data:

$$\sum x = 24, \sum y = 44, \sum xy = 306, \sum x^2 = 164, \sum y^2 = 574 \text{ and } n = 4.$$

Sol. The given data may be written as

$$\sum x_i = 24, \sum y_i = 44, \sum x_i y_i = 306, \sum x_i^2 = 164, \sum y_i^2 = 574,\ n = 4$$

$$\therefore \quad b_{yx} = \frac{\sum x_i y_i - \frac{\left(\sum x_i\right)\left(\sum y_i\right)}{n}}{\left\{\sum x_i^2 - \frac{\left(\sum x_i\right)^2}{n}\right\}} = \frac{\left\{306 - \frac{24 \times 44}{4}\right\}}{\left\{164 - \frac{(24)^2}{4}\right\}}$$

$$= \frac{306 - 264}{164 - 144} = \frac{42}{20} = 2.1. \quad \textbf{Ans.}$$

Example 22: Find the regression coefficient b_{xy}, between x and y for the following data:

$$\sum x = 30, \sum y = 42, \sum xy = 199, \sum x^2 = 184, \sum y^2 = 318 \text{ and } n = 6.$$

Sol. The given equation can be written as:

$$\sum x_i = 30, \sum y_i = 42, \sum x_i y_i = 199, \sum x_i^2 = 184, \sum y_i^2 = 318 \quad n = 6$$

$$\therefore \quad b_{xy} = \frac{\sum x_i y_i - \frac{\left(\sum x_i\right)\left(\sum y_i\right)}{n}}{\left\{\sum y^2_i - \frac{\left(\sum y_i\right)^2}{n}\right\}}$$

$$= \frac{\left\{199 - \frac{30 \times 42}{6}\right\}}{\left(318 - \frac{42 \times 42}{6}\right)} = \frac{199 - 210}{318 - 294} = \frac{-11}{24} = -0.46. \quad \textbf{Ans.}$$

Example 23: Find the following observations (x, y), find the regression coefficient b_{yx} and b_{xy} and hence find the correlation coefficient between x and y.

(1, 2), (2, 4), (3, 8), (4, 7), (5, 10), (6, 5), (7, 14), (8, 16), (9, 2), (10, 20).

Sol. Here n = 10. We may prepare the table, given below:

x_i	y_i	x_i^2	y_i^2	$x_i y_i$
1	2	1	4	2
2	4	4	16	8
3	8	9	64	24
4	7	16	49	28
5	10	25	100	50
6	5	36	25	30
7	14	49	196	98
8	16	64	256	128
9	2	81	4	18
10	20	100	400	200
$\sum x_i = 55$	$\sum y_i = 88$	$\sum x_i^2 = 385$	$\sum y_i^2 = 1114$	$\sum x_i y_i = 586$

$$\therefore \qquad b_{yx} = \frac{\sum x_i y_i - \dfrac{(\sum x_i)(\sum y_i)}{n}}{\left\{\sum x^2{}_i - \dfrac{(\sum x_i)^2}{n}\right\}} = \frac{\left\{586 - \dfrac{55 \times 88}{10}\right\}}{\left\{385 - \dfrac{(55)^2}{10}\right\}} = \frac{102}{82.5} = 1.24$$

and
$$b_{xy} = \frac{\sum x_i y_i - \dfrac{(\sum x_i)(\sum y_i)}{n}}{\left\{\sum y^2{}_i - \dfrac{(\sum y_i)^2}{n}\right\}} = \frac{\left\{586 - \dfrac{55 \times 88}{10}\right\}}{\left(1114 - \dfrac{88 \times 88}{10}\right)} = \frac{102}{339.6} = 0.30$$

Now $b_{yx} \cdot b_{xy} = \left(r \cdot \dfrac{\sigma_y}{\sigma_x}\right) . \left(r \cdot \dfrac{\sigma_x}{\sigma_y}\right) = r^2$, where r is the coefficient of correlation.

$$\therefore \qquad r = \sqrt{b_{yx} \cdot b_{xy}} = \sqrt{1.24 \times 0.30} = 0.609.$$

Thus $\qquad b_{yx} = 1.24, b_{xy} = 0.30$ and $r = 0.609.$ **Ans.**

4.12 SOME PROPERTIES OF REGRESSION COEFFICIENTS

Let the regression coefficient of y on $x = b_{yx}$; the regression coefficient of x on y = b_{xy}; and the correlation coefficient between x and y = r. Then, we have the following results.

Theorem 1: Prove that $r = \sqrt{b_{yx} . b_{xy}}$

Proof: We have $b_{yx} = r . \dfrac{\sigma_y}{\sigma_x}$ and $b_{xy} = r . \dfrac{\sigma_x}{\sigma_y}$

$\therefore \qquad b_{yx}.b_{xy} = r^2$ or $r = \sqrt{b_{yx}.b_{xy}}$.

Remark: Clearly we can say that, the correlation coefficient is the geometric mean between the two regression coefficients.

Theorem 2: Prove that r, b_{yx} and b_{xy} are of the same sign.

Proof: We know that $b_{yx} = r.\dfrac{\sigma_y}{\sigma_x}$ and $b_{xy} = r.\dfrac{\sigma_x}{\sigma_y}$

Since σ_x and σ_y are both positive, it follows from the two equations, given above that r, b_{yx} and b_{xy} have the same sign as r.

Hence r, b_{yx} and b_{xy} are always same sign.

Theorem 3: Prove that the arithmetic mean of regression coefficients is greater than the correlation coefficient.

Proof: Clearly the required result is true,

If $\dfrac{1}{2}(b_{yx} + b_{xy}) > r$ *i.e.*, if $\dfrac{1}{2}\left[r.\dfrac{\sigma_y}{\sigma_x} + r.\dfrac{\sigma_x}{\sigma_y}\right] > r$

i.e. if $\sigma_y^2 + \sigma_x^2 > 2\sigma_x\sigma_y$

i.e. if $\left(\sigma_y^2 + \sigma_x^2\right) - 2\sigma_x\sigma_y > 0$

i.e. if $\left(\sigma_y - \sigma_x\right)^2 > 0$, which is true.

Hence, the required result is true.

Theorem 4: Let θ be the angle between the regression line of y on x and the regression line of x on y.

Then, prove that $$\tan\theta = \left\{\frac{(1-r^2)}{r} \cdot \frac{\sigma_x.\sigma_y}{(\sigma_x^2 + \sigma_y^2)}\right\}.$$

Proof: The equation of the line of regression of x on y is

$$(x - \bar{x}) = r \cdot \frac{\sigma_x}{\sigma_y}(y - \bar{y}) \qquad \text{...(1)}$$

and the equation of the line of regression of y on x is

$$(y - \bar{y}) = r \cdot \frac{\sigma_y}{\sigma_x}(x - \bar{x}) \qquad \text{...(2)}$$

Let m_1 and m_2 be the slopes of (1) and (2) respectively.

Then, $$m_1 = \frac{\sigma_y}{r.\sigma_x} \text{ and } m_2 = \frac{r.\sigma_y}{\sigma_x}$$

$$\therefore \qquad \tan\theta = \frac{(m_1 - m_2)}{(1 + m_1 m_2)}$$

$$= \frac{\frac{\sigma_y}{r.\sigma_x} - \frac{r.\sigma_y}{\sigma_x}}{1 + \frac{(\sigma_y)^2}{(\sigma_x)^2}}$$

or $$\tan\theta = \left\{ \frac{(1-r^2)}{r} \cdot \frac{\sigma_x.\sigma_y}{(\sigma_x^2 + \sigma_y^2)} \right\}$$

Corollary 1: If $r = 0$, we get $\tan \theta = \infty$ or $\theta = \frac{\pi}{2}$, *i.e.*, two lines of regression one perpendicular each other.

Corollary 2: If $r = \pm 1$, we get $\tan \theta = 0 \Rightarrow \theta = 0$, since θ is acute *i.e.*, the lines of regression coincide and there is a perfect correlation between the two variates x and y.

Example 24: The line of regression of x on y and of y on x are respectively

$$x = 19.13 - 0.87y \text{ and } y = 11.64 - 0.50x.$$

Find (*a*) The mean of x-series;

(*b*) The mean of y-series;

(*c*) The correlation coefficient between x and y.

Sol. Let the mean of x-series be $\bar{x}$ and that of y-series be $\bar{y}$.

Since the line of regression pass through ($\bar{x}$, $\bar{y}$), we have

$$\bar{x} = 19.13 - 0.87\bar{y} \text{ or } \bar{x} + 0.87\bar{y} = 19.13 \quad ...(1)$$

and $$\bar{y} = 11.64 - 0.50\bar{x} \text{ or } 0.50\bar{x} + \bar{y} = 11.64 \quad ...(2)$$

On solving (1) and (2), we get $\bar{x} = 15.94$ and $\bar{y} = 3.67$.

$\therefore$ mean of x-series = 15.94.

and mean of y-series = 3.67.

Now, the line of regression of y on x is

$$y = 11.94 - 0.50x \qquad \therefore\ b_{yx} = -0.50$$

Also, the line of regression of x on y is

$$x = 19.13 - 0.87y \qquad \therefore\ b_{xy} = -0.87.$$

$\therefore$ $$r = \pm\sqrt{b_{yx}.b_{xy}} = \pm\sqrt{(-0.50)(-0.87)} = \pm\sqrt{.435} = -0.66. \quad \textbf{Ans.}$$

Clearly, r is taken as negative, since each one of b_{yx} and b_{xy} is negative.

Example 25: Out of the following two regression lines, find the line of regression of x on y:

$$2x + 3y = 7 \text{ and } 5x + 4y = 9.$$

Sol. Let $2x + 3y = 7$ be the regression line of x on y.

Then, $5x + 4y = 9$ is the regression line of y on x.

$\therefore$ $$2x + 3y = 7 \text{ and } 5x + 4y = 9$$

$\Rightarrow$ $$x = -\frac{3}{2}y + \frac{7}{2} \text{ and } y = -\frac{5}{4}x + \frac{9}{4} \Rightarrow b_{xy} = -\frac{3}{2} \text{ and } b_{yx} = -\frac{5}{4}$$

$$\Rightarrow \qquad r = \pm\sqrt{b_{yx} \cdot b_{xy}} = -\sqrt{\left(-\frac{3}{2}\right)\left(-\frac{5}{4}\right)} = -\sqrt{\frac{15}{8}} < -1, \text{ which is impossible.}$$

$\therefore$ Our choice of regression line is incorrect.

Hence, the regression line of x on y is $5x + 4y = 9$. **Ans.**

Example 26: Find the correlation coefficient between x and y, when the lines of regression are:

$$2x - 9y + 6 = 0 \text{ and } x - 2y + 1 = 0.$$

Sol. Let the line of regression of x on y be $2x - 9y + 6 = 0$.

Then the line of regression of y on x is $x - 2y + 1 = 0$.

$$\therefore \qquad 2x - 9y + 6 = 0 \text{ and } x - 2y + 1 = 0.$$

$$\Rightarrow \qquad x = \frac{9}{2}y - 3 \text{ and } y = \frac{1}{2}x + \frac{1}{2} \Rightarrow b_{xy} = \frac{9}{2} \text{ and } b_{yx} = \frac{1}{2}$$

$$\Rightarrow \qquad r = \pm\sqrt{b_{yx}.b_{xy}} = \pm\sqrt{\frac{9}{2} \times \frac{1}{2}} = \frac{3}{2} > 1, \text{ which is impossible.}$$

$\therefore$ Our choice of regression line is incorrect.

Hence, the regression line of x on y is $x - 2y + 1 = 0$.

and, the regression line of y on x is $2x - 9y + 6 = 0$.

$\therefore$ $x - 2y + 1 = 0$ and $2x - 9y + 6 = 0$

$$\Rightarrow \qquad x = 2y - 1 \text{ and } y = \frac{2}{9}x + \frac{2}{3} \Rightarrow b_{xy} = 2 \text{ and } b_{yx} = \frac{2}{9}$$

$$\Rightarrow \qquad r = \sqrt{b_{yx}.b_{xy}} = \sqrt{2 \times \frac{2}{9}} = \frac{2}{3}.$$

Hence, the correlation coefficient between x and y is $\frac{2}{3}$.

Example 27: The equation of the two lines of regression are:

$$3x + 12y = 19 \text{ and } 3y + 9x = 46.$$

Find (*a*) The mean of x-series,

(*b*) The mean of y-series,

(*c*) Regression coefficients b_{yx} and b_{xy},

(*d*) Correlation coefficient between x and y.

Sol. Let the mean of x-series be $\bar{x}$ and the mean of y-series be $\bar{y}$. Then, each of the given lines passes through $(\bar{x}, \bar{y})$

$$3\bar{x} + 12\bar{y} = 19 \qquad \text{...(1)}$$

and

$$9\bar{x} + 3\bar{y} = 46 \qquad \text{...(2)}$$

On solving (1) and (2), we get $\bar{x} = 5$ and $\bar{y} = \frac{1}{3}$

$\therefore$ mean of x-series = 5.

and mean of y-series = $\frac{1}{3}$

Now, let the line of regression of x on y be $3x + 12y = 19$.
Then the line of regression of y on x is $3y + 9x = 46$.

$\therefore \quad 3x + 12y = 19$ and $3y + 9x = 46$

$$\Rightarrow \quad x = -4y + \frac{19}{3} \text{ and } y = -3x + \frac{46}{3} \Rightarrow b_{xy} = -4 \text{ and } b_{yx} = -3$$

$$\therefore \quad r = \pm\sqrt{b_{yx}.b_{xy}} = \pm\sqrt{(-4)(-3)} = -2\sqrt{3} < -1, \text{ which is impossible.}$$

$\therefore$ Our choice of regression line is incorrect.

Consequently, the regression line of x on y is $3y + 9x = 46$.

and, the regression line of y on x is $3x + 12y = 19$.

$\therefore$ $3y + 9x = 46$ and $3x + 12y = 19$.

$$\Rightarrow \quad x = -\frac{1}{3}y + \frac{46}{9} \text{ and } y = -\frac{1}{4}x + \frac{19}{12} \quad \Rightarrow \quad b_{xy} = -\frac{1}{3}, b_{yx} = -\frac{1}{4}$$

$$\therefore \quad r = \pm\sqrt{b_{yx}.b_{xy}} = \pm\sqrt{\left(-\frac{1}{3}\right)\left(\frac{-1}{4}\right)} = -\frac{1}{2\sqrt{3}} = -\frac{\sqrt{3}}{6}. \quad \text{[as, } b_{xy} \text{ and } b_{yx} \text{ have the same sign]}$$

Example 28: You are given the following data:

Series	x	y
Mean	18	100
Standard deviation	14	20

Correlation coefficient between x and y is 0.8.

Find the two regression lines.

Estimate the value of y, when $x = 70$.

Estimate the value of x, when $y = 90$.

Sol. It being given that

$$\bar{x} = 18, \bar{y} = 100, \sigma_x = 14, \sigma_y = 20 \text{ and } r = 0.8.$$

$\therefore$ The line of regression of y on x is

$$(y - \bar{y}) = r.\frac{\sigma_y}{\sigma_x}(x - \bar{x})$$

or $$(y - 100) = \left(\frac{0.8 \times 20}{14}\right)(x - 18)$$

or $$y = 1.14x + 79.41$$

When $x = 70$, we have $\quad y = (1.14 \times 70 + 79.41) = 159.21$

and, the line of regression of x on y is

$$(x - \bar{x}) = r.\frac{\sigma_x}{\sigma_y}(y - \bar{y})$$

or $$(x - 18) = 0.8 \times \frac{14}{20}(y - 100) \text{ or } x = 0.56y - 38$$

When $y = 90$, we have $\quad x = (0.56 \times 90 - 38) = 12.4.$

4.12.1 To find b_{yx} and b_{xy} and Using Assumed Mean

Let the assumed mean of x-series and y-series be A and B respectively. Then, taking $dx_i = (x_i - A)$ and $dy_i = (y_i - B)$, we have

$$b_{yx} = \frac{\sum (dx_i . dy_i) - \frac{(\sum dx_i)(\sum dy_i)}{n}}{\left\{\sum (dx_i)^2 - \frac{(\sum dx_i)^2}{n}\right\}}$$

and

$$b_{xy} = \frac{\sum (dx_i . dy_i) - \frac{(\sum dx_i)(\sum dy_i)}{n}}{\left\{\sum (dy_i)^2 - \frac{(\sum dy_i)^2}{n}\right\}}$$

Example 29: Find the regression coefficients and hence the equation of the two lines of regression from the following data:

Age of husband (x)	25	22	28	26	35	20	22	40	20	18
Age of wife (y)	18	15	20	17	22	14	16	21	15	14

Hence estimate

(a) The age of wife, when the age of husband is 30.

(b) The age of husband, when the age of wife is 19.

Sol. We have:

$$\bar{x} = \frac{\sum x_i}{n} = \frac{256}{10} = 25.6 \approx 26 \text{ and } \bar{y} = \frac{\sum y_i}{n} = \frac{172}{10} = 17.2 \approx 17$$

Let assumed means of x-series and y-series be 26 and 17 respectively. Then, we may prepare the table given below.

x_i	y_i	$dx_i = (x_i - 26)$	$dy_i = (y_i - 17)$	$(dx_i)^2$	$(dy_i)^2$	$dx_i\, dy_i$
25	18	–1	1	1	1	–1
22	15	–4	–2	16	4	8
28	20	2	3	4	9	6
26	17	0	0	0	0	0
35	22	9	5	81	25	45
20	14	–6	–3	36	9	18
22	16	–4	–1	16	1	4
40	21	14	4	196	16	56
20	15	–6	–2	36	4	12
18	14	–8	–3	64	9	24
$\Sigma x_i = 256$	$\Sigma y_i = 172$	$\Sigma dx_i = -4$	$\Sigma dy_i = 2$	$\Sigma(dx_i)^2 = 450$	$\Sigma(dy_i)^2 = 78$	$\Sigma dx_i\, dy_i = 172$

$$\therefore \quad b_{xy} = \frac{\sum (dx_i.dy_i) - \frac{(\sum dx_i)(\sum dy_i)}{n}}{\left\{\sum (dy_i)^2 - \frac{(\sum dy_i)^2}{n}\right\}} = \frac{172 - \frac{(-4)(2)}{10}}{\left\{78 - \frac{(2)^2}{10}\right\}}$$

$$= \frac{172 + 0.8}{78 - 0.4} = \frac{172.8}{77.6} = 2.226$$

and

$$b_{yx} = \frac{\sum (dx_i.dy_i) - \frac{(\sum dx_i)(\sum dy_i)}{n}}{\left\{\sum (dx_i)^2 - \frac{(\sum dx_i)^2}{n}\right\}} = \frac{172 - \frac{(-4)(2)}{10}}{\left\{450 - \frac{16}{10}\right\}} = \frac{172 + 0.8}{450 - 1.6} = \frac{172.8}{448.4} = 0.385$$

$\therefore$ The equation of the line of regression of y on x is

$$(y - \bar{y}) = b_{yx}.(x - \bar{x}) \text{ or } (y - 17.2) = 0.385\,(x - 25.6)$$

Now, when $x = 30$, we get

$$y - 17.2 = 0.385\,(30 - 25.6) \text{ or } y = 19 \text{ (nearly)}$$

$\therefore$ When the age of husband is 30 years, the estimated age of the wife is 19 years.
Again, the equation of the line of regression of x on y is

$$(x - \bar{x}) = b_{xy}(y - \bar{y}) \text{ or } (x - 25.6) = 2.26(y - 17.2)$$

Thus, when $y = 19$, we get $x = 30$(nearly).
So, when the age of wife is 19 years, the estimate age of husband is 30 years.

4.13 ERROR OF PREDICTION

The deviation of the predict value from the observed value is known as the standard error of prediction. It is given by

$$E_{yx} = \sqrt{\frac{\sum (y - y_p)^2}{n}}$$

where y is the actual value and y_p the predicted value.

Theorem: Prove that

(*a*) $E_{yx} = \sigma_y.\sqrt{(1 - r^2)}$, (*b*) $E_{xy} = \sigma_x.\sqrt{(1 - r^2)}$.

Proof: The equation of the line of regression of y on x is

$$(y - \bar{y}) = r.\frac{\sigma_y}{\sigma_x}(x - \bar{x})$$

$$\therefore \quad y_p = \bar{y} + r.\frac{\sigma_y}{\sigma_x}(x - \bar{x})$$

So $$E_{yx} = \sqrt{\frac{\sum(y-y_p)^2}{n}} = \left[\frac{1}{n}\sum\left\{y-\bar{y}-r.\frac{\sigma_y}{\sigma_x}(x-\bar{x})\right\}^2\right]^{\frac{1}{2}}$$

$$= \left[\frac{1}{n}\left\{\sum\{(y-\bar{y})^2 + \frac{r^2.\sigma_y^2.(x-\bar{x})^2}{\sigma_x^2} - \frac{2r.\sigma_y}{\sigma_x}.(x-\bar{x})(y-\bar{y})\}\right\}\right]^{\frac{1}{2}}$$

$$= \left\{\frac{\sum(y-\bar{y})^2}{n} + \frac{r^2.\sigma_y^2}{\sigma_x^2}\frac{\sum(x-\bar{x})^2}{n} - \frac{2r.\sigma_y}{\sigma_x}.\frac{\sum(x-\bar{x})(y-\bar{y})}{n}\right\}^{\frac{1}{2}}$$

$$= \left\{\sigma_y^2 + \frac{r^2.\sigma_y^2}{\sigma_x^2}.\sigma_x^2 - \frac{2r.\sigma_y}{\sigma_x}.r.\sigma_x.\sigma_y\right\}^{\frac{1}{2}}$$

$$= (\sigma_y^2 - r^2.\sigma_y^2)^{\frac{1}{2}} = \sigma_y.\sqrt{1-r^2}$$

Similarly, (*b*) may be proved.

Example 30: For the data given below, find the standard error of estimate of y on x.

x	1	2	3	4	5
y	2	5	3	8	7

Sol. We leave it to the reader to find the line of regression of y on x. This is, $y = 1.3x + 1.1$.

So, $y_p = 1.3x + 1.1$. Now we prepare the table given below.

x	y	$y_p = 1.3x + 1.1$	$(y - y_p)$	$(y - y_p)^2$
1	2	2.4	–0.4	0.16
2	5	3.7	1.3	1.69
3	3	5	–2	4
4	8	6.3	1.7	2.89
5	7	7.6	–0.6	0.36
				$\Sigma(y - y_p)^2 = 9.10$

$$\therefore \quad E_{yx} = \sqrt{\frac{\sum(y-y_p)^2}{n}} = \sqrt{\frac{9.10}{5}} = 1.349.$$

Example 31: Obtain the line of regression of y on x for the data given below:

x :	1.53	1.78	2.60	2.95	3.42
y :	33.50	36.30	40.00	45.80	53.50

Sol.

x	y	x^2	xy
1.53	33.50	2.3409	51.255
1.78	36.30	2.1684	64.614
2.60	40.00	6.76	104.000
2.95	45.80	8.7025	135.11
3.42	53.50	11.6964	182.97
$\Sigma x = 12.28$	$\Sigma y = 209.1$	$\Sigma x^2 = 32.6682$	$\Sigma xy = 537.949$

We know that the line of regression of y on x is

$$y - \bar{y} = b_{yx}(x - \bar{x}) \qquad \text{...(1)}$$

where
$$b_{yx} = \frac{n\Sigma xy - \Sigma x \Sigma y}{n\Sigma x^2 - (\Sigma x)^2} = \frac{5 \times 537.949 - 12.28 \times 209.1}{5 \times 32.6682 - (12.28)^2}$$

$$= \frac{121.997}{12.543} = 9.726$$

and
$$\bar{x} = \frac{\Sigma x}{n} = \frac{12.28}{5} = 2.456, \quad \bar{y} = \frac{\Sigma y}{n} = \frac{209.1}{5} = 41.82$$

$\therefore$ From (1), we obtain

$$y - 41.82 = 9.726\ (x - 2.456) = 9.726x - 23.887$$

$\Rightarrow$ $y = 17.932 + 9.726x.$ **Ans.**

Example 32: The following data regrading the heights (y) and weights (x) of 100 college students are given:

$\Sigma x = 15000$, $\Sigma x^2 = 2272500$, $\Sigma y = 6800$, $\Sigma y^2 = 463025$ and

$\Sigma xy = 1022250$, find the equation of regression line of height on weight.

Sol. We known that

$$y - \bar{y} = b_{y_x}(x - \bar{x}) \qquad \text{...(1)}$$

Now
$$\bar{x} = \frac{15000}{100} = 150, \quad \bar{y} = \frac{6800}{100} = 68$$

and
$$b_{y_x} = \frac{n\Sigma xy - \Sigma x \Sigma y}{n\Sigma x^2 - (\Sigma x)^2} = \frac{100 \times 1022250 - 15000 \times 6800}{100 \times 2272500 - (15000)^2} = \frac{225000}{2250000} = 0.1$$

Example 33: If the tangent of the angle between the lines of regression is 0.6 and $\sigma_x = \sigma_{y/2}$ find the correlation coefficient

Sol. $\tan\theta = \left[\left(\frac{1-r^2}{r}\right)\frac{\sigma_x\sigma_y}{\sigma_x^2+\sigma_y^2}\right]$

$$\therefore \qquad \tan\theta = \left(\frac{1-r^2}{r}\right)\cdot\left(\frac{2\sigma_x^2}{5\sigma_x^2}\right) \qquad (\because 2\sigma_x = \sigma_y)$$

Given $\qquad \tan\theta = 0.6$

$$\therefore \qquad 0.6 = \frac{1-r^2}{r}\left(\frac{2}{5}\right)$$

$$\Rightarrow \qquad 3r = 2(1-r^2)$$

$$\Rightarrow \qquad 2r^2 + 3r - 2 = 0$$

$(2r-1)(r+2) = 0 \quad \therefore \quad r = 0.5$ (other value is not possible)

Example 34: The variable x and y are connected by the equation $ax + by + c = 0$. Show that $r_{xy} = -1$ or 1 according as a and b are of the same sign or of opposite sign.

Sol. We have $ax+by+c=0 \Rightarrow y = -\frac{a}{b}x - \frac{c}{b} \quad \therefore b_{yx} = -\frac{a}{b}$

Again writing $ax + by + c = 0$ in the form

$$x = -\frac{b}{a}y - \frac{c}{a} \quad \therefore \quad b_{xy} = -\frac{b}{a}$$

Now $\qquad r^2 = b_{yx}b_{xy} = \left(-\frac{a}{b}\right)\left(-\frac{b}{a}\right) = 1$

If a and b are if some signs, then $r = -1$ (since b_{xy} and b_{yx} are positive).

If a and b are if opposite signs, then $r = 1$ (since b_{yx} and b_{xy} are positive).

$\therefore$ From (1), we get

$$y - 68 = 0.1\,(x - 150) = 0.1x - 15$$

$$\Rightarrow \qquad y = 0.1x + 53. \quad \textbf{Ans.}$$

Example 35: In a partially destroyed laboratory record, only the lines of regression of y on x and x on y are available as

$$4x - 5y + 33 = 0 \quad \text{and} \quad 20x - 9y = 107$$

respectively. Calculate $\bar{x}, \bar{y}$ and the coefficient of correlation between x and y.

Sol. Let $\bar{x}$ and $\bar{y}$ are the mean of x and y series respectively. Then each satisfies the given equation of line of regression.

$$\therefore \qquad 4\bar{x} - 5\bar{y} + 33 = 0 \qquad ...(1)$$

$$20\bar{x} - 9\bar{y} = 107 \qquad ...(2)$$

On solving (1) and (2), we get $\bar{x} = 13$ and $\bar{y} = 17$

$\therefore$ Mean of x series *i.e.*, $\bar{x} = 13$

and mean of y series *i.e.*, $\bar{y} = 17$

Next of the regression line of y on x is

$$4x - 5y + 33 \Rightarrow y = \frac{33}{5} + \frac{4}{5}x$$

$$\therefore \qquad b_{yx} = \frac{4}{5}$$

And the regression line of x on y is

$$20x - 9y = 107 \Rightarrow \frac{107}{20} + \frac{9}{20}y$$

$$\therefore \qquad b_{xy} = \frac{9}{20}$$

Hence $$r = \pm\sqrt{b_{yx}b_{xy}} = \pm\sqrt{\frac{4}{5} \times \frac{9}{20}} = \pm\sqrt{\frac{36}{100}} = +\frac{6}{10} = 0.6. \quad \textbf{Ans.}$$

Example 36: A study of prices of a certain commodity at Hapur and Kanpur yields the following data:

	Hapur (Rs.)	Kanpur (Rs.)
Average price/kilo	2.463	2.797
Standard deviation	0.326	0.207

'r' between prices at Hapur and Kanpur is 0.774.

Estimate, from the above data, the most likely price at Hapur corresponding to the price of Rs. 3.052 per kilo at Kanpur.

Sol. Let $$\bar{x} = 2.463, \bar{y} = 2.797, \sigma_x = 0.326, \sigma_y = 0.207$$

and $$r = 0.774$$

$$\therefore \qquad b_{xy} = r\frac{\sigma_x}{\sigma_y} = 0.774 \times \frac{0.326}{0.207} = 1.218956$$

Regression line of x on y is given by

$$x - \bar{x} = b_{xy}(y - \bar{y})$$

$$\Rightarrow \quad x - 2.463 = 1.218956(y - 2.797)$$

$$\Rightarrow \quad x = 1.218956y - 0.9464213$$

when $y = 3.052$

$$x = 1.218956 \times 3.052 - 0.9464213$$

$$= 2.7738. \quad \textbf{Ans.}$$

Example 37: For 10 observations on price (x) and supply (y), the following data were obtained (in appropriate units)

$$\Sigma x = 130, \Sigma y = 220, \Sigma x^2 = 2288, \Sigma y^2 = 5506 \text{ and } \Sigma xy = 3467$$

Obtain the two lines of regression and estimate the supply when price is 16 units.

Sol. We know that the regression line of y on x and regression line of x on y are respectively

$$y - \bar{y} = b_{yx}(x - \bar{x}) \quad ...(1)$$

and

$$x - \bar{x} = b_{xy}(y - \bar{y}) \quad ...(2)$$

$$\therefore \quad \bar{x} = \frac{\Sigma x}{n} = \frac{130}{10} = 13, \bar{y} = \frac{\Sigma y}{n} = \frac{220}{10} = 22$$

$$b_{yx} = \frac{n\Sigma xy - \Sigma x \Sigma y}{n\Sigma x^2 - (\Sigma x)^2} = \frac{10 \times 3467 - 130 \times 220}{10 \times 2288 - (130)^2} = \frac{6070}{5980} = 1.015$$

Similarly

$$b_{xy} = \frac{n\Sigma xy - \Sigma x \Sigma y}{n\Sigma y^2 - (\Sigma y)^2} = \frac{6070}{6660} = 0.9114$$

From equations (1) and (2), we get

$$y - 22 = 1.015(x - 13) \Rightarrow y = 1.015 + 8.805$$

and

$$x - 13 = 0.9114(y - 22) \Rightarrow x = 0.9114y - 7.0508 \quad \textbf{Ans.}$$

when $x = 16$ units,

$$y = 1.015 \times 16 + 8.805 = 25.045 \text{ units.} \quad \textbf{Ans.}$$

Example 38: Partially destroyed laboratory record of an analysis of correlation data, the following results are only eligible variance of $x = 9$, Regression lines are $8x - 10y + 66 = 0$ and $40x - 18y = 214$. What were

(*i*) Themean value of x and y,

(*ii*) The correlation coefficient of x and y,

(*iii*) Standard deviation of y. [*U.T.U. 2008*]

Sol. (*i*) We have $8x - 10y = -66$, $40x - 18y = 214$

Since the two regression lines pass through $(\bar{x}, \bar{y})$

$$\therefore \quad 8\bar{x} - 10\bar{y} = -66 \quad ...(i)$$

$$40\bar{x} - 18\bar{y} = 214 \quad ...(ii)$$

Solving equations (*i*) and (*ii*), we get

$$\bar{x} = 13$$

$$\bar{y} = 17. \quad \textbf{Ans.}$$

(*ii*) Let the equation of regression line of y on x is

$$8x - 10y = -66 \Rightarrow y = 6.6 + 0.8x$$

$$\therefore \quad b_{xy} = 0.8$$

and the equation of regression line of x on y is

$$40x - 18y = 214 \Rightarrow x = 5.35 + 0.45y$$

$$\therefore \quad b_{xy} = 0.45$$

Correlation coefficient $\quad r^2 = b_{yx} \cdot b_{xy} = 0.8 \times 0.45 = 0.36$

$$\Rightarrow \quad r = \pm 0.6$$

Since both b_{yx} and b_{xy} are positive so r is positive.

$$\Rightarrow \quad r = 0.6. \quad \textbf{Ans.}$$

(*iii*) We have $\quad r_x = 9$

$$\therefore \quad b_{yx} = r\frac{\sigma y}{\sigma x}$$

$$0.8 = 0.6 \times \frac{\sigma_y}{9} \Rightarrow \sigma_y = \frac{0.8 \times 9}{0.6}$$

$$= \sigma_y = 12. \quad \textbf{Ans.}$$

PROBLEM SET 4.2

1. Find the correlation coefficient between x and y from the given data:

x	78	89	97	69	59	79	68	57
y	125	135	156	112	107	138	123	108

[**Ans.** 0.96]

2. If $z = ax + by$ and r is the correlation coefficient between x and y, show that $\sigma_z^2 = a^2\sigma_x^2 + b^2\sigma_y^2 + 2ab\sigma_x\sigma_y$.

3. Find the coefficient of correlation between x and y from the following table:

x	10	14	18	22	26	30
y	18	12	24	6	30	36

[**Ans.** 0.6]

4. The following table gives according to age the frequency of the marks obtained by 100 students in an intelligence test:

Age (years) → Marks ↓	18	19	20	21	Total
10–20	4	2	2	–	8
20–30	5	4	6	4	19
30–40	6	8	10	11	35
40–50	4	4	6	8	22
50–60	–	2	4	4	10
60–70	–	2	3	1	6
Total	19	22	31	28	100

Calculate the coefficient of correlation between age and intelligence. [**Ans.** 0.25]

5. Calculate the coefficient of correlation for the following data:

Husband's age (years) x	23	27	28	28	29	30	31	33	35	36
Wife's age (years) y	18	20	22	27	21	29	27	29	28	29

[**Ans.** 0.82]

6. Two ladies Seema and Payal were asked to ranked 7 different types of lipsticks. The ranks given by them are given below:

Lipsticks	A	B	C	D	E	F	G
Seema	2	1	4	3	5	7	6
Payal	1	3	2	4	5	6	7

Calculate Spearman's rank correlation coefficient. [**Ans.** 0.786]

7. Calculate the coefficient of correlation or the following data:

Sales	50	50	55	60	65	65	65	60	60	60
Express	11	13	14	16	16	15	15	14	13	13

[**Ans.** 0.751]

8. If $\Sigma x = 71$, $\Sigma y = 70$, $\Sigma x^2 = 555$, $\Sigma y^2 = 526$ and $\Sigma xy = 527$ find r_{xy}.

9. In two sets of variables x and y with 50 observations each of the following data were observed: $\bar{x}=10, \bar{y}=6, \sigma_x=3, \sigma_y=2, r=0.3$. But on subsequent verification it was found that one value of $x = 10$ and one value of $y = 6$ were found inaccurate and hence weeded out. With the remaining 49 pairs of values how is the original value of r affected?

[**Ans.** $\bar{x}=13, \bar{y}=17, \sigma_y=3, r=0.6$]

10. From the following data of marks obtained by 10 students in Physics and Chemistry calculate the rank correlation coefficient

Physics (P)	35	56	50	65	44	38	44	50	15	26
Chemistry (C)	50	35	70	25	35	58	75	60	55	35

[**Ans.** – 0.139]

11. Find the rank correlation coeffiient for the following indices of supply and price of an article.

Supply index	124	100	105	112	102	93	99	115	123	104	99	113	121	103	101
Price index	80	100	102	91	100	111	109	100	89	104	111	102	98	111	123

[**Ans.** – 0.76]

12. Following are the marks obtained by 10 students in the first three semesters in three Engineering Mathematics papers out of 75.

Semester I (Maths I)	60	55	75	45	69	45	72	39	35	45
Semester II (Maths II)	70	58	73	49	60	49	60	55	60	48
Semester III (Maths III)	55	61	68	40	58	60	50	38	50	60

Calculate the rank correlation coefficient for all three pairs of these three subjects.

13. Two judges in a beauty contest rank the ten competitors in the following order.

Judge I	4	3	4	2	7	9	8	10	5
Judge II	1	6	7	5	8	10	9	3	2

[**Ans.** 0.224]

14. The $I.Q^s$ of a group of 6 persons were measured and then they sat for a certain examination. Their $I.Q^s$ and examination marks were as follows:

Exam mrks	70	60	80	60	10	20
I-Q	110	100	140	12	80	90

Compute the coefficient of correlation and rank correlation. [**Ans.** 0.882; 0.97]

Problems Related to Regression:

15. If the regression coefficient are 0.8 and 0.2, what would be the value of coefficient of correlation? [**Ans.** 0.4]

16. The two variables x and y have the regression lines $3x + 2y - 26 = 0$ and $6x + y - 31 = 0$. Find

(*i*) The mean values of x and y.

(*ii*) The correlation coefficient between x and y.

(*iii*) The variance of y if the variance of x is 25.

$\left[\textbf{Ans.}\ (i)\ 5/3, 21\ (ii) -\frac{1}{2}\ (iii)\ 15\right]$

17. From the following data find the regression lines and correlation coefficeint between x and y.

x	25	28	35	32	31	36	29	38	34	32
y	43	46	49	41	36	32	31	30	33	39

[**Ans.** $y = 57.28 - 0.61x$, $x = 39.58 - 0.21y$, $r = -0.36$]

18. If F is the pull required to lift a load W by means of a pulley block, fit a linear law of the form $F = mW + C$ connecting F and W, using the data

W	50	70	100	120
F	12	15	21	25

where F and W are in kg wt. Compute F when $W = 150$ kg wt.

19. The following table shows the age x and blood pressure y of 12 women. (*i*) Find the correlation coefficient between x and y. (*ii*) Determine the regression equation of y on x. (*iii*) Estimate the blood pressure of a women whose Age is 45 years.

Age (x)	56	42	72	36	63	47	55	49	38	42	68	60
B. Pressure (y)	147	125	160	118	149	128	150	145	115	140	152	155

[**Ans.** (*i*) 0.8961, (*ii*) $y = 1.138x + 80.78$ (*iii*) 132]

20. The average daily wage for working class in Madras is Rs. 12 and for that in Delhi is Rs. 18; their respective standard deviations are Rs. 2 and Rs. 3 and the correlation coefficient is 0.67. Find the most likely wage in Delhi corresponding to the wage of Rs. 20 in Madras.

[**Ans.** 26.04]

21. (*i*) Find the regression line of y on x for the following data:

x	1	3	4	6	8	9	11	14
y	1	2	4	4	5	7	8	9

(*ii*) Derive regression line for the following data

$\Sigma x = 30$, $\Sigma x^2 = 190$, $\Sigma xy = 192$, $\Sigma y = 30$, $\Sigma y^2 = 190$, $n = 5$

[**Ans.** $7x - 11y + 6 = 0$ (*ii*) $y - 1.2x + 1.2 = 0$, $x - 1.2y + 1.2 = 0$]

22. Find σ_y and r from the following data. Regression lines $3x = y$; $4y = 3x$ and S.D. of $x = 2$

[**Ans.** $r = 0.5$, $\sigma_y = 3$]

23. The following results were obtained from marks in applied Mechanics and Engineering Mathematics in an examination:

Mean	*Applied Mechanics* (x)	*Engg. Mathematics* (y)
	47.5	39.5
S.D.	16.8	10.8

$r = 0.95$. Find both the regression equations. Also estimate the value of y for $x = 30$.

[**Ans.** $y = 0.6107x + 10.49$, $x = 1.477y - 10.8415$, $y = 28.81$]

24. Calculate linear regression coefficients from the following: [*U.P.T.U. (MCA) 2001*]

x	1	2	3	4	5	6	7	8
y	3	7	10	12	14	17	20	24

[**Ans.** 2.7976, 0.3540]

25. From a partially destroyed laboratory record only the following are availale. $x = 4y + 5$ and $y = kx + 4$ are the regression lines of x on y and y on x respectively. Show that $0 \le k \le \frac{1}{4}$. If $k = \frac{1}{16}$ find the means of the two variables x and y and also find the coefficient between them. **[Ans.** $\bar{x} = 28, \bar{y} = 5.57, r = 0.5$]

4.14 PROBABILITY

Historically, probability theory began with the study of games of chance, such as roulette and cards, Apart from games, uncertainty prevails in business and other aspects of life.

Probability is a concept which numerically measure the degree of uncertainty and therefore, of certainty of the occurrence of events.

If an event A can happen in m ways and fail n ways, all these being equally likely to occur, then the happing of A is

$$P(A) = \frac{\text{Number of favourable cases}}{\text{Total number of mutually exclusive and equally likely cases}}$$

$$= \frac{m}{m+n} = p, \text{ say}$$

and that of its failing is defined as P (not A) $= \frac{n}{m+n} = q, \text{ say}$

$$\Rightarrow \qquad p + q = \frac{m}{m+n} + \frac{n}{m+n} = \frac{m+n}{m+n} = 1$$

$$\Rightarrow \qquad p\ (A) + p\ (\text{not } A) = 1$$

4.14.1 Probability Distribution

We have discussed frequency distribution, which are based on actual observations. We shall now discuss another important type of distribution known as *Theoretical or Expected frequency distribution*. Often we can mathematically deduce the expected frequency distribution of certain population on the basis of theoretical considerations. Such distributions are known as *Theoretical distribution or probability distributions*. In this chapter, we shall discuss three important theoretical (or probability) distributions, *viz*., (1) **Binomial distribution**, (2) **Poisson distribution** and (3) **Normal distribution**. Of these, the first two are discrete probability distribution and the third is a continuous probability distribution.

4.15 BINOMIAL DISTRIBUTION (OR BERNOULLI DISTRIBUTION)

Binomial distribution is the most fundamental and important discrete probability distribution in statistics and is defined as $P(r) = {}^nC_r p^r q^{n-r}$ ($r = 0, 1, 2, 3, \ldots\ldots, n$), which gives the probability of r successes in a series of n independent trails $(r \le n)$ where p is the constant probability of success in a single trial and $q = 1 - p$.

If n and p are constants, the above function $P(r)$ represents a discrete probability distribution given by

r	0	1	2	....	n
$P(r)$	q^n	${}^nC_1p^1q^{n-1}$	${}^nC_2p^2q^{n-2}$	....	p^n

It is called Binomial distribution, because for $r = 0, 1, 2,\ldots, n$, it gives the successive terms of the Binomial expansion $(p+q)^n = q^n + {}^nC_1p^1q^{n-1} + {}^nC_2p^2q^{n-2} + \ldots\ldots + p^n$

Clearly $\sum_{r=0}^{n} P(r) = (p+q)^n = 1^n = 1 \quad [\because q = 1-p \text{ or } p+q=1]$

4.15.1 Conditions of Binomial Distribution

The binomials distribution holds under the following conditions:

(1) Trials are repeated under identical conditions for a fixed number of items, say n times.

(2) There are only two possible outcomes, *viz*., success or failure for each trial.

(3) The probability of successes in each trial remains constant and does not change from trial to trial.

(4) If two independent random variables x and y follow binomial distribution with parameters (n_1, p) and (n_2, p) respectively their sum $(x + y)$ also follows binomial distribution with parameters $(n_1 + n_2, p)$.

4.15.2 Properties of Binomial Distribution

(1) Binomial distribution in a discrete probability distribution in the random variable r (the no. of successes) assumes the values $0, 1, 2,\ldots, n$ where n is finite.

(2) Mean = np.variance = npq, S.D. $(\sigma) = \sqrt{npq}$, Skewness $= \dfrac{(q-p)}{\sqrt{npq}}$ and kurtosis $= \dfrac{(1-6pq)}{npq}$, where $q = 1 - p$.

(3) The mode of binomial distribution in that value of x which occurs with the largest probability. It may have either one or two modes.

(4) The trials are independent *i.e.*, the probability of an event in a trial is not affected by the results of any other trial.

***Extension*:** If n independent trials are repeated N times, we get N sets of n trails and the expected frequency of x successes is $N.{}^nC_rp^rq^{n-r}$. The expected frequencies of $0, 1, 2,\ldots.n$, successes are the successive terms of the binomial expansion of $N(p + q)^n$.

4.15.3 Mean of the Binomial Distribution

For the binomial distribution, $P(r) = {}^nC_rp^rq^{n-r}$

Mean $\mu = \sum_{r=0}^{n} rP(r) = \sum_{r=0}^{n} r.{}^nC_rq^{n-r}p^r$

$$= 0 + 1.{}^nC_1q^{n-1}p + 2.{}^nC_2q^{n-2}p^2 + 3.{}^nC_3q^{n-3}p^3 + \ldots\ldots + n.{}^nC_np^n$$

$$= nq^{n-1}p + 2.\frac{n(n-1)}{2.1}q^{n-2}p^2 + 3.\frac{n(n-1)(n-2)}{3.2.1}q^{n-3}p^3 + ... + np^n$$

$$= nq^{n-1}p + n(n-1)q^{n-2}p^2 + \frac{n(n-1)(n-2)}{2.1}q^{n-3}p^3 + ... + np^n$$

$$= np\left[q^{n-1} + (n-1)q^{n-2}p^2 + \frac{(n-1)(n-2)}{2.1}q^{n-3}p^3 + ...p^{n-1}\right]$$

$$= np\left[{}^{n-1}C_0 q^{n-1} + {}^{n-1}C_1 q^{n-2}p^2 + {}^{n-1}C_2 q^{n-3}p^3 + ... + {}^{n-1}C_{n-1}p^{n-1}\right]$$

$$= np(p+q)^{n-1} = np \qquad (\text{as } p + q = 1)$$

which is the required mean of the Binomial distribution.

4.15.4 Variance of the Binomial Distribution

Variance $\sigma^2 = \sum_{r=0}^{n} r^2 P(r) - \mu^2 = \sum_{r=0}^{n} [r + r(r-1)]P(r) - \mu^2$

$$= \sum_{r=0}^{n} rP(r) + \sum_{r=0}^{n} [r(r-1)]P(r) - \mu^2 = \mu + \sum_{r=2}^{n} r(r-1)\,{}^nC_r q^{n-r}p^r - \mu^2$$

(Since the contribution due to $r = 0$ and $r = 1$ is zero)

$$= \mu + \left[2.1\,.\,{}^nC_2 q^{n-2}p^2 + 3.2\,.\,{}^nC_3 q^{n-3}p^3 + ... + n(n-1)\,{}^nC_n p^n\right] - \mu^2$$

$$= \mu + \left[2.1\frac{n(n-1)}{2.1}q^{n-2}p^2 + 3.2\,.\,\frac{n(n-1)(n-2)}{3.2.1}\,.\,q^{n-3}p^3 + ... + n(n-1)p^n\right] - \mu^2$$

$$= \mu + [n(n-1)q^{n-2}p^2 + n(n-1)(n-2)\,.\,q^{n-3}p^3 + ... + n(n-1)p^n] - \mu^2$$

$$= \mu + n(n-1)p^2(q+p)^{n-2} - \mu^2 = \mu + n(n-1)p^2 - \mu^2 \qquad [\because q + p = 1]$$

$$= np[1 + (n-1)p - np] = np[1-p] = npq \qquad [\because \mu = np]$$

which is the required variance of the Binomial distribution

4.15.5 Recurrence or Recursion Formula for the Binomial Distribution

[*U.T.U. 2008*]

In the Binomial distribution,

$$P(r) = {}^nC_r p^r q^{n-r} = \frac{n!}{(n-r)!r!}q^{n-r}p^r$$

$$P(r+1) = {}^nC_{r+1}p^{r+1}q^{n-r-1} = \frac{n!}{(n-r-1)!(r+1)!}p^{r+1}q^{n-r-1}$$

$$\therefore \frac{P(r+1)}{P(r)} = \frac{(n-r)!}{(n-r-1)!} \times \frac{r!}{(r+1)!} \times \frac{p}{q} = \frac{(n-r)\times(n-r-1)!}{(n-r-1)!} \times \frac{r!}{(r+1)\times r!} \times \frac{p}{q} = \frac{n-r}{r+1}.\frac{p}{q}$$

$$\Rightarrow \qquad \boxed{P(r+1) = \frac{n-r}{r+1} \cdot \frac{p}{q} P(r)}$$

which is the required recurrence formula.

Theorem: Moment generating function of a binomial distribution about the origin is $(q + pe^t)^n$.

Proof:
$$M_{x=0}(t) = E(e^{xt}) = \sum_{x=0}^{n} e^{xt} \cdot p(x) = \sum_{x=0}^{n} (e^{tx} \cdot {}^nC_x p^x q^{n-x})$$
$$= \sum_{x=0}^{n} {}^nC_x (pe^t)^x q^{n-x} = (q + pe^t)^n$$

Example 1: Find the mean of binomial distribution. Also find the moment generating function of the binomial distribution about its mean.

Sol. Mean = np [Sub-Art. 4.15.3, Page 338]

And moment generating function about mean is

$$M_{x=np}(t) = E[e^{t(x-np)}] = \sum_{x=0}^{n} e^{xt} . e^{-npt} p(x)$$
$$= e^{-npt} \sum_{x=0}^{\infty} e^{xt} p(x) = e^{-npt} (q + pe^t)^n \qquad \text{| By Theorem 4.1}$$
$$= \left[qe^{-pt} + pe^{-pt} \cdot e^t\right]^n$$
$$= \left[qe^{-pt} + pe^{(1-p)t}\right]^n$$
$$= \left[qe^{-pt} + pe^{qt}\right]^n. \quad \textbf{Ans.}$$

Example 2: Six coins are tossed simultaneously. What is the probability of getting (*a*) 2 heads (*b*) at least two heads and (*c*) more than two heads?

Sol. Let getting a head by tossing a coin be a success.

Then $p = 1/2$, $q = 1 - p = 1/2$, here $n = 6$

Then probability $\quad P(r) = {}^nC_r p^r q^{n-r} = {}^6C_r \left(\frac{1}{2}\right)^r \left(\frac{1}{2}\right)^{6-r}$ ...(1)

(*a*) Putting $r = 2$ in (1), we get the probability of (2) of 2 heads given by

$$P(2) = {}^6C_2 \left(\frac{1}{2}\right)^2 \left(\frac{1}{2}\right)^{6-2} = 15 . \frac{1}{4} . \frac{1}{16} = \frac{15}{64}$$

(*b*) Probability of at least two heads

$$= 1 - \{P(0) + P(1)\} = 1 - \left\{ {}^6C_0 \left(\frac{1}{2}\right)^0 \left(\frac{1}{2}\right)^{6-0} + {}^6C_1 \left(\frac{1}{2}\right)^1 \left(\frac{1}{2}\right)^{6-1} \right\}$$

$$= 1 - \left\{1 . 1 . \frac{1}{64} + 6 . \frac{1}{2} . \frac{1}{32}\right\} = 1 - \frac{7}{64} = \frac{57}{64}.$$

(*c*) Probability of getting more than 2 heads

= 1 – (Probability of getting not more than 2 heads)

= 1 – $\{p(0) + p(1) + p(2)\}$ = 1 – {1/64 + 6/64 + 15/64}

= 1 – 22/64 = 21/32. **Ans.**

Example 3: Five unbaised coins are tossed and number of heads are noted. The experiment is repeated 64 times and the following distribution is obtained.

No. of heads	0	1	2	3	4	5	Total
Frequencies	3	6	24	26	4	1	64

Fit binomial distribution using by recurrence relation.

Sol. Here $n = 5$ and $N = 64$

Since the coins are unbaised $p = \frac{1}{2} = q$ so that $\frac{p}{q} = 1$

Now $\quad P(0) = q^n = \left(\frac{1}{2}\right)^5 = \frac{1}{32}.$

Hence $\quad f(0) = Nq^n = 64\left(\frac{1}{32}\right) = 2.$

Using the recurrence formula $P(r+1) = \left(\frac{n-r}{r+1}\right), \frac{p}{q} p(r)$, we get $P(1) = \left(\frac{5-0}{0+1}\right) P(0) = \frac{5}{32}.$

$f(1) = 64(5/32) = 10.$

Hence

r	*Probabilities* $p(r)$	*Expected frequencies* $f(r) = N.P(r)$	*Observed frequencies*
0.	$P(0) = \frac{1}{32}$	$64 \times \frac{1}{32} = 2$	3
1.	$P(1) = \frac{5}{1} P(0) = \frac{5}{32}$	$64 \times \frac{5}{32} = 10$	6
2.	$P(2) = \frac{4}{2} P(1) = \frac{10}{32}$	$64 \times \frac{10}{32} = 20$	24
3.	$P(3) = \frac{3}{3} P(2) = \frac{10}{32}$	$64 \times \frac{10}{32} = 20$	26
4.	$P(4) = \frac{2}{4} P(3) = \frac{5}{32}$	$64 \times \frac{5}{32} = 10$	4
5.	$P(5) = \frac{1}{5} P(4) = \frac{1}{32}$	$64 \times \frac{1}{32} = 2$	1
Total		64	64

Example 4: Eight coins are tossed simultaneously. Show that the probability of getting at least 6 heads is 37/256.

Sol. Let the occurrence of head by tossing a coin be a success. Then p = constant probability of success = 1/2, $q = 1 - p$.

The probability of r successes is given by $P(r) = {}^nC_r p^r q^{n-r} = {}^8C_r \left(\frac{1}{2}\right)^r \left(\frac{1}{2}\right)^{8-r}$

∴ The required probability of getting at least 6 heads (*i.e.*, at least 6 successes)

$$= f(6) + f(7) + f(8)$$

$$= {}^8C_6(1/2)^6(1/2)^{8-6} + {}^8C_7(1/2)^7(1/2)^{8-7} + {}^8C_8(1/2)^8(1/2)^{8-8}$$

$$= 28 \times \frac{1}{256} + 8 \times \frac{1}{256} + 1 \times \frac{1}{256} = \frac{37}{256}. \quad \textbf{Ans.}$$

Example 5: Fit a binomial distribution to the following freqeuncy data:

x =	0	1	3	4
f	28	62	10	4

Sol. First of all we find arithmatic mean from the following table:

x	f	fx
0	28	0
1	62	62
3	10	30
4	4	16
	$\Sigma f = 104$	$\Sigma fx = 108$

∴ $\quad \bar{x} = \frac{108}{104} = 1.038 \Rightarrow np = 1.038$ or $4 = 1.038 \quad$ (As $n = 4$)

$$\Rightarrow p = .2595$$

∴ $\quad q = 1 - p = 1 - 0.2595 = 0.7405$

Also $\quad N = \Sigma f = 104$

Hence binomial distribution $= N(p + q)^n$

$$= 104\ (.2595 + .7405)^4. \quad \textbf{Ans.}$$

Example 6: If the probability of the defective bolts be $\frac{1}{10}$, find the following for the Binomial Distribution of the defective bolts in a total 400.

(*a*) the mean (*b*) the standard deviation (*c*) the moment coefficient of skewness.

Sol. Here $n = 400$, $p = 1/10=0.1$, $q = 1 - p = 0.9$

(*a*) Mean of the Binomial Distribution $= np = 400 \times 0.1 = 40$.

(*b*) $\text{S.D.}(\sigma) = \sqrt{npq} = \sqrt{400 \times 0.1 \times 0.9} = \sqrt{36} = 6.$

(*c*) Moment coefficient of skewness $= \dfrac{q-p}{\sqrt{npq}} = \dfrac{0.9-0.1}{6} = \dfrac{0.8}{6} = 0.133.$ **Ans.**

Example 7: What is the probability of guessing correctly at least six of the ten answers in a TRUE FALSE objective test?

Sol. We have $P(r) = {}^nC_r p^r q^{n-r}$, where $q = 1 - p$.

Here $n = 10$, $p = 1/2$, $q = 1 - 1/2, = 1/2$, $r = 6, 7, 8, 9, 10$

Hence the required probability

$$= f(6) + f(7) + f(8) + f(9) + f(10)$$

$$= {}^{10}C_6\left(\frac{1}{2}\right)^6\left(\frac{1}{2}\right)^{10-6} + {}^{10}C_7\left(\frac{1}{2}\right)^7\left(\frac{1}{2}\right)^{10-7} + {}^{10}C_8\left(\frac{1}{2}\right)^8\left(\frac{1}{2}\right)^{10-8}$$

$$+ {}^{10}C_9\left(\frac{1}{2}\right)^9\left(\frac{1}{2}\right)^{10-9} + {}^{10}C_{10}\left(\frac{1}{2}\right)^{10}\left(\frac{1}{2}\right)^{10-10}$$

$$= \left(\frac{1}{2}\right)^{10}\left\{{}^{10}C_6 + {}^{10}C_7 + {}^{10}C_8 + {}^{10}C_9 + {}^{10}C_{10}\right\}$$

$$= \left(\frac{1}{2}\right)^{10}\left\{{}^{10}C_6 + {}^{10}C_7 + {}^{10}C_8 + {}^{10}C_9 + 1\right\}$$

$$= \frac{1}{1024}(210 + 120 + 45 + 10 + 1) = \frac{386}{1024} = \frac{193}{512}. \quad \textbf{Ans.}$$

Example 8: Assume that on the average one telephone number out of fifteen called between 2 p.m. and 3 p.m. on week-days is busy. What is the probability that if 6 randomly selected telephone numbers are called (*a*) not more than 3, (*b*) at least three of them will be busy?

Sol. p, the probability of a telephone number being busy between 2 p.m. and 3 p.m. on week-days $= \dfrac{1}{15}$.

$$q = 1 - \frac{1}{15} = \frac{14}{15},\ n = 6$$

The probability that not more than three will be busy

$$= P(r \le 3) = P(0) + P(1) + P(2) + P(3)$$

$$= {}^6C_0\left(\frac{14}{15}\right)^6 + {}^6C_1\left(\frac{14}{15}\right)^5\left(\frac{1}{15}\right) + {}^6C_2\left(\frac{14}{15}\right)^4\left(\frac{1}{15}\right)^2 + {}^6C_3\left(\frac{14}{15}\right)^3\left(\frac{1}{15}\right)^3$$

$$= \frac{(14)^3}{(15)^6}[2744 + 1176 + 210 + 20] = \frac{2744 \times 4150}{(15)^6} = 0.9997$$

The probability that at least three of them will be busy $= P(r \geq 3)$

$$= P(3) + P(4) + P(5) + P(6)$$

$$= {}^6C_3\left(\frac{14}{15}\right)^3\left(\frac{1}{15}\right)^3 + {}^6C_4\left(\frac{14}{15}\right)^2\left(\frac{1}{15}\right)^4 + {}^6C_5\left(\frac{14}{15}\right)\left(\frac{1}{15}\right)^5 + {}^6C_6\left(\frac{1}{15}\right)^6 = 0.005. \quad \textbf{Ans.}$$

Example 9: Six dice are thrown 729 times. How many times do you expect at lest three dice to show a five or six?

Sol. p = The chance of getting 5 or 6 with one die $= \frac{2}{6} = \frac{1}{3}$

$$q = 1 - \frac{1}{3} = \frac{2}{3}, n = 6, N = 729$$

Since dice are in sets of 6 and there are 729 sets.

The expected number of times at least three showing five or six

$$= N.P(r \geq 3)$$

$$= 729[P(3) + P(4) + P(5) + P(6)]$$

$$= 729\left[{}^6C_3\left(\frac{2}{3}\right)^3\left(\frac{1}{3}\right)^3 + {}^6C_4\left(\frac{2}{3}\right)^2\left(\frac{1}{3}\right)^4 + {}^6C_5\left(\frac{2}{3}\right)\left(\frac{1}{3}\right)^5 + {}^6C_6\left(\frac{1}{3}\right)^6\right]$$

$$= \frac{729}{3^6}[160 + 60 + 12 + 1] = 233. \quad \textbf{Ans.}$$

Example 10: (*a*) Determine the Binomial distribution for which the mean is 4 and and deviation is $\sqrt{3}$.

(*b*) Point out the fallacy, if any, in the following statement : "The mean of a Binomial distrbution is 10 and S.D. is 4".

Sol. (*a*) For the Binomial distribution.

$$np = \text{Mean} = 4 \quad \text{...(1)}$$

$$\sqrt{npq} = \text{Standard deviation} = \sqrt{3}, \quad \text{or} \quad npq = 3 \quad \text{...(2)}$$

Substituting $np = 4$ in (2), we get

$$4 \,.\, q = 3, \quad \text{or} \quad q = \frac{3}{4}.$$

$$\therefore \quad p = 1 - q = 1 - \frac{3}{4} = \frac{1}{4}$$

Substituting $p = \frac{1}{4}$ in (1), we get

$$n.\ \frac{1}{4} = 4, \text{ or, } n = 16.$$

Hence the required probability of Binomial distribution is

$$P(r) = {}^nC_r p^r q^{n-r}, \text{ where } q = 1 - p$$

$$= {}^{16}C_r \cdot \left(\frac{1}{4}\right)^r \cdot \left(\frac{3}{4}\right)^{16-r}, \text{ where } r = 0, 1, 2, 3, \ldots, 16.$$

(*b*) For the given Binomial distribution,

$$np = \text{Mean} = 10 \qquad \ldots(1)$$

and $$\sqrt{npq} = \text{Standard deviation} = 4, \quad \text{or} \quad npq = 16 \qquad \ldots(2)$$

Dividing (2) by (1), we get

$$q = \frac{16}{10} = 1.6 > 1, \text{ which is absurd, since } 0 < q < 1.$$

Hence the statement can not be true.

Example 11: (*a*) Out of 800 families with 4 children each, how many families would be expected to have (1) 2 boys and 2 girls (2) at least one boy (3) no girl (4) at most two girls?

Assume equal probabilities for boys and girls.

(*b*) Out of 800 families with 5 children each, how many would you expect to have (1) 3 boys (2) 5 girls (3) either 2 or 3 boys?

Assume equal probabilities for boys and girls.

Sol. (*a*) Since probabilities for boys and girls are equal

P = probability of having a boy = $\frac{1}{2}$, q = probability for having a girl = $\frac{1}{2}$, $n = 4$, $N = 800$

(1) The expected number of families having 2 boys and 2 girls

$$= 800 \cdot {}^4C_2 \left(\frac{1}{2}\right)^2 \left(\frac{1}{2}\right)^2 = 800 \times 6 \times \frac{1}{16} = 300.$$

(2) The expected number of families having at least one boy

$$= 800 \cdot \left[{}^4C_1 \left(\frac{1}{2}\right)^3 \left(\frac{1}{2}\right) + {}^4C_2 \left(\frac{1}{2}\right)^2 \left(\frac{1}{2}\right)^2 + {}^4C_3 \left(\frac{1}{2}\right) \left(\frac{1}{2}\right)^3 + {}^4C_4 \left(\frac{1}{2}\right)^4 \right]$$

$$= 800 \times \frac{1}{16} [4 + 6 + 4 + 1] = 750.$$

(3) The expected number of families having no girls *i.e.*, having 4 boys

$$= 800 \cdot {}^4C_4 \left(\frac{1}{2}\right)^4 = 50.$$

(4) The expected number of families having at most 2 girls *i.e.*, having at least 2 boys

$$= 800 \cdot \left[{}^4C_2 \left(\frac{1}{2}\right)^2 \left(\frac{1}{2}\right)^2 + {}^4C_3 \left(\frac{1}{2}\right) \left(\frac{1}{2}\right)^3 + {}^4C_4 \left(\frac{1}{2}\right)^4 \right]$$

$$= 800 \times \frac{1}{16} [6 + 4 + 1] = 550. \quad \textbf{Ans.}$$

(*b*) Here, $n = 5$,

Probability of having a boy, $p = \frac{1}{2}$

Probability of having a girl, $q = \frac{1}{2}$

No. of families, $N = 800$.

(1) The probability having 3 boys out of 5 children

$$= P(r = 3)$$

$$= {}^5C_3\left(\frac{1}{2}\right)^3\left(\frac{1}{2}\right)^{5-3} = \frac{5!}{3!2!}\left(\frac{1}{8}\right)\left(\frac{1}{4}\right) = \frac{10}{32} = 0.3125.$$

$\therefore$ The expected number of families $= 800 \times 0.3125 = 250.$

(2) The probability of having 5 girls out 5 children.

$$= P(r = 0) \qquad \text{(Since there is no boy)}$$

$$= {}^5C_0\left(\frac{1}{2}\right)^0\left(\frac{1}{2}\right)^{5-0} = \left(\frac{1}{2}\right)^5 = \frac{1}{32} = 0.03125.$$

$\therefore$ The expected number of families $= 800 \times 0.03125 = 25$

(3) The probability of having either 2 or 3 boys out 5 children.

$$= P(r = 2) + P(r = 3)$$

$$= {}^5C_2\left(\frac{1}{2}\right)^2\left(\frac{1}{2}\right)^{5-2} + {}^5C_3\left(\frac{1}{2}\right)^3\left(\frac{1}{2}\right)^{5-3}$$

$$= 10\left(\frac{1}{4}\right)\left(\frac{1}{8}\right) + 10\left(\frac{1}{8}\right)\left(\frac{1}{4}\right) = \frac{20}{32} = 0.625.$$

$\therefore$ The expected number of families $= 800 \times 0.625 = 500.$ **Ans.**

Example 12: In a Binomial distribution consisting of 5 independent trials, probabilities of 1 and 2 successes are 0.4096 and 0.2048 respectively. Find the parameter 'p' of the distribution.

Sol. For a Binomial distribution, the probability of r successes in a series of n independent trials is given by

$$P(r) = {}^nC_r p^r q^{n-r}$$

Here $n = 5$; $\quad \therefore P(r) = {}^5C_r p^r q^{5-r}$

$\therefore \quad P(1) = {}^5C_1 \cdot p.q^{5-1} = 5pq^4$ and $P(2) = {}^5C_2 \cdot p^2.q^{5-2} = 10p^2q^3$, where $q = 1 - p$, it is given.

$$\therefore \qquad 5pq^4 = 0.4096 \qquad \text{...(1)}$$

and

$$10p^2q^3 = 0.2048 \qquad \text{...(2)}$$

Dividing (2) by (1), we get

$$\frac{10p^2q^3}{5pq^4} = \frac{0.2048}{0.4096}, \quad \text{or} \quad \frac{2p}{q} = \frac{1}{2}, \quad \text{or} \quad 4p = q = 1 - p,$$

or

$$5p = 1, \qquad \text{or} \quad p = 1/5 = 0.2. \qquad \textbf{Ans.}$$

Example 13: If the probability of hitting a target is 10% and 10 shots are fixed independently. What is the probability that the greatest will be hit at least once?

Sol. Here, $p = \frac{10}{100} = \frac{1}{10}, q = 1 - p = 1 - \frac{1}{10} = \frac{9}{10}$

Probability that the greatest will be hit at least once

$$= P(r \geq 1) = 1 - P(r = 0)$$

$$= 1 - [{}^nC_0 p^0 q^n] = 1 - \left[{}^{10}C_0 \left(\frac{1}{10}\right)^0 \left(\frac{1}{10}\right)^{10}\right]$$

$$= 1 - \left(\frac{9}{10}\right)^{10} = 1 - 0.3486 = 0.6513. \quad \textbf{Ans.}$$

Example 14: A policeman fires 6 bullets on a dacoit. The probability that the dacoit will be killed by a bullet is 0.6. What is the probability that dacoit is still alive?

Sol. Here, n = no. of bullets fired = 6

$$p = 0.6$$

$$q = 1 - p = 0.4$$

Probability that the dacoit is still alive (not killed)

$$= P(r = 0)$$

$$= {}^nC_0 p^0 q^{n-0} = {}^6C_0 (.6)^0 (.4)^6 = (.4)^6 = 0.004096. \quad \textbf{Ans.}$$

Example 15: The overall percentage of failures in a certain examination is 40. What is the probability that out of a group of 6 candidates at least 4 passed the examination?

Sol. Let the event of a candidate passing the examination 'be a success'. Then

q = Probability of failure of a candidate in the examination = 40% (given)= $\frac{40}{100} = \frac{2}{5}$

$\therefore$ p = Probability of success (*i.e.* passing in the examination) in a single trial

$$= 1 - q = 1 - \frac{2}{5} = \frac{3}{5}$$

Here n = Number of independent trials = No. of candidates in the group = 6.

The probability in Binomial distribution is

$$P(r) = {}^nC_r p^r q^{n-r} = {}^6C_r \left(\frac{3}{5}\right)^r \left(\frac{2}{5}\right)^{6-r} \quad ...(1)$$

$\therefore$ The required probability = the probability of at least 4 success in 6 independent trial

$$= f(4) + f(5) + f(6)$$

$$= {}^6C_4 \left(\frac{3}{5}\right)^4 \left(\frac{2}{5}\right)^{6-4} + {}^6C_5 \left(\frac{3}{5}\right)^5 \left(\frac{2}{5}\right)^{6-5} + {}^6C_6 \left(\frac{3}{5}\right)^6 \left(\frac{2}{5}\right)^{6-6}$$

[using (1)]

$$= 15 \times \frac{81}{625} \times \frac{4}{25} + 6 \times \frac{243}{3125} \times \frac{2}{5} + 1 \times \frac{729}{15625} \times 1$$

$$= \frac{4860 + 2916 + 729}{15625} = \frac{8505}{15625} = \frac{1701}{3125}. \quad \textbf{Ans.}$$

Example 16: In sampling a large number of ports manufactured by a machine, the mean number of defectives in a sample of 20 is 2. Out of 1000 such samples, how many would be expected to contain at least 3 defective parts.

Sol. Mean no. of defectives = 2.

$$\mu = np = 2 \qquad \text{...(1)}$$

But $\qquad n = 20$ (given)

$$p = \frac{2}{20} = \frac{1}{10} = 0.1 \quad \text{[From (1)]}$$

so that, $\qquad q = 1 - 0.1 = 0.9$

No. of samples, $N = 1000$

The probability of having at least 3 defectives in a sample of 20 parts $= P(r \geq 3)$

$$= 1 - P(r < 3)$$

$$= 1 - [P(0) + P(1) + P(2)]$$

$$= 1 - \left[{}^{20}C_0 (0.1)^0 (0.9)^{20} + {}^{20}C_1 (0.1)^1 (0.9)^{19} + {}^{20}C_2 (0.1)^2 (0.9)^{18} \right]$$

$$= 1 - [0.1216 + 0.2702 + 0.2852]$$

$$= 1 - 0.677 = 0.323$$

$\therefore$ No. of sample expected $= 1000 \times 0.323 = 323$. **Ans.**

Example 17: If 10% of the bolts produced by a machine are defective, determine the probability that out of 10 bolts chosen at random.

(*i*) 1 (*ii*) none (*iii*) at most 2 bolts will be defective.

Sol. Here, $\quad p(\text{defective}) = \frac{10}{100} = \frac{1}{10}$

$\therefore \quad q(\text{non-defective}) = 1 - \frac{1}{10} = \frac{9}{10}$

Also, n =10, (n is no. of bolts chosen).

The probability in Binomial distribution is

$$P(r) = {}^nC_r p^r q^{n-r} \qquad \text{...(1)}$$

(*i*) Here $r = 1$

$$P(1) = {}^{10}C_1 \left(\frac{1}{10}\right)^1 \left(\frac{9}{10}\right)^{10-1} \qquad \text{using (1)}$$

$$= 10\left(\frac{1}{10}\right)\left(\frac{9}{10}\right)^9 = (.9)^9 = 0.3874$$

(*ii*) Here $r = 0$

$$P(0) = {}^{10}C_0 \left(\frac{1}{10}\right)^0 \left(\frac{9}{10}\right)^{10-0} \qquad \text{using (1)}$$

$$=\left(\frac{9}{10}\right)^{10}=0.3486$$

(*iii*) Probability, that at most 2 bolts will be defective $=P(r\le 2)$

$$=P(0)+P(1)+P(2) \qquad ...(2)$$

Now, $$P(2)={}^{10}C_2\left(\frac{1}{10}\right)^2\left(\frac{9}{10}\right)^{10-2} \quad \text{using (1)}$$

$$=\frac{10!}{2!8!}\left(\frac{1}{100}\right)\left(\frac{9}{10}\right)^8=45\left(\frac{1}{100}\right)(.43046)=0.1937$$

$\therefore$ From (2)

Required probability $=P(0)+P(1)+P(2)$

$=0.3486+0.3874+0.1937=0.9297.$ **Ans.**

Example 18: A product is 0.5 per cent defective and is picked in cartons of 100. What percentage of it contains not more than 3 defectives?

Sol. p(defective) = 0.5 per cent = 0.005

$\therefore$ $q(\text{non-defective})=1-0.005=0.995$

No. of packed products in a carton = 100

$\therefore$ $n=100$

Probability that not more than 3 defectives are there in a carton

$$P(r\le 3)=P(0)+P(1)+P(2)+P(3)$$

$$={}^{100}C_0(0.005)^0(.995)^{100}+{}^{100}C_1(0.005)^1(.995)^{99}+{}^{100}C_2(0.005)^2(.995)^{98}+{}^{100}C_3(0.005)^3(.995)^{97}$$

$$=0.6057+0.3044+0.07571+0.01242=0.99823$$

Hence required percentage = 99.823. **Ans.**

Example 19: A student is given a true-false examination with 8 questions. If he corrects at least 7 questions, he passes the examination. Find the probability that he will pass given that he guesses all questions.

Sol. Here, n = no. of questions asked = 8.

$$p=\frac{1}{2}$$

$$q=\frac{1}{2}$$ (Since the question can either be true or false)

Probability that he will pass $= P(r \geq 7) = P(7) + P(8)$

$$= {}^8C_7\left(\frac{1}{2}\right)^7\left(\frac{1}{2}\right)^{8-7} + {}^8C_8\left(\frac{1}{2}\right)^8\left(\frac{1}{2}\right)^{8-8} = 8\left(\frac{1}{2}\right)^7\left(\frac{1}{2}\right)^1 + 1.\left(\frac{1}{2}\right)^8$$

$$= \left(\frac{1}{2}\right)^8(8+1) = \frac{9}{256} = 0.03516. \quad \textbf{Ans.}$$

Example 20: The probability that a man aged 60 will live to be 70 is 0.65. What is the probability that out of 10 men, now 60, at least 7 will livc to be 70?

Sol. The probability that a man aged 60 will live to be 70 = p = 0.65

$\therefore \quad q = 1 - p = 1 - 0.65 = 0.35$

Number of men (n) =10.

Probability that at least 7 men will live to 70.

$$= P(7 \text{ or } 8 \text{ or } 9 \text{ or } 10)$$
$$= P(7) + P(8) + P(9) + P(10)$$
$$= 10C_7q^3p^7 + 10C_8q^2p^8 + 10C_9qp^9 + 10C_{10}p^{10}$$
$$= 120(0.35)^3(0.65)^7 + 45(0.35)^2(0.65)^8 + 10(0.35)(0.65)^9 + (0.65)^{10}$$
$$= (0.65)^7\left[120(0.35)^3 + 45(0.35)^2(0.65)^8 + 10(0.35)(0.65)^9 + (0.65)^{10}\right]$$
$$= (0.65)^7 \times 125\left[120\times(0.7)^3 + 45\times(0.7)^2(0.13) + 10(0.07)(0.13)^2 + (0.1)\times(0.13)^3\right]$$
$$= 0.04901\times 125[0.04116 + 0.028665 + 0.011830 + 0.002197]$$
$$= 6.12625\times 0.0833852 = 0.05137. \quad \textbf{Ans.}$$

Example 21: (*i*) Assuming that 20% of the population of a city are literate, so that the chance of an individual being literate is 1/5 and assuming that 100 investigators each take 10 individuals to see whether they are literate, how many investigators would you expect to report 3 or less were literate? (*ii*) Assuming half the population of a town consumes chocolates and that 100 investigators each take 10 individuals to see whether they are consumers, how many investigators would you expect to report that three people or less were consumers?

Sol. (*i*) $p = \frac{1}{5},\ n = 10$

$$P(3 \text{ or less}) = P(0 \text{ or } 1 \text{ or } 2 \text{ or } 3) = P(0) + P(1) + P(2) + P(3)$$

$$= {}^{10}C_0\left(\frac{1}{5}\right)^0\left(\frac{4}{5}\right)^{10} + {}^{10}C_1\left(\frac{1}{5}\right)^1\left(\frac{4}{5}\right)^9 + {}^{10}C_2\left(\frac{1}{5}\right)^2\left(\frac{4}{5}\right)^8 + {}^{10}C_3\left(\frac{1}{5}\right)^3\left(\frac{4}{5}\right)^7$$

$$= \left(\frac{4}{5}\right)^{10} + \frac{10}{5}\left(\frac{4}{5}\right)^9 + \frac{45}{25}\left(\frac{4}{5}\right)^8 + \frac{120}{125}\left(\frac{4}{5}\right)^7$$

$$P(3 \text{ or less}) = \left(\frac{4}{5}\right)^7 \left[(0.8)^3 + 2(0.8)^2 + 1.8(0.8) + 0.96\right]$$

$$= 0.2097152[0.512 + 1.28 + 1.44 + 0.96]$$

$$= 0.2097152 \times 4.192 = 0.879126118.$$

Required number of investigators = 0.879126118 × 100=88 (approx.)

(*ii*) The chance for an individual to be consumer = $p = \frac{1}{2}$

The chance of not being a consumer $= q = 1 - \frac{1}{2} = \frac{1}{2}$

$$P(r \le 3) = P(0) + P(1) + P(2) + P(3)$$

$$= q^{10} + {}^{10}C_1 q^9 p^1 + {}^{10}C_2 q^8 p^2 + {}^{10}C_3 q^7 p^3$$

$$= \left(\frac{1}{2}\right)^{10} + 10\left(\frac{1}{2}\right)^9\left(\frac{1}{2}\right) + 45\left(\frac{1}{2}\right)^8\left(\frac{1}{2}\right)^2 + 120\left(\frac{1}{2}\right)^7\left(\frac{1}{2}\right)^3$$

$$= \left(\frac{1}{2}\right)^{10}(1 + 10 + 45 + 120) = \frac{176}{1024}$$

Number of investigators to report that three or less people were consumers of chocolate

$$= \frac{176}{1024} \times 100 = 17.2$$

Hence, 17 investigators would report that 3 or less people are consumers.

Example 22: Five coins are tossed 3200 times. Find the expected frequencies of the distribution of heads and tails and calculate the result. Fit a binomial distribution and calculate the mean number of heads and standard deviation.

Sol. Let getting a head by tossing a coin be a success. The p = constant probability of success $= \frac{1}{2}, q = 1 - p = \frac{1}{2}$.

The expected frequencies are the successive terms of the binomial expansion of $N(p+q)^n$.

Here n = 5, N = 3200, $p = \frac{1}{2}$, $q = \frac{1}{2}$.

The Binomial expansion of $N(p+q)^n$, *i.e.* of $3200\left(\frac{1}{2} + \frac{1}{2}\right)^n$

$$= 3200\left[\left(\frac{1}{2}\right)^5 + {}^5C_1\left(\frac{1}{2}\right)^{5-1}\left(\frac{1}{2}\right) + {}^5C_2\left(\frac{1}{2}\right)^{5-2}\left(\frac{1}{2}\right)^2 + {}^5C_3\left(\frac{1}{2}\right)^{5-3}\left(\frac{1}{2}\right)^3\right.$$

$$\left. + {}^5C_4\left(\frac{1}{2}\right)^{5-4}\left(\frac{1}{2}\right)^4 + {}^5C_5\left(\frac{1}{2}\right)^{5-5}\left(\frac{1}{2}\right)^5\right]$$

$$= 3200\left[\left(\frac{1}{2}\right)^5 + 5\left(\frac{1}{2}\right)^4 + 10\left(\frac{1}{2}\right)^5 + 10\left(\frac{1}{2}\right)^5 + 5\left(\frac{1}{2}\right)^5 + \left(\frac{1}{2}\right)^5\right]$$

$$= 3200 \times \left(\frac{1}{2}\right)^5 (1+5+10+10+5+1)$$

$$= 100(1+5+10+10+5+1) = 100 + 500 + 1000 + 1000 + 500 + 100$$

Success(s) (No. of heads)	0	1	2	3	4	5	Total
Failures (No. of tails)	5	4	3	2	1	0	
Frequecy (f)	100	500	1000	1000	500	100	3200

Mean number of success $(\bar{x}) = \dfrac{\sum fx}{\sum f} = \dfrac{8000}{3200} = 2.5$

Otherwise $\quad$ Mean $= np = 5 \times \dfrac{1}{2} = 2.5$

$$\text{S.D. } (\sigma) = \sqrt{npq} = \sqrt{5 \times \frac{1}{2} \times \frac{1}{2}} = \frac{\sqrt{5}}{2} = 1.12 \quad \text{(approx.)}$$

PROBLEM SET 4.3

1. Find the probability of getting 4 heads in 6 tosses of a fair coin. $\left[\textbf{Ans. } \dfrac{15}{64}\right]$

2. Ten coins are tossed simultaneously. Find the probability of getting at least seven heads. $\left[\textbf{Ans. } \dfrac{11}{64}\right]$

3. Ten per cent of screws produced in a certain factory turn out to be defective. Find the probability that in a sample of 10 screws chosen at random, exactly two will be defective. [**Ans.** 0.1937]

4. A die is thrown 8 times and it is required to find the probability that 3 will show (*i*) exactly 2 times. (*ii*) at least seven times (*iii*) at least once.

   $\left[\textbf{Ans. } (i)\ \dfrac{28 \times 5^6}{6^8}\quad (ii)\ \dfrac{41}{6^8}\quad (iii)\ 1 - \left(\dfrac{5}{6}\right)^8\right]$

5. Six dice are thrown 729 times. How many times do you expect at least three dice to show a five or a six? [**Ans.** 233]

6. Assuming a Binomial distribution, find the probability of obtaining at least two 'six' in rolling a fair die 4 times. $\left[\textbf{Ans. } \dfrac{171}{1296}\right]$

7. A box contains 10 screws, 3 of which are defective. Two screws are drown at random without replacement. Find the probability that none of the two screws is defective.

$$\left[\textbf{Ans. } \frac{7}{15}\right]$$

8. Find the Binomial distribution whose mean is 5 and variance is $\frac{10}{3}$.

$$\left[\textbf{Ans. } {}^{15}C_r\left(\frac{1}{3}\right)^r\left(\frac{2}{3}\right)^{15-r}\right]$$

9. A and B take turns in throwing dice, the first to throw 10 being awarded the prize. Show that if A has the first throw, their chance of winning are in the ratio 12 : 11.

10. The latest poll indicates that for Americans, randomly selected the prob. That they are conservative is 0.55 probability. That they are liberal is 0.30 probability. That they are middle of road is 0.15. Assume the probability as accurate, answer the following pertaining to a randomly chosen group of 10 Americans.

(*i*) Probability of 4 of them being liberal

(*ii*) Probability of none being conservative.

(*iii*) Probability of 2 of them being middle of road.

[**Ans.** (*i*) 0.2, (*ii*) 0.00034, (*iii*) 0.2759]

11. A bag contains 5 white, 7 red and 8 black balls. If your balls are thrown, one by one, with replacement, what is the probability that

(*i*) None is white

(*ii*) All are white

(*iii*) At least one is white

(*iv*) Only 2 are white

$$\left[\textbf{Ans. } (i)\ \frac{81}{256}\quad (ii)\ \frac{1}{256}\quad (iii)\ \frac{175}{256}\quad (iv)\ \frac{27}{428}\right]$$

12. The probability that a bomb dropped from a plane will strike the target is $\frac{1}{5}$. If six bombs are dropped, find the probability that

(*i*) Exactly two will strike the target and

(*ii*) At least two will strike the target.

[**Ans.** (*i*) 0.245 (*ii*) 0.345]

13. Five dice were thrown together 96 times. The number of times 4, 5, 6 was actually thrown is given below. Calculate the expected frequencies.

No of dice	0	1	2	3	4	5
Frequencies	1	10	24	35	18	8

[**Ans.** 3, 15, 30, 30, 15, 3]

14. Out of 2000 families with 4 children each, how many would you expect to have (*i*) at least one boy (*ii*) exactly 2 boys (*iii*) 1 or 2 girls (*iv*) no girls. (Assume that the prob. of a girl child is 1/2).

[**Ans.** (*i*) 1875 (*ii*) 750 (*iii*) 1250 (*iv*) 125]

15. If on the average rain falls on 10 days in every thirty days find the probability that (*i*) rain falls on at least 3 days of a given week (*ii*) first 3 days of a given week be dry and the remaining wet.

$$\left[\textbf{Ans. } (i)\sum_{r=3}^{9}{}^{7}C_r\left(\frac{1}{3}\right)^r\left(\frac{2}{3}\right)^{7-r}\quad (ii)\left(\frac{2}{3}\right)^3\left(\frac{1}{3}\right)^4\right]$$

4.16 POISSON DISTRIBUTION

Poisson distribution is a discrete (countably infinite) probability distribution and is defined by the probability functions.

$$P(r) = \frac{e^{-m} m^r}{r!} \quad (r = 1, 2, 3......) \qquad ...(1)$$

where m is a positive constant called the parameter of the distributive and $e = 2.71828$ approximately.

It was discovered by a French mathematician S.D. Poisson in 1837. It is widely used in statistical work. Poisson distribution is the limiting form of the Binomial distribution when the number of trials n becomes sufficiently large and the probability p of success in a trial very small, where $np = m$ is a positive constant.

Examples of Poisson Distribution

(*i*) The number of telephone calls per minute at a switch board.
(*ii*) The number of printing mistakes per page in a large text.
(*iii*) The number of cars passing a certain point in one minute.
(*iv*) The number of particles emitted by a radioactive substance.
(*v*) The number of persons born blind per year in a large city.
(*vi*) The number of typographical errors made by a typist in a large number of pages.
(*vii*) The number of defective electric bulbs manufactured by a reputed company.

4.16.1 Properties of Poisson Distribution

1. Poisson distribution is a discrete probability distribution in which the random variable x assumes the infinite set of values. 0, 1, 2, 3,....
2. Mean = m = parameter of the distribution,

   Variance = m, $S.D.(\sigma) = \sqrt{m}$, Skewness $= \frac{1}{\sqrt{m}}$ and Kurtosis $= \frac{1}{m}$.
3. The mode of Poisson distribution is that value of x which occurs with the largest probability. It may have either one or two modes. If m is not an integer, the mode is the integral value between $m - 1$ and m. But if m is an integer, then there are two modes which are $m - 1$, m.
4. If x and y be two independent Poisson variates with parameter m_1 and m_2 respectively, then their sum $x + y$ is also a Poisson variate with parameter $m_1 + m_2$.
5. The first, second, and third raw moments are respectively m, $m^2 + m$, $m^3 + 3m^2 + m$.

4.16.2 Poisson Distribution as a Limit of the Binomial Distribution

If in the Binomial distribution, the number of trials n becomes very large and the probability p of success in a trial very small where $np = m$ remains constant (*i.e.* if $n \to \infty, p \to 0$, where $np = m$ = constant). Then the binomial distribution tends to Poisson distribution as shown below.

In case of Binomial distribution, the probability $P(r)$ of r successes in a series of n independent trials is given by

$$p(r) = {}^nC_r q^{n-r} p^r = {}^nC_r (1-p)^{n-r} p^r$$

$$\left(\text{Since mean} = m = np, p = \frac{m}{n}\right)$$

$$= {}^nC_r\left(1-\frac{m}{n}\right)^{n-r}\left(\frac{m}{n}\right)^r \qquad (m \text{ is constant})$$

$$= \frac{n(n-1)(n-2)\ldots\ldots(n-\overline{r-1})}{r!}\left(\frac{m}{n}\right)^r\left(1-\frac{m}{n}\right)^{n-r}$$

$$= \frac{\frac{n}{n}\left(\frac{n}{n}-\frac{1}{n}\right)\left(\frac{n}{n}-\frac{2}{n}\right)\ldots\ldots\left(\frac{n}{n}-\frac{r-1}{n}\right)m^r\left(1-\frac{m}{n}\right)^n}{r!\left(1-\frac{m}{n}\right)^r}$$

$$= \frac{1\left(1-\frac{1}{n}\right)\left(1-\frac{2}{n}\right)\ldots\ldots\left(1-\frac{r-1}{n}\right)m^r\left(1-\frac{m}{n}\right)^n}{r!\left(1-\frac{m}{n}\right)^r}$$

As $n \to \infty$, then $\left(1-\frac{1}{n}\right)\left(1-\frac{2}{n}\right)\ldots\ldots\left(1-\frac{r-1}{n}\right)$ tends to 1.

Also $\left(1-\frac{m}{n}\right)^r \to 1$ as $n \to \infty$

and $$\lim_{n\to\infty}\left(1-\frac{m}{n}\right)^n = \lim_{n\to\infty}\left[\left(1-\frac{m}{n}\right)^{-n/m}\right]^{-m} = e^{-m}$$

$\therefore$ $$P(r) = \frac{m^r}{r!}e^{-m}$$

$$\boxed{P(r) = \frac{e^{-m}.m^r}{r!}} \text{ where } r = 0, 1, 2, 3.....$$

4.16.3 Mean of the Poisson Distribution

For Poisson distribution

$$P(r) = \frac{m^r e^{-m}}{r!}$$

$$\text{Mean } (\mu) = \sum_{r=0}^{\infty} rP(r) = \sum_{r=0}^{\infty}\frac{re^{-m}m^r}{r!}$$

$$= e^{-m}\sum_{r=0}^{\infty}\frac{m^r}{(r-1)!}$$

$$= e^{-m}\left(m + \frac{m^2}{1!} + \frac{m^3}{2!} + ...\right)$$

$$= e^{-m} m\left(1 + \frac{m}{1!} + \frac{m^2}{2!} + ...\right)$$

$$= e^{-m}.m.e^{m} = m$$

∴ Mean of Poission distribution = m

4.16.4 Variance of the Poisson Distribution

$$\text{Variance } (\sigma)^2 = \sum_{r=0}^{\infty} r^2 P(r) - \mu^2$$

$$= \sum_{r=0}^{\infty} r^2 P(r) - m^2 = e^{-m} \sum_{r=0}^{\infty} \frac{r^2 m^r}{r!} - m^2$$

$$= e^{-m}\left[\frac{1^2 m}{1!} + \frac{2^2 m^2}{2!} + \frac{3^2 m^3}{3!} + \frac{4^2 m^4}{4!} ......\right] - m^2$$

$$= me^{-m}\left[1 + \frac{2m}{1!} + \frac{3m^2}{2!} + \frac{4m^3}{3!} ......\right] - m^2$$

$$= me^{-m}\left[1 + \frac{(1+1)m}{1!} + \frac{(1+2)m^2}{2!} + \frac{(1+3)m^3}{3!} + ...\right] - m^2$$

$$= me^{-m}\left[\left(1 + \frac{m}{1!} + \frac{m^2}{2!} + \frac{m^3}{3!} + ...\right) + \left(\frac{m}{1!} + \frac{2m^2}{2!} + \frac{3m^3}{3!}\right) + ...\right] - m^2$$

$$= me^{-m}\left[e^m + m\left(1 + \frac{m}{1!} + \frac{m^2}{2!} + ...\right)\right] - m^2$$

$$= me^{-m}\left[e^m + me^m\right] - m^2 = m$$

∴ Variance of Poisson distribution is m.
Hence standard deviation of Poisson distribution is $\sqrt{m}$.

4.16.5 Recurrence Formula for Poisson Distribution

By Poisson distribution

$$P(r) = \frac{e^{-m} m^r}{r!} \quad ...(1)$$

$$P(r+1) = \frac{e^{-m} m^{r+1}}{(r+1)!} \quad ...(2)$$

dividing (2) by (1), we get

$$\frac{P(r+1)}{P(r)}=\frac{e^{-m}m^{r+1}}{(r+1)!}\cdot\frac{r!}{e^{-m}m^{r}}=\frac{m}{r+1}$$

$$\boxed{P(r+1)=\frac{m}{r+1}P(r)}$$, which is required recurrence formula.

Example 1: A random variable x follows Poisson distribution with parameter 4. Find the probabilities that x assumes the values (*i*) 0, 1, 2, 4; (*ii*) less than 2; (*iii*) at least 3{Given that $e^{-4}=0.0183$ }.

Sol. For the Poisson distribution with parameter *m*, the probability $P(x)$ of *x* success is

$$P(x)=\frac{e^{-m}m^{x}}{x!}=\frac{e^{-4}4^{x}}{x!} \qquad [\because m=4]$$

(*i*) Here x = 0, 1, 2, 4

$$P(0)=\frac{e^{-4}4^{0}}{0!}=\frac{e^{-4}1}{1}=0.0183, \qquad P(2)=\frac{e^{-4}4^{2}}{2!}=\frac{0.0183\times 16}{2}=0.1464$$

$$P(1)=\frac{e^{-4}4^{1}}{1!}=\frac{0.0183\times 4}{1}=0.0732,\ P(4)=\frac{e^{-4}4^{4}}{4!}=\frac{0.0183\times 32}{3}=0.1952$$

(*ii*) Here 'less than 2' means 'either 0 or 1'. *i.e.*, x = 0, or 1.

Hence the required probability $=P(0)+P(1)=0.0183+0.0732=0.0915.$

(*iii*) Here "at least 3 means either 3 or 4 or 5, or ...."

Hence the required probability that *x* assumes at least 3 is

$$=P(3)+P(4)+P(5)+.........$$

$$=1-\left[P(0)+P(1)+P(2)`\right] \qquad \left[\because P(0)+P(1)+P(2)+P(3)+.....=1\right]$$

$$=1-(0.0183+0.0732+0.1464)=1-0.2379=0.7621.$$

Example 2: In a Poisson distribution, if [*U.T.U. 2008*]

$P(r = 2) = 9P(r = 4) + 90(\mathbf{r} = 6)$, then find $P(r = 1)$

Sol. We apply the formula $P(r)=\dfrac{e^{-m}\cdot m^{r}}{\lfloor r}$, we get

$$\frac{e^{-m}\cdot m^{2}}{\lfloor 2}=9\cdot\frac{e^{-m}\cdot m^{4}}{\lfloor 4}+90\cdot\frac{e^{-m}\cdot m^{6}}{\lfloor 6}$$

or
$$\frac{m^{2}\cdot e^{-m}}{2}=\frac{9\cdot m^{4}e^{-m}}{4\cdot 3\cdot 2\cdot 1}+90\cdot\frac{m^{6}}{6\cdot 5\cdot 4\cdot 3\cdot 2}e^{-m}$$

or
$$\frac{1}{2}=\frac{3m^{4}}{8}+\frac{m^{4}}{8}$$

or $\quad m^4 + 3m^2 - 4 = 0$

or $\quad (m^2 - 1)(m^2 + 4) = 0$

$\Rightarrow \quad m = \pm 1 \quad$ or $\quad m = 1$

Hence $\quad P(r = 1) = \dfrac{e^{-1} \cdot 1^1}{\lfloor 1} = \dfrac{1}{e}$. **Ans.**

Example 3: If 5% of the electric bulbs manufactures by a company are defective, use Poisson distribution to find the probability that in a sample of 100 bulbs (*i*) none is defective, (*ii*) 5 bulbs will be defective {Given $e^{-5} = 0.007$}

Sol. Let the occurrence of a defective bulb be a success.

Then $p = \dfrac{5}{100} = 0.05$ (small) and $n = 100$ (large).

$\therefore m = np = 100 \times 0.05 = 5$ which is the parameter of Poisson distribution.

For Poisson distribution with parameter $m = 5$, the probability of r successes is

$$P(r) = \frac{e^{-m} m^r}{r!} = \frac{e^{-5} 5^r}{r!}$$

(*i*) Since none is defective.

$\therefore \quad r = 0$

Hence the required probability that some is defective $= P(0) = \dfrac{e^{-5} 5^0}{1!} = \dfrac{e^{-5} 1}{1} = 0.007$

(*ii*) Since 5 bulbs will be defective

$\therefore \quad r = 5$

Hence the required probability that 5 bulbs will be defective

$$P(5) = \frac{e^{-5} 5^5}{5!} = \frac{0.007 \times 625}{24} = 0.182. \quad \textbf{Ans.}$$

Example 4: If the variance of the Poisson distribution is 2. Find the probabilities for $r = 1, 2, 3, 4$ from the recurrence relation of the Poisson distribution. Also find $P(r \geq 4)$.

Sol. Variance = m = 2, Mean = 2.

$$P(r+1) = \frac{m}{r+1} P(r) \quad \text{(Recurrence Relation)}$$

Now, $\quad P(r+1) = \dfrac{2}{r+1} P(r) \qquad \left[P(r) = \dfrac{e^{-m} \cdot m^r}{r!} \right.$, $\left. P(0) = e^{-m} = e^{-2} = 0.1353 \right.$

If $r = 0$, $\quad P(1) = \dfrac{2}{0+1} P(0) = \dfrac{2}{0+1}(0.1353) = 0.2706$

If $r = 1$, $\quad P(2) = \dfrac{2}{1+1} P(1) = \dfrac{2}{2}(0.2706) = 0.2706$

If $r = 2$, $$P(3) = \frac{2}{2+1}P(2) = \frac{2}{3}(0.2706) = 0.1804$$

If $r = 3$, $$P(4) = \frac{2}{3+1}P(3) = \frac{2}{4}(0.1804) = 0.0902$$

$$P(r \geq 4) = P(4) + P(5) + P(6) + \ldots\ldots$$
$$= 1 - \left[P(0) + P(1) + P(2) + P(3)\right]$$
$$= 1 - [0.1353 + 0.2706 + 0.2706 + 0.1804]$$
$$= 1 - 0.8569 = 0.1431.$$ **Ans.**

Example 5: Between the hours of 2 p.m. and 4 p.m., the average number of phone calls per minute coming into the switch board of a company is 2.5. Find the probability that during one particular minute there will be no phone call at all. {Given $e^{-2} = 0.13523$ and $e^{-0.5} = 0.60650$ }

Sol. Let the random variable r represents the number of phone calls received per minute between 2 p.m. and 4 p.m. Then r follows Poisson distribution with parameter, say m. For a Poisson distribution, m = parameter = mean, or average. $\therefore m = 2.5$.

The probability of a phone call is

$$P(r) = \frac{e^{-m}m^r}{r!} = \frac{e^{-2.5}2.5^r}{r!}$$

Here $r = 0$,

$$\therefore P(0) = \frac{e^{-2.5}(2.5)^0}{0!} = e^{-2.5} = e^{-2+(-0.5)} = e^{-2} \times e^{-0.5} = 0.13524 \times 0.60650 = 0.082023.$$

Hence the required probability that there will be no phone call $= P(0) = 0.082023$. **Ans.**

Example 6: The number of arrivals of customers during any day follows Poisson distribution with a mean of 5. What is the probability that the total number of customers on two days elected at random is less than 2?

Sol. $$m = 5,\ P(r) = \frac{e^{-m}m^r}{r!}.P(r) = \frac{e^{-5}(5)^r}{r!}$$

If the number of customers on two days < 2 = 1 or 0.

No. of customers

First day	*Second day*	*Total*
0	0	0
0	1	1
1	0	1

Required Probability $= P(0)P(0) + P(0)P(1) + P(1)P(0)$

$$= \frac{e^{-5}(5)^0}{0!} \cdot \frac{e^{-5}(5)^0}{0!} + \frac{e^{-5}(5)^0}{0!} \cdot \frac{e^{-5}(5)^1}{1!} + \frac{e^{-5}(5)^1}{1!} \frac{e^{-5}(5)^0}{0!}$$

$$= e^{-5}e^{-5} + e^{-5}e^{-5}.5. + e^{-5}.5.e^{-5}$$

$$= e^{-10}[1+5+5] = 11e^{-10} = 11 \times 4.54 \times 10^{-5}$$

$$= 4.994 \times 10^{-4}.$$ **Ans.**

Example 7: Using Poisson distribution, find the probability that the ace of spades will be drawn from a pack of well shuffled cards at least once in 104 consecutive trials.

Sol. Probability of the ace of spades $= p = \frac{1}{52}$, $n = 104$

$$m = np = 104 \times \frac{1}{52} = 2$$

$$P(r) = e^{-m}\frac{m^r}{r!} = e^{-2}.\frac{2^r}{r!} = \frac{1}{e^2}\frac{2^r}{r!}$$

P(At least once) $= P(1) + P(2) + P(3) + ... + P(104) = 1 - P(0)$

$$= 1 - \frac{1}{e^2} \times \frac{2^0}{0!} = 1 - \frac{1}{e^2} = 1 - 0.135 = 0.865.$$ **Ans.**

Example 8: Records show that the probability is 0.00002 that a car will have a flat tyre while driving over a certain bridge. Use the Poisson probability distribution to determine the probability that among 20,000 cars driven over the bridge, and more than one will have a flat tyre.

$\left[\text{Given } e^{-0.2} = 0.819,\ e^{-0.4} = 0.670\right]$

Sol. Here $p = 0.00002$ and $n = 20{,}000$, so that m = parameter $= np = 0.00002 \times 20{,}000 = 0.4$

$\therefore$ The Poisson probability function is

$$P(r) = \frac{e^{-m}m^r}{r!} = \frac{e^{-0.4}(0.4)^r}{r!},\ \text{where } r = 0, 1, 2, 3.$$

The required probability that not more than one will have a flat tyre $= P(0) + P(1)$.

$$= \frac{e^{-0.4}(0.4)^0}{1!} + \frac{e^{-0.4}(0.4)^1}{2!} = \frac{0.670 \times 1}{1} + \frac{0.670 \times 0.4}{2}$$

$$= 0.670 + 0.2680 = 0.9380 = 0.938.$$ **Ans.**

Example 9: Find the probability that at most 5 defective bolts will be found in a box of 200 bolts, if its is known that 2 per cent of such bolts are expected to be defective. (You may take the distribution to be Poisson). Take $e^{-4} = 0.0183$.

Sol. Let the occurrence of a defective bolt be a success.

Then p = Probability that a bolt drawn at random is defective = $\frac{2}{100} = 0.02$ which is small and n = total no. of bolts in the box = 200 which is large.

Also $m = np = 200 \times 0.02 = 4$ which is a finite constant.

$\therefore$ The number of defective bolts (*i.e.*, r) approximately follow Poisson distribution. The probability $P(r)$ of r success (*i.e.*, defectives) is given by

$$P(r) = \frac{e^{-m} m^r}{r!} = \frac{e^{-4}(4)^r}{r!}.$$

By "at most 5" we mean "either 0, or 1 or 2 or 3, or 5 *i.e.* r = 0, or 1 or 2 or 3 or 4 or 5.

Hence the required probability that at most 5 bolts are defective.

$$= P(0) + P(1) + P(2) + P(3) + P(4) + P(5)$$

$$= \frac{e^{-4}.4^0}{0!} + \frac{e^{-4}.4^1}{1!} + \frac{e^{-4}.4^2}{2!} + \frac{e^{-4}.4^3}{3!} + \frac{e^{-4}.4^4}{4!} + \frac{e^{-4}.4^5}{5!}$$

$$= e^{-4}\left(1 + \frac{4}{1} + \frac{16}{2} + \frac{64}{6} + \frac{256}{24} + \frac{1024}{120}\right)$$

$$= 0.0183\left(1 + 4 + 8 + \frac{32}{3} + \frac{32}{3} + \frac{128}{15}\right) = 0.0183 \times \frac{643}{15} = 7.844. \textbf{ Ans.}$$

Example 10: The probability that a man aged 45 years will die within a year is 0.012. What is the probability that out of 10 such men at least 9 will reach their forty-sixth birthday?

$\left[\text{Given } e^{-0.12} = 0.88692\right]$

Sol. Since the probability $p = 0.012$ that a man aged 45 years will be within a year is very small, we have to use Poisson distribution, m = parameter = $np = 10 \times 0.012 = 0.12$.

For the Poisson distribution, the probability $P(r)$ is given by

$$P(r) = \frac{e^{-m} m^r}{r!} = \frac{e^{-0.12}(0.12)^r}{r!},$$

where r is the no. of men that will die within a year.

$$P(0) = \frac{e^{-0.12}(0.12)^0}{0!} = \frac{0.88682 \times 1}{1} = 0.88682$$

and $$P(1) = \frac{e^{-0.12}(0.12)^1}{1!} = \frac{0.88682 \times 0.12}{1} = 0.10643$$

Hence the required probability that at least 9 out of 10 men will reach their 46th birthday.

= Probability that 9 will not die + Probability that 10 will not die

= Probability that 1 man will die + Probability that nobody will die

= $P(1) + P(0) = 0.10643 + 0.88682 = 0.98325$. **Ans.**

Example 11: If the probability that an individual suffers a bad reaction from a certain injection is 0.001, determine the probability that out of 2000 individuals. (*a*) exactly 3 (*b*) more than 2 individuals (*c*) None (*d*) More than one individual will suffer a bad reaction.

Sol.
$$p = 0.001,\ n = 2000$$
$$m = np = 2000 \times 0.001 = 2$$
$$P(r) = \frac{e^{-m} m^r}{r!} = \frac{e^{-2}(2)^r}{r!} = \frac{1}{e^2} \times \frac{2^r}{r!}$$

(*a*)
$$P(3) = \frac{1}{e^2} \cdot \frac{2^3}{3!} = \frac{1}{(2.718)^2} \times \frac{8}{6} = (.135) \times \frac{4}{3} = 0.18. \quad \textbf{Ans.}$$

(*b*) P (more than 2) = $P(3) + P(4) + P(5) + \ldots\ldots + P(2000)$
$$= 1 - [P(0) + P(1) + P(2)]$$
$$= 1 - \left[\frac{e^{-2}2^0}{0!} + \frac{e^{-2}2^1}{1!} + \frac{e^{-2}2^2}{2!}\right]$$
$$= 1 - e^{-2}[1 + 2 + 2] = 1 - \frac{5}{e^2}$$
$$= 1 - 5 \times 0.135 = 1 - 0.675 = 0.325. \quad \textbf{Ans.}$$

(*c*) P(none) = $P(0) = \frac{e^{-2}2^0}{0!} = 0.135$. **Ans.**

(*d*) P(More than 1) = $P(2) + P(3) + P(4) + \ldots\ldots + P(2000)$
$$= 1 - [P(0) + P(1)]$$
$$= 1 - \left[\frac{e^{-2}2^0}{0!} + \frac{e^{-2}2^1}{1!}\right] = 1 - 3e^{-2} = 1 - 3 \times 0.135 = 1 - 0.405$$
$$= 0.595. \quad \textbf{Ans.}$$

Example 12: A manufacturer knows that the razor blades he makes contain on an average 0.5% of defectives. He packs them in packet of 5. What is the probability that a packet picked a random will contain 3 or more faulty blades?

Sol.
$$p = 0.5\% = 0.005, \quad n = 5$$
$$m = np = 5 \times 0.005 = 0.025$$
$$P(r) = \frac{e^{-m} m^r}{r!} = \frac{e^{-0.025}(0.025)^r}{r!}$$
$$P(3 \text{ or more}) = P(3) + P(4) + P(5)$$

$$= \frac{e^{-0.025}(0.025)^3}{3!} + \frac{e^{-0.025}(0.025)^4}{4!} + \frac{e^{-0.025}(0.025)^5}{5!}$$

$$= \frac{e^{-0.025}(0.025)^3}{5!}\left[20 + 5(0.025) + (0.025)^2\right]$$

$$= \frac{0.975 \times .000015625 \times 20.125625}{120}$$

$$= 0.000002555. \quad \textbf{Ans.}$$

Example 13: Assuming that the probability of a fatal accident in a factory during the year is $\frac{1}{1200}$. Calculate the probability that in a factory employing 300 workers, there will be at least two fatal accident in a year. $\left(e^{-0.25} = 0.7788\right)$

Sol. Here p = Probability of a fatal accident during the year = $\frac{1}{1200}$, and n = 300.

Then m = The parameter of Poisson distribution = np = $300 \times \frac{1}{1200} = 0.25$

For Poisson distribution with parameter m = 0.25, the probability of r successes (*i.e.* r fatal accident) is

$$P(r) = \frac{e^{-m}m^r}{r!} = \frac{e^{-0.25}(0.25)^r}{r!} \qquad ...(1)$$

$\therefore$ The required probability that there will be at least two fatal accident in a year

$$= P(2) + P(3) + P(4) + ... = 1 - \left[P(0) + P(1)\right]$$

$$= 1 - \left[\frac{e^{-0.25}(0.25)^0}{0!} + \frac{e^{-0.25}(0.25)^1}{1!}\right]$$

$$= 1 - \left[e^{-0.25} + e^{-0.25}(0.25)\right]$$

$$= 1 - e^{-0.25}(1 + 0.25) = 1 - 0.7788 \times 1.25 = 1 - 0.9735 = 0.0265. \quad \textbf{Ans.}$$

Example 14: In a certain factory producing cycle tyres, where is a small chance of 1 in 500 tyres to be defective. The tyres are supplied in lots of 10. Using Poisson distribution calculate the approximate number of lots containing no defective, one defective and two defective tyres respectively, in a consignment of 10,000 lots.

Sol. Here $p = \frac{1}{500}$, and n = 10.

The m = The parameter of Poisson distribution = np = $10 \times \frac{1}{500} = 0.02$

$$P(r) = \frac{e^{-m}m^r}{r!}$$

(a) Probability of no defective $= P(0) = \dfrac{e^{-0.02}(.02)^0}{0!} = e^{-.02} = 0.9802$

Number of lots containing no defective $= 10{,}000 \times 0.9802 = 9802$ lots.

(b) Probability of one defective $= P(1) = \dfrac{e^{-0.02}(.02)^1}{1!} = 0.9802 \times 0.02 = 0.019604$

Number of lots containing one defective $= 10{,}000 \times 0.019604 = 196$ lots.

(c) Probability of two defective $= P(2) = \dfrac{e^{-0.02}(.02)^2}{2!} = 0.9802 \times 0.0002 = 0.00019604$

Number of lots containing two defective = 10,000 × 0.00019604 = 2 lots. **Ans.**

Example 15: In a certain factory turning out razor blades, there is a small chance of 0.02 for any blade to be defective. The blades are supplied in packets of 10. Calculate the approximate number of packets containing no defective, one defective and two defective blades in a consignment of 10, 000 packets. (Given : $e^{-0.02} = 0.9802$)

Sol. Here $p = 0.002$, $n = 10$ and $m = np = 10 \times .002 = 0.02$, $N = 10000$

∴ Probability of having no defective

$$P(0) = \frac{e^{-0.02} \cdot (.02)^0}{0!} = e^{-0.02} = 0.9802$$

Number of packets $= N \times 0.9802 = 10000 \times 0.9802 = 9802$. **Ans.**

Next, probability of having one defective

$$p(1) = \frac{e^{-0.02} \cdot (0.02)^1}{1!} = 0.9802 \times 0.02 = 0.019604$$

∴ Number of packets having one defective in the consignment

$= 0.019604 \times 10000 = 196$. **Ans.**

and Probability of having two defective blades

$$p(2) = \frac{e^{-0.02}(0.02)^2}{2!} = \frac{(0.980198)\times(0.0004)}{2} = 0.000196$$

Example 16: Fit a Poisson distribution to the following data:

x	0	1	2	3	4	Total
f	123	59	14	3	1	200

Sol. First of all mean $m = \dfrac{\Sigma fx}{\Sigma f} = \dfrac{59+28+9+4}{200} = \dfrac{100}{200} = 0.5$

and $p(r) = \dfrac{e^{-.5}(.5)^r}{r!}$

$\therefore \quad P(0) = e^{-0.5} = 6065.$

Hence $\quad f(0) = NP(0) = 200\ (.6065) = 121.3$

Using the recurrence formula $P(r+1) = \left(\frac{m}{r+1}\right)P(r)$, we have following table:

r	$P(r+1) = \left(\frac{m}{r+1}\right)P(r)$	Expected frequencies $f(x) = N.\ p(r)$	Observed frequencies
0	$P(0) = 0.6065$	$200 \times 0.6065 = 121.3$	121
1	$P(1) = \left(\frac{0.5}{1}\right)P(0) = 0.5(0.6065) = 0.3033$	$200 \times 0.3033 = 60.66$	61
2	$P(2) = \left(\frac{0.5}{2}\right)P(1) = \frac{0.5}{2} \times 0.3033 = 0.0758$	$200 \times 0.0758 = 15.16$	15
3	$P(3) = \left(\frac{0.5}{3}\right)P(2) = \frac{0.5}{3} \times 0.0758 = 0.0126$	$200 \times 0.0126 = 2.52$	3
4	$P(4) = \left(\frac{0.5}{4}\right)P(2) = \frac{0.5}{4} \times 0.0126 = 0.00157$	$200 \times 0.00157 = 0.315$	0

Example 17: The distribution of the number of road accidents per day in a city is Poisson with mean 4. Find the number of days out of 100 days when there will be

(*i*) no accident (*ii*) at least 2 accident (*iii*) atmost 3 accidents (*iv*) between 2 and 5 accidnets.

Sol. Mean $m = 4$, $N = 100$

(*i*) $P(r=0) = \frac{e^{-4} \cdot 4^0}{0!} = e^{-4} = 0.01831$

$\therefore$ Required number of days = 100 × 0.1831 = 1.831 ≈ 2. **Ans.**

(*ii*) $P(r \geq 2) = 1 - P(r<2) = 1 - [P(r=0) + P(r=1)] = 1 - \left[e^{-4} + \frac{e^{\cdot -4} 4}{1!}\right] = 1 - 5e^{-4} = 0.90842$

$\therefore$ Required number of days = 100 × 0.90.842 = 90.842 ≈ 91. **Ans.**

(*iii*) $P(r \leq 3) = P(0) + P(1) + P(2) + P(3) = \frac{e^{-4} 4^0}{0!} + \frac{e^{-4}(4)^1}{1!} + \frac{e^{-4}(4)^2}{2!} + \frac{e^{-4}(4)^3}{3!}$

$$= e^{-4}\left(1 + 4 + 8 + \frac{32}{3}\right) = 0.43347$$

(*iv*) $P(2 < r < 5) = P(3) + P(4) = \frac{e^{-4}(4)^3}{3!} + \frac{e^{-4}(4)^4}{4!} = \left(\frac{64}{6} + \frac{256}{24}\right)e^{-4} = 0.3907$

$\therefore$ Number of days = 100 × 0.3907 = 39.07 ≈ 39. **Ans.**

Example 18: Find the moment generating function of the poisson distribution about $x = 0$.

Sol. Moment generating function about $x = 0$ is

$$M_0(t) = E(e^{tr}) = \sum_r e^{tr} \cdot \frac{e^{-\lambda} \cdot \lambda^r}{r!}$$

$$= e^{-\lambda} \sum_{r=0}^{\infty} \frac{(\lambda e^t)^r}{r!} = e^{-\lambda} e^{\lambda e^t}$$

$$= e^{\lambda(e^t - 1)}. \quad \textbf{Ans.}$$

Similarly the m.g.f. about mean is $e^{\lambda(e^t - 1 - m)}$.

Example 19: A manufacture knows that the condensers he makes contain on an average 1% of defectives. He packs them in boxes of 100. What is the probability that a box packed a random will contain 4 or more faulty condensers?

Sol. $p = 1\% = 0.01,\ n = 100,\ m = np = 100 \times 0.01 = 1$

$$P(r) = \frac{e^{-m} m^r}{r!} = \frac{e^{-1}(1)^r}{r!} = \frac{e^{-1}}{r!}$$

P(4 or more faulty condensers) $= P(4) + P(5) + ... + P(100)$

$$= 1 - \left[P(0) + P(1) + P(2) + P(3)\right] = 1 - \left[\frac{e^{-1}}{0!} + \frac{e^{-1}}{1!} + \frac{e^{-1}}{2!} + \frac{e^{-1}}{3!}\right]$$

$$= 1 - e^{-1}\left[1 + 1 + \frac{1}{2} + \frac{1}{6}\right]$$

$$= 1 - \frac{8}{3e} = 1 - 0.981 = 0.019. \quad \textbf{Ans.}$$

Example 20: An insurance company found that only 0.01% of the population is involved in a certain type of accident each year. If its 1000 policy holders were randomly selected from the population, what is the probability that not more that two of its clients are involved in such an accident next year? (Given that $e^{-0.1} = 0.9048$)

Sol. $$p = 0.01\% = \frac{1}{100} \times \frac{1}{100} = \frac{1}{10000}, n = 1000$$

$$m = np = (1000) \times \frac{1}{10,000} = \frac{1}{10} = 0.1$$

$$P(r) = \frac{e^{-m} m^r}{r!}$$

P(not more that 2) = P (0, 1 and 2) $= P(1) + P(1) + P(2)$

$$= \frac{e^{-0.1}(0.1)^0}{0!} + \frac{e^{-0.1}(0.1)^1}{1!} + \frac{e^{-0.1}(0.1)^2}{2!}$$

$$= e^{-0.1}\left[1 + 0.1 + \frac{0.01}{2}\right]$$

$$= 0.9048 \times 1.105 = 0.9998. \quad \textbf{Ans.}$$

Example 21: A car hire firm two cars which it hires out day by day. The number of demands for a car on each day is distributed as a Poisson distribution with mean 1.5. Calculate the number of days in a year on which (*i*) car is not used (*ii*) the number of days in a year some demand is refused (given that $e^{-1.5}$ = 0.2231).

Sol. m =1.5

(*i*) If the car is not used, then demand $(r) = 0$

$$P(r) = \frac{e^{-m} m^r}{r!}$$

$$P(0) = \frac{e^{-1.5}(1.5)^0}{0!} = e^{-1.5} = 0.2231$$

Number of days in a year when there is no demand of car = $365 \times 0.2231 = 81.4315$ or 81 days. **Ans.**

(*ii*) Some demand is refused if the number of demands is more that two *i.e.*, $r > 2$.

$$P(r > 2) = P(3) + P(4).... = 1 - [P(0) + P(1) + P(2)]$$

$$= 1 - \left[\frac{e^{-1.5}(1.5)^0}{0!} + \frac{e^{-1.5}(1.5)^1}{1!} + \frac{e^{-1.5}(1.5)^2}{2!}\right]$$

$$= 1 - \left[e^{-1.5} + e^{-1.5} \times 1.5 + e^{-1.5} \times 1.125\right]$$

$$= 1 - e^{-1.5}[1 + 1.5 + 1.125] = 1 - e^{-1.5} \times 3.625$$

$$= 1 - 0.2231 \times 3.625 = 1 - 0.8087375$$

$$= 0.1912625$$

Number of days in a year when some demand of car is refused

$$= 365 \times 0.1912625 = 69.81 = 70 \text{ days.} \quad \textbf{Ans.}$$

PROBLEM SET 4.4

1. Six coins are tossed 6400 times. Using the Poisson distribution, determine the approximate probability of getting six head x times.

$$\left[\textbf{Ans. } \frac{(100)^2 . e^{-100}}{x!}\right]$$

2. Using Poisson distribution, find the probability that the ace of spades will be drawn from a pack of well-stuffed cards at least once in 104 consecutive trials. (Given e^{-2} = 0.136)

[**Ans.** 0.8647]

3. Find the probability that at most 5 defective fuses will be found in a box of 200 fuses if experience show that 2 per cent of each fuses are defective. **[Ans.** 0.785]

4. If the probability of a bad reaction from a certain injection is 0.0002, determine the chance that out of 1000 individuals more than two will get a bad reaction. **[Ans.** 0.0012]

5. An insurance company finds that 0.005% of the population dies from a certain kind of accident each year. What is the probability that the company must pay off no more than 3 of 10,000 insured risks against such incident in a given year? **[Ans.** 0.0016]

6. Wireless sets are manufactured with 25 solder joints each on the average 1 joint in 500 is defective. How many sets can be expected to be free from defective joints in a consignment of 10000 sets? **[Ans.** 9512]

7. The number of accident during a year in a factory has the Poisson distribution with mean 1.5. The accident during different years are assumed independent. Find the probability that only 2 accidents take place during 2 years time. **[Ans.** 0.224]

8. The random variable x has a Poisson distribution, if

$P(x=1)=0.01487,\ P(x=2)=0.0446.$ Then find $P(x=3)$. **[Ans.** 0.08922]

9. Let x be the number of cars, passing a certain point, per minute at a particular time. Assuming that x has a Poisson distribution with mean 0.5, find the probability of observing 3 or fewer cars during any given minute. **[Ans.** 0.998]

10. For a Poisson distribution with mean m show that

$$\mu_{r+1} = mr\mu_{r-1} + m\frac{d\mu_r}{dm}$$

where,
$$\mu_r = \sum_{x=0}^{\infty}(x-m)^r \frac{e^{-m}m^x}{x!}$$

11. If a random variable has a Poisson distribution such that $P(1)=P(2)$, find

(*i*) Mean of the distribution (*ii*) $P(4)$. $\left[\textbf{Ans. } (i)\ 2,\ (ii)\ \frac{2}{3e^2}\right]$

12. If $P(x=0)=P(x=1)=a$ in a Poisson distribution, show that $ae = 1$.

13. The probability that an evening college student will be graduate is 0.4. Determine the probability that out of 5 students (*a*) none (*b*) one and (*c*) atleast one will be graduate.

[Ans. (*a*) 0.08 (*b*) 0.26 (*c*) 0.92]

14. Suppose 3% of bolts made by a machine are defective, the defects occurring at random during production. If bolts are packaged 50 per box, find (*a*) exact probability and (*b*) Poisson approximation to it, that a given box will contain 5 defective.

[Ans. (*a*) 0.013074 (*b*) 0.01412]

15. The frequency of accident per shift in a factory is shown in the following table.

Accident / shift	0	1	2	3	4	Total
Frequency	192	100	24	3	1	320

Find the expected frequencies. **[Ans.** 193, 97, 24, 1]

16. In a certain factory producing razor blades there is a small chance of $\frac{1}{500}$ for any blade to be defective. The blades are supplied in packets of 10. Use Poisson distribution to calculate the approximate number of packets containing (*i*) no defective (*ii*) one defective (*iii*) two defective blades respectively in a consignment of 10000 packets.

[**Ans.** (*i*) 42 (*ii*) 4 (*iii*) 3

17. The probability that a man aged 50 years will die with a year is 0.01125. What is the probability that of 12 such men atleast eleven will reach their 51st birth day.

[**Ans.** 0.9916]

4.17 NORMAL DISTRIBUTION (OR GAUSSIAN DISTRIBUTION)

Normal (or Gaussian) distribution is the most important continuous probability distribution on statistics and is defined by the probability density function (or simply density function).

$$f(x) = \frac{1}{\sigma\sqrt{2\pi}} . e^{-\left[(x-m)^2/2\sigma^2\right]}, (-\infty < x < \infty) \quad ...(1)$$

where m = mean, σ = standard deviation, $\pi = 3.1416$ or $\frac{22}{7}$ and e = 2.71828 approximately.

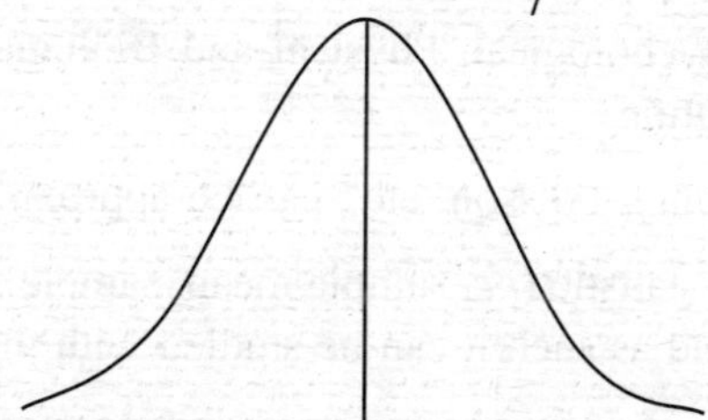

The curve representing the Normal distribution (1) is called the normal curve and the total area bounded by the cure and the x-axis is one, *i.e.*,

$$\int_{-\infty}^{\infty} f(x)dx = 1.$$

The area under the curve between the ordinates $x = a$ and $x = b$, where $a < b$, represents the probability that x lies between a and b, *i.e.* $P(a < x < b)$.

Thus, $P(a < x < b)$ = area under the normal curve between the vertical lines $x = a$ and $x = b$ which is $\int_a^b f(x)dx$.

The curve is symmetrical about the mean (m) and bell-shaped and the two tails on the right and the left sides of the mean (m) extends to infinity. Clearly, Mean = Median = Mode.

4.17.1 Standard Normal Distribution

If we take $\frac{x-m}{\sigma} = z$ in (1), we obtain the Standard Normal Distribution with mean zero and S.D. $\sigma = 1$ defined by the density function.

$$p(z) = \frac{1}{\sqrt{2\pi}} . e^{-(z^2/2)}, (-\infty < x < \infty)$$

Here z is called the standard normal variable with mean 0 and standard deviation 1 and in short, we say that $p(z)$ is $N(0,1)$.

4.17.2 Properties of Normal Distribution

1. Normal distribution is a continuous probability distribution having two parameters m and σ.
2. Standard Deviation $= \sigma$, Skewness = 0, Kurtosis = 0, Mean = Median = Mode = m.
3. The normal curve is bell-shaped and symmetrical about the line $x = m$, and the two tails of the curve on both sides of $x = m$ extends to infinity.
4. The quartiles Q_1 and Q_2 are equidistant from the mean m and $Q_1 = m - 0.67\,\sigma$, $Q_2 = m + 0.67\,\sigma$ approximately. Quartile deviation = $0.67\,\sigma$ and Mean deviation = $0.80\,\sigma$.
5. All odd order central moments are zero, *i.e.* $m_1 = m_2 = m_3 .... = 0$ Mean $(m) = A + m$.

Even order moments are given by $m_{2r} = 1, 3, 5....(2r-1)\,\sigma^{2r}; r = 1, 2, 3......$ clearly $m_2 = 1.\sigma^2 = \sigma^2$,

$m_4 = 1.3\sigma^4 = 3\sigma^4$ etc.

4.17.3 Importance of the Normal Distribution

Normal distribution plays a very important role in Statistical Theory and in Sampling Theory. It has been found that:

1. Data obtained from Psychological, Physical and Biological measurements approximately follow Normal distribution.
2. Distribution like Binomial, Poisson, etc., can be approximated to Normal distribution.
3. For large samples, any statistic (*i.e.* sample mean, sample S.D., etc.) approximately follows Normal distribution and as such it can be studied with the help of normal curve.
4. Normal curve is used to find confidence limits of the population parameters.
5. Normal distribution are largely applied in statistical quality control in industry for finding, control limits.
6. The theory of errors of observations in physical measurements are based on Normal distribution.

4.17.4 Area under the Normal Curve

By taking $z = \dfrac{x - m}{\sigma}$, standard normal curve is formed. The total area under the curve is 1.

The area under the curve is divided into two equal parts by $z = 0$. The area between the ordinate $z = 0$ and any other ordinate can be noted from the supplied table. It should be noted that mean for the Normal distribution is 0.

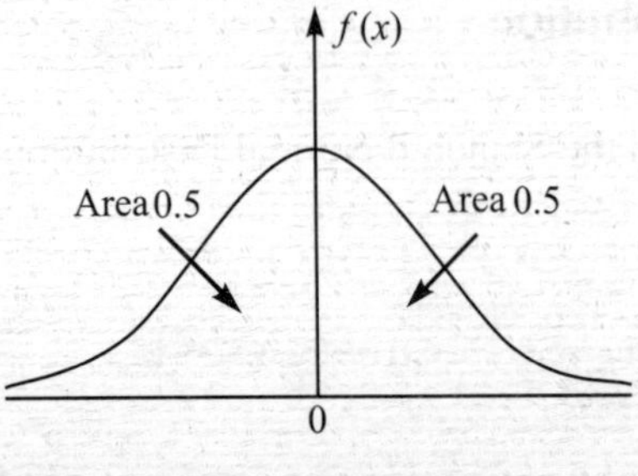

$f(x)$

Area 0.5 Area 0.5

0

TABLE: Normal Table

Areas under the standard normal

$$\textbf{Curve} = \frac{1}{\sqrt{2\pi}} \int_0^z e^{-\frac{z^2}{2}} dz$$

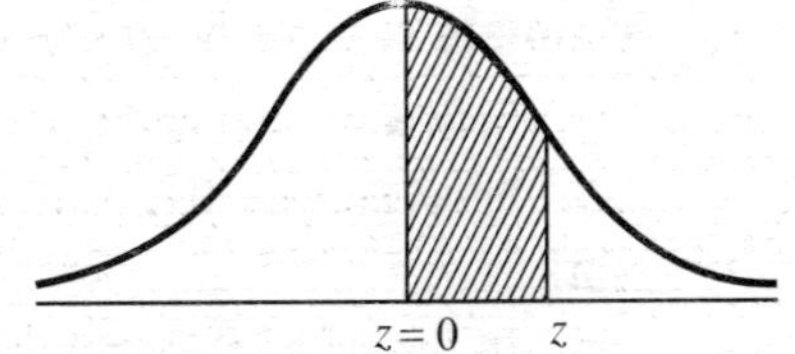

z	.00	.01	.02	.03	.04	.05	.06	.07	.08	.09
0.0	.0000	.0040	.0080	.0120	.0160	.0199	.0239	.0279	.0319	.0359
0.1	.0398	.0438	.0478	.0517	.0557	.0596	.0636	.0675	.0714	.0754
0.2	.0793	.0832	.0871	.0910	.0948	.0987	.1026	.1064	.1103	.1141
0.3	.1179	.1217	.1255	.1293	.1331	.1368	.1406	.1443	.1480	.1517
0.4	.1554	.1591	.1628	.1664	.1700	.1736	.1772	.1808	.1844	.1879
0.5	.1915	.1950	.1985	.2019	.2054	.2088	.2123	.2157	.2190	.2224
0.6	.2257	.2291	.2324	.2357	.2389	.2422	.2454	.2485	.2517	.2549
0.7	.2580	.2611	.2642	.2673	.2704	.2734	.2764	.2794	.2823	.2852
0.8	.2881	.2910	.2939	.2967	.2995	.3023	.3051	.3078	.3106	.3133
0.9	.3159	.3186	.3212	.3238	.3264	.3289	.3315	.3340	.3365	.3389
1.0	.3413	.3438	.3461	.3485	.3508	.3531	.3554	.3577	.3599	.3621
1.1	.3643	.3665	.3686	.3708	.3729	.3749	.3770	.3790	.3810	.3830
1.2	.3849	.3869	.3888	.3907	.3925	.3944	.3962	.3980	.3997	.4015
1.3	.4032	.4049	.4066	.4082	.4099	.4115	.4131	.4147	.4162	.4177
1.4	.4192	.4207	.4222	.4236	.4251	.4255	.4279	.4292	.4306	.4319
1.5	.4332	.4345	.4357	.4370	.4382	.4394	.4406	.4418	.4429	.4441
1.6	.4452	.4463	.4474	.4484	.4495	.4505	.4515	.4525	.4535	.4545
1.7	.4554	.4564	.4573	.4582	.4591	.4599	.4608	.4616	.4625	.4633
1.8	.4641	.4649	.4656	.4664	.4671	.4678	.4686	.4693	.4699	.4706
1.9	.4713	.4719	.4726	.4732	.4738	.4744	.4750	.4756	.4761	.4767
2.0	.4772	.4778	.4783	.4788	.4793	.4798	.4803	.4808	.4812	.4817
2.1	.4821	.4826	.4830	.4834	.4838	.4842	.4846	.4850	.4854	.4857
2.2	.4861	.4864	.4868	.4871	.4875	.4878	.4881	.4884	.4887	.4890
2.3	.4893	.4896	.4898	.4901	.4904	.4906	.4909	.4911	.4913	.4916
2.4	.4918	.4920	.4922	.4925	.4927	.4929	.4931	.4932	.4934	.4936
2.5	.4938	.4940	.4941	.4943	.4945	.4946	.4948	.4949	.4951	.4952
2.6	.4953	.4955	.4956	.4957	.4959	.4930	.4961	.4962	.4963	.4964
2.7	.4965	.4966	.4967	.4968	.4969	.4970	.4971	.4972	.4973	.4974
2.8	.4974	.4975	.4976	.4977	.4977	.4978	.4979	.4979	.4980	.4981
2.9	.4981	.4982	.4982	.4983	.4984	.4984	.4985	.4985	.4986	.4986
3.0	.4987	.4987	.4987	.4988	.4988	.4989	.4989	.4989	.4990	.4990
3.1	.4990	.4991	.4991	.4991	.4992	.4992	.4992	.4992	.4993	.4993

Example 1: Find the area under the normal curve in the following cases using table.

(*i*) between $z = 0$ and $z = 1.4$ (*ii*) between $z = -0.56$ and $z = 0$

(*iii*) between $z = -0.45$ and $z = 2.42$ (*iv*) between $z = 0.75$ and $z = 1.95$

Sol.

(*i*) In the table, proceed downward under column marked z on top until 1.4 is reached, and then proceed right, to column marked 0 on top.

The result obtained is 0.4192 which is the required area. Here 0.4192 represents the probability that x lies between 0 and 1.4 *i.e.* $P(0 \le x \le 1.4) = 0.4192$. **Ans.**

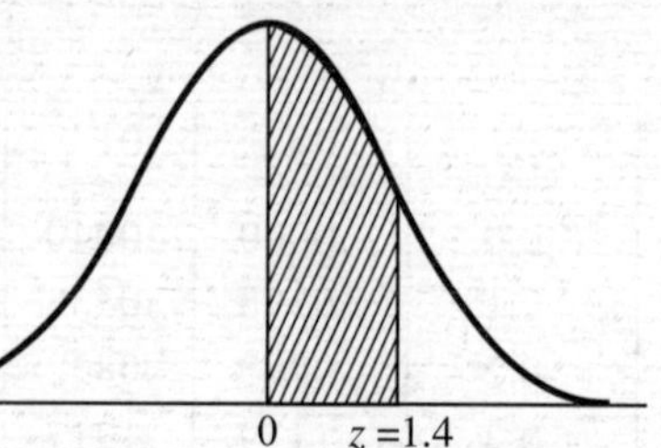

(*ii*) Required area = (area between $z = -0.56$ and $z = 0$) = (area between $z = 0$ and $z = 0.56$) (By Symmetry)

To find the area between $z = 0$ and $z = 0.56$ from the table proceed downward under column marked z unit 0.5 is reached. The proceed right to column marked 6 on top. The result obtained is 0.2123 which is the required area and it represents the probability that x lies between –0.56 and 0 *i.e.*, $P(-0.56 \le z \le 0) = 0.2123$. **Ans.**

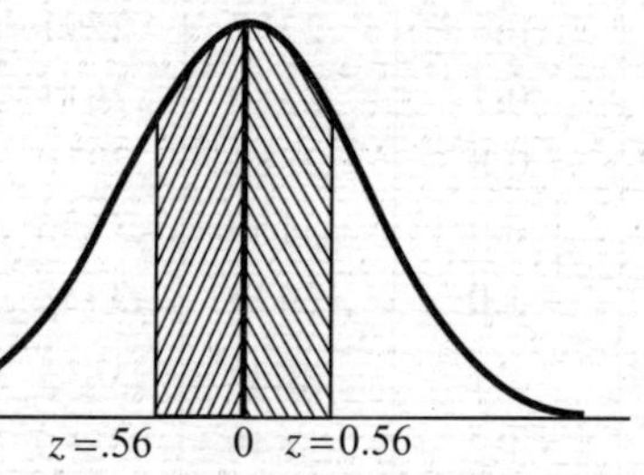

(*iii*) Required area = (area between $z = -0.45$ and $z = 0$) + (area between $z = 0$ and $z = 2.46$)
= (area between $z = 0$ and $z = 0.45$) + (area between $z = 0$ and $z = 2.42$) (By Symmetry)
= 0.1736 + 0.4922 (from table) = 0.6658. $\left[\text{Clearly, } P(-0.45 \le z \le 2.42) = 0.6658\right]$

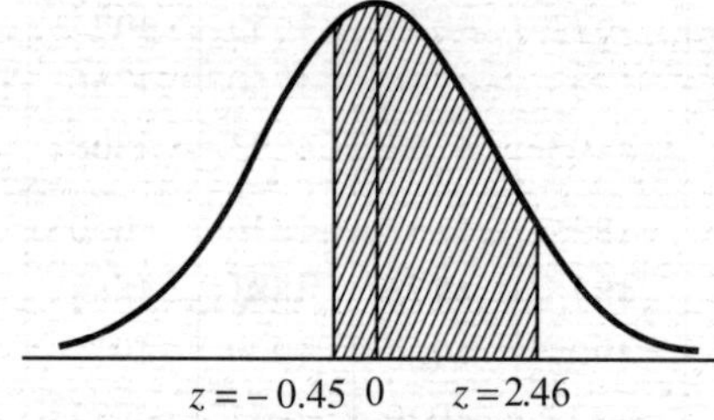

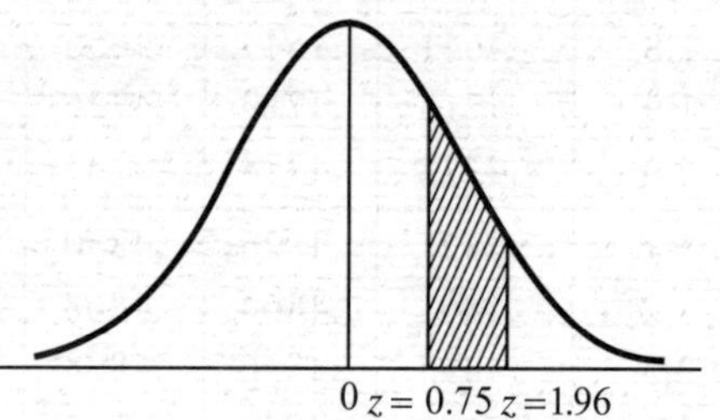

(*iv*) Required area = (area between $z = 0$ and $z = 1.96$) – (area between $z = 0$ and $z = 0.75$)
= 0.4750 – 0.2734(from table) = 0.2016. **Ans.**

Example 2: Find the probability that the standard normal variate is (*i*) less than –0.8, (*ii*) greater than 1.76, (*iii*) greater than 2.13 but less than –1.24.

Sol. (*i*) Since the standard normal curve is symmetrical about $z = 0$;

$\therefore$ (area under the normal curve to the left of $z = 0$) = (area under the normal curve to the right of $z = 0$) $= \frac{1}{2} = 0.5$.

Required probability $P(z<-0.8)$

= area under the normal curve to the left of $z = -0.8$

= (area to the left $z = 0$) – (area between $z = -0.8$ and $z = 0$)

= 0.5 – (area between $z = 0$ and $z = 0.8$)

= 0.5 – 0.2881=0.2119. **Ans.**

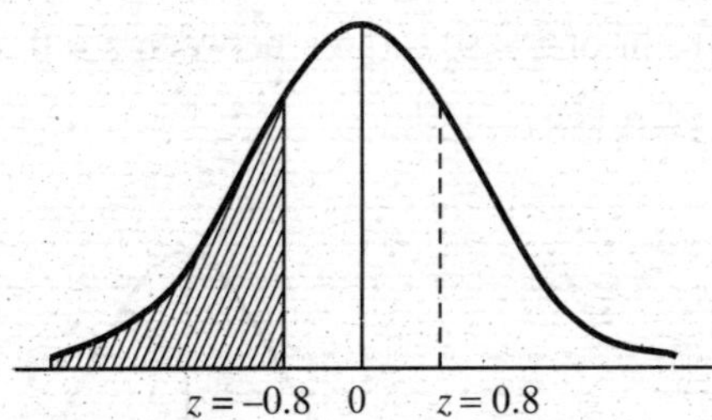

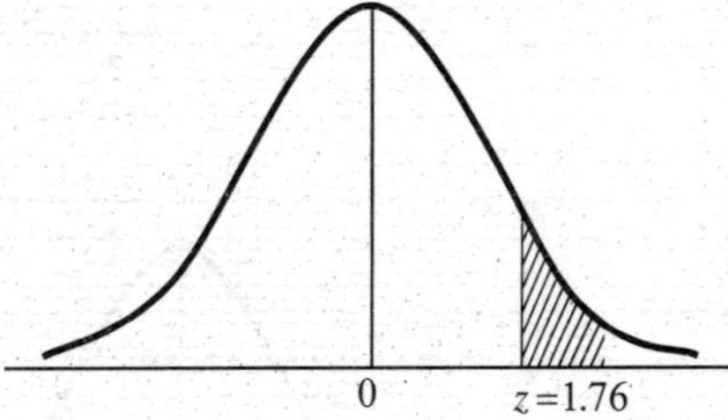

(*ii*) Required probability $P(z > 1.76)$=area under the normal curve to the right of $z = 1.76$ = (area to the right of $z = 0$) – (area between $z = 0$ and $z = 1.76$) = 0.5 – 0.4607=0.0392. **Ans.**

(*iii*) Required Probability

$= P(z > 2.13 \text{ and } z < -1.24)$

= total area under the standard normal curve
 – (area between $z = -1.24$ and $z = 0$)
 – (area between $z = 0$ and $z = 2.15$)

= 1 – (area between $z = 0$ and $z = 1.24$)
 – (area between $z = 0$ and $z = 2.13$)

= 1 – 0.3925 – 0.4834 (using table)

= 0.1241. **Ans.**

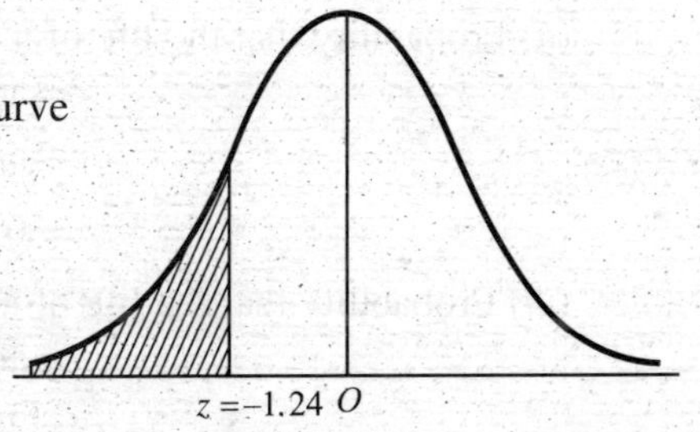

Example 3: A sample of 100 dry battery cells tested to find the length of life produced the following results: $\bar{x}$ = 12 hours, $\sigma = 3$ hours.

Assuming that the data are normally distributed, what percentage of battery cells are expected to have life: (*i*) more than 15 hours, (*ii*) less than 6 hours and (*iii*) between 10 and 14 hours?

Given z	2.5	2	1	0.67
Area	0.4938	0.4772	0.3413	0.2487

(*U.P.T.U. 2003*)

Sol. We know that standardized value of x is $z = \dfrac{x-m}{\sigma}$.

Here m = mean =12 hours, σ = S.D. = 3 hours.

$\therefore$ z = (Standardized value of 15) $= \dfrac{15-12}{3} = +1$;

z = (Standardized value of 6) $= \dfrac{6-12}{3} = -2$;

$$z = \text{(Standardized value of 10)} = \frac{10-12}{3} = -0.67;$$

$$z = \text{(Standardized value of 14)} = \frac{14-12}{3} = 0.67$$

∴ Probability that the life of a battery cell is more than 15 hours

= area under the standard normal curve to be right of $z = 1$

=(area to the right of $z = 0$) – (area between $z = 0$ and $z = 1$)

$= 0.5 - 0.3413 = 0.1587 = 15.87\%$

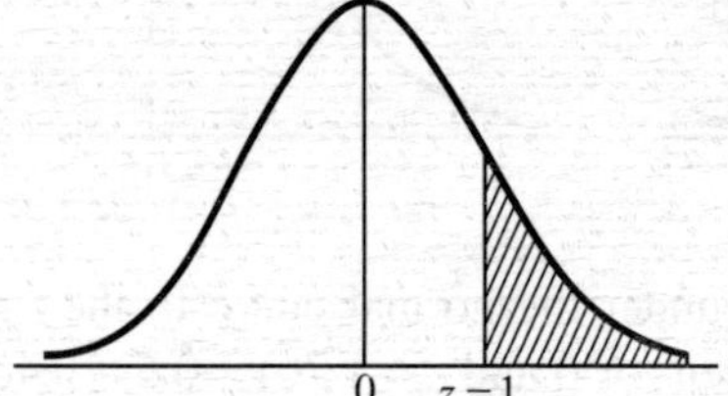

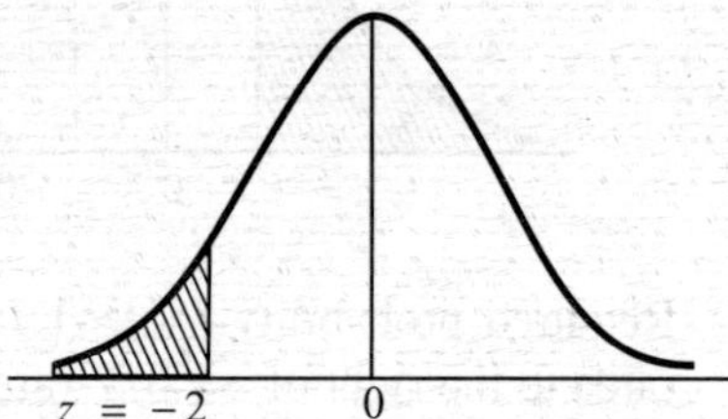

(*ii*) Probability that the life of a battery cell is less than 6 hours

= area under the normal curve to the left of $z = -2$

= area to the right of $z = +2$ [By symmetry]

$= 0.5 - 0.4772 = 0.0228 = 2.28\%$

(*iii*) Probability that the life of a battery cell lies between 10 and 14 hours

= area under the normal curve between $z = -0.67$ and $z = 0.67$

= 2(area between $z = 0$ and $z = 0.67$)

$= 2 \times 0.2487 = 0.4974 = 49.74\%$

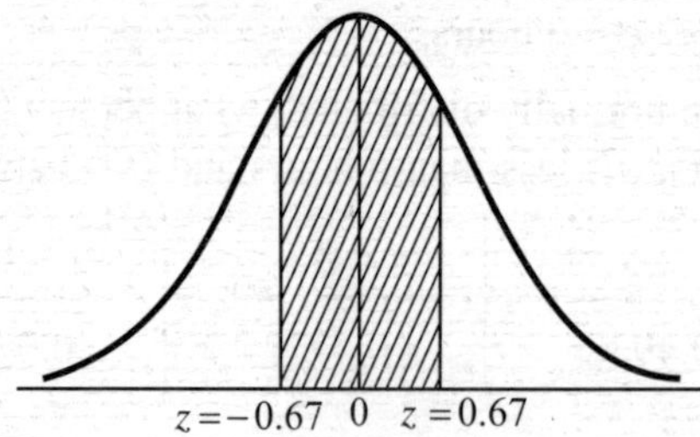

Hence the required percentage of battery cell with life (*i*) more than 15 hours is 15.87, (*ii*) less than 6 hours is 2.28 and (*iii*) between 10 and 14 hours = 49.74.

Example 4: In a sample of 1000 cases, the mean of a certain test is 14 and standard deviation is 2.5. Assuming the distribution to be normal, find

(*i*) How many students score between 12 and 15?

(*ii*) How many score above 18?

(*iii*) How many score below 8?

(*iv*) How many score 16?

Sol. n = 1000, m = 14 and S.D. = 2.5

(*i*) $$z_1 = \frac{x-m}{\sigma} = \frac{12-14}{2.5} = -0.8$$

$$z_2 = \frac{x-m}{\sigma} = \frac{15-14}{2.5} = 0.4$$

z=−0.8 0 z=0.4

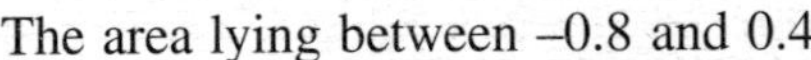

The area lying between –0.8 and 0.4

=Area from 0 to –0.8 + Area from 0 to 0.4

= 0.2881 + 0.1554 = 0.4435

The required number of students $= z = 1000 \times 0.4435 = 443.5 = 444(\text{say})$

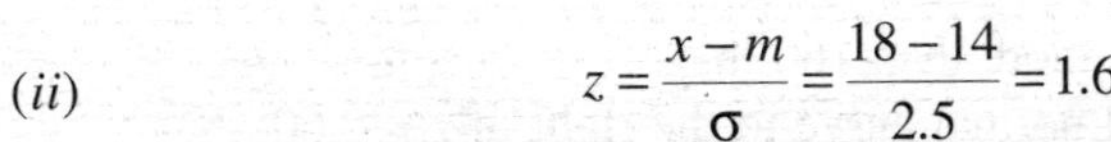

(*ii*) $$z = \frac{x-m}{\sigma} = \frac{18-14}{2.5} = 1.6$$

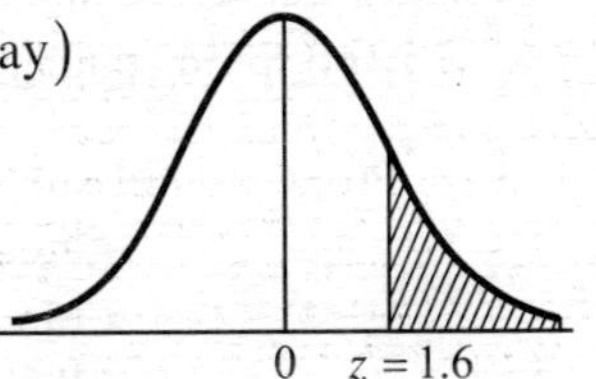

The area right to 1.6 = 0.5 – Area between 0 and 1.6

= 0.5 – 0.4452 = 0.0548

The required number of students $= 1000 \times 0.0548 = 54.8 = 55(\text{say})$

(*iii*) $$z = \frac{x-m}{\sigma} = \frac{8-14}{2.5} = -2.4$$

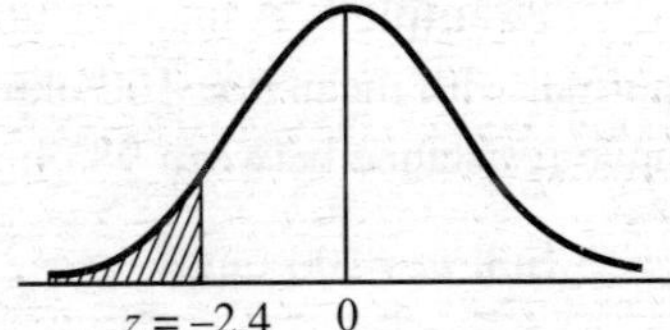

The area to the left –2.4 = 0.5 – Area between 0 and –2.4

= 0.5 – 0.4918 = 0.0082

The required number of students $= 1000 \times 0.0082 = 8.2 = 8(\text{say})$

(*iv*) Area between 15.5 and 16.5

$$z_1 = \frac{x-m}{\sigma} = \frac{15.5-14}{2.5} = 0.6$$

$$z_2 = \frac{x-m}{\sigma} = \frac{16.5-14}{2.5} = 1$$

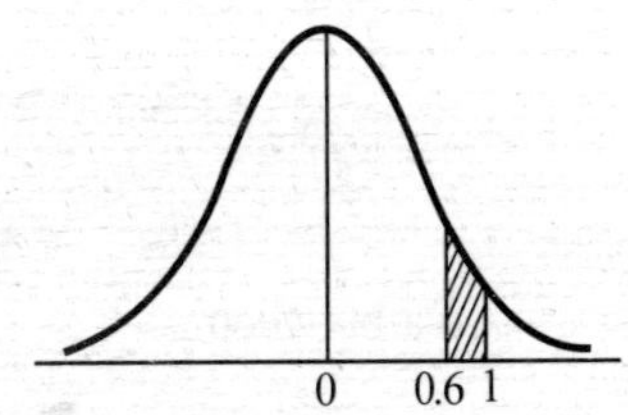

The area between 0.6 and 1 = 0.3413 – 0.2257 = 0.1156

The required number of students $= 1000 \times 0.1156 = 115.6 = 116(\text{say})$

Example 5: The mean weight of 500 male students at a certain college is 151 lb. and the standard deviation is 15 lb. Assuming that the weights are normally distributed, find how many students weight (*i*) between 119.5 and 155.5 lb. (*ii*) more than 160 lb.

Sol. Here m = mean =151 lb., S.D. =15 lb.

We have $z = \dfrac{x-m}{\sigma}$ = Value of x in standard unit (or standardized value of x)

(*i*) 119.5 lb. in standard units = $\dfrac{119.5-151}{15}=-2.10$

and 155.5 lb. in standard units = $\dfrac{155.5-151}{15}=0.30$

∴ Proportion of students whose weights lie between 119.5 and 155.5 lb.

= (area under the standard normal curve between $z = -2.10$ and $z = 0.30$)
= (area between $z = -2.10$ and $z = 0$) + (area between $z = 0$ and $z = 0.30$)
= (area between $z = 0$ and $z = 2.10$) + (area between $z = 0$ and $z = 0.30$)
= 0.4821 + 0.1179 (using table) = 0.6000.

Hence the required number of students whose weights lie between 119.5 and 155.5 lb. $= 0.6000\times 500 = 300$.

(*ii*) 160 lb. in standard units = $\dfrac{160-151}{15}=\dfrac{9}{15}=.60$

∴ Proportion of students whose weights more than 160 lb.

= area under the standard normal curve to the right of $z = 0.60$
= (area to the right of $z = 0$) – (area between $z = 0$ and $z = 0.60$)
= 0.5 – 0.2257 (using table) = 0.2743.

Hence, the required no. of students weighing more than 160 lb. = $0.2743\times 500=137$ (approximately).

Example 6: A manufacturer knows from experience that the resistance of resistors he produces is normal with mean m = 100 ohms and standard deviation $\sigma = 2$ ohms. What percentage of resistors will have resistance between 98 ohms and 102 ohms?

Sol. m = 100 ohms, $\sigma = 2$ ohms

$$x_1 = 98, x_2 = 102$$

$$z_1 = \frac{x_1 - m}{\sigma} = \frac{98-100}{2} = -1$$

$$z_2 = \frac{x_2 - m}{\sigma} = \frac{102-100}{2} = +1$$

$z_1=-1$ 0 $z_2=+1$

Area between $z_1 = -1,\ z_2 = +1$

= (Area between $z = 0$ and $z = -1$) + (Area between $z = 0$ and $z = +1$)

= 2(Area between $z = 0$ and $z = +1$) $= 2\times 0.3413 = 0.6826$

Percentage of resistors having resistance between 98 ohms and 102 ohms = 68.26. **Ans.**

Example 7: In a normal distribution, 31% of the items are under 45 and 8% are over 64. Find the mean and standard deviation of the distribution. (*U.T.U. 2008*)

Sol. Let m be the mean and σ the S.D.

If $x = 45,\ z = \dfrac{45-m}{\sigma}$

If $\qquad x = 64, \quad z = \dfrac{64 - m}{\sigma}$

19% 42%

$z = -0.496 \quad 0 \quad z = 1.4$

Area between 0 and $z = \dfrac{45 - m}{\sigma} = 0.50 - 0.31 = 0.19$

[From the table, for the area 0.19, $z = 0.496$]

$$z = \frac{45 - m}{\sigma} = -0.496 \qquad \text{...(1)}$$

Area between $z = 0$ and $z = \dfrac{64 - m}{\sigma} = 0.5 - 0.06 = 0.42$

[From the table, for the area 0.42, $z = 1.405$]

$$z = \frac{64 - m}{\sigma} = 1.405 \qquad \text{...(2)}$$

Solving (1) and (2), we get

$$m = 50, \ \sigma = 10. \quad \textbf{Ans.}$$

Example 8: The mean of the inner diameter (in each) of a sample of 200 tubes produced by a machine is 0.502 inch and the standard deviation is 0.005 inch. The purpose for which these tubes are intended allows a maximum tolerance is the diameter of 0.496 to 0.508 (*i.e.* otherwise the tubes are considered defective). What percentage of the tubes produced by the machine to defective if the diameter are found to be normally distributed? [Area under the standard normal curve between $z = 0$ and $z = 1.2$ is 0.3849]

Sol. Here m = mean of the inner diameter = 0.502 inch, σ = S.D. = 0.005 inch.

The tolerance limits of non-defective tubes are 0.496 inch and 0.508 inch.

$$z_1 \text{ (Standardized value of 0.496)} = \frac{x - m}{\sigma} = \frac{0.496 - 0.502}{0.005} = -1.2$$

$$z_2 \text{ (Standardized value of 0.508)} = \frac{x - m}{\sigma} = \frac{0.508 - 0.502}{0.005} = +1.2$$

$\therefore$ Proportion of non-defective tubes

= Area under the standard normal curve between $z = -1.2$ and $z = +1.2$

= 2(Area between $z = 0$ and $z = 1.2$) $= 2 \times 0.3849 = 0.7698 = 76.98\%$

Hence the required percentage of defective tubes =100% – 76.98% = 23.02%.

Example 9: Assume mean height of soldiers to be 68.22 inches with a variance of 10.8 inches square. How many soldiers in a regiment of 1,000 would you expect to be over 6 feet tall, given that the area under the standard normal curve between $x = 0$ and $x = 0.35$ is 0.1368.

and between $x = 0$ and $x = 1.15$ is 0.3746

Sol. x = 6 feets = 72 inches

$$\therefore \qquad z = \frac{x - m}{\sigma} = \frac{72 - 68.22}{\sqrt{10.8}} = 1.15$$

$$p(x > 72) = p(z > 1.15) \;= 0.5 - P(0 \le z \le 1.15)$$

$$= 0.5 - 0.3746 = 0.1254$$

Hence expected number of soldiers = 1000 × 0.1254 = 125.4 = 125 (approx.). **Ans.**

Example 10: The income of a group of 10,000 persons was found to be normally distributed with mean Rs. 750 p.m. and S.D. = Rs. 50. Show that of this group. about 95% had income exceeding Rs. 668 only 5% had income exceeding Rs. 832. Also find the lowest income among the richest 100.

Sol. $m = 750$, $\sigma = 50$

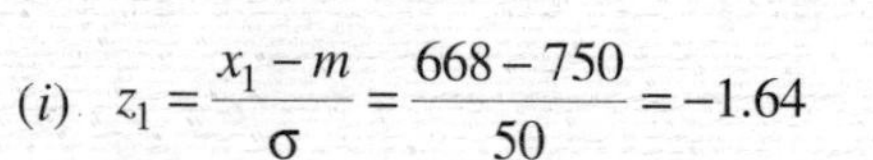

(*i*) $z_1 = \dfrac{x_1 - m}{\sigma} = \dfrac{668 - 750}{50} = -1.64$

$$P(x_1) = 0.5 + P(-1.64 \le z \ge 0) = 0.5 + P(0 \le z \le 1.64) = 0.5 + 0.4495 = 0.9495$$

∴ Required percentage of persons having income exceeding Rs. 668 = 95% (approx.). **Ans.**

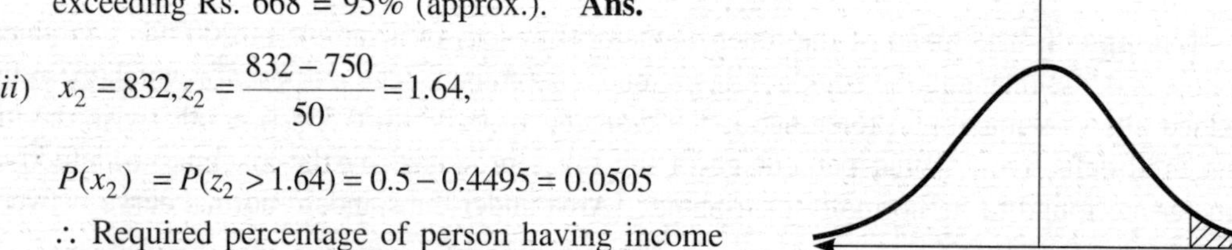

(*ii*) $x_2 = 832, z_2 = \dfrac{832 - 750}{50} = 1.64,$

$$P(x_2) \;= P(z_2 > 1.64) = 0.5 - 0.4495 = 0.0505$$

∴ Required percentage of person having income exceeding Rs. 832 = 5%. **Ans.**

(*iii*) Let x be the lowest income among the richest 100 persons

∴ The area between O and z = 0.49, (By Table) z = 2.33

Thus, $\dfrac{x - m}{\sigma} = 2.33 \Rightarrow \dfrac{x - 750}{50} = 2.33$

$$\Rightarrow x = 866.5. \quad \textbf{Ans.}$$

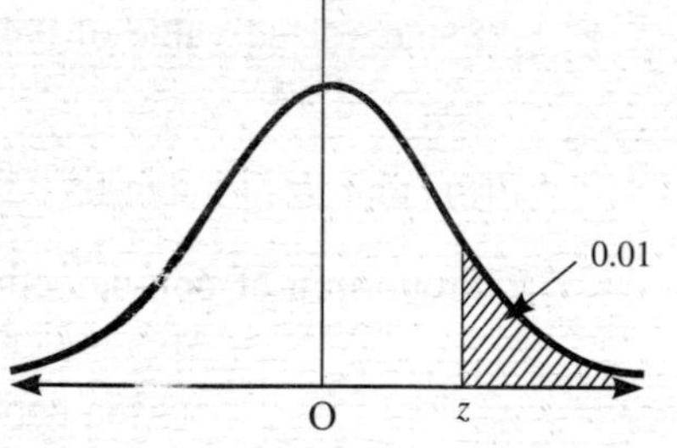

Example 11: The mean of a Normal distribution is 50 and 5% of the values are greater than 60. Find the standard deviation of the distribution.

(Given: The area under standard normal curve between $z = 0$ and $z = 1.64$ is 0.45)

Sol. Here m = mean = 50, σ = S.D. = ?, $z = \dfrac{x - m}{\sigma}$ = Standardized value of x, and $P(x > 60) = \dfrac{5}{100} = 0.05$.

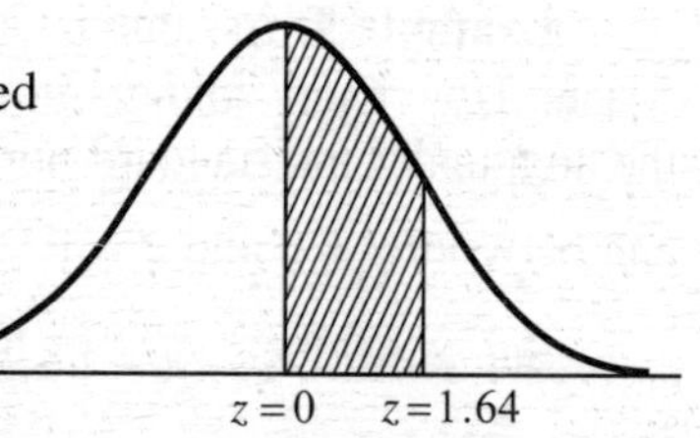

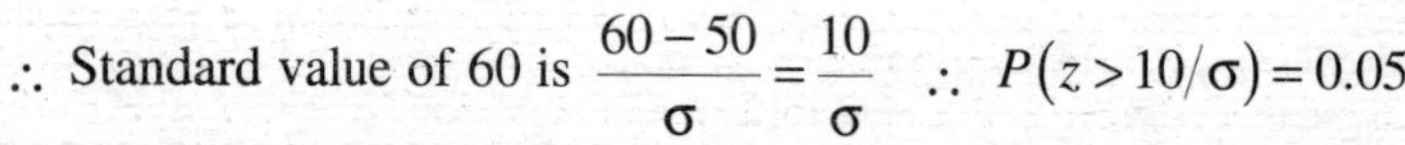

∴ Standard value of 60 is $\dfrac{60 - 50}{\sigma} = \dfrac{10}{\sigma}$ ∴ $P(z > 10/\sigma) = 0.05$

By the condition of the problem, 0.05 must be the area under the standard normal curve to the right of $z = \frac{10}{\sigma}$.

Again, given that the area between $z = 0$ and $z = 1.64$ is 0.45

$\therefore$ area to the right of $z = 1.64$ = (area to the right of $z = 0$) – (area between $z = 0$ and $z = 1.64$)

$= 0.5 - 0.45 = 0.05.$

Thus 0.05 is the area to the right of $z = \frac{10}{\sigma}$ as well as to the right of $z = 1.64$.

Hence $\frac{10}{\sigma} = 1.64$, or $\sigma = \frac{10}{1.64} = \frac{1000}{164} = 6$ (approx.).

Example 12: In a test on 2000 electric bulbs, it was found that the life of a particular make, was normally distributed with an average life of 2040 hours and S.D. of 60 hours, estimated the number of bulbs likely to burn for

(*i*) more than 2150 hours. (*ii*) less than 1950 hours.

(*iii*) more than 1920 hours and but less than 2160 hours.

Sol. Mean $m = 2040$, S.D. $(\sigma) = 60$ and $N = 2000$

(*i*) $x_1 = 2150$ $\therefore$ $z_1 = \frac{2150 - 2040}{60} = 1.83$

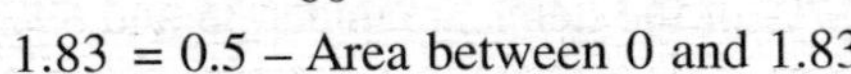

The area right to 1.83 = 0.5 – Area between 0 and 1.83

$= 0.5 - 4664 = 0.0336$

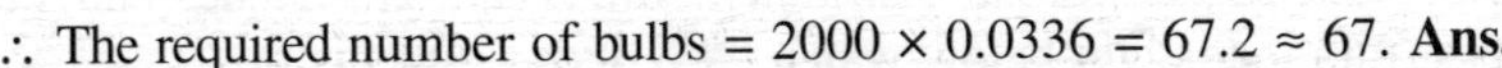

$\therefore$ The required number of bulbs = $2000 \times 0.0336 = 67.2 \approx 67$. **Ans.**

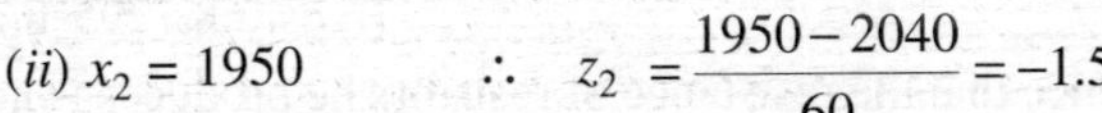

(*ii*) $x_2 = 1950$ $\therefore$ $z_2 = \frac{1950 - 2040}{60} = -1.5$

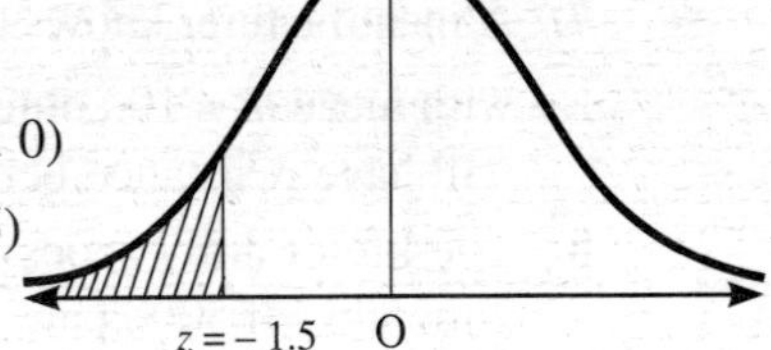

The area left to 1.5 = 0.5 – (Area between – 1.5 and 0)

= 0.5 – (Area between 0 and 1.5)

$= 0.5 - 0.4332 = 0.0668$

$\therefore$ The required number of bulbs = $2000 \times 0.0668 = 133.6 = 133$. **Ans.**

(*iii*) $x_1 = 1920,\quad x_2 = 2160$

$$z_1 = \frac{1920 - 2040}{60},\quad z_2 = \frac{2160 - 2040}{60}$$

$\therefore$ $z_1 = -2,\quad z_2 = 2$

Area between $z = -2$ and $z = 2$

= Area between ($z = 0$ and $z = -2$)

+ Area between ($z = 0$ and $z = 2$)

= $2 \times$ Area ($z = 0$ and $z = 2$)

$= 2 \times .4772 = 0.9544$

Hence the number of bulbs $= 2000 \times .9544 = 1908.8$

≈ 1909. **Ans.**

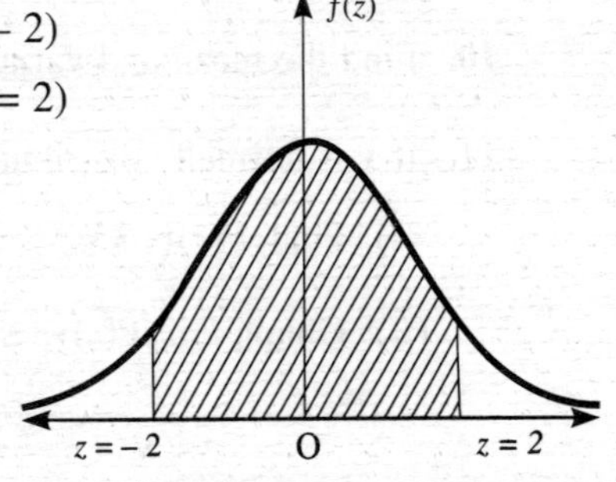

PROBLEM SET 4.5

1. A sample of 100 day battery cells tested to find the length of life produced the following results:

$$m = 12 \text{ hours}, \quad \sigma = 3\text{hours}$$

Assuming the data to be normal distributed, what percentage of battery cells are expected to have life.

(*i*) more that 15 hours (*ii*) less than 6 hours (*iii*) between 10 and 14 hours.

[**Ans.** (*i*) 15.87% (*ii*) 2.28% (*iii*) 44.70%]

2. The mean height of 500 students is 151 cm. and the standard deviation is 15 cm. Assuming that the heights are normally distributed, find how many students have height between 120 and 155 cm? [**Ans.** 294 (approximately)]

3. A large number of measurement is normally distributed with mean 65.5″ and S.D. of 6.2″. Find the percentage of measurements that fall between 54.87″ and 68.8″. [**Ans.** 66.01%]

4. The mean yield per plot of a crop is 17 kg. and standard deviation is 3 kg. If distribution of yield per plot is normal, find the percentage of plot giving yields:

(*i*) Between 15.5 kg. and 20 kg. and

(*ii*) More that 20 kg. [**Ans.** (*i*) 53.28%, (*ii*) 15.87%]

5. In a normal distribution, 7% of the items are under 35 and 89% are under 63. What are the mean and standard deviation of the distribution? [**Ans.** M = 50.3, $\sigma = 10.33$]

6. If the height of 300 students are normal distributed with mean 64.5 inches and standard deviation 3.3 inches, find the height below which 99% of the students lie.

[**Ans.** 68.7295 inches]

7. A manufacturer knows from experience that the resistance of resistors he produces is normal with mean m = 100 ohms and standard deviation $\sigma = 2$ ohms. What percentage of resistors will have resistance between 98 ohm and 1.2 ohms? [**Ans.** 68.26]

8. The life of army shoes is normally distributed with mean 8 months and standard deviation 2 months. If 5000 pairs are insured how many pairs would be expected to need replacement after 12 months? [Given that $P(z \geq 2) = 0.0228$ and $z = \dfrac{x-m}{\sigma}$. [**Ans.** 4886 pairs of shoes]

9. The lifetime of radio tubes manufactured in a factory is known to have an average value of 10 years. Find the probability that the lifetime of a tube taken randomly (*i*) exceeds 15 years, (*ii*) is less than 5 years, assuming that the exponential probability law is followed.

[**Ans.** (*i*) 0.2231 (*ii*) 0.3935]

10. Find the mean and variance of the density function $f(x) = \lambda e^{-\lambda x}$. $\left[\textbf{Ans. } \dfrac{1}{\lambda}, \dfrac{1}{\lambda^2}\right]$

11. If x is normally distributed with mean 1 and variance 4

(*i*) Find $P(r)(-3 \leq r \leq 3)$

(*ii*) Obtain k if $P(r)(r \leq k) = 0.90$. [**Ans.** (*i*) 0.8185 (*ii*) 3.56]

12. The income of a group of 10000 persons was formed to be normally distributed with mean Rs. 750 p.m. and standard deviation of Rs. 50 show that, of this group, about 95% had income exceeding Rs. 668 and only 5% had income exceeding Rs. 832. Also find the lowest income among the richest 100. *(U.P.T.U. 2005)*

[Ans. Rs. 866]

13. The weekly wages of 1000 workers are normally distributed around a mean of Rs. 70 and with a standard deviation of Rs. 5. Estimate the number of workers whose weekly wages will be (*i*) between Rs. 70 and 72 (*ii*) between Rs. 69 and Rs. 72 (*iii*) more than Rs. 75 (*iv*) less than Rs. 63 (*v*) more than Rs. 80. **[Ans.** (*i*) 155 (*ii*) 235 (*iii*) 159 (*iv*) 81 (*v*) 23]

14. A random variable has a normal distribution with 10 as S.D. Find its mean of the probability that the random variable takes on a value less than 80.5 is 0.3264. **[Ans.** $m = 45$]

15. At a certain examination 10% of the students who appeared for the paper in statistics got less than 30 marks and 97% of the students got less than 62 marks. Assuming the distribution to be normal find the mean and standard deviation.

[Ans. 43.04, 10.03]

UNIT V

Curve Fitting and Solution of Equation

5.1 CURVE FITTING

In many branches of applied mathematics and engineering sciences we come across experiments and problems, which involve two variables. For example, it is known that the speed v of a ship varies with the horsepower p of an engine according to the formula $p = a + bv^3$. Here a and b are the constants to be determined. For this purpose we take several sets of readings of speeds and the corresponding horsepowers. The problem is to find the best values for a and b using the observed values of v and p. Thus, the general problem is to find a suitable relation or law that may exist between the variables x and y from a given set of observed values (x_i, y_i), $i = 1, 2, \ldots\ldots\ldots, n$. Such a relation connecting x and y is known as empirical law. For above example, $x = v$ and $y = p$.

The process of finding the equation of the curve of best fit, which may be most suitable for predicting the unknown values, is known as curve fitting. Therefore, curve fitting means an exact relationship between two variables by algebraic equations. There are following methods for fitting a curve.

I. Graphic method
II. Method of group averages
III. Method of moments
IV. Principle of least square.

Out of above four methods, we will only discuss and study here principle of least square.

5.2 PRINCIPLE OF LEAST SQUARES

The graphical method has the drawback in that the straight line drawn may not be unique but principle of least squares provides a unique set of values to the constants and hence suggests a curve of best fit to the given data. The method of least square is probably the most systematic procedure to fit a unique curve through the given data points.

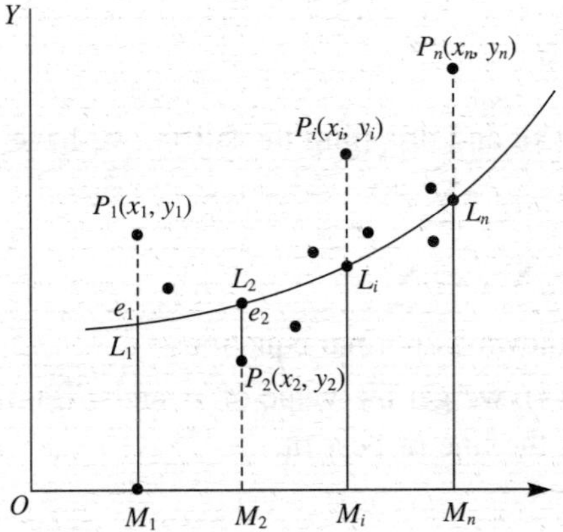

Let the curve $y = a + bx + cx^2 + ...kx^{m-1}$ ...(1)

be fitted to the set of n data points (x_1, y_1), (x_2, y_2), (x_3, y_3),......, (x_n, y_n). At $(x = x_i)$ the observed (or experimental) value of the ordinate is $y_i = P_iM_i$ and the corresponding value on the fitting curve (i) is $a + bx_i + cx_i^2 + ...kx_i^m = L_iM_i$ which is the expected or calculated value. The difference of the observed and the expected value is $P_iM_i - L_iM_i = e_i$(say) this difference is called error at $(x = x_i)$ clearly some of the error $e_1, e_2, e_3,.....,e_i,.......,e_n$ will be positive and other negative. To make all errors positive we square each of the errors *i.e.* $S = e_1^2 + e_2^2 + e_3^2 + ... + e_i^2 + ... + e_n^2$ the curve of best fit is that for which e's are as small as possible *i.e.* S, the sum of the square of the errors is a minimum this is known as the principle of least square. The theoretical values for $x_1, x_2, ..., x_n$ may be $y_{\lambda_1}, y_{\lambda_2}, ...y_{\lambda_n}$.

5.3 FITTING OF STRAIGHT LINE

Let a straight line $y = a + bx$ ...(1)

which is fitted to the given data points (x_1, y_1), (x_2, y_2), (x_3, y_3),......, (x_n, y_n).

Let y_{λ_1} be the theoretical value for x_1 then $e_1 = y_1 - y_{\lambda_1}$

$$\Rightarrow \quad e_1 = y_1 - (a + bx_1)$$

$$\Rightarrow \quad e_1^2 = (y_1 - a - bx_1)^2$$

Now we have
$$S = e_1^2 + e_2^2 + e_3^2 + ... + e_n^2$$

$$S = \sum_{i=1}^{n} e_i^2$$

$$S = \sum_{i=1}^{n} (y_i - a - bx_i)^2$$

By the principle of least squares, the value of S is minimum therefore,

$$\frac{\partial S}{\partial a} = 0 \qquad ...(2)$$

and $$\frac{\partial S}{\partial b} = 0 \qquad ...(3)$$

On solving equations (2) and (3), and dropping the suffix, we have

$$\sum y = na + b\sum x \qquad ...(4)$$

$$\sum xy = a\sum x + b\sum x^2 \qquad ...(5)$$

The equation (3) and (4) are known as normal equations.

On solving equations (3) and (4), we get the value of a and b. Putting the value of a and b in equation (1), we get the equation of the line of best fit.

5.4 FITTING OF PARABOLA

Let a parabola $y = a + bx + cx^2$ ...(1)

which is fitted to a given data (x_1, y_1), (x_2, y_2), (x_3, y_3) ,......, (x_n, y_n).

Let y_λ be the theoretical value for x_1 then $e_1 = y_1 - y_\lambda$

$$\Rightarrow \qquad e_1 = y_1 - (a + bx_1 + cx_1^2)$$

$$\Rightarrow \qquad e_1^2 = (y_1 - a - bx_1 - cx_1^2)^2$$

Now we have $$S = \sum_{i=1}^{n} e_i^2$$

$$S = \sum_{i=1}^{n} (y_i - a - bx_i - cx_i^2)^2$$

By the principle of least squares, the value of S is minimum, therefore

$$\frac{\partial S}{\partial a} = 0, \frac{\partial S}{\partial b} = 0 \text{ and } \frac{\partial S}{\partial c} = 0 \qquad ...(2)$$

Solving equation (2) and dropping suffix, we have

$$\sum y = na + b\sum x + c\sum x^2 \qquad ...(3)$$

$$\sum xy = a\sum x + b\sum x^2 + c\sum x^3 \qquad ...(4)$$

$$\sum x^2 y = a\sum x^2 + b\sum x^3 + c\sum x^4 \qquad ...(5)$$

The equation (3), (4) and (5) are known as normal equations.

On solving equations (3), (4) and (5), we get the values of a,b and c. Putting the values of a, b and c in equation (1), we get the equation of the parabola of best fit.

5.5 CHANGE OF SCALE

When the magnitude of the variable in the given data is large number then calculation becomes very much tedious then problem is further simplified by taking suitable scale when the value of x are given at equally spaced intervals.

Let h be the width of the interval at which the values of x are given and let the origin of x and y be taken at the point x_0, y_0 respectively, then putting

$$u = \frac{(x - x_0)}{h} \text{ and } v = y - y_0$$

If m is odd then, $$u = \frac{x - (\text{middle term})}{\text{interval}(h)}$$

But if m is even then, $$u = \frac{x - (\text{middle of two middle term})}{\frac{1}{2}(\text{interval})}$$

Example 1: Find the best-fit values of a and b so that $y = a + bx$ fits the data given in the table.

x:	0	1	2	3	4
y:	1	1.8	3.3	4.5	6.3

Sol. Let the straight line is $y = a + bx$ ...(1)

x	y	xy	x^2
0	1	0	0
1	1.8	1.8	1
2	3.3	6.6	4
3	4.5	13.5	9
4	6.3	25.2	16
$\sum x = 10$	$\sum y = 16.9$	$\sum xy = 47.1$	$\sum x^2 = 30$

Normal equations are, $\sum y = na + b\sum x$ ...(2)

$$\sum xy = a\sum x + b\sum x^2 \quad ...(3)$$

Here $n = 5$, $\sum x = 10$, $\sum y = 16.9$, $\sum xy = 47.1$, $\sum x^2 = 30$

Putting these values in normal equations, we get

$$16.9 = 5a + 10b$$

$$47.1 = 10a + 30b$$

On solving these two equations, we get

$$a = 0.72, \quad b = 1.33.$$

So required line $y = 0.72 + 1.33x$. **Ans.**

Example 2: Fit a straight line to the given data regarding x as the independent variable.

x	1	2	3	4	5	6
y	1200	900	600	200	110	50

Sol. Let the straight line obtained from the given data by $y = a + bx$ ...(1)

Then the normal equations are $\sum y = na + b\sum x$ ...(2)

$$\sum xy = a\sum x + b\sum x^2 \quad ...(3)$$

x	y	x^2	xy
1	1200	1	1200
2	900	4	1800
3	600	9	1800
4	200	16	800
5	110	25	550
6	50	36	300
$\sum x = 21$	$\sum y = 3060$	$\sum x^2 = 91$	$\sum xy = 6450$

Putting all values in the equations (2) and (3), we get

$$3060 = 6a + 21b$$

$$6450 = 21a + 91b$$

Solving these equations, we get

$$a = 1361.97 \quad \text{and} \quad b = -243.42$$

Hence the fitted equation is $y = 1361.97 - 243.42x$. **Ans.**

Example 3: Fit a straight line to the following data:

x	71	68	73	69	67	65	66	67
y	69	72	70	70	68	67	68	64

Sol. Here we from the following table:

x	y	xy	x^2
71	69	4899	5041
68	72	4896	4624
73	70	5110	5329
69	70	4830	4761
67	68	4556	4489
65	67	4355	4225
66	68	4488	4356
67	64	4288	4489
$\Sigma x = 546$	$\Sigma y = 548$	$\Sigma xy = 37422$	$\Sigma x^2 = 37314$

Let the equation of straight line to be fitted be

$$y = a + bx \quad ...(1)$$

And the normal equations are

$$\Sigma y = an + b\Sigma x \quad ...(2)$$

$$\Sigma xy = a\Sigma x + b\Sigma x^2 \quad ...(3)$$

$$\Rightarrow \quad 8a + 546b = 548$$

$$546a + 37314b = 37422$$

Solving these equations, we get

$$a = 39.5454,\ b = 0.4242$$

Hence from (1)

$$y = 39.5454 + 0.4242x. \quad \textbf{Ans.}$$

Example 4: Find the least square polynomial approximation of degree two to the data.

x	0	1	2	3	4
y	−4	−1	4	11	20

Also compute the least error.

Sol. Let the equation of the polynomial be $y = a + bx + cx^2$ ...(1)

x	y	xy	x^2	x^2y	x^3	x^4
0	−4	0	0	0	0	0
1	−1	−1	1	−1	1	1
2	4	8	4	16	8	16
3	11	33	9	99	27	81
4	20	80	16	320	64	256
$\sum x = 10$	$\sum y = 30$	$\sum xy = 120$	$\sum x^2 = 30$	$\sum x^2y = 434$	$\sum x^3 = 100$	$\sum x^4 = 354$

The normal equations are,

$$\sum y = na + b\sum x + c\sum x^2 \quad ...(2)$$

$$\sum xy = a\sum x + b\sum x^2 + c\sum x^3 \quad ...(3)$$

$$\sum x^2y = a\sum x^2 + b\sum x^3 + c\sum x^4 \quad ...(4)$$

Here $n = 5, \sum x = 10, \sum y = 30, \sum xy = 120, \sum x^2 = 30, \sum x^2y = 434, \sum x^3 = 100,$ $\sum x^4 = 354.$

Putting all these values in (2), (3) and (4), we get

$$30 = 5a + 10b + 30c \quad ...(5)$$

$$120 = 10a + 30b + 100c \quad ...(6)$$

$$434 = 30a + 100b + 354c \quad ...(7)$$

On solving these equations, we get $a=-4$, $b=2$, $c=1$. Therefore required polynomial is $y=-4+2x+x^2$, errors = 0. **Ans.**

Example 5: Fit a second degree curve of regression of y on x to the following data:

x	1	2	3	4
y	6	11	18	27

Sol. We form the following table:

x	y	x^2	x^3	x^4	xy	x^2y
1	6	1	1	1	6	6
2	11	4	8	16	22	44
3	18	9	27	81	54	162
4	27	16	64	256	108	432
$\Sigma x = 10$	$\Sigma y = 62$	$\Sigma x^2 = 30$	$\Sigma x^3 = 100$	$\Sigma x^4 = 354$	$\Sigma xy = 190$	$\Sigma x^2y = 644$

The equation of second degree parabola is given by

$$y = a + bx + cx^2 \qquad \text{...(1)}$$

And the normal equations are

$$\Sigma y = an + b\Sigma x + c\Sigma x^2 \qquad \text{...(2)}$$

$$\Sigma xy = a\Sigma x + b\Sigma x^2 + c\Sigma x^3 \qquad \text{...(3)}$$

$$\Sigma x^2 y = a\Sigma x^2 + b\Sigma x^3 + c\Sigma x^4 \qquad \text{...(4)}$$

$$\Rightarrow \quad \left.\begin{aligned} 4a + 10b + 30c &= 62 \\ 10a + 30b + 100c &= 190 \\ 30a + 100b + 354c &= 644 \end{aligned}\right\} \Rightarrow a = 3, b = 2, c = 1$$

Hence $\qquad y = 3 + 2x + x^2$. **Ans.**

Example 6: By the method of least squares, find the straight line that best fits the following data:

x	1	2	3	4	5	6	7
y	14	27	40	55	68	77	85

Sol. The equation of line is

$$y = a + bx \qquad \text{...(1)}$$

The normal equations are $\Sigma y = an + b\Sigma x$ ...(2)

and $$\Sigma xy = a\Sigma x + b\Sigma x^2 \qquad \text{...(3)}$$

Now we from the following table:

x	y	xy	x^2
1	14	14	1
2	27	54	4
3	40	120	9
4	55	220	16
5	68	340	25
6	77	462	36
7	85	595	49
$\Sigma x = 28$	$\Sigma y = 356$	$\Sigma xy = 1805$	$\Sigma x^2 = 140$

∴ From equations (2) and (3), we get

$$7a + 28b = 356$$

$$28a + 140b = 1805$$

On solving these equations, we get

$$a = -3.5714$$

$$b = 13.6071$$

∴ $$y = -3.5714 + 13.6071x. \quad \textbf{Ans.}$$

Example 7: Find the least squares fit of the form $y = a_0 + a_1 x^2$ to the following data

x	−1	0	1	2
y	2	5	3	0

Sol. We have $$y = a_0 + a_1 x^2$$

By principle of least squares

$$s = \sum_i \{y_i - (a_0 + a_1 x_i^2\}^2$$

$$\frac{\partial s}{\partial a} = \sum_i 2\{y_i - (a_0 + a_1 x_i^2\}(-x_i^2) = 0$$

⇒ $$\Sigma x^2 y = a_0 \Sigma x^2 + a_1 \Sigma x^4 \quad \text{(Drop suffix)} \qquad ...(1)$$

and $$\frac{\partial s}{\partial a_0} = \sum -2\{y_i - (a_0 + a_1 x_i^2)\} = 0$$

$\Rightarrow$ $$\Sigma y = a_0 n + a_1 \Sigma x^2 \quad \text{(Drop suffix)} \qquad ...(2)$$

Now, we form the following table:

x	y	x^2	x^2y	x^4
–1	2	1	2	1
0	5	0	0	0
1	3	1	3	1
2	0	4	0	16
$\Sigma x = 2$	$\Sigma y = 10$	$\Sigma x^2 = 6$	$\Sigma x^2y = 5$	$\Sigma x^4 = 18$

From equations (1) and (2), we get

$$6a_0 + 18a_1 = 5 \qquad ...(3)$$

and $$4a_0 + 6a_1 = 10 \qquad ...(4)$$

Solving the equations (3) and (4), we get

$$a_1 = -\ 1.111,\ a_0 = 4.166$$

$\therefore$ The equation is given by

$$y = 4.166 - 1.111x^2.$$ **Ans.**

Example 8: Fit a second-degree parabola to the following data taking x as the independent variable.

x	1	2	3	4	5	6	7	8	9
y	2	6	7	8	10	11	11	10	9

Sol. The equation of second-degree parabola is given by $y = a + bx + cx^2$ and the normal equations are:

$$\left.\begin{aligned} \sum y &= na + b\sum x + c\sum x^2 \\ \sum xy &= a\sum x + b\sum x^2 + c\sum x^3 \\ \sum x^2 y &= a\sum x^2 + b\sum x^3 + c\sum x^4 \end{aligned}\right\} \qquad ...(1)$$

Here $n = 9$. The various sums are appearing in the table as follows:

x	y	xy	x^2	x^2y	x^3	x^4
1	2	2	1	2	1	1
2	6	12	4	24	8	16
3	7	21	9	63	27	81
4	8	32	16	128	64	256
5	10	50	25	250	125	625
6	11	66	36	396	216	1296
7	11	77	49	539	343	2401
8	10	80	64	640	512	4096
9	9	81	81	729	729	6561
$\sum x=45$	$\sum y=74$	$\sum xy=421$	$\sum x^2=284$	$\sum x^2y=2771$	$\sum x^3=2025$	$\sum x^4=15333$

Putting these values of $\sum x, \sum y, \sum x^2, \sum xy, \sum x^2y, \sum x^3$ and $\sum x^4$ in equation (1) and solving the equations for a, b and c, we get

$$a=-0.923\ ;\ b=3.520\ ;\ c=-0.267\cdot$$

Hence the fitted equation is

$$y=-0.923+3.53x-0.267x^2. \textbf{ Ans.}$$

Example 9: Show that the line of fit to the following data is given by $y=0.7x+11.28$.

x	0	5	10	15	20	25
y	12	15	17	22	24	30

Sol. Here $n=6$ (even)

Let $x_0=12.5,\ h=5,\ y_0=20$ (say)

Then, $u=\dfrac{x-12.5}{2.5}$ and $v=y-20$, we get

x	y	u	v	uv	u^2
0	12	−5	−8	40	25
5	15	−3	−5	15	9
10	17	−1	−3	3	1
15	22	1	2	2	1
20	24	3	4	12	9
25	30	5	10	50	25
		$\sum u=0$	$\sum v=0$	$\sum uv=122$	$\sum u^2=70$

The normal equations are,

$$0=6a+0\cdot b \quad\Rightarrow\quad a=0$$

$$122=0\cdot a+70b \quad\Rightarrow\quad b=1.743$$

Thus line of fit is $v = 1.743u$.

or $$y - 20 = (1.743)\left(\frac{x - 12.50}{2.5}\right) = 0.69x - 8.715$$

or $$y = 0.7x + 11.285. \quad \textbf{Ans.}$$

Example 10: Fit a second-degree parabola to the following data by least squares method.

x	1929	1930	1931	1932	1933	1934	1935	1936	1937
y	352	356	357	358	360	361	361	360	359

Sol. Taking $x_0 = 1933,\ y_0 = 357$ then $u = \dfrac{(x - x_0)}{h}$

Here $h = 1$

Taking $u = x - x_0$ and $v = y - y_0$, therefore, $u = x - 1933$ and $v = y - 357$

x	$u = x - 1933$	y	$v = y - 357$	uv	u^2	u^2v	u^3	u^4
1929	–4	352	–5	20	16	–80	–64	256
1930	–3	356	–1	3	9	–9	–27	81
1931	–2	357	0	0	4	0	–8	16
1932	–1	358	1	–1	1	1	–1	1
1933	0	360	3	0	0	0	0	0
1934	1	361	4	4	1	4	1	1
1935	2	361	4	8	4	16	8	16
1936	3	360	3	9	9	27	27	81
1937	4	359	2	8	16	32	64	256
Total	$\sum u = 0$		$\sum v = 11$	$\sum uv = 51$	$\sum u^2 = 60$	$\sum u^2v = -9$	$\sum u^3 = 0$	$\sum u^4 = 708$

Then the equation $y = a + bx + cx^2$ is transformed to $v = A + Bu + Cu^2$ ...(1)

Normal equations are:

$$\sum v = 9A + B\sum u + C\sum u^2 \qquad \Rightarrow 11 = 9A + 60C$$
$$\sum uv = A\sum u + B\sum u^2 + C\sum u^3 \qquad \Rightarrow B = 17/20$$
$$\sum u^2v = A\sum u^2 + B\sum u^3 + C\sum u^4 \qquad \Rightarrow -9 = 60A + 708C$$

On solving these equations, we get $A = \dfrac{694}{231} = 3,\ B = \dfrac{17}{20} = 0.85$ and $C = -\dfrac{247}{924} = -0.27$

$\therefore$ $$v = 3 + 0.85u - 0.27u^2$$

$\Rightarrow$ $$y - 357 = 3 + 0.85(x - 1933) - 0.27(x - 1933)^2$$

$\Rightarrow$ $$y = -1010135.08 + 1044.69x - 0.27x^2. \quad \textbf{Ans.}$$

Example 11: Fit second degree parabola to the following:

x	0	1	2	3	4
y	1	1.8	1.3	2.5	6.3

Sol. Here $n = 5$ (odd) therefore $x_0 = 2$, $h = 1$, $y_0 = 0$ (say)

Now let $u = x - 2$, $v = y$ and the curve of fit be $v = a + bu + cu^2$.

x	y	u	v	uv	u^2	u^2v	u^3	u^4
0	1	−2	1	−2	4	4	−8	16
1	1.8	−1	1.8	−1.8	1	1.8	−1	1
2	1.3	0	1.3	0	0	0	0	0
3	2.5	1	2.5	2.5	1	2.5	1	1
4	6.3	2	6.3	12.6	4	25.2	8	16
Total		0	12.9	11.3	10	33.5	0	34

Hence the normal equations are,

$$\sum v = 5a + b\sum u + c\sum u^2$$

$$\sum uv = a\sum u + b\sum u^2 + c\sum u^3$$

$$\sum u^2v = a\sum u^2 + b\sum u^3 + c\sum u^4$$

On putting the values of $\sum u, \sum v$ etc. from the table in these, we get

$$12.9 = 5a + 10c,\ 11.3 = 10b,\ 33.5 = 10a + 34c.$$

On solving these equations, we get

$$a = 1.48,\ b = 1.13 \text{ and } c = 0.55$$

Therefore the required equation is $v = 1.48 + 1.13u + 0.55u^2$.

Again substituting $u = x - 2$ and $v = y$, we get

$$y = 1.48 + 1.13(x-2) + 0.55(x-2)^2$$

or $$y = 1.42 - 1.07x + 0.55x^2.$$ **Ans.**

5.6 FITTING OF AN EXPONENTIAL CURVE

Suppose an exponential curve of the form

$$y = ae^{bx}$$

Taking logarithm on both the sides, we get

$$\log_{10} y = \log_{10} a + bx \log_{10} e$$

i.e., $$Y = A + Bx \quad ...(1)$$

where $$Y = \log_{10} y,\ A = \log_{10} a \text{ and } B = b\log_{10} e.$$

The normal equations for (1) are,

$$\sum Y = nA + B\sum x$$

$$\sum xY = A\sum x + B\sum x^2$$

On solving the above two equations, we get A and B

then $$a = \text{antilog } A,\ b = \frac{B}{\log_{10} e}$$

5.7 FITTING OF THE CURVE y = ax + bx²

Error of estimate for ith point (x_i, y_i) is

$$e_i = (y_i - ax_i - bx_i^2)$$

We have, $$S = \sum_{i=1}^{n} e_i^2$$

$$= \sum_{i=1}^{n} (y_i - ax_i - bx_i^2)^2$$

By the principle of least square, the value of S is minimum

$$\therefore \quad \frac{\partial S}{\partial a} = 0 \text{ and } \frac{\partial S}{\partial b} = 0$$

Now $$\frac{\partial S}{\partial a} = 0$$

$$\Rightarrow \quad \sum_{i=1}^{n} 2(y_i - ax_i - bx_i^2)(-x_i) = 0$$

or $$\sum_{i=1}^{n} x_i y_i = a\sum_{i=1}^{n} x_i^2 + b\sum_{i=1}^{n} x_i^3 \quad ...(1)$$

and $$\frac{\partial S}{\partial b} = 0$$

$$\Rightarrow \quad \sum_{i=1}^{n} 2(y_i - ax_i - bx_i^2)(-x_i^2) = 0$$

or $$\sum_{i=1}^{n} x_i^2 y_i = a\sum_{i=1}^{n} x_i^3 + b\sum_{i=1}^{n} x_i^4 \quad ...(2)$$

Dropping the suffix i from (1) and (2), then the normal equations are,

$$\sum xy = a\sum x^2 + b\sum x^3$$

$$\sum x^2 y = a\sum x^3 + b\sum x^4$$

5.8 FITTING OF THE CURVE $y = ax + \frac{b}{x}$

Error of estimate for ith point (x_i, y_i) is

$$e_i = (y_i - ax_i - \frac{b}{x_i})$$

We have,

$$S = \sum_{i=1}^{n} e_i^2$$

$$= \sum_{i=1}^{n} (y_i - ax_i - \frac{b}{x_i})^2$$

By the principle of least square, the value of S is minimum

$$\therefore \quad \frac{\partial S}{\partial a} = 0 \text{ and } \frac{\partial S}{\partial b} = 0$$

Now

$$\frac{\partial S}{\partial a} = 0$$

$$\Rightarrow \quad \sum_{i=1}^{n} 2(y_i - ax_i - \frac{b}{x_i})(-x_i) = 0$$

or

$$\sum_{i=1}^{n} x_i y_i = a\sum_{i=1}^{n} x_i^2 + nb \qquad ...(1)$$

and

$$\frac{\partial S}{\partial b} = 0$$

$$\Rightarrow \quad \sum_{i=1}^{n} 2(y_i - ax_i - \frac{b}{x_i})(-\frac{1}{x_i}) = 0$$

or

$$\sum_{i=1}^{n} \frac{y_i}{x_i} = na + b\sum_{i=1}^{n} \frac{1}{x_i^2} \qquad ...(2)$$

Dropping the suffix i from (1) and (2), then the normal equations are,

$$\sum xy = nb + a\sum x^2$$

$$\sum \frac{y}{x} = na + b\sum \frac{1}{x^2}$$

where n is the number of pair of values of x and y.

5.9 FITTING OF THE CURVE $y = \frac{c_0}{x} + c_1\sqrt{x}$

Error of estimate for ith point (x_i, y_i) is

$$e_i = (y_i - \frac{c_0}{x_i} - c_1\sqrt{x_i})$$

We have,
$$S = \sum_{i=1}^{n} e_i^2$$
$$= \sum_{i=1}^{n} (y_i - \frac{c_0}{x_i} - c_1\sqrt{x_i})^2$$

By the principle of least square, the value of S is minimum

$$\therefore \quad \frac{\partial S}{\partial c_0} = 0 \text{ and } \frac{\partial S}{\partial c_1} = 0$$

Now
$$\frac{\partial S}{\partial c_0} = 0$$

$$\Rightarrow \quad \sum_{i=1}^{n} 2(y_i - \frac{c_0}{x_i} - c_1\sqrt{x_i})(-\frac{1}{x_i}) = 0$$

or
$$\sum_{i=1}^{n} \frac{y_i}{x_i} = c_0 \sum_{i=1}^{n} \frac{1}{x_i^2} + c_1 \sum_{i=1}^{n} \frac{1}{\sqrt{x_i}} \qquad \text{...(1)}$$

and
$$\frac{\partial S}{\partial c_1} = 0$$

$$\Rightarrow \quad \sum_{i=1}^{n} 2(y_i - \frac{c_0}{x_i} - c_1\sqrt{x_i})(-\sqrt{x_i}) = 0$$

or
$$\sum_{i=1}^{n} y_i\sqrt{x_i} = c_0 \sum_{i=1}^{n} \frac{1}{\sqrt{x_i}} + c_1 \sum_{i=1}^{n} x_i \qquad \text{...(2)}$$

Dropping the suffix i from (1) and (2), then the normal equations are,

$$\sum \frac{y}{x} = c_0 \sum \frac{1}{x^2} + c_1 \sum \frac{1}{\sqrt{x}}$$

$$\sum y\sqrt{x} = c_0 \sum \frac{1}{\sqrt{x}} + c_1 \sum x.$$

Example 12: Find the curve of best fit of the type $y = ae^{bx}$ to the following data by the method of least squares:

x	1	5	7	9	12
y	10	15	12	15	21

Sol. The curve to be fitted is $y = ae^{bx}$ or $Y = A + Bx$, where $Y = \log_{10} y$, $A = \log_{10} a$, and $B = b \log_{10} e$.

Therefore the normal equations are:

$$\sum Y = 5A + B\sum x$$
$$\sum xY = A\sum x + B\sum x^2$$

x	y	$Y = \log_{10} y$	x^2	xY
1	10	1.0000	1	1
5	15	1.1761	25	5.8805
7	12	1.0792	49	7.5544
9	15	1.1761	81	10.5849
12	21	1.3222	144	15.8664
$\sum x = 34$		$\sum Y = 5.7536$	$\sum x^2 = 300$	$\sum xY = 40.8862$

Substituting the values of $\sum x$, etc. and calculated by means of above table in the normal equations, we get

$$5.7536 = 5A + 34B$$

and
$$40.8862 = 34A + 300B$$

On solving these equations, we obtain,

$$A = 0.9766\,;\; B = 0.02561$$

Therefore $\quad a = \text{antilog}_{10} A = 9.4754\,;\; b = \dfrac{B}{\log_{10} e} = 0.059$

Hence the required curve is $y = 9.4754e^{0.059x}$. **Ans.**

Example 13: For the data given below, find the equation to the best fitting exponential curve of the form $y = ae^{bx}$.

x	1	2	3	4	5	6
y	1.6	4.5	13.8	40.2	125	300

Sol. $y = ae^{bx}$

On taking log both the sides, $\log y = \log a + bx \log e$ which is of the form $Y = A + Bx$, where $Y = \log y$, $A = \log a$ and $B = b \log e$.

x	y	$Y = \log y$	x^2	xY
1	1.6	0.2041	1	0.2041
2	4.5	0.6532	4	1.3064
3	13.8	1.1399	9	3.4197
4	40.2	1.6042	16	6.4168
5	125	2.0969	25	10.4845
6	300	2.4771	36	14.8626
$\sum x = 21$		$\sum Y = 8.1754$	$\sum x^2 = 91$	$\sum xY = 36.6941$

Normal equations are: $\sum Y = 6A + B\sum x$

$$\sum xY = A\sum x + B\sum x^2$$

Therefore from these equations, we have

$$8.1754 = 6A + 21B$$

$$36.6941 = 21A + 91B$$

$\Rightarrow$ $$A = -0.2534, B = 0.4617$$

Therefore, $a = \text{antilog}A = \text{antilog}(-0.2534) = \text{antilog}(\bar{1}.7466) = 0.5580.$

and $$b = \frac{B}{\log e} = \frac{0.4617}{0.4343} = 1.0631$$

Hence required equation is $y = 0.5580e^{1.0631x}$. **Ans.**

Example 14: Given the following experimental values:

x	0	1	2	3
y	2	4	10	15

Fit by the method of least squares a parabola of the type $y = a + bx^2$.

Sol. Error of estimate for ith point (x_i, y_i) is $e_i = (y_i - a - bx_i^2)$

By the principle of least squares, the values of a and b are such that

$$S = \sum_{i=1}^{n} e_i^2 = \sum_{i=1}^{n} (y_i - a - bx_i^2)^2 \text{ is minimum.}$$

Therefore normal equations are given by

$$\frac{\partial S}{\partial a} = 0 \Rightarrow \sum y = na + b\sum x^2 \qquad ...(1)$$

and $$\frac{\partial S}{\partial b} = 0 \Rightarrow \sum x^2 y = a\sum x^2 + b\sum x^4 \qquad ...(2)$$

x	y	x^2	x^2y	x^4
0	2	0	0	0
1	4	1	4	1
2	10	4	40	16
3	15	9	135	81
Total	$\sum y = 31$	$\sum x^2 = 14$	$\sum x^2 y = 179$	$\sum x^4 = 98$

Here $n = 4$.

From (1) and (2), $31 = 4a + 14b$ and $179 = 14a + 98b$

Solving for a and b, we get

$$a = 2.71 \text{ and } b = 1.44$$

Hence the required curve is $y = 2.71 + 1.44x^2$. **Ans.**

Example 15: By the method of least square, find the curve $y = ax + bx^2$ that best fits the following data:

x	1	2	3	4	5
y	1.8	5.1	8.9	14.1	19.8

Sol. Error of estimate for ith point (x_i, y_i) is $e_i = (y_i - ax_i - bx_i^2)$

By the principle of least squares, the values of a and b are such that

$$S = \sum_{i=1}^{n} e_i^2 = \sum_{i=1}^{n} (y_i - ax_i - bx_i^2)^2 \text{ is minimum.}$$

Therefore normal equations are given by

$$\frac{\partial S}{\partial a} = 0 \Rightarrow \sum_{i=1}^{n} x_i y_i = a\sum_{i=1}^{n} x_i^2 + b\sum_{i=1}^{n} x_i^3 \quad \text{and} \quad \frac{\partial S}{\partial b} = 0 \Rightarrow \sum_{i=1}^{n} x_i^2 y_i = a\sum_{i=1}^{n} x_i^3 + b\sum_{i=1}^{n} x_i^4$$

Dropping the suffix i, normal equations are

$$\sum xy = a\sum x^2 + b\sum x^3 \quad \text{...(1)}$$

and

$$\sum x^2 y = a\sum x^3 + b\sum x^4 \quad \text{...(2)}$$

x	y	x^2	x^3	x^4	xy	x^2y
1	1.8	1	1	1	1.8	1.8
2	5.1	4	8	16	10.2	20.4
3	8.9	9	27	81	26.7	80.1
4	14.1	16	64	256	56.4	225.6
5	19.8	25	125	625	99	495
Total		$\sum x^2 = 55$	$\sum x^3 = 225$	$\sum x^4 = 979$	$\sum xy = 194.1$	$\sum x^2 y = 822.9$

Substituting these values in equations (1) and (2), we get

$$194.1 = 55a + 225b \text{ and } 822.9 = 225a + 979b$$

$$\Rightarrow \quad a = \frac{83.85}{55} \approx 1.52$$

and

$$b = \frac{317.4}{664} \approx 0.49$$

Hence required parabolic curve is $y = 1.52x + 0.49x^2$. **Ans.**

Example 16: Fit an exponential curve of the form $y = ab^x$ to the following data:

x	1	2	3	4	5	6	7	8
y	1.0	1.2	1.8	2.5	3.6	4.7	6.6	9.1

Sol. $y = ab^x$ takes the form $Y = A + Bx$, where $Y = \log y$; $A = \log a$ and $B = \log b$.

Hence the normal equations are given by

$$\sum Y = nA + B\sum x \text{ and } \sum xY = A\sum x + \sum x^2.$$

x	y	$Y = \log y$	xY	x^2
1	1.0	0.0000	0.000	1
2	1.2	0.0792	0.1584	4
3	1.8	0.2553	0.7659	9
4	2.5	0.3979	1.5916	16
5	3.6	0.5563	2.7815	25
6	4.7	0.6721	4.0326	36
7	6.6	0.8195	5.7365	49
8	9.1	0.9590.	7.6720	64
$\sum x = 36$	$\sum y = 30.5$	$\sum Y = 3.7393$	$\sum xY = 22.7385$	$\sum x^2 = 204$

Putting the values in the normal equations, we obtain

$3.7393 = 8A + 36B$ and $22.7385 = 36A + 204B$

$\Rightarrow$ $B = 0.1407$ and $A = 0.1656$

$\Rightarrow$ $b = \text{antilog} B = 1.38$ and $a = \text{antilog}\, A = 0.68$.

Thus the required curve of best fit is $y = (0.68)(1.38)^x$. **Ans.**

Example 17: Fit a curve $y = ab^x$ to the following data:

x	2	3	4	5	6
y	144	172.8	207.4	248.8	298.5

Sol. Given equation $y = ab^x$ reduces to $Y = A + Bx$ where $Y = \log y$, $A = \log a$ and $B = \log b$.

The normal equations are,

$$\sum \log y = n \log a + \log b \sum x$$

$$\sum x \log y = \log a \sum x + \log b \sum x^2$$

The calculations of $\sum x$, $\sum \log y$, $\sum x^2$ and $\sum x \log y$ are substitute in the following tabular form.

x	y	x^2	$\log y$	$x\log y$
2	144	4	2.1584	4.3168
3	172.8	9	2.2375	6.7125
4	207.4	16	2.3168	9.2672
5	248.8	25	2.3959	11.9795
6	298.5	36	2.4749	14.8494
20		90	11.5835	47.1254

Putting these values in the normal equations, we have

$$11.5835 = 5\log a + 20\log b$$

$$47.1254 = 20\log a + 90\log b.$$

Solving these equations and taking antilog, we have a = 100, b = 1.2 approximate. Therefore equation of the curve is $y = 100(1.2)^x$. **Ans.**

Example 18: Derive the least square equations for fitting a curve of the type $y = ax^2 + (b/x)$ to a set of n points. Hence fit a curve of this type to the data.

x	1	2	3	4
y	−1.51	0.99	3.88	7.66

Sol. Let the n points are given by (x_1, y_1), (x_2, y_2), (x_3, y_3), ……, (x_n, y_n). The error of estimate for the ith point (x_i, y_i) is $e_i = [y_i - ax_i^2 - (b/x_i)]$.

By the principle of least square, the values of a and b are such so that the sum of the square of error S, *viz.*,

$$S = \sum_{i=1}^{n} e_i^2 = \sum\left(y_i - ax_i^2 - \frac{b}{x_i}\right)^2 \text{ is minimum.}$$

Therefore the normal equations are given by

$$\frac{\partial S}{\partial a} = 0, \frac{\partial S}{\partial b} = 0$$

or
$$\sum_{i=1}^{n} y_i x_i^2 = a\sum_{i=1}^{n} x_i^4 + b\sum_{i=1}^{n} x_i \text{ and } \sum_{i=1}^{n}\frac{y_i}{x_i} = a\sum_{i=1}^{n} x_i + b\sum_{i=1}^{n}\frac{1}{x_i^2}$$

These are the required least square equations.

x	y	x^2	x^4	$\frac{1}{x}$	$\frac{1}{x^2}$	yx^2	$\frac{y}{x}$
1	−1.51	1	1	1	1	−1.51	−1.51
2	0.99	4	16	0.5	0.25	3.96	0.495
3	3.88	9	81	0.3333	0.1111	34.92	1.2933
4	7.66	16	256	0.25	0.0625	122.56	1.0943
10			354		1.4236	159.93	1.1933

Putting the values in the above least square equations, we get

$$159.93 = 354a + 10b \text{ and } 1.1933 = 10a + 1.4236b.$$

Solving these, we get $a = 0.509$ and $b = -2.04$.

Therefore, the equation of the curve fitted to the above data is $y = 0.509x^2 - \dfrac{2.04}{x}$. **Ans.**

Example 19: Fit the curve $pv^\gamma = k$ to the following data:

p(kg/cm^2)	0.5	1	1.5	2	2.5	3
v (litres)	1620	1000	750	620	520	460

Sol. Given $\qquad pv^\gamma = k$

$$v = \left(\frac{k}{p}\right)^{1/\gamma} = k^{1/\gamma}p^{-1/\gamma}$$

Taking log both the sides, we get

$$\log v = \frac{1}{\gamma}\log k - \frac{1}{\gamma}\log p$$

which is of the form $Y = A + BX$

where $Y = \log v$, $X = \log p$, $A = \dfrac{1}{\gamma}\log k$ and $B = -\dfrac{1}{\gamma}$.

p	v	X	Y	XY	X^2
0.5	1620	–0.30103	3.20952	–0.96616	0.09062
1	1000	0	3	0	0
1.5	750	0.17609	2.87506	0.50627	0.03101
2	620	0.30103	2.79239	0.84059	0.09062
2.5	520	0.39794	2.716	1.08080	0.15836
3	460	0.47712	2.66276	1.27046	0.22764
Total		$\sum X = 1.05115$	$\sum Y = 17.25573$	$\sum XY = 2.73196$	$\sum X^2 = 0.59825$

Here $n = 6$

Normal equations are,

$$17.25573 = 6A + 1.05115B$$

$$2.73196 = 1.05115A + 0.59825B$$

On solving these, we get

$$A = 2.99911 \text{ and } B = -0.70298$$

$\therefore$ $$\gamma = -\frac{1}{B} = \frac{1}{0.70298} = 1.42252$$

Again $$\log k = \gamma A = 4.26629$$

$\therefore$ $$k = \text{antilog}(4.26629) = 18462.48$$

Hence, the required curve is $pv^{1.42252} = 18462.48$. **Ans.**

Example 20: For the data given below, find the equation to the best fitting exponential curve of the form $y = ae^{bx}$.

x	1	2	3	4	5	6
y	1.6	4.5	13.8	40.2	125	300

Sol. Given $y = ae^{bx}$, taking log we get $\log y = \log a + bx \log_{10} e$ which is of the $Y = A + Bx$,

where $$Y = \log y,\ A = \log a \text{ and } B = \log_{10} e.$$

Put the values in the following tabular form, also transfer the origin of x series to 3, so that $u = x - 3$.

x	y	$\log y = Y$	$x - 3 = u$	uY	u^2
1	1.6	0.204	−2	−0.408	4
2	4.5	0.653	−1	−0.653	1
3	13.8	1.140	0	0	0
4	40.2	1.604	1	1.604	1
5	125.0	2.094	2	4.194	4
6	300	2.477	3	7.431	9
Total		8.175	3	12.168	19

In case $Y = A + B.u,$ then normal equations are given by

$$\sum Y = nA + B\sum u \quad \Rightarrow \quad 8.175 = 6A + 3B \qquad ...(1)$$

$$\sum uY = A\sum u + B\sum u^2 \quad \Rightarrow \quad 12.168 = 3A + 19B \qquad ...(2)$$

Solving (1) and (2), we get

$$A = 1.13 \text{ and } B = 0.46$$

Thus equation is $Y = 1.13 + 0.46u,$ *i.e.* $Y = 1.13 + 0.46(x - 3)$

or $$Y = 0.46x - 0.25$$

Which gives $\log a = -0.25$ *i.e.* $\text{antilog}(-0.25) = \text{antilog}(\bar{1}.75) = 0.557$

$$b = \frac{B}{\log_{10} e} = \frac{.46}{0.4343} = 1.06$$

Hence, the required equation of the curve is $y = (0.557)e^{1.06x}$. **Ans.**

PROBLEM SET 5.1

1. Fit a straight line to the given data regarding x as the independent variable:

x	1	2	3	4	6	8
y	2.4	3.1	3.5	4.2	5.0	6.0

[**Ans.** $y = 2.0253 + 0.502x$]

2. Fit a straight line $y = a + bx$ to the following data by the method of least square:

x	0	1	3	6	8
y	1	3	2	5	4

[**Ans.** $1.6 + 0.38x$]

3. Find the least square approximation of the form $y = a + bx^2$ for the following data:

x	0	0.1	0.2	0.3	0.4	0.5
y	1	1.01	0.99	0.85	0.81	0.75

[**Ans.** $y = 1.0032 - 1.1081x^2$]

4. Fit a second degree parabola to the following data:

x	0.0	1.0	2.0
y	1.0	6.0	17.0

[**Ans.** $y = 1 + 2x + 3x^2$]

5. Fit a second degree parabola to the following data:

x	1.0	1.5	2.0	2.5	3.0	3.5	4.0
y	1.1	1.3	1.6	2.0	2.7	3.4	4.1

[**Ans.** $y = 1.04 - 0.193x + 0.243x^2$]

6. Fit a second degree parabola to the following data by the least square method:

x	1	2	3	4	5
y	1090	1220	1390	1625	1915

[**Ans.** $y = 27.5x^2 + 40.5x + 1024$]

7. Fit a parabola $y = a + bx + cx^2$ to the following data:

x	2	4	6	8	10
y	3.07	12.85	31.47	57.38	91.29

[**Ans.** $y = 0.34 - 0.78x + 0.99x^2$]

8. Determine the constants a and b by the method of least squares such that $y = ae^{bx}$ fits the following data:

x	2	4	6	8	10
y	4.077	11.084	30.128	81.897	222.62

[**Ans.** $y = 1.49989e^{0.50001x}$]

9. Fit a least square geometric curve $y = ax^b$ to the following data:

x	1	2	3	4	5
y	0.5	2	4.5	8	12.5

[**Ans.** $y = 0.5012x^{1.9977}$]

10. A person runs the same race track for five consecutive days and is timed as follows:

Day(x)	1	2	3	4	5
Time(y)	15.3	15.1	15	14.5	14

Make a least square fit to the above data using a function $a+\frac{b}{x}+\frac{c}{x^2}$.

$$\left[\textbf{Ans. } y=13.0065+\frac{6.7512}{x}+\frac{4.4738}{x^2}\right]$$

11. Use the method of least squares to fit the curve $y=\frac{c_0}{x}+c_1\sqrt{x}$ to the following table of values:

x	0.1	0.2	0.4	0.5	1	2
y	21	11	7	6	5	6

$$\left[\textbf{Ans. } y=\frac{1.97327}{x}+3.28182\sqrt{x}\right]$$

12. Using the method of least square to fit a parabola $y=a+bx+cx^2$ in the following data:

$(x,y):(-1,2),(0,0),(0,1),(1,2)$ $\left[\textbf{Ans. } y=\frac{1}{2}+\frac{3}{2}x^2\right]$

13. The pressure of the gas corresponding to various volumes V is measured, given by the following data:

$V(\text{cm}^3)$	50	60	70	90	100
$p(\text{kgcm}^{-2})$	64.7	51.3	40.5	25.9	78

Fit the data to the equation $pV^{\gamma}=c$. [**Ans.** $pV^{0.28997}=167.78765$]

14. Employ the method of least squares to fit a parabola $y = a + bx + cx^2$ in the following data: (x, y): $(-1, 2)$, $(0, 0)$, $(0, 1)$, $(1, 2)$ [**Ans.** $y = 0.5 + 1.5x^2$]

15. Fit a second degree parabola in the following data: [*U.T.U. 2008*]

x	0.0	1.0	2.0	3.0	4.0
y	1.0	4.0	10.0	17.0	30.0

[**Ans.** $y = 1 + 2x + 3x^2$]

16. Fit at least square quadratic curve to the following data:

x	1	2	3	4
y	1.7	1.8	2.3	3.2

, estimate $y(2.4)$

[**Ans.** $y = 2 - 0.5x + 0.2x^2$ and $y(2.4) = 1.952$]

17. Fit an exponential curve by least squares

x	1	2	5	10	20	30	40	50
y	98.2	91.7	81.3	64.0	36.4	32.6	17.1	11.3

Estimate y when $x = 25$. **[Ans.** $y = 100(0.96)^x$, $y(25) = 33.9$]

18. Fit the curve $y = a + \frac{b}{x}$ to the following data

x	1	2	3	4
y	3	1.5	6	7.5

Estimate y when $x = 2.25$. $\left[\textbf{Ans. } y = 1.3 + \frac{1.7}{x}, y(2.25) = 4.625\right]$

5.10 POLYNOMIAL

If $$f(x) = a_0x^n + a_1x^{n-1} + a_2x^{n-2} + \ldots\ldots\ldots\ldots + a_{n-1}x + a_n$$

Then the above relation is called the polynominal of nth order in x.

5.10.1 Degree of Polynomial

The highest power of x occurring in the given polynomial is called degree of polynomial.

The constant $c = cx^0$ is called a polynomial of degree zero.

The polynomial $f(x) = ax + b, a \neq 0$ is of degree one and is called a linear polynomial.

The polynomial $f(x) = ax^2 + bx + c,\ a \neq 0$ is of degree two and is called a quadratic polynomial.

The polynomial $f(x) = ax^3 + bx^2 + cx + d, a \neq 0$ is of degree three and is called a cubic polynomial.

The polynomial $f(x) = ax^4 + bx^3 + cx^2 + dx + e, a \neq 0$ is of degree four and is called biquadratic polynomial.

5.11 DESCARTE'S RULE OF SIGNS

The number of positive roots of the equation $f(x) = 0$ cannot exceed the number of changes of sign in $f(x)$, and the number of negative roots cannot exceed the number of changes of sign of $f(-x)$.

Existence of imaginary roots: If an equation of the nth degree has at most p positive roots and at most q negative roots, then it has at least $n - (p + q)$ imaginary roots.

Example 1: Apply Descarte's Rule of signs to discuss the nature of the roots of the equation $x^4 + 15x^2 + 7x - 11 = 0$.

Sol. The given equation is $f(x) = x^4 + 15x^2 + 7x - 11 = 0$

Signs of $f(x)$ are + + + − [from + to −]

It has one change of sign and hence it must have one +ve root.

$$f(-x)=(-x)^4+15(-x)^2+7(-x)-11=0$$

or $$f(-x)=x^4+15x^2-7x-11 \qquad =0$$

Signs of $f(-x)$ are + + – – [from + to –]

It has only one change in sign and hence it must have one –ve root.

Thus the equation has two real roots , one +ve and one –ve and hence the other two roots must be imaginary.

Example 2: Show that $x^7-3x^4+2x^3-1=0$ has at least four imaginary roots.

Sol. The given equation is

$$f(x)=x^7-3x^4+2x^3-1=0 \qquad \text{[from + to – or – to +]}$$

Signs of $f(x)$ + – + –

$\therefore f(x)=0$ has 3 changes of sign. Therefore, it cannot have more than three positive roots.

Also $$f(-x) = (-x)^7-3(-x)^4+2(-x)^3-1=0$$

or $$f(-x)=-x^7-3x^4-2x^3-1=0$$

Signs of $f(x)$ are – – – –

$\therefore f(-x)=0$ has no changes in sign. Therefore the given equation has no negative root.

Thus the given equation cannot have more than 3 + 0 = 3 real roots. But the given equation has 7 roots. Hence the given equation has 7 – 3 = 4 imaginary roots.

Example 3: Find the least positive number of imaginary roots of the equation

$$f(x)=x^9-x^5+x^4+x^2+1=0$$

Sol. The given equation is $f(x)=x^9-x^5+x^4+x^2+1=0$

Signs of $f(x)$ are + – + + +

$\therefore$ $f(x)=0$ has two changes of signs, and hence 2 is the max. number of +ve root.

$$f(-x)=(-x)^9-(-x)^5+(-x)^4+(-x)^2+1=0$$

or $$f(-x)=-x^9+x^5+x^4+x^2+1=0$$

Signs of $f(-x)$ are – + + + +

$\therefore f(-x)=0$ has only one changes of sign and hence it has only one –ve root or $f(x)=0$ has only one –ve root.

Thus the max. number of real roots is 2 + 1=3 and the equation being of 9th degree and it will have at least 9 – 3 = 6 imaginary roots.

5.12 CARDON'S METHOD

Case 1. When Cubic is of the Form $x^3 + qx + r = 0$

The given cubic is $x^3 + qx + r = 0$ ...(1)

Let $x = u + v$ be a root of (1)

Cubing, $x^3 = u^3 + v^3 + 3uv(u+v) = u^3 + v^3 + 3uvx$

or $x^3 - 3uvx - (u^3 + v^3) = 0$ ...(2)

Comparing (1) and (2), $uv = -\frac{1}{3}q$ or $u^3v^3 = -\frac{1}{27}q^3$ and $u^3 + v^3 = -r$

$\therefore$ $u^3$and v^3 are the roots of the equation

$$t^2 - (u^3 + v^3)\,t + u^3v^3 = 0 \quad \text{or} \quad t^2 + rt - \frac{1}{27}q^3 = 0 \qquad ...(3)$$

Solving (3),
$$t = \frac{-r \pm \sqrt{r^2 + \frac{4q^3}{27}}}{2}$$

Let
$$u^3 = \frac{-r + \sqrt{r^2 + \frac{4q^3}{27}}}{2};\; v^3 = \frac{-r - \sqrt{r^2 + \frac{4q^3}{27}}}{2}$$

Now, the three cube roots of u^3 are $u, u\omega, u\omega^2$ and those of v^3 are $v, v\omega, v\omega^2$, where ω and ω^2 are imaginary cube root of unity.

Since $x = u + v$

To find x, we have to add a cube root of u^3 and a cube of v^3 in such a manner that their product is real.

$\therefore$ The three values of x are $u + v,\ u\omega + v\omega^2,\ u\omega^2 + v\omega$ $(\because \omega^3 = 1)$

Example 4: Use Cardon's method to solve

$$x^3 - 27x + 54 = 0.$$

Sol. Let $x = u + v$

Cubing, $x^3 = (u+v)^3 = u^3 + v^3 + 3uv(u+v) = u^3 + v^3 + 3uvx$

$\Rightarrow$ $x^3 - 3uvx - (u^3 + v^3) = 0$

Comparing with the given equation,we get

$$uv = 9 \Rightarrow u^3v^3 = 729 \quad \text{(on cubing)}$$

and $u^3 + v^3 = -54$

$\therefore$ u^3 and v^3 are the root of $t^2 - (u^3 + v^3)t + u^3v^3 = 0$

$$\Rightarrow \quad t^2 + 54t + 729 = 0 \Rightarrow (t+27)^2 = 0$$

$$\Rightarrow \quad t = -27, -27$$

Let $u^3 = -27$ and $v^3 = -27$

So that, $u = -3, -3\omega, -3\omega^2$ and $v = -3, -3\omega, -3\omega^2$

To find x, we have to add a cube root of u^3 and a cube root of v^3 in such a way that their product is real.

$$\therefore \quad x = (-3-3), (-3\omega - 3\omega^2), (-3\omega^2 - 3\omega) \qquad (\because \omega^3 = 1)$$

$$= -6, -3(\omega + \omega^2), -3(\omega^2 + \omega) = -6, 3, 3 \qquad (\because 1 + \omega + \omega^2 = 0)$$

Hence the required roots are –6, 3, 3. **Ans.**

Example 5: Solve by Cardon's method $x^3 - 15x - 126 = 0$.

Sol. Let $x = u + v$

Cubing, $x^3 = (u+v)^3 = u^3 + v^3 + 3uv(u+v) = u^3 + v^3 + 3uvx$

$$\Rightarrow \quad x^3 - 3uvx - (u^3 + v^3) = 0$$

Comparing with the given equation, we get

$$uv = 5 \Rightarrow u^3v^3 = 125 \text{ (on cubing)}$$

and $u^3 + v^3 = 126$

$\therefore$ u^3 and v^3 are the roots of $t^2 - (u^3 + v^3)t + u^3v^3 = 0$

$$\Rightarrow \quad t^2 - 126t + 125 = 0 \quad \Rightarrow \quad (t-1)(t-125) = 0 \quad \Rightarrow \quad t = 1, 125$$

Let $u^3 = 1$ and $v^3 = 125$

So that, $u = 1, \omega, \omega^2$ and $v = 5, 5\omega, 5\omega^2$

To find x, we have to add a cube root of u^3 and a cube root of v^3 in such a way that their product is real.

$$\therefore \quad x = 1 + 5, \omega + 5\omega^2, \omega^2 + 5\omega \qquad (\because \omega^3 = 1)$$

$$= 6, \left\{\left(\frac{-1+i\sqrt{3}}{2}\right) + 5\left(\frac{-1-i\sqrt{3}}{2}\right)\right\}, \left\{\left(\frac{-1-i\sqrt{3}}{2}\right) + 5\left(\frac{-1+i\sqrt{3}}{2}\right)\right\}$$

$$= 6, -3 - 2\sqrt{3}i, -3 + 2\sqrt{3}i = 6, -3 \pm 2\sqrt{3}i$$

Hence, the required roots are $6, -3 \pm 2\sqrt{3}i$. **Ans.**

Example 6: Solve $x^3 - 6x - 9 = 0$ by Cardon's method.

Sol. Let $x = u + v$

Cubing, $$x^3 = (u+v)^3 = u^3 + v^3 + 3uv(u+v)$$

$$= u^3 + v^3 + 3uvx$$

$\Rightarrow$ $x^3 - 3uvx - (u^3 + v^3) = 0$

Comparing, we get

$$uv = 2 \Rightarrow u^3v^3 = 8 \text{ (on cubing)}$$

and $u^3 + v^3 = 9$

$\therefore$ u^3, v^3 are the roots of $t^2 - (u^3+v^3)t + u^3v^3 = 0$

$\Rightarrow$ $$t^2 - 9t + 8 = 0 \Rightarrow (t-1)(t-8) = 0 \Rightarrow t = 1, 8$$

Let $u^3 = 1$ and $v^3 = 8$ so that

$$u = 1, \omega, \omega^2 \text{ and } v = 2, 2\omega, 2\omega^2$$

To find x, we have to add a cube root of u^3 and a cube root of v^3 in such that their product is real.

$\therefore$ $$x = 1 + 2, \omega + 2\omega^2, \omega^2 + 2\omega \qquad (\because \omega^3 = 1)$$

$$= 3, \left\{\left(\frac{-1+i\sqrt{3}}{2}\right) + 2\left(\frac{-1-i\sqrt{3}}{2}\right)\right\}, \left\{\left(\frac{-1-i\sqrt{3}}{2}\right) + 2\left(\frac{-1+i\sqrt{3}}{2}\right)\right\}$$

$$= 3, \frac{-3-i\sqrt{3}}{2}, \frac{-3+i\sqrt{3}}{2} \Rightarrow x = 3, \frac{-3 \pm i\sqrt{3}}{2}.$$ **Ans.**

Example 7: Solve the cubic equation $x^3 - 18x - 35 = 0$

Sol. Let $x = u + v$

Cubing, $$x^3 = (u+v)^3 = u^3 + v^3 + 3uv(u+v)$$

$$= u^3 + v^3 + 3uvx$$

$\Rightarrow$ $x^3 - 3uvx - (u^3 + v^3) = 0$

Comparing, we get $uv = 6 \Rightarrow u^3v^3 = 216$ (on cubing)

and $u^3 + v^3 = 35$

$\therefore$ u^3 and v^3 are the roots of $t^2 - (u^3+v^3)t + u^3v^3 = 0$

$\Rightarrow$ $$t^2 - 35t + 216 = 0 \Rightarrow (t-8)(t-27) = 0 \Rightarrow t = 8, 27$$

Let $u^3 = 8$ & $v^3 = 27$ so that

$$u = 2, 2\omega, 2\omega^2 \text{ and } v = 3, 3\omega, 3\omega^2$$

To find x, we have to add a cube root of u^3 and a cube root of v^3 in such that their product is real.

$$\therefore \quad x = 2+3,\ 2\omega+3\omega^2,\ 2\omega^2+3\omega \qquad (\because \omega^3 = 1)$$

$$= 5, \left\{2\left(\frac{-1+i\sqrt{3}}{2}\right)+3\left(\frac{-1-i\sqrt{3}}{2}\right)\right\}, \left\{2\left(\frac{-1-i\sqrt{3}}{2}\right)+3\left(\frac{-1+i\sqrt{3}}{2}\right)\right\}$$

$$= 5, \frac{-5-i\sqrt{3}}{2}, \frac{-5+i\sqrt{3}}{2}$$

$$\Rightarrow \quad x = 5, \frac{-5\pm i\sqrt{3}}{2}. \quad \textbf{Ans.}$$

Case 2. When the Cubic Equation is of the Form $a_0x^3 + a_1x^2 + a_2x + a_3 = 0$

Then, first of all we remove the term containing x^2.

This is done by diminishing the roots of the given equation by $\frac{-a_1}{na_0}$ or Sum of root/No. of roots. Where n is 3. We proceed with the help of following examples:

Example 8: Solve by Cardon's method $x^3 + 6x^2 + 9x + 4 = 0$

Sol. Equationing $x^3 + 6x^2 + 9x + 4 = 0$ ...(1)

Equating with $a_0x^3 + a_1x^2 + a_2x + a_3 = 0$, we get ...(2)

$$a_0 = 1, a_1 = 6$$

Then $$h = -\frac{a_1}{3a_0} = -\frac{6}{3} = -2$$

$$h = -2.$$

Now remove the x^2 terms ,we have

−2	1	6	9	4
		−2	−8	−2
	1	4	1	(2)
		−2	−4	
	1	2	(−3)	
		−2		
	1	(0)		

The transformed equation is $y^3 - 3y + 2 = 0$ ...(3)

where $$y = x + 2.$$

Let $y = u + v$ be a solution of (3), then we get

$$y^3 - 3uvy - (u^3 + v^3) = 0 \qquad ...(4)$$

Equating (3) and (4), we get $uv = 1$

$$\Rightarrow \quad u^3v^3 = 1 \text{ and } u^3 + v^3 = -2$$

Let us consider an equation whose roots are u^3 and v^3

$$t^2 - st + p = 0 \quad s = \text{sum of roots}, \; p = \text{product of roots}$$

$$t^2 + 2t + 1 = 0 \;\Rightarrow\; (t+1)^2 = 0$$

$\Rightarrow$ $$t = -1, -1$$

Let $u^3 = -1$ and $v^3 = -1$ so that

$$u = -1, -\omega, -\omega^2 \text{ and } v = -1, -\omega, -\omega^2$$

$\therefore$ $$y = u + v = (-1-1), (-\omega - \omega^2), (-\omega^2 - \omega)$$

$$= -2, 1, 1 \qquad (\because 1 + \omega + \omega^2 = 0)$$

But $$x = y - 2$$

$$= -2 - 2, 1 - 2, 1 - 2 = -4, -1, -1$$

$\therefore$ $x = -4, -1, -1$. **Ans.**

Example 9: Solve by Cardon's method $x^3 - 15x^2 - 33x + 847 = 0$.

Sol. The given equation $x^3 - 15x^2 - 33x + 847 = 0$ ...(1)

Equating with $a_0x^3 + a_1x^2 + a_2x + a_3 = 0$, we get

$$a_0 = 1, a_1 = -15$$

$\therefore$ $$h = -\frac{a_1}{3a_0} = \frac{15}{3} = 5$$

Now remove the x^2 term, using synthetic division, we have

5	1	–15	–33	847
		5	–50	–415
	1	–10	–83	(432)
		5	–25	
	1	–5	(–108)	
		5		
	1	(0)		

Transformed equation is

$$y^3 - 108y + 432 = 0 \qquad ...(2)$$

where $$y = x - 5.$$

Let $$y = u + v$$

$\therefore$ $$y^3 - 3uvy - (u^3 + v^3) = 0 \qquad ...(3) \text{ (on cubing)}$$

Comparing (2) and (3),

$$uv = 36 \Rightarrow u^3v^3 = (6)^6 \text{ and } u^3 + v^3 = -432$$

$\therefore$ u^3 and v^3 are the roots of $t^2 + 432t + (6)^6 = 0$

$$\Rightarrow \quad t^2 + 2.(6)^3 t + (6)^6 = 0 \Rightarrow (t+216)^2 = 0 \Rightarrow t = -216, -216$$

Let $\quad u^3 = -216$ and $v^3 = -216$ so that

$$u = -6, -6\omega, -6\omega^2 \text{ and } v = -6, -6\omega, -6\omega^2$$

$\therefore \quad y = u + v = (-6-6), (-6\omega - 6\omega^2), (-6\omega^2 - 6\omega)$

$$= -12, 6, 6 \qquad \left(\begin{array}{r} \because \omega + \omega^2 + 1 = 0 \\ \omega + \omega^2 = -1 \end{array}\right)$$

But $\quad x = y + 5 = -12 + 5, 6 + 5, 6 + 5$

$\therefore \quad x = -7, 11, 11.$ **Ans.**

Example 10: Solve $x^3 - 3x^2 + 12x + 16 = 0$

Sol. Here $a_0 = 1, a_1 = -3 \quad \therefore \; h = -\dfrac{a_1}{3a_0} = \dfrac{3}{3} = 1$

Now remove the term x^2, using synthetic division

1	1	–3	12	16
		1	–2	10
	1	–2	10	(26)
		1	–1	
	1	–1	(9)	
		1		
	1	(0)		

The transformed equation is

$$(x-1)^3 + 9(x-1) + 26 = 0$$

Let $\quad y = x - 1$, then $y^3 + 9y + 26 = 0$ ...(1)

Let $\quad y = u + v$ ...(2)

$$\Rightarrow \quad y^3 - 3uv(y) - (u^3 + v^3) = 0$$

Comparing (1) and (2), we get

$$uv = -3 \text{ and } u^3 + v^3 = -26$$

$$\Rightarrow \quad u^3v^3 = -27$$

Now let us have an equation whose roots are u^3 and v^3

$$t^2-(-26)t+(-27)=0 \Rightarrow t^2+26t-27=0$$

$$\Rightarrow \quad (t+27)(t-1)=0$$

$$\Rightarrow \quad t=1,-27 \quad i.e., \quad u^3=1 \text{ and } v^3=-27$$

So that $\quad u=1,\omega,\omega^2$ and $v=-3,-3\omega,-3\omega^3$

$$y=u+v=(1-3),(\omega-3\omega^2),(\omega^2-3\omega)$$

$$=-2,\left\{\frac{-1+i\sqrt{3}}{2}-3\left(\frac{-1-i\sqrt{3}}{2}\right)\right\},\left\{\frac{-1-i\sqrt{3}}{2}-3\left(\frac{-1+i\sqrt{3}}{2}\right)\right\}$$

$$=-2,\left\{(1+2i\sqrt{3})\right\},\left\{(1-2i\sqrt{3})\right\}$$

$$\therefore \quad x=y+1$$

$$\Rightarrow \quad x=(-2+1),(1+2i\sqrt{3}+1),(1-2i\sqrt{3}+1)$$

$$=-1,(2+2i\sqrt{3}),(2-2i\sqrt{3})$$

$$=-1,2(1\pm i\sqrt{3}). \quad \textbf{Ans.}$$

Example 11: Solve $x^3-6x^2+6x-5=0$ by Cardon's method.

Sol. The given equation $\quad x^3-6x^2+6x-5=0$ ...(1)

Comparing with $a_0x^3+a_1x^2+a_2x+a_3=0$, we get

$$a_0=1, a_1=-6$$

$$\therefore \quad h=-\frac{a_1}{3a_0}=\frac{6}{3}=2$$

Now remove the x^2 term, using synthetic division

2	1	–6	6	–5
		2	–8	–4
	1	–4	–2	(–9)
		2	–4	
	1	–2	(–6)	
		2		
	1	(0)		

Transformed equation is $\quad y^3-6y-9=0$ ...(2)

where $$y = x - 2$$

Let $y = u + v$ be the solution of (2), then

$$y^3 = u^3 + v^3 + 3uv\,(u + v)$$

$$y^3 = u^3 + v^3 + 3uv\,(y)$$

$$y^3 - 3uv(y) - (u^3 + v^3) = 0 \qquad ...(3)$$

Comparing (2) and (3), we have

$$3uv = 6 \;\Rightarrow\; uv = 2 \Rightarrow u^3v^3 = 8$$

and $$u^3 + v^3 = 9$$

Now let us have an equation whose roots are u^3 and v^3

$$t^2 - st + p = 0 \text{ ; } s = \text{sum of roots, } p = \text{product of the roots.}$$

$$t^2 - 9t + 8 = 0$$

$$t = 1,8 \quad i.e., \; u^3 = 1 \text{ and } v^3 = 8 \text{ so that}$$

$$u = 1, \omega, \omega^2 \text{ and } v = 2, 2\omega, 2\omega^2$$

$\therefore$ $$y = u + v$$

$$= (1+2), (\omega + 2\omega^2), (\omega^2 + 2\omega)$$

$$= 3, \left\{\left(\frac{-1+i\sqrt{3}}{2}\right) + 2\left(\frac{-1-i\sqrt{3}}{2}\right)\right\}, \left\{\left(\frac{-1-i\sqrt{3}}{2}\right) + 2\left(\frac{-1+i\sqrt{3}}{2}\right)\right\}$$

$$= 3, \frac{-3-i\sqrt{3}}{2}, \frac{-3+i\sqrt{3}}{2}.$$

But $$x = y + 2$$

$$= 5, \frac{1-i\sqrt{3}}{2}, \frac{1+i\sqrt{3}}{2}$$

$$= 5, \frac{1}{2}(1 \pm i\sqrt{3}). \quad \textbf{Ans.}$$

Example 12: Solve the equation $x^3 - 3x^2 + 3x - 1 = 0$ by Cardon's method.

Sol. Here, $a_0 = 1, a_1 = -3$

$\therefore$ $$h = -\frac{a_1}{3a_0} = \frac{3}{3} = 1$$

Now remove the x^2 term, using synthetic division, we have

$$\begin{array}{c|cccc} 1 & 1 & -3 & 3 & -1 \\ & & 1 & -2 & 1 \\ \hline & 1 & -2 & 1 & (0) \\ & & 1 & -1 & \\ \hline & 1 & -1 & (0) & \\ & & 1 & & \\ \hline & 1 & (0) & & \end{array}$$

Transformed equation is $y^3 = 0$ ...(1)

where $y = x - 1.$

From (1), $y = 0, 0, 0$

$\therefore$ $x = y + 1 = 1, 1, 1$

Hence the required roots are 1, 1, 1. **Ans.**

Example 13: Solve the cubic equation $x^3 + x^2 - 16x + 20 = 0.$

Sol. Given equation is $x^3 + x^2 - 16x + 20 = 0$

Here, $a_0 = 1, a_1 = 1$

$$\therefore \qquad h = -\frac{a_1}{3a_0} = -\frac{1}{3}$$

Now remove the x^2 term, using synthetic division, we have

$$\begin{array}{c|cccc} -1/3 & 1 & 1 & -16 & 20 \\ & & -1/3 & -2/9 & \frac{146}{27} \\ \hline & 1 & 2/3 & \frac{-146}{9} & \left(\frac{686}{27}\right) \\ & & -1/3 & -1/9 & \\ \hline & 1 & 1/3 & \left(\frac{-147}{9}\right) & \\ & & -1/3 & & \\ \hline & 1 & (0) & & \end{array}$$

Transformed equation is $y^3 - \frac{147}{9}y + \frac{686}{27} = 0$ ...(1)

where $y = x + \frac{1}{3}.$

Let $y = u + v$

Cubing, $$y^3 = u^3 + v^3 + 3uvy$$

$$\Rightarrow \quad y^3 - 3uvy - (u^3 + v^3) = 0 \qquad ...(2)$$

Comparing with (1), we get

$$uv = \frac{49}{9} \Rightarrow u^3v^3 = \left(\frac{7}{3}\right)^6 \text{ and } u^3 + v^3 = -\frac{686}{27}$$

$\therefore$ u^3 and v^3 are the roots of $t^2 + \frac{686}{27}t + \left(\frac{7}{3}\right)^6 = 0$

$$t^2 + 2\left(\frac{7}{3}\right)^3 t + \left(\frac{7}{3}\right)^6 = 0$$

$$\Rightarrow \quad \left\{t + \left(\frac{7}{3}\right)^3\right\}^2 = 0 \Rightarrow t = -\left(\frac{7}{3}\right)^3, -\left(\frac{7}{3}\right)^3$$

Let $$u^3 = -\left(\frac{7}{3}\right)^3 \text{ and } v^3 = -\left(\frac{7}{3}\right)^3$$

So that $$u = \frac{-7}{3}, -\frac{7}{3}\omega, -\frac{7}{3}\omega^2 \text{ and } v = -\frac{7}{3}, -\frac{7}{3}\omega, -\frac{7}{3}\omega^2$$

$\therefore$ $$y = u + v = \left(-\frac{7}{3} - \frac{7}{3}\right), \left(-\frac{7}{3}\omega - \frac{7}{3}\omega^2\right), \left(-\frac{7}{3}\omega^2 - \frac{7}{3}\omega\right) = -\frac{14}{3}, \frac{7}{3}, \frac{7}{3}$$

Now, $$x = y - \frac{1}{3} = \frac{-14}{3} - \frac{1}{3}, \frac{7}{3} - \frac{1}{3}, \frac{7}{3} - \frac{1}{3} = -5, 2, 2$$

Hence required roots are –5, 2, 2. **Ans.**

Example 14: Solve the cubic equation $x^3 + 6x^2 - 12x + 32 = 0.$

Sol. Given equation is $x^3 + 6x^2 - 12x + 32 = 0$

Here, $a_0 = 1, a_1 = 6$

$\therefore$ $$h = -\frac{a_1}{3a_0} = -\frac{6}{3} = -2$$

Now remove the x^2 term, using synthetic division method, we have

–2	1	6	– 12	32
		–2	– 8	40
	1	4	– 20	(72)
		–2	– 4	
	1	2	(–24)	
		–2		
	1	(0)		

Transformed equation is

$$y^3 - 24y + 72 = 0 \qquad ...(1)$$

where $\qquad y = x + 2$

Let $\qquad y = u + v$

Cubing, $\qquad y^3 = u^3 + v^3 + 3uvy$

$$\Rightarrow \quad y^3 - 3uvy - (u^3 + v^3) = 0 \qquad ...(2)$$

Comparing (1) and (2), we get

$$uv = 8 \;\Rightarrow\; u^3v^3 = 512 \text{ and } u^3 + v^3 = -72$$

$\therefore$ u^3 and v^3 are the roots of $t^2 + 72t + 512 = 0$

$\Rightarrow \qquad t = -8, -64$

Let $\qquad u^3 = -8$ and $v^3 = -64$

So that $\qquad u = -2, -2\omega, -2\omega^2$ & $v = -4, -4\omega, -4\omega^2$

$$\therefore \qquad y = u + v = (-2 - 4), (-2\omega - 4\omega^2), (-2\omega^2 - 4\omega)$$

$$= -6, 3 + i\sqrt{3}, 3 - i\sqrt{3}$$

Now, $\qquad x = y - 2 = -8, 1 + i\sqrt{3}, 1 - i\sqrt{3}$

Hence the required roots are $-8, 1 \pm i\sqrt{3}$. **Ans.**

Example 15: Solve $x^3 + 3ax^2 + 3(a^2 - bc)x + a^3 + b^3 + c^3 - 3abc = 0$ by Cardon's method.

Sol. Given equation is $x^3 + 3ax^2 + 3(a^2 - bc)x + a^3 + b^3 + c^3 - 3abc = 0$

Here $a_0 = 1$, $a_1 = 3a$

$$\therefore \qquad h = -\frac{a_1}{n.a_0} = -\frac{3a}{3.1} = -a$$

Now remove the x^2 term, using synthetic division, we have

$-a$	1	$3a$	$3(a^2 - bc)$	$a^3 + b^3 + c^3 - 3abc$
		$-a$	$-2a^2$	$-a^3 \quad + 3abc$
	1	$2a$	$a^2 - 3bc$	$(b^3 + c^3)$
		$-a$	$-a^2$	
	1	a	$-3bc$	
		$-a$		
	1	(0)		

Transformed equation is

$$y^3 - 3bcy + (b^3 + c^3) = 0 \quad \text{...(1)}$$

where $y = x + a$

Let $y = u + v$

Cubing, $y^3 = u^3 + v^3 + 3uvy$

$\Rightarrow \quad y^3 - 3uvy - (u^3 + v^3) = 0$

Comparing with (1), we get

$$uv = bc \;\Rightarrow\; u^3v^3 = b^3c^3 \;\&\; u^3 + v^3 = -(b^3 + c^3)$$

$\therefore$ u^3 and v^3 are the roots of

$$t^2 + (b^3 + c^3)t + b^3c^3 = 0 \Rightarrow (t + b^3)(t + c^3) = 0$$

$\Rightarrow \quad t = -b^3, -c^3$

Let $u^3 = -b^3$ and $v^3 = -c^3$

So that $u = -b, -b\omega, -b\omega^2$ and $v = -c, -c\omega, -c\omega^2$

$\therefore \quad y = u + v = -b - c, -b\omega - c\omega^2, -b\omega^2 - c\omega$

$\therefore \quad x = y - a = -a - b - c, -a - b\omega - c\omega^2, -a - b\omega^2 - c\omega.$ **Ans.**

Example 16: Solve $27x^3 + 54x^2 + 198x - 73 = 0$ by Cardon's method.

Sol. Given equation $27x^3 + 54x^2 + 198x - 73 = 0$

Equating with $a_0x^3 + a_1x^2 + a_2x^3 + a_3 = 0$, we get

$$a_0 = 27, a_1 = 54$$

$$\therefore \quad h = -\frac{a_1}{3.a_0} = -\frac{54}{3.27} = -\frac{2}{3}$$

Now remove the x^2 term, using synthetic division, we have

−2/3	27	54	198	−73
		−18	−24	−116
	27	36	174	(−189)
		−18	−12	
	27	18	(162)	
		−18		
	27	(0)		

Transformed equation is

$$27y^3 + 162y - 189 = 0$$

$$\Rightarrow \qquad y^3 + 6y - 7 = 0 \qquad \text{...(1)}$$

where $y = x + \frac{2}{3}$.

Let $y = u + v$.

Cubing, $$y^3 = u^3 + v^3 + 3uvy$$

$$\Rightarrow \qquad y^3 - 3uvy - (u^3 + v^3) = 0$$

Comparison with (1) gives

$$uv = -2 \Rightarrow u^3v^3 = -8 \text{ and } u^3 + v^3 = 7$$

$\therefore$ u^3 and v^3 are the roots of $t^2 - 7t - 8 = 0$

$$\Rightarrow \qquad (t+1)(t-8) = 0 \Rightarrow t = -1, 8$$

Let $$u^3 = -1 \Rightarrow u - 1, -\omega, -\omega^2$$

and $$v^3 = 8 \Rightarrow v = 2, 2\omega, 2\omega^2$$

$$\therefore \qquad y = u + v = (-1+2), (-\omega + 2\omega^2), (-\omega^2 + 2\omega)$$

$$= 1, \left\{-\left(\frac{-1+i\sqrt{3}}{2}\right) + 2\left(\frac{-1-i\sqrt{3}}{2}\right)\right\}, \left\{-\left(\frac{-1-i\sqrt{3}}{2}\right) + 2\left(\frac{-1+i\sqrt{3}}{2}\right)\right\}$$

$$= 1, \frac{-1-3i\sqrt{3}}{2}, \frac{-1+3i\sqrt{3}}{2}$$

Now, $$x = y - \frac{2}{3} = 1 - \frac{2}{3}, \left(\frac{-1-3\sqrt{3}i}{2}\right) - \frac{2}{3}, \left(\frac{-1+3\sqrt{3}i}{2}\right) - \frac{2}{3}$$

$$= \frac{1}{3}, \frac{-7-9\sqrt{3}i}{6}, \frac{-7+9\sqrt{3}i}{6}$$

$$\therefore \qquad x = \frac{1}{3}, \frac{-7 \pm 9\sqrt{3}i}{6}. \qquad \textbf{Ans.}$$

Example 17: Solve the equation $46x^3 + 72x^2 + 18x - 11 = 0$ by Cardon's method.

Sol. First we remove the second term of the given equation by diminishing its roots by

$$h = -\frac{a_1}{na_0} = -\frac{72}{3.46} = -\frac{12}{23}$$

$-\frac{12}{23}$	46	72	18	-11
		-24	$-\frac{576}{23}$	$\frac{1944}{(23)^2}$
	46	48	$-\frac{162}{23}$	$\left(\frac{-3875}{(23)^2}\right)$
		-24	$-\frac{288}{23}$	
	46	24	$\left(-\frac{450}{23}\right)$	
		-24		
	46	(0)		

$\therefore$ the transformed equation is

$$46y^3 - \frac{450}{23}y - \frac{3875}{(23)^2} = 0 \qquad \text{...(1)}$$

where $$y = x - \left(-\frac{12}{23}\right).$$

Now to remove the fractional coefficients, we multiply the roots by 46, and get the following transformed equation

$$z^3 - 900z - 15500 = 0 \qquad \text{...(2)}$$

where $$z = 46y = 46\left(x + \frac{12}{23}\right) = 46x + 24$$

$$\therefore \qquad x = \frac{z - 24}{46} \qquad \text{...(3)}$$

Let $z = u + v$ be the solution of the cubic (2).

Cubing both sides, we get

$$z^3 - 3uvz - (u^3 + v^3) = 0 \qquad \text{...(4)}$$

Comparing (2) and (4),we get

$$uv = 300 \qquad \Rightarrow \qquad u^3v^3 = 27000000$$

and $$u^3 + v^3 = 15500$$

$\therefore u^3$ and v^3 are the roots of the quadratic

$$t^2 - 15500t + 27000000 = 0$$

$$t = \frac{15500 \pm \sqrt{(15500)^2 - 4 \times 1 \times 27000000}}{2}$$

$$= \frac{15500 \pm 500\sqrt{(31)^2 - 4 \times 108}}{2}$$

$$= 250(31 \pm 23) = 250 \times 54, 250 \times 8$$

Let $u^3 = 250 \times 54$ and $v^3 = 250 \times 8$

$$u = 15.(2)^{2/3} \text{ and } v = 10.(2)^{1/3}$$

The roots of the cubic (2) are

$$z = u + v = 15.(2)^{2/3} + 10.(2)^{1/3}$$

and $$z = \omega u + \omega^2 v = \omega.15(2)^{2/3} + \omega^2.10(2)^{1/3}$$

and $$z = \omega^2 u + \omega v = \omega^2.15(2)^{2/3} + \omega.10(2)^{1/3}$$

$\therefore$ From (3) the roots of the given cubic are

$$x = \frac{1}{46}\left[15(2)^{2/3} + 10(2)^{1/3} - 24\right], \frac{1}{46}\left[15\omega(2)^{2/3} + 10\omega^2(2)^{1/3} - 24\right]$$

and $\frac{1}{46}\left[15\omega^2(2)^{2/3} + 10\omega(2)^{1/3} - 24\right]$. **Ans.**

Case 3. When Cubic Equation is of the Form $x^3 + qx^2 + r = 0$

Here the term containing x is absent, we put $x = \frac{1}{y}$ so that the given equation transforms to the form of cubic given in case 1.

When three values of y are found, replace y by $\frac{1}{x}$ and obtain the required roots of the given cubic equation.

Note: In Cardon's solution if $u^3 = x + iy$ and $v^3 = x + iy$, then we can use $x = r\cos\theta$ and $y = r\sin\theta$.

Example 18: Solve $9x^3 + 6x^2 - 1 = 0$ by Cardon's method.

Sol. Put $x = \frac{1}{y}$ then given equation reduces to

$$9\left(\frac{1}{y}\right)^3 + 6\left(\frac{1}{y}\right)^2 - 1 = 0$$

$$\Rightarrow \quad \frac{9}{y^3} + \frac{6}{y^2} - 1 = 0$$

$\Rightarrow \qquad y^3 - 6y - 9 = 0 \qquad ...(1)$

Let $\qquad y = u + v$

Cubing, $\qquad y^3 = u^3 + v^3 + 3uvy$

$\Rightarrow \qquad y^3 - 3uvy - (u^3 + v^3) = 0$

Comparing, $\qquad uv = 2 \qquad \Rightarrow \quad u^3v^3 = 8$ and $u^3 + v^3 = 9$

$\therefore$ u^3 and v^3 are the roots of $t^2 - 9t + 8 = 0$

$\Rightarrow \qquad (t-1)(t-8) = 0 \qquad \Rightarrow \quad t = 1, 8$

Let $\qquad u^3 = 1$ and $v^3 = 8$

So that $\qquad u = 1, \omega, \omega^2$ and $v = 2, 2\omega, 2\omega^2$

$$\therefore \qquad y = u + v = (1+2), (\omega + 2\omega^2), (\omega^2 + 2\omega)$$

$$= 3, \left\{\left(\frac{-1+i\sqrt{3}}{2}\right) + 2\left(\frac{-1-i\sqrt{3}}{2}\right)\right\}, \left\{\left(\frac{-1-i\sqrt{3}}{2}\right) + 2\left(\frac{-1+i\sqrt{3}}{2}\right)\right\}$$

$$\Rightarrow \qquad y = 3, \frac{-3-i\sqrt{3}}{2}, \frac{-3+i\sqrt{3}}{2}.$$

But $\qquad x = \dfrac{1}{y}$

$$\therefore \qquad x = \frac{1}{3}, \frac{2}{-3-i\sqrt{3}}, \frac{2}{-3+i\sqrt{3}}. \qquad \textbf{Ans.}$$

Example 19: Solve by Cardon's method $28x^3 - 9x^2 + 1 = 0.$

Sol. Put $x = \dfrac{1}{y}$, then given equation reduces to $y^3 - 9y + 28 = 0.$

Let $\qquad y = u + v$

Cubing, $\qquad y^3 = u^3 + v^3 + 3uvy$

$\Rightarrow \qquad y^3 - 3uvy - (u^3 + v^3) = 0$

Comparing, $\qquad uv = 3 \qquad \Rightarrow \quad u^3v^3 = 27 \,\&\, u^3 + v^3 = -28$

$\therefore$ u^3 and v^3 are the roots of $t^2 + 28t + 27 = 0$

$\Rightarrow \qquad (t+1)(t+27) = 0 \qquad \Rightarrow \quad t = -1, -27$

Let $\qquad u^3 = -1$ and $v^3 = -27$ so that

$\qquad u = -1, -\omega, -\omega^2$ and $v = -3, -3\omega, -3\omega^2$

$\therefore \quad y = u + v = (-1-3), (-\omega - 3\omega^2), (-\omega^2 - 3\omega)$

$$= -4, \left\{-\left(\frac{-1+i\sqrt{3}}{2}\right) - 3\left(\frac{-1-i\sqrt{3}}{2}\right)\right\}, \left\{-\left(\frac{-1-i\sqrt{3}}{2}\right) - 3\left(\frac{-1+i\sqrt{3}}{2}\right)\right\}$$

$$= -4, 2+i\sqrt{3}, 2-i\sqrt{3}$$

$\therefore \quad x = \frac{1}{y} = -\frac{1}{4}, \frac{1}{2+i\sqrt{3}}, \frac{1}{2-i\sqrt{3}} = -\frac{1}{4}, \frac{2-i\sqrt{3}}{7}, \frac{2+i\sqrt{3}}{7} = -\frac{1}{4}, \frac{2 \pm i\sqrt{3}}{7}.$ **Ans.**

Example 20: Show that the roots of the cubic

$$x^3 - 3a^2x - 2a^3 \cos 3A = 0 \text{ are } 2a \cos A,\ 2a \cos\left(\frac{2\pi}{3} \pm A\right).$$

Sol. The given equation is

$$x^3 - 3a^2x - 2a^3 \cos 3A = 0 \quad \text{...(1)}$$

Let $x = u + v$ be the solution of (1)

Cubing on both sides $x^3 - 3uvx - (u^3 + v^3) = 0$ ...(2)

Comparing (1) and (2), we get

$$uv = a^2 \quad \Rightarrow \quad u^3v^3 = a^6$$

and $$u^3 + v^3 = 2a^3 \cos 3A$$

$\therefore$ u^3 and v^3 are the roots of the quadratic

$$t - 2a^3 \cos 3At + a^6 = 0$$

$\therefore \quad t = \dfrac{2a^3 \cos 3A \pm \sqrt{4a^6 \cos^2 3A - 4a^6}}{2} = a^3[\cos 3A \pm i \sin 3A]$

Let $$u^3 = a^3[\cos 3A + i \sin 3A] \Rightarrow u = a[\cos 3A + i \sin 3A]^{1/3}$$

and $$v^3 = a^3[\cos 3A - i \sin 3A] \Rightarrow v = a[\cos 3A - i \sin 3A]^{1/3}$$

$\therefore \quad x = u + v = a[\cos 3A + i \sin 3A]^{1/3} + a[\cos 3A - i \sin 3A]^{1/3}$

$$= a[\{\cos(2n\pi + 3A) + i \sin(2n\pi + 3A)\}^{1/3} + \{\cos(2n\pi + 3A) - i \sin(2n\pi + 3A)\}^{1/3}]$$

$$= a\left[2\cos\left(\frac{2n\pi + 3A}{3}\right)\right] \quad \text{where} \quad n = 0, 1, 2.$$

Hence the roots of the given cubic are

$$2a\cos A, 2a\cos\left(\frac{2\pi}{3} + A\right) \text{ and } 2a\cos\left(\frac{4\pi}{3} + A\right)$$

or $$2a\cos A, 2a\cos\left(\frac{2\pi}{3} + A\right) \text{ and } 2a\cos\left\{2\pi - \left(\frac{2\pi}{3} - A\right)\right\}$$

or $$2a\cos A, 2a\cos\left(\frac{2\pi}{3}+A\right) \text{ and } 2a\cos\left(\frac{2\pi}{3}-A\right)$$

or $$2a\cos A, 2a\cos\left(\frac{2\pi}{3}\pm A\right).$$ **Proved.**

Example 21: Prove that the roots of the equation $x^3-3x+1=0$ are $2\cos\left(\frac{2\pi}{9}\right)$, $2\cos\left(\frac{8\pi}{9}\right)$, $2\cos\left(\frac{14\pi}{9}\right)$.

Sol. The given equation is

$$x^3-3x+1=0 \qquad ...(1)$$

Let $x=u+v$ be the solution of (1)

Cubing on both sides

$$x^3-3uvx-(u^3+v^3)=0 \qquad ...(2)$$

Comparing (1) and (2), we get

$3uv=3 \quad \Rightarrow \quad uv=1 \quad \Rightarrow \quad u^3v^3=1$ and $u^3+v^3=-1$

$\therefore$ u^3 and v^3 are the roots of the quadratic

$$t^2+t+1=0$$

$$t=\frac{-1\pm i\sqrt{3}}{2}$$

$$\therefore \quad u^3=-\frac{1}{2}+i\frac{\sqrt{3}}{2} \text{ and } v^3=-\frac{1}{2}-i\frac{\sqrt{3}}{2}$$

$$u=\left(-\frac{1}{2}+i\frac{\sqrt{3}}{2}\right)^{1/3}, v=\left(-\frac{1}{2}-i\frac{\sqrt{3}}{2}\right)^{1/3}$$

$$\therefore \quad x=u+v$$

$$=\left(-\frac{1}{2}+i\frac{\sqrt{3}}{2}\right)^{1/3}+\left(-\frac{1}{2}-i\frac{\sqrt{3}}{2}\right)^{1/3}$$

Putting $-\frac{1}{2}=r\cos\theta$ and $\frac{\sqrt{3}}{2}=r\sin\theta$ then, we get

$r^2=1 \quad \Rightarrow \quad r=1$ and

$\tan\theta=-\sqrt{3} \quad \Rightarrow \theta=\frac{2\pi}{3}$

$$\therefore \quad x=r^{1/3}\left[(\cos\theta+i\sin\theta)^{1/3}+(\cos\theta-i\sin\theta)^{1/3}\right]$$

$$x = \left[\left\{\cos\left(\frac{2n\pi+\theta}{3}\right)+i\sin\left(\frac{2n\pi+\theta}{3}\right)\right\}+\left\{\cos\left(\frac{2n\pi+\theta}{3}\right)-i\sin\left(\frac{2n\pi+\theta}{3}\right)\right\}\right]$$

$$= 2\cos\left(\frac{2n\pi}{3}+\frac{2\pi}{9}\right) \quad \text{where } n = 0,1,2$$

Hence the roots are

$$2\cos\left(\frac{2\pi}{9}\right), 2\cos\left(\frac{8\pi}{9}\right), 2\cos\left(\frac{14\pi}{9}\right). \qquad \textbf{Proved.}$$

Example 22: Show that the roots of the cubic $8a^3x^3 - 6ax + 2\sin 3A = 0$ are

$$\frac{1}{a}\sin A, \frac{1}{a}\sin\left(\frac{\pi}{3}-A\right), -\frac{1}{a}\sin\left(\frac{\pi}{3}+A\right).$$

Sol. The given cubic equation can be written as

$$x^3 - \frac{3}{4a^2}x + \frac{1}{4a^3}\sin 3A = 0 \qquad \text{...(1)}$$

Let $x = u + v$ be the solution of (1)
Cubing on both sides

$$x^3 - 3uvx - (u^3+v^3) = 0 \qquad \text{...(2)}$$

Comparing (1) and (2), we get

$$uv = \frac{1}{4a^2} \quad \text{or} \quad u^3v^3 = \frac{1}{64a^6}$$

and $$u^3 + v^3 = -\frac{\sin 3A}{4a^3}$$

$\therefore$ u^3 and v^3 are the roots of the quadratic

$$t^2 + \frac{\sin 3A}{4a^3}t + \frac{1}{64a^6} = 0$$

$\therefore$ $$t = \frac{\dfrac{-\sin 3A}{4a^3} \pm \sqrt{\left(\dfrac{\sin^2 3A}{16a^6} - 4.1.\dfrac{1}{64a^6}\right)}}{2}$$

$$= \frac{1}{8a^3}[-\sin 3A \pm i\cos 3A]$$

$$= \frac{1}{8a^3}\left[\cos\left(\frac{\pi}{2}+3A\right) \pm i\sin\left(\frac{\pi}{2}+3A\right)\right]$$

Let $$u^3 = \frac{1}{8a^3}\left[\cos\left(\frac{\pi}{2}+3A\right) + i\sin\left(\frac{\pi}{2}+3A\right)\right]$$

and $$v^3 = \frac{1}{8a^3}\left[\cos\left(\frac{\pi}{2}+3A\right) - i\sin\left(\frac{\pi}{2}+3A\right)\right]$$

$\therefore$ $$x = u + v$$

$$= \frac{1}{2a}\left[\left[\left\{\cos\left(\frac{\pi}{2}+3A\right)+i\sin\left(\frac{\pi}{2}+3A\right)\right\}\right]^{1/3} + \left[\left\{\cos\left(\frac{\pi}{2}+3A\right) - i\sin\left(\frac{\pi}{2}+3A\right)\right\}\right]^{1/3}\right]$$

$$= \frac{1}{2a}\left[2\cos\frac{1}{3}\left(2n\pi+\frac{\pi}{2}+3A\right)\right], \quad \text{where } n = 0, 1, 2.$$

$\therefore$ The roots of the given cubic are

$$\frac{1}{a}\cos\frac{1}{3}\left(\frac{\pi}{2}+3A\right), \frac{1}{a}\cos\frac{1}{3}\left(2\pi+\frac{\pi}{2}+3A\right) \text{ and } \frac{1}{a}\cos\frac{1}{3}\left(4\pi+\frac{\pi}{2}+3A\right)$$

i.e., $$\frac{1}{a}\cos\left(\frac{\pi}{6}+A\right), \frac{1}{a}\cos\left(\frac{5\pi}{6}+A\right) \& \frac{1}{a}\cos\left(\frac{3\pi}{2}+A\right)$$

i.e., $$\frac{1}{a}\cos\left[\frac{\pi}{2}-\left(\frac{\pi}{3}-A\right)\right], \frac{1}{a}\cos\left[\frac{\pi}{2}+\left(\frac{\pi}{3}+A\right)\right] \& \frac{1}{a}\cos\left[\frac{\pi}{2}+(\pi+A)\right]$$

i.e., $$\frac{1}{a}\sin\left(\frac{\pi}{3}-A\right), -\frac{1}{a}\sin\left(\frac{\pi}{3}+A\right) \& \frac{1}{a}\sin A$$

Hence the roots of the given cubic equation are $\frac{1}{a}\sin A, \frac{1}{a}\sin\left(\frac{\pi}{3}-A\right)$ and $-\frac{1}{a}\sin\left(\frac{\pi}{3}+A\right)$.

Proved.

Example 23: Prove that the roots of the cubic equation $x^3 - 12x - 8\sqrt{2} = 0$ are

$$4\cos\frac{\pi}{12}, 4\cos\frac{3\pi}{4}, 4\cos\frac{17\pi}{12}.$$

Sol. The given cubic equation is $x^3 - 12x - 8\sqrt{2} = 0$

Let $x = u + v$.

Cubing, $x^3 - 3uvx - (u^3 + v^3) = 0$

Comparing, $uv = 4 \quad \Rightarrow \quad u^3v^3 = 64 \; \& \; u^3 + v^3 = 8\sqrt{2}$

$\therefore$ u^3 and v^3 are the roots of the quadratic

$$t^2 - 8\sqrt{2}t + 64 = 0$$

$\Rightarrow$ $$t = \frac{8\sqrt{2} \pm \sqrt{128-256}}{2} = \frac{8\sqrt{2} \pm 8\sqrt{2}i}{2} = 4\sqrt{2} \pm 4\sqrt{2}i$$

Let $u^3 = 4\sqrt{2} + 4\sqrt{2}i$ and $v^3 = 4\sqrt{2} - 4\sqrt{2}i$

Let $4\sqrt{2} = r\cos\theta$ and $4\sqrt{2} = r\sin\theta$

So that $r^2 = 64 \quad \Rightarrow r = 8$

Also $\tan\theta = 1 \quad \Rightarrow \theta = \dfrac{\pi}{4}$

$\therefore \quad u^3 = r(\cos\theta + \sin\theta) = r[\cos(2k\pi + \theta) + i\sin(2k\pi + \theta)]$

$$\Rightarrow \quad u = r^{1/3}\left[\cos\left(\frac{2k\pi + \theta}{3}\right) + i\sin\left(\frac{2k\pi + \theta}{3}\right)\right] \quad \text{(using De Moivre's theorem)}$$

Similarly,
$$v = r^{1/3}\left[\cos\left(\frac{2k\pi + \theta}{3}\right) - i\sin\left(\frac{2k\pi + \theta}{3}\right)\right]$$

$$x = u + v = 2r^{1/3}\cos\left(\frac{2k\pi + \theta}{3}\right); (k = 0, 1, 2)$$

$$\Rightarrow \quad x = 2(8)^{1/3}\cos\left(\frac{2k\pi + \pi/4}{3}\right); k = 0, 1, 2$$

$$= 4\cos\left(\frac{8k\pi + \pi}{12}\right); k = 0, 1, 2$$

$$\Rightarrow \quad x = 4\cos\frac{\pi}{12}, 4\cos\frac{3\pi}{4}, 4\cos\frac{17\pi}{12}.$$ **Proved.**

PROBLEM SET 5. 2

Solve the following cubic equations by Cardon's method:

1. $x^3 - 21x - 344 = 0$ $\left[\textbf{Ans. } 8, -4 \pm i3\sqrt{3}\right]$

2. $x^3 - 18x - 35 = 0$ $\left[\textbf{Ans. } 5, \frac{1}{2}(-5 \pm i\sqrt{3})\right]$

3. $x^3 - 3x^2 + 3 = 0$ $\left[\textbf{Ans. } 2\cos\frac{2\pi}{9}, 2\cos\frac{8\pi}{9}, 2\cos\frac{14\pi}{9}\right]$

4. $x^3 - 15x^2 - 357x + 5491 = 0$ $\left[\textbf{Ans. } -19, 17, 17\right]$

5. $64x^3 - 144x^2 + 108x - 27 = 0$ $\left[\textbf{Ans. } \frac{3}{4}, \frac{3}{4}, \frac{3}{4}\right]$

6. $2x^3 + 3x^2 + 3x + 1 = 0$ $\left[\textbf{Ans. } -\frac{1}{2}, -\frac{1}{2} \pm i\frac{\sqrt{3}}{2}\right]$

7. $x^3 - 3x^2 + 12x + 16 = 0$ $\left[\textbf{Ans. } 1, 2 \pm 2i\sqrt{3}\right]$

8. $2x^3 + 5x^2 + x - 2 = 0$ [**Ans.** $-1, -2, 1/2$]

9. $x^3 + 3x - 14 = 0$ $\left[\textbf{Ans. } 2, -1 \pm i\sqrt{6}\right]$

10. $x^3 - 12x - 65 = 0$ $\left[\textbf{Ans. } 5, \frac{-5 \pm 3\sqrt{3}i}{2}\right]$

11. $x^3 - 3abx + (a^3 + b^3) = 0$ $\left[\textbf{Ans. } -(a-b), \frac{(a+b) \pm (a-b)\sqrt{3}i}{2}\right]$

12. If the cubic equation $a_0x^3 + 3a_1x^2 + 3a_2x + a_3 = 0$ has two roots equal to α, show that $-\alpha = \frac{H_2}{H_1} = \frac{H_1}{H}$, where $H = a_0a_2 - a_1^2, 2H_1 = a_0a_3 - a_1a_2$ and $H_2 = a_1a_3 - a_2^2$.

5.13 DESCARTE'S METHOD FOR SOLVING BIQUADRATIC EQUATION

Consider a biquadratic equation of the form

$$x^4 + ax^2 + bx + c = 0 \quad \text{...(1)}$$

Suppose the biquadratic (1) can be split up into two quadratic factors $x^2 + lx + m$ and $x^2 - lx + n$

Then $x^4 + ax^2 + bx + c = (x^2 + lx + m)(x^2 - lx + n)$

Equating the coefficients of like power of x, we get

$$m + n - l^2 = a \qquad \therefore \quad m + n = l^2 + a \quad \text{...(2)}$$

$$ln - lm = b \qquad \therefore \quad m - n = \frac{b}{l} \quad \text{...(3)}$$

$$mn = c$$

Since $(n+m)^2 - (n-m)^2 = 4mn$

$$\therefore \quad (l^2 + a)^2 - \frac{b^2}{l^2} = 4c \quad \text{or} \quad l^6 + 2al^4 + (a^2 - 4c)l^2 - b^2 = 0 \quad \text{...(4)}$$

It is cubic in l^2. Solving (4) as a cubic in l^2 by trial method, m and n can be obtained from (2) and (3) thus the biquadratic equation reduces to the quadratic equation $x^2 + lx + m = 0, x^2 - lx + n = 0$ which gives four value of x.

Example 1: Solve the biquadratic equation $x^4 - 3x^2 - 42x - 40 = 0$ by Descarte's method.

Sol. The given equation is

$$x^4 - 3x^2 - 42x - 40 = 0$$

Let $x^4 - 3x^2 - 42x - 40 = (x^2 + lx + m)(x^2 - lx + n)$

Equating the coefficients of like power of x, we get

$$m + n - l^2 = -3 \qquad \therefore \quad m + n = l^2 - 3 \quad \text{...(1)}$$

$$ln - lm = -42 \qquad \therefore \quad n - m = \frac{-42}{l} \quad \text{...(2)}$$

$$mn = -40$$

Since $(n+m)^2-(n-m)^2=4mn$

$$\therefore \quad (l^2-3)^2-\frac{(42)^2}{l^2}=-160 \quad \text{or} \quad l^6-6l^4+169l^2-1764=0$$

By trial $l^2=9$ or $l=3$;

From (1) and (2),

$$m+n=6$$

$$m-n=14$$

$\therefore$ $m=10$ and $n=-4$

Therefore, given biquadratic equation can be written as

$$(x^2+3x+10)(x^2-3x-4)=0$$

$$\therefore \quad x=4,\ -1,\ \frac{-3\pm i\sqrt{31}}{2}. \quad \textbf{Ans.}$$

Example 2: Solve $x^4-5x^2-6x-5=0$.

Sol. The given equation is $x^4-5x^2-6x-5=0$.

Let $x^4-5x^2-6x-5=(x^2+lx+m)(x^2-lx+n)$.

Equating the coefficients of like power of x on both sides, we have

$$m+n-l^2=-5 \qquad \therefore \ m+n=l^2-5 \qquad \text{...(1)}$$

$$ln-lm=-6 \qquad \therefore \ n-m=\frac{-6}{l} \qquad \text{...(2)}$$

$$mn=-5$$

Since $(n+m)^2-(n-m)^2=4mn$

$$\therefore \quad (l^2-5)^2-\frac{36}{l^2}=-20 \quad \text{or} \quad l^6-10l^4+45l^2-36=0$$

By trial $l^2=1$ or $l=1$;

From (1) and (2), we get

$$m+n=-4$$

$$m-n=6$$

$\therefore$ $m=1$ and $n=-5$

Therefore, given biquadratic equation can be written as

$$(x^2+x+1)(x^2-x-5)=0$$

$$\therefore \quad x=\frac{-1\pm\sqrt{1-4}}{2},\frac{1\pm\sqrt{1+20}}{2}$$

Hence $$x=\frac{-1\pm i\sqrt{3}}{2};\ \frac{1\pm\sqrt{21}}{2}. \quad \textbf{Ans.}$$

Example 3: Solve $x^4 - 8x^2 - 24x + 7 = 0$.

Sol. The given equation is $x^4 - 8x^2 - 24x + 7 = 0$

Let $x^4 - 8x^2 - 24x + 7 = (x^2 + lx + m)(x^2 - lx + n)$

Equating the coefficients of like power of x on both sides, we have

$$m + n - l^2 = -8 \qquad \therefore \quad m + n = l^2 - 8 \qquad \text{...(1)}$$

$$ln - lm = -24 \qquad \therefore \quad n - m = \frac{-24}{l} \qquad \text{...(2)}$$

$$mn = 7$$

Since $\quad (n + m)^2 - (n - m)^2 = 4mn$

$$\therefore \quad (l^2 - 8)^2 - \frac{(-24)^2}{l^2} = 28 \quad \text{or} \quad l^6 - 16l^4 + 36l^2 - 576 = 0$$

By trial $\quad l^2 = 16 \quad$ or $\quad l = 4$;

From (1) and (2), we get

$$m + n = 8$$

$$n - m = -6$$

$$\therefore \quad m = 7 \text{ and } n = 1$$

Therefore, given biquadratic equation can be written as

$$x^4 - 8x^2 - 24x + 7 = (x^2 + 4x + 7)(x^2 - 4x + 1)$$

$$\therefore \quad x = \frac{-4 \pm \sqrt{16 - 28}}{2}, \frac{4 \pm \sqrt{16 - 4}}{2}$$

Hence $\quad x = -2 \pm \sqrt{3}i, 2 \pm \sqrt{3}$. **Ans.**

Example 4: Solve $x^4 + 8x^3 + 9x^2 - 8x - 10 = 0$.

Sol. Given equation is $x^4 + 8x^3 + 9x^2 - 8x - 10 = 0$

Sum of roots = –8, number of roots = 4

$$\therefore \quad h = \frac{-8}{4} = -2$$

Now removing the term of x^3 by synthetic diviosion method, we have

–2	1	8	9	–8	–10
		–2	–12	6	4
	1	6	–3	–2	(–6)
		–2	–8	22	
	1	4	–11	(20)	
		–2	–4		
	1	2	(–15)		
		–2			
	1	(0)			

$\therefore$ Transformed equation is $y^4 - 15y^2 + 20y - 6 = 0$ ...(1)

where $y = x - (-2) = x + 2$.

Let $y^4 - 15y^2 + 20y - 6 = (y^2 + ly + m)(y^2 - ly + n)$

Comparing, we get

$$m + n - l^2 = -15 \quad \text{or} \quad m + n = l^2 - 15$$

$$ln - lm = -20 \quad \text{or} \quad n - m = \frac{20}{l}$$

$$mn = -6$$

Since $(n + m)^2 - (n - m)^2 = 4mn$

$\therefore$ $(l^2 - 15)^2 - \dfrac{400}{l^2} = -24$ or $l^6 - 30l^4 + 249l^2 - 400 = 0$

By trial $l^2 = 16$ or $l = 4$;

From (1) and (2), we get

$$m + n = 1$$

$$n - m = 5$$

$\therefore$ $m = -2$ and $n = 3$

Therefore, given biquadratic equation can be written as

$$(y^2 + 4y - 2)(y^2 - 4y + 3) = 0$$

$\therefore$ $y = \dfrac{-4 \pm \sqrt{16+8}}{2}, \dfrac{4 \pm \sqrt{16-12}}{2}$

$$= -2 \pm \sqrt{6},\ 1, 3 \ \text{ but } \ x = y - 2$$

Hence $x = -4 \pm \sqrt{6}, -1, 1.$ **Ans.**

Example 5: Solve the biquadratic equation $x^4 - 8x^3 - 12x^2 + 60x + 63 = 0$ by Descarte's method.

Sol. We have $x^4 - 8x^3 - 12x^2 + 60x + 63 = 0$

Here $a_1 = -8$ and $a_0 = 1$

$$h = -\frac{a_1}{na_0} = \frac{8}{4.1} = 2$$

Now removing the term of x^3 by synthetic diviosion method, we have

2	1	–8	–12	60	63
		2	–12	–48	24
	1	–6	–24	12	(87)
		2	–8	–64	
	1	–4	–32	(–52)	
		2	–4		
	1	–2	(–36)		
		2			
	1	(0)			

$\therefore$ Transformed equation is $z^4 - 36z^2 - 52z + 87 = 0$

where $$z = x - 2.$$

Let $z^4 - 36z^2 - 52z + 87 = (z^2 + lz + m)(z^2 - lz + n)$ ...(1)

Comparing the coefficient of like power of both sides, we get

$$m + n - l^2 = -36 \quad \text{or} \quad m + n = l^2 - 36$$

$$ln - lm = -52 \quad \text{or} \quad n - m = \frac{-52}{l}$$

$$mn = 87$$

Since $(n+m)^2 - (n-m)^2 = 4mn$

$\therefore$ $(l^2 - 36)^2 - \dfrac{(-52)^2}{l^2} = 348$ or $l^6 - 72l^4 + 948l^2 - 2704 = 0$

By trial $l^2 = 4$ or $l = 2;$

$\therefore$ We get $m + n = -32$

$$n - m = -26$$

$\therefore$ $m = -3$ and $n = -29$

Therefore given biquadratic equation can be written as

$$(z^2 + 2z - 3)(z^2 - 2z - 29) = 0$$

$\therefore$ $z = -3, 1, 1 \pm \sqrt{30}$

But $x = z + 2$

$$= -1, 3, 3 \pm \sqrt{30}. \quad \textbf{Ans.}$$

Example 6: Solve the equation $x^4 - 6x^3 - 9x^2 + 66x - 22 = 0$ by Descarte's method.

Sol. To remove the second term, we have to diminish the roots by 3/2 which will not be easy, so we first multiply the roots by 2.

Multiplying the roots by 2, reduced equation is

$y^4 - 12y^3 - 36y^2 + 528y - 352 = 0$ ...(1)

where $y = 2x$.

Now diminish the root by 3

3	1	–12	–36	528	–352
		3	–27	–189	1017
	1	–9	–63	339	(665)
		3	–18	–243	
	1	–6	–81	(96)	
		3	–9		
	1	–3	(–90)		
		3			
	1	(0)			

$\therefore$ Transformed equation is

$$z^4 - 90z^2 + 96z + 665 = 0 \quad ...(2)$$

where $z = y - 3 = 2x - 3$.

Let $z^4 - 90z^2 + 96z + 665 = (z^2 + lz + m)(z^2 - lz + n)$ ...(3)

Comparing the coefficient of like power x of both sides, we get

$$m + n - l^2 = -90 \quad \text{or} \quad m + n = l^2 - 90$$

$$ln - lm = 96 \quad \text{or} \quad n - m = \frac{96}{l}$$

$$mn = 665$$

Since $(n+m)^2 - (n-m)^2 = 4mn$

$$\therefore \quad (l^2 - 90)^2 - \frac{(96)^2}{l^2} = 2660 \quad \text{or} \quad l^6 - 180l^4 + (20 \times 272)l^2 - (96)^2 = 0$$

By trial $l^2 = 36$ or $l = 6$

$\therefore$ We get

$$m + n = -54$$

$$n - m = 16$$

$$\therefore \quad m = -35 \;\&\; n = -19$$

Therefore, from equation (3) we have,

$$(z^2 + 6z - 35)(z^2 - 6z - 19) = 0$$

$$\Rightarrow \quad z = \frac{-6 \pm \sqrt{36 + 140}}{2}, \frac{6 \pm \sqrt{36 + 76}}{2}$$

$$= -3 \pm 2\sqrt{11}, 3 \pm 2\sqrt{7}$$

$$2x - 3 = -3 \pm 2\sqrt{11}, 3 \pm 2\sqrt{7}$$

Hence, $x = \pm\sqrt{11}, 3 \pm \sqrt{7}$. **Ans.**

Example 7: Solve $x^4 - 2x^3 - 5x^2 + 10x - 3 = 0$.

Sol. The given equation is $x^4 - 2x^3 - 5x^2 + 10x - 3 = 0$ ...(1)

To remove the second term, we have diminish the roots by 2/4 *i.e.*1/2 which will be difficult, so we first multiply the roots by 2.

Multiplying the roots of (1) by 2, we get

$$y^4 - 2.2y^3 - 4.5y^2 + 8.10y - 16.3 = 0$$

or

$$y^4 - 4y^3 - 20y^2 + 80y - 48 = 0 \quad ...(2)$$

where $y = 2x$.

Sum of roots of (2) = 4

Now diminish the root of (2) by 4/4 *i.e.*, by 1

1	1	–4	–20	80	–48
		1	–3	–23	57
	1	–3	–23	57	(9)
		1	–2	–25	
	1	–2	–25	(32)	
		1	–1		
	1	–1	(–26)		
		1			
	1	(0)			

$\therefore$ Transformed equation is $z^4 - 26z^2 + 32z + 9 = 0$ ...(3)

where $z = y - 1 = 2x - 1$.

Let $z^4 - 26z^2 + 32z + 9 = (z^2 + lz + m)(z^2 - lz + n)$

Equating the coefficient of like power of z on both sides, we get

$$m + n - l^2 = -26 \quad \text{or} \quad m + n = l^2 - 26$$

$$ln - lm = 32 \quad \text{or} \quad n - m = \frac{32}{l}$$

$$mn = 9$$

Since $(n+m)^2 - (n-m)^2 = 4mn$

$$\therefore \quad (l^2 - 26)^2 - \frac{(36)^2}{l^2} = 36 \quad \text{or} \quad l^6 - 52l^4 + 640l^2 - 1024 = 0$$

By trial $l^2 = 16$ or $l = 4$

$\therefore$ We get

$$m + n = -10$$

$$n - m = 8$$

$$\therefore \quad m = -9 \text{ and } n = -1$$

Therefore from equation (3), we have

$$(z^2 + 4z - 9)(z^2 - 4z - 1) = 0$$

$$\Rightarrow \quad z = \frac{-4 \pm \sqrt{16+36}}{2}, \frac{4 \pm \sqrt{16+4}}{2} = \frac{-4 \pm 2\sqrt{13}}{2}; \frac{4 \pm 2\sqrt{5}}{2}$$

$$= -2 \pm \sqrt{13};\ 2 \pm \sqrt{5}$$

$$2x - 1 = -2 \pm \sqrt{13};\ 2 \pm \sqrt{5}$$

Hence, $x = \frac{-1 \pm \sqrt{13}}{2}; \frac{3 \pm \sqrt{5}}{2}$. **Ans.**

Example 8: Solve the biquadratic equation $x^4 + 2x^3 - 7x^2 - 8x + 12 = 0$.

Sol. Here $a_1 = 2$ and $a_0 = 1$

$$h = -\frac{a_1}{na_0} = \frac{-2}{4.1} = -\frac{1}{2}$$

Multiplying the roots of the given equation by 2, we get the transformed equation

$$y^4 + 2 . 2y^3 - 2^2 . 7y^2 - 2^3 . 8y + 2^4 . 12 = 0$$

$$y^4 + 4y^3 - 28y^2 - 64y + 192 = 0 \qquad ...(1)$$

where, $y = 2x$.

Now $h = \frac{-4}{4.1} = -1$ removing the second term, using synthetic division method, we have

−1	1	4	−28	−64	192
		−1	−3	31	33
	1	3	−31	33	(225)
		−1	−2	33	
	1	2	−33	(0)	
		−1	−1		
	1	1	(−34)		
		−1			
	1	(0)			

$\therefore$ Transformed equation is $z^4 - 34z^2 + 225 = 0$ ...(2)

where $z = 2x + 1$.

Now, $z^4 - 34z^2 + 225 = (z^2 - 25)(z^2 - 9)$

$\therefore$ From equation (2), we have

$$(z^2 - 25)(z^2 - 9) = 0 \qquad \Rightarrow \quad z = \pm 5, \pm 3$$

$\therefore$ $2x + 1 = \pm 5, \pm 3$

or $x = \frac{5-1}{2}, \frac{-5-1}{2}, \frac{3-1}{2}, \frac{-3-1}{2}$

or $x = 2, -3, 1, -2$

Hence the required roots are $1, \pm 2, -3$. **Ans.**

Example 9: Solve the biquadratic equation $2x^4 + 6x^3 - 3x^2 + 2 = 0$.

Sol. Here the term containing x is missing

Put $x = \frac{1}{y}$, we get

$$2y^4 - 3y^2 + 6y + 2 = 0 \qquad \text{...(1)}$$

Let $\quad 2y^4 - 3y^2 + 6y + 2 = 2(y^2 + ly + m)(y^2 - ly + n)$

Comparing the coefficient of like power y of both sides, we get

$$2(m + n - l^2) = -3 \quad \text{or} \quad m + n = l^2 - 3/2$$

$$2(ln - lm) = 6 \quad \text{or} \quad n - m = \frac{3}{l}$$

$$mn = 1$$

Since $\quad (n + m)^2 - (n - m)^2 = 4mn$

$$\therefore \quad (l^2 - \frac{3}{2})^2 - \frac{9}{l^2} = 4 \quad \text{or} \quad 4l^6 - 12l^4 - 7l^2 - 36 = 0$$

By trial $\quad l^2 = 4 \quad$ or $\quad l = 2$

$\therefore$ We get

$$m + n = 5/2$$

$$n - m = 3/2$$

$$\therefore \quad m = 1/2 \text{ and } n = 2$$

Therefore, from equation (1), we have

$$2(y^2 + 2y + 1/2)(y^2 - 2y - 2) = 0$$

$$\Rightarrow \quad y = \frac{-2 \pm \sqrt{2}}{2}, 1 \pm i$$

Since $\quad x = \frac{1}{y} = \frac{2}{-2 \pm \sqrt{2}}, \frac{1}{1 \pm i}$

$$= -(2 \pm \sqrt{2}), \frac{1}{2}(1 \pm i). \quad \textbf{Ans.}$$

5.14 FERRARI'S METHOD

This method is applied in solving the biquadratic equation and is illustrated by solving the following examples:

Example 10: Solve the biquadratic equation $x^4 - 2x^3 - 5x^2 + 10x - 3 = 0$ by Ferrari's method.

Sol. We have $\quad x^4 - 2x^3 - 5x^2 + 10x - 3 = 0$

$$x^4 - 2x^3 + x^2 = 5x^2 - 10x + 3 + x^2$$

$$x^4 - 2x^3 + x^2 = 6x^2 - 10x + 3$$

$$(x^2-x)^2=6x^2-10x+3$$

Now $$(x^2-x+\lambda)^2=6x^2-10x+3+\lambda^2+2\lambda(x^2-x)$$

$$(x^2-x+\lambda)^2=x^2(6+2\lambda)-x(10+2\lambda)+(\lambda^2+3) \quad ...(1)$$

Here choose λ so that RHS of (1) is a perfect square *i.e.*, $b^2-4ac=0$.

$$(10+2\lambda)^2-4(6+2\lambda)(\lambda^2+3)=0$$

$$(\lambda+5)^2-(6+2\lambda)(\lambda^2+3)=0$$

$$2\lambda^3+5\lambda^2-4\lambda-7=0$$

By inspection, $\lambda=-1$

Putting $\lambda=-1$ in (1), we get

$$(x^2-x-1)^2=4x^2-8x+4$$

$\Rightarrow$ $$(x^2-x-1)^2=(2x-2)^2$$

$\Rightarrow$ $$(x^2-x-1)^2-(2x-2)^2=0$$

$\Rightarrow$ $$(x^2-x-1+2x-2)(x^2-x-1-2x+2)=0$$

$\Rightarrow$ $$(x^2+x-3)(x^2-3x+1)=0$$

$\therefore$ $$x=\frac{-1\pm\sqrt{1+12}}{2} \text{ or } \frac{3\pm\sqrt{9-4}}{2}$$

Hence the roots are $\frac{-1\pm\sqrt{13}}{2}, \frac{3\pm\sqrt{5}}{2}$. **Ans.**

Example 11: Solve the equation $x^4-10x^3+35x^2-50x+24=0$ by Ferrari's method.

[*U.T.U. 2008*]

Sol. We have $x^4-10x^3+35x^2-50x+24=0$ ...(1)

$$x^4-10x^3=-35x^2+50x-24$$

$$x^4-10x^3+25x^2=-35x^2+50x-24+25x^2$$

$$x^4-10x^3+25x^2=-10x^2+50x-24$$

$$(x^2-5x)^2=-10x^2+50x-24$$

Now $$(x^2-5x+\lambda)^2=-10x^2+50x-24+\lambda^2+2\lambda(x^2-5x)$$

$$(x^2-5x+\lambda)^2=x^2(2\lambda-10)+x(50-10\lambda)+(\lambda^2-24) \quad ...(2)$$

Here choose λ so that RHS of (2) is a perfect square *i.e.*, $b^2-4ac=0$.

$$(50-10\lambda)^2-4(2\lambda-10)(\lambda^2-24)=0$$

$$(25-5\lambda)^2-(2\lambda-10)(\lambda^2-24)=0$$

$$2\lambda^3 - 35\lambda^2 + 202\lambda - 385 = 0$$

By inspection, $\lambda = 5$

Putting $\lambda = 5$ in (2), we get

$$(x^2 - 5x + 5)^2 = 1 \quad \Rightarrow \quad (x^2 - 5x + 5)^2 - 1^2 = 0$$

$$\Rightarrow \quad (x^2 - 5x + 6)(x^2 - 5x + 4) = 0$$

$$\Rightarrow \quad (x - 2)(x - 3)(x - 1)(x - 4) = 0$$

$$\Rightarrow \quad x = 1, 2, 3, 4.$$ **Ans.**

Example 12: Solve the biquadratic equation $2x^4 + 6x^3 - 3x^2 + 2 = 0$ by Ferrari's method.

Sol. Multiplying the roots of the given equation by 2, we have

$$\Rightarrow \quad z^4 + 6z^3 - 6z^2 + 16 = 0, \qquad \text{where } z = 2x.$$

$$\Rightarrow \quad z^4 + 6z^3 = 6z^2 - 16$$

$$\Rightarrow \quad z^4 + 6z^3 + 9z^2 = 15z^2 - 16 \quad (\text{add } 9z^2)$$

$$\Rightarrow \quad (z^2 + 3z)^2 = 15z^2 - 16 \qquad \text{...(1)}$$

Now, $$(z^2 + 3z + \lambda)^2 = (z^2 + 3z)^2 + \lambda^2 + 2\lambda(z^2 + 3z)$$

$$= 15z^2 - 16 + \lambda^2 + 2\lambda z^2 + 6\lambda z$$

$$= z^2(2\lambda + 15) + 6z\lambda + (\lambda^2 - 16) \qquad \text{...(2)}$$

Here choose λ so that RHS of (2) is a prefect square *i.e.*, $b^2 - 4ac = 0$.

$$36\lambda^2 = 4(2\lambda + 15)(\lambda^2 - 16)$$

$$9\lambda^2 = 2\lambda^3 + 15\lambda^2 - 32\lambda - 240$$

$$\Rightarrow \quad \lambda^3 + 3\lambda^2 - 16\lambda - 120 = 0$$

By inspection, $\lambda = 5$

Putting $\lambda = 5$ in (2), we get

$$(z^2 + 3z + 5)^2 = 25z^2 + 30z + 9 = (5z + 3)^2$$

$$\Rightarrow \quad (z^2 + 3z + 5)^2 - (5z + 3)^2 = 0$$

$$\Rightarrow \quad (z^2 - 2z + 2)(z^2 + 8z + 8) = 0$$

$$\Rightarrow \quad z = -4 \pm 2\sqrt{2}, 1 \pm i$$

Hence $$x = z/2 = -2 \pm \sqrt{2}, \ \frac{1}{2}(1 \pm i).$$ **Ans.**

Example 13: Solve the equation $x^4 - 12x^3 + 41x^2 - 18x - 72 = 0$ by Ferrari's method.

Sol. We have $x^4 - 12x^3 + 41x^2 - 18x - 72 = 0$ ...(1)

$$x^4 - 12x^3 = -41x^2 + 18x + 72$$

$$x^4 - 12x^3 + 36x^2 = -41x^2 + 36x^2 + 18x + 72$$

or $$(x^2 - 6x)^2 = -5x^2 + 18x + 72$$

Now, $$(x^2 - 6x + \lambda)^2 = -5x^2 + 18x + 72 + \lambda^2 + 2\lambda(x^2 - 6x)$$

$$= x^2(2\lambda - 5) + x(18 - 12\lambda) + (\lambda^2 + 72) \qquad ...(2)$$

Here choose λ so that RHS of (2), is a prefect square i.e., $b^2 - 4ac = 0$.

$$(18 - 12\lambda)^2 - 4(2\lambda - 5)(\lambda^2 + 72) = 0$$

$$2\lambda^3 - 41\lambda^2 + 252\lambda - 441 = 0 \quad \text{which gives } \lambda = 3$$

Therefore, from (2), we get

$$(x^2 - 6x + 3)^2 = (x - 9)^2$$

$\Rightarrow$ $$(x^2 - 6x + 3)^2 - (x - 9)^2 = 0$$

$\Rightarrow$ $$(x^2 - 5x - 6)(x^2 - 7x + 12) = 0$$

$\therefore$ $$x = -1, 3, 4 \text{ and } 6. \quad \textbf{Ans.}$$

PROBLEM SET 5.3

Solve the following biquadratic equations by Descarte's method:

1. $x^4 - 2x^2 + 8x - 3 = 0.$ $\left[\textbf{Ans. } -1 \pm \sqrt{2}, 1 \pm i\sqrt{2}\right]$

2. $x^4 - 5x^2 - 6x - 5 = 0.$ $\left[\textbf{Ans. } \frac{-1 \pm i\sqrt{3}}{2}, \frac{1 \pm \sqrt{21}}{2}\right]$

3. $x^4 - 8x^2 - 24x + 7 = 0.$ $\left[\textbf{Ans. } -2 \pm i\sqrt{3}, 2 \pm \sqrt{3}\right]$

4. $x^4 + 11x^2 + 10x + 50 = 0.$ $\left[\textbf{Ans. } -1 \pm 2i, 1 \pm 3i\right]$

5. $x^4 - 15x^2 + 20x - 6 = 0.$ $\left[\textbf{Ans. } 1, 3, -2 \pm \sqrt{6}\right]$

6. $x^4 - 10x^2 - 20x - 16 = 0.$ $\left[\textbf{Ans. } 4, -2, -1 \pm i\right]$

7. $x^4 + 8x^3 + 9x^2 - 8x - 10 = 0.$ $\left[\textbf{Ans. } \pm 1, -4 \pm \sqrt{6}\right]$

8. $x^4 - 10x^3 + 35x^2 - 50x + 24 = 0.$ [**Ans.** 1, 2, 3, 4]

9. $x^4 - 4x^3 + 9x^2 - 12x + 18 = 0.$ $\left[\textbf{Ans. } \pm i\sqrt{3}, 2 \pm i\sqrt{3}\right]$

10. $x^4 - 8x^3 - 12x^2 + 60x + 63 = 0.$ $\left[\textbf{Ans. } 3, -1, 3 \pm \sqrt{30}\right]$

11. $z^4 + 4z^3 + 12z^2 + 8z + 95 = 0.$ $\left[\textbf{Ans. } -3 \pm i\sqrt{10}, 1 \pm i\right]$

12. $x^4 - 6x^3 + 3x^2 + 22x - 6 = 0.$ $\left[\textbf{Ans. } 1 \pm \sqrt{7}, 2 \pm \sqrt{3}\right]$

13. $x^4 + 12x - 5 = 0.$ $\left[\textbf{Ans. } -1 \pm \sqrt{2}, 1 \pm 2i\right]$

Solve the following equations by Ferrari's method.

14. $x^4 - 3x^2 - 42x - 40 = 0.$ $\left[\textbf{Ans. } 4, -1, \dfrac{-3 \pm \sqrt{31}i}{3}\right]$

15. $x^4 - 10x^2 - 20x - 16 = 0.$ $[\textbf{Ans. } 4, -2, 1 \pm i]$

16. $x^4 + 2x^3 - 7x^2 - 8x + 12 = 0.$ $[\textbf{Ans. } -2, -3, 1, 2]$

17. $x^4 - 8x^3 - 12x^2 + 60x + 63 = 0.$ $\left[\textbf{Ans. } -1, 3, 3 \pm \sqrt{30}\right]$

18. $x^4 - 10x^3 + 44x^2 - 104x + 96 = 0.$ $\left[\textbf{Ans. } 2, 4, 2 \pm 2\sqrt{2}i\right]$